Modern Cold Spray

Julio Villafuerte
Editor

Modern Cold Spray

Materials, Process, and Applications

Springer

Editor
Julio Villafuerte
Corporate—Supersonic Spray
Technologies
CenterLine Windsor Ltd
Windsor
Ontario
Canada

ISBN 978-3-319-16771-8 ISBN 978-3-319-16772-5 (eBook)
DOI 10.1007/978-3-319-16772-5

Library of Congress Control Number: 2015938600

Springer Cham Heidelberg New York Dordrecht London

Printed on acid-free paper

Springer International Publishing AG Switzerland is part of Springer Science+Business Media (www.springer.com)

Preface

Education, conceived as the sharing of knowledge between individuals and generations of individuals, represents a powerful tool by which human kind improves their understanding of the world around us towards, ideally at least, the creation of better societies where all individuals prosper, live happily, and in harmony with each other and their environments. Books are instruments of such principle, where the most current information can be perpetually stored to be shared among current and future generations of individuals.

This book is intended for engineers, thermal spray applicators, manufacturing, business visionaries, and professionals who, through their innovative input, can make impact on the future of manufacturing. Over the years, there have been many different names for the technology, main focus of this book, many being better descriptors of its science and operating principles; however, the name "cold spray" appears to stick better than most; therefore, we have decided to refer to the technology accordingly. This book is about the evolution of the cold spray technology, the current state of the art, and its practical use. More than anything, cold spray is a solid-state material consolidation method; it is its ability of producing unique material consolidates in ways that are not attainable with other thermal methods, the main reason for its important place in modern manufacturing.

Since its wide commercial inception at the beginning of the twenty-first century, there have been a number of excellent books written on the subject of cold spray. The persistent dynamic evolution of this technology requires that a continuous flow of renewed information becomes available to be shared among readers and individuals interested in the topic. This book contains cutting edge information on the cold spray technology at the time of its publication; however, without any promise that it includes absolutely all current developments and applications available in the marketplace.

At the time of this publication, cold spraying was commercially used for dimensional restoration of difficult-to-repair materials, such as cast iron and magnesium alloys used in heavy equipment and noncritical aircraft components. The process was also reportedly being commercially used to create thermally and/or electrically conductive surfaces in nonconductive substrates. At the same time, similar to the nanotechnology buzzword of the 2000s, there was wide spread speculation on the potential benefit of using cold spray technology for additive manufacturing, another novel technique in which components could be fabricated by systematically consolidating layers of materials, following a digital model.

This book is comprised of 12 chapters written by well-known authors from the academic, research, and industrial world, who have already spent significant part of their professional careers on various aspects of the cold spray technology. Our hope is that, by reading this book, readers can get the inspiration necessary to further improve the process and nurture more real-world applications.

I want to take this opportunity to sincerely thank all of the contributing authors for their valuable contributions; without these contributions this book would not be what it is supposed to be: a source of well-founded knowledge of cold spray technology for current and future generations. Their time and mental efforts to convey this information in written form is greatly appreciated. I also want to express my special thanks to CenterLine Windsor Ltd for providing me with a nurturing environment so that I could pursue this project. Sincere appreciation also goes to Ania Levinson and Abira Sengupta of Springer for their assistance with coordinating and final editing of the book. Their patience, especially when awaiting for long overdue drafts, is greatly appreciated.

Last but not the least, I want to express my deep gratitude to my friends, Ionel Botef, Vladimir Franjo, and Kofi Adomako for their emotional support during tough times in this phase of my life. To my girlfriend, Sandra, for her unconditional support during the production of this book. Her continuous encouragement helped me overcome my fears and frustrations. Finally, I want to dedicate this book to my three boys, Julian, Nico, Marcos, and their mother, Marcela, for what they represent in my life. Children are the future, who trespass our own mortality. I wish they can improve our society, whereby the good aspects of modern technology are fully exploited more for the good of all than for the greed of a few.

Windsor, On, Canada Julio Villafuerte
February 2015

Contents

Contributors

J. Abelson Donaldson Company, Inc., Minneapolis, MN, USA

W. Birtch Corporate—Supersonic Spray Technologies, CenterLine Windsor Ltd, Windsor, ON, Canada

I. Botef School of Mechanical, Industrial, and Aeronautical Engineering, University of the Witwatersrand, Johannesburg, South Africa

V. K. Champagne U.S. Army Reasearch Laboratory, Aberdeen Proving Ground, MD, USA

D. Christoulis CeraMetal Surface Engineering S.A., Koropi, Greece

T. J. Eden The Applied Research Laboratory, The Pennsylvania State University, Reston, VA, USA

R. Ghelichi Department of Mechanical Engineering, Massachusetts Institute of Technology, Cambridge, MA, USA

D. Goldbaum Automotive and Surface Transportation, National Research Council of Canada, Boucherville, QC, Canada

D. Helfritch TKC Global, Hearndon, VA, USA

T Hussain Division of Materials, Mechanics and Structure, Faculty of Engineering, University of Nottingham, Nottingham, UK

E. Irissou Automotive and Surface Transportation, National Research Council of Canada, Boucherville, QC, Canada

H. Jahed Mechanical and Mechatronics Engineering Department, University of Waterloo, Waterloo, ON, Canada

M. Jeandin MINES ParisTech, Paris, France

B. Jodoin University of Ottawa, Ottawa, ON, Canada

P. King CSIRO, Clayton, VIC, Australia

P.K. Koh SIM University, Singapore, Singapore

H. Koivuluoto University of Tampere, Tampere, Finland

J.-G. Legoux Automotive and Surface Transportation, National Research Council of Canada, Boucherville, QC, Canada

C.-J. Li School of Materials Science & Engineering, Xi'an Jiaotong University, Xi'an, Shaanxi, PRC

C. Moreau Department of Mechanical and Industrial Engineering, Concordia University, Montreal, QC, Canada

D. Poirier Automotive and Surface Transportation, National Research Council of Canada, Boucherville, QC, Canada

L. Pouliot Tecnar Automation Ltd., St-Bruno, QC, Canada

C. Sarafoglou School of Naval Architecture and Marine Engineering, National Technical University of Athens, Athens, Greece

O. Stier Corporate Technology, Siemens AG, Berlin, Germany

S. Vezzu Veneto Nanotech, Venezia, Italy

J. Villafuerte Corporate—Supersonic Spray Technologies, CenterLine Windsor Ltd, Windsor, ON, Canada

J. Wang Corporate—Supersonic Spray Technologies, CenterLine Windsor Ltd, Windsor, ON, Canada

D. E. Wolfe The Applied Research Laboratory, The Pennsylvania State University, Reston, VA, USA

M. Yandouzi University of Ottawa, Ottawa, ON, Canada

S Yue Department of Mining and Materials Engineering, McGill University, Montreal, QC, Canada

Chapter 1
Overview

I. Botef and J. Villafuerte

1.1 Introduction

Cold spray is one of many names for describing a solid-state process that consolidates metals by exposing a substrate to a high-velocity jet (300–1200 m/s) of small particles (5–40 μm) accelerated by a supersonic gas jet (Papyrin et al. 2007). The powder particles intended to form the deposit are injected into the gas stream which accelerates such particles as it expands in the divergent section of a DeLaval nozzle. The temperature of the gas stream is always below the melting point of the feedstock material, and, as the gas expands to supersonic velocity, its pressure and temperature also decrease. After exiting the nozzle and upon impact with the substrate, the solid particles plastically deform and create a combination of metallurgical and mechanical bonding with the surrounding material. This results in the formation of a coating or freestanding shape in the solid state. However, material deposition *only* takes place if the impact velocity of the particles exceeds a defined material- and temperature-dependent threshold, namely critical velocity (Li et al. 2010). Figure 1.1 compares the coating thickness capabilities of the cold spray process to other surfacing technologies including surface conversion/modification and deposition methods. The cold spray process covers a vast range of possible coating thicknesses.

The details of the aspects highlighted above as well many more aspects related to the cold spray process will be presented in subsequent chapters of this book. In the following sections of this chapter, we review the beginnings of thermal and cold spray technologies, the advantages and limitations of cold spray, and a comparison with other coating processes. This is followed by a discussion on the need for a goal-orientated surface engineering approach as it pertains to cold spray technology.

I. Botef (✉)
School of Mechanical, Industrial, and Aeronautical Engineering, University of the Witwatersrand, Johannesburg, South Africa
e-mail: ionel.botef@wits.ac.za

J. Villafuerte
Corporate—Supersonic Spray Technologies, CenterLine Windsor Ltd., Windsor, ON, Canada
e-mail: julio.villafuerte@cntrline.com

J. Villafuerte (ed.), *Modern Cold Spray*, DOI 10.1007/978-3-319-16772-5_1

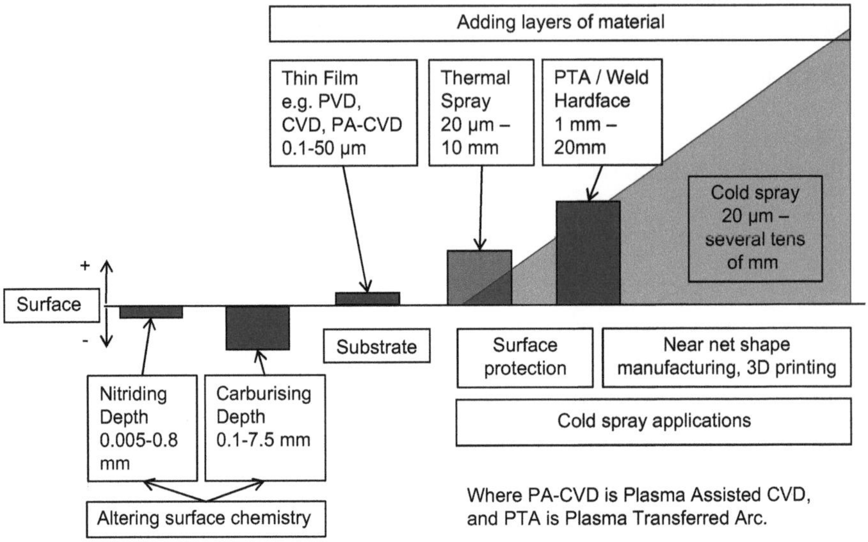

Fig. 1.1 Cold spray in the surface engineering space. *CS* cold spray, *PVD* physical vapor deposition, *CVD* chemical vapor deposition

1.2 Beginnings

The birth of new technologies is often associated with the result of rigorous methodical research and analysis; however, this is not always the case. Historically, chance discoveries led to new great ideas that eventually directed important further scientific investigations of the observed phenomena. In this respect, some examples include the discovery of penicillin by Alexander Fleming in 1928 while studying *Staphylococcus* (the bacteria that causes food poisoning), Teflon by Roy Plunkett in 1938 in an attempt to make a new refrigerant, the microwave effect by P. L. Spencer in 1945 while walking past a radar tube and noticing that the chocolate bar in his pocket melted, or cellophane by Jacques Brandenberger in 1908 while trying to apply a clear, protective film to a cloth making it waterproof.

Innovation can be the result of "a happy accident" sometimes referred to as "serendipity" (Walpole 1754). However, as Louis Pasteur (1854) pointed out, "where observation is concerned, luck favours only the prepared mind" (Dusek 2006, p. 35) suggesting that "luck" was harnessed to some discoveries but it was actually because of the scientist who actively created the conditions for the discovery by planning and carefully controlling the conditions of a systematic experimental research. Moreover, it was the "prepared mind" that properly interpreted the importance of the unforeseen incident, employed creative analogies, and so used the incident constructively.

The discovery of the thermal spray and cold spray processes discussed in the following sections could be mentioned in the same context. Therefore, with the current demands on scientific research to solve critical problems, the unexpected chance events should not be discounted.

1.2.1 Thermal Spraying

Myth has it that Dr. Max Ulrick Schoop developed in the early 1900s in Zurich, Switzerland, the concept of thermal spraying when playing "soldiers" with his son and observing the deformation of lead pellets being fired from a toy cannon against a brick wall. As a good observer, Schoop recognized the potential of metallic splats and, in about 1912, he developed a simple device based on the principle that if a wire rod were fed into an intense, concentrated flame, it would melt and, with the flame surrounded by a stream of compressed gas, the molten metal would become atomized and readily propelled onto a surface to create a coating (Knight 2008). Schoop sold the rights of his process to Metallizator, a German company which, by the early 1920s, made the "metallizing" process available throughout Europe and the USA, and so promoted the spraying of railroads, navy ship tanks, coal barges, and even the Panama Canal's emergency gates (Hermanek 2013).

Whatever the rationale, it can be stated that the pioneer work in the early 1900s of Schoop resulted in the discovery and development of metal spraying and subsequently the thermal spray technology represented by a group of coating processes in which finely divided metallic or nonmetallic materials are deposited in a molten or semi-molten condition to form a coating (Davis 2004).

During World War II, the thermal spray market exploded as war products were needed to last in battle and harsh conditions. In the 1960s, the detonation gun (D-gun) was developed, followed by induction plasma spray in 1965, vacuum plasma spray in 1973, and high-velocity oxy-fuel spray (HVOF) in the 1980s (Knight 2008). More recent new thermal spray developments include high-velocity air fuel (HVAF), low-velocity flame spray, and the suspension or solution precursor plasma spray (SPS or SPPS), a process which has been used to create thermal barrier coatings (TBCs) with novel low thermal conductivity columnar microstructures, and also an economical alternative to the electron beam physical vapor deposition (EB-PVD) process (Xie et al. 2006; Sampath et al. 2012).

1.2.2 Cold Spraying

According to NASA, when flying insects get in the way of an airplane's wing during takeoff or landing, it is not just the bugs that suffer; those little blasts of bug guts disrupt the laminar flow of air over the airplane's wings, creating more drag on the airplane and contributing to increased fuel consumption. That is why the "bug team" at NASA recently ran several flight tests of coatings that may one day reduce the amount of bug contamination on the wings of commercial aircraft (Atkinson 2013).

In this context, there is a large variety of two-phase flows, which, regarding the physical state, could be classified in gas–particle (solid or liquid), liquid–solid, and liquid–liquid mixtures. Out of these, the special case of gas–particle flows is of major importance in aerospace: (i) flame ignition and stability in turbojet engines depend to a large extent on the dynamics and evaporation of the solid small droplets of kerosene spray injected in the combustion chamber (Murrone and Villedieu 2011); (ii) in solid propellant rockets, the addition of small (1–100 μm) solid aluminum particles increase the temperature of the burnt gases, suppress pressure oscillations in the combustion chamber, but when exhausted with the rest of the combustion products, may cause additional erosion in the rocket nozzle wall and increase heat transfer (Rudinger 1976); and (iii) the presence of super-cooled water droplets in the air lead to ice accretion which may deposit on aircraft surfaces, depending on their size and relative velocity (Murrone and Villedieu 2011).

Hence, due to the importance of two-phase flows in aerospace optimal design and safety operations, it was normal for the Russian scientists at the Institute of Theoretical and Applied Mechanics of the Siberian Branch of the Russian Academy of Science (ITAM SB RAS) in Novosibirsk to perform, in the mid-1980s, wind tunnel experiments and study the influence of particles on flow structure and their interaction with a body (Papyrin et al. 2007).

However, apart from other results, the original carefully planned and controlled conditions of wind tunnel experiments at ITAM SB RAS also led to the observation for the first time of deposition of aluminum on a body in a "cold" (280 K) supersonic two-phase flow with aluminum particle at a velocity of 400–450 m/s (Papyrin 2007).

This was the "happy accident" of observing a new phenomenon. However, it was the scientist who actively created the conditions for discovery and the "prepared minds" of scientists, such as Professor Papyrin and his colleagues, who properly identified and understood the importance of the unforeseen incident, used creative analogies, and, using the incident constructively, transformed it into what we call today the cold spray process.

The scientific and practical importance of the discovery of the cold spray process stimulated further experiments for a more detailed study of the observed phenomenon and the establishment of the basic physical principles of the process. The wide spectrum of research conducted include, in no particular order: experimental studies, modeling of the process, gas dynamics inside and outside the supersonic nozzle, optimization of the nozzle, the impact of a supersonic jet on a substrate, deformation of particles and bonding mechanisms, coating properties, and equipment and applications development (Papyrin et al. 2007).

All these studies led to a large number of initial patents in Russia; then, as the process became known and appreciated, new cold spray patents were filed in other countries around the world, as discussed in Chap. 12. Also, the cold spray process received important support by a number of consortiums formed, for example, in 1994–1995 between companies such as Ford Motor Company, General Motors, General Electric Aircraft Engines, and Pratt & Whitney Division of United Technologies and then, in 2000–2003, the Cooperative Research and Development Agreement (CRADA) consortium formed between companies such as Alcoa, ASB Industries, Sandia National Laboratories/DOE, Daimler Chrysler, Ford Motor Company, Jacobs Chuck Manufacturing Company, Ktech Corporation, Pratt & Whitney, Praxair, and Siemens-Westinghouse (Papyrin et al. 2007; Irissou et al. 2008).

Although the cold spray process, as we know it today, had been recognized for more than 20 years, its true commercial development only started in the early 2000s (Irissou et al. 2008). Since this time, the number of research publications on cold spraying has grown exponentially, a situation which also led to some sort of patent race in the field. In this respect, it is important to mention that way back in 1900, H. S. Thurston filed a patent for a method of applying metal particles upon a metal plate by a blast of pressurized gas; that in 1958, Rocheville filed a patent to protect a device that essentially used the method patented by Thurston but used a DeLaval-type nozzle to accelerate the gas and the fine powder particles; and, because sometimes the novelty of a patent application seems disputable, the decision in 2007 of the Supreme Court of the USA restrained to a greater degree the possibility of issuing patents that combine elements from different preexisting patents (Irissou et al. 2008). A comprehensive state of the art of the patenting of cold spray technology, at the time of the writing this book, is presented in Chap. 12.

1.3 Cold Spray: Advantages and Limitations

Like any other material consolidation techniques, the cold spray process has its own advantages and limitations. The main attribute of the cold spray process is the fact that it is a solid-state process, which results in many unique characteristics (Karthikeyan 2007). The most obvious limitation arises from the inherent plastic deformation of particles, which leads to loss of ductility at the expense of strength (Ogawa et al. 2008). The latter, however, may indeed be an advantage for certain applications. There are many ways in which advantages and limitations could be presented; this chapter divides these topics into three categories, namely:

1. Deposited material property advantages
2. Manufacturing advantages
3. Process limitations

1.3.1 Deposited Material Property Advantages

Cold spray coating property advantages are summarized in Fig. 1.2. Chapter 4 "Coating Properties" explores in more detail all factors related to the properties of cold spray coatings. Many of these properties are interrelated; for example, cold-sprayed characteristics such as high density, low porosity, and no oxidation maximize both thermal and electrical conductivity, and, depending on the corrosion environment, may be conducive to improved corrosion resistance.

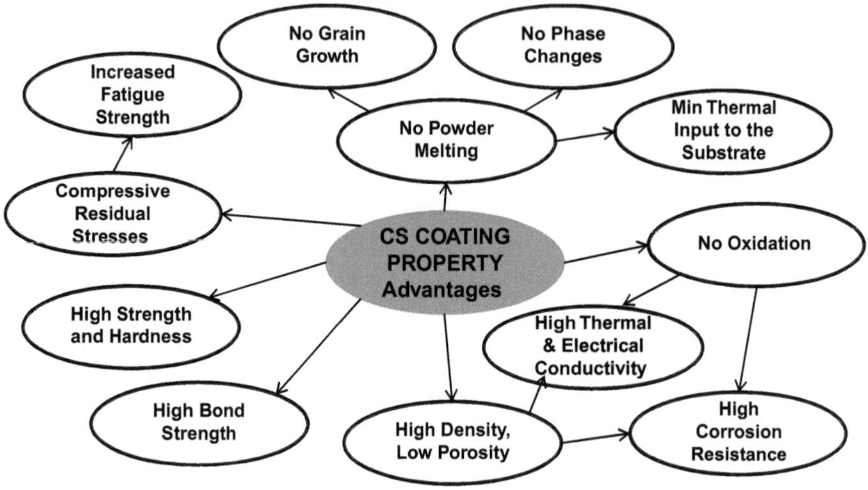

Fig. 1.2 Cold spray coating property advantages

1.3.1.1 No Powder Melting

The key physical difference between cold spraying and more conventional thermal spraying methods is that, in cold spray, material consolidation occurs entirely in the solid state (Papyrin 2007) which requires enough impact energy (particle velocity) to stimulate bonding by rapid plastic deformation. Therefore, to achieve higher gas flow velocities in the DeLaval nozzle, the compressed carried gas is often preheated. However, while preheated temperatures as high as 1000 °C or more may be used, the fact that contact time of spray particles with the hot gas is quite short and that the gas rapidly cools as it expands in the diverging section of the nozzle, the temperature of the particles actually remains substantially below the initial gas preheat temperature, hence, below the melting temperature of the spray material (Grujicic et al. 2003).

1.3.1.2 No Grain Growth

During other material consolidation processes, such as powder metallurgy and conventional thermal spray processes, grain recrystallization and coarsening is a fact of life which is unacceptable in many instances (Kim et al. 2005). In contrast, during cold spraying, the net heat input into the material is low enough that extensive grain growth and recrystallization do not typically happen; this is largely beneficial as the consolidated material may be able to retain desirable mechanical and physical properties of the feedstock material (Al-Mangour et al. 2013), such as fatigue strength, which strongly depends on surface microstructure and grain size (Ghelichi et al. 2012). After cold spraying, the grain microstructure of the consolidated material remains largely equaxial contrary to the splat-like microstructure typical of conventional thermal spray (Luzin et al. 2011). Others have even reported fine grain microstructures with ultimate tensile strength and hardness always higher than the equivalent properties in equivalent wrought materials- attributed to the high degree of plastic deformation during cold spraying (Karthikeyan 2007; Phani et al. 2007; Koivuluoto et al. 2008; Al-Mangour et al. 2013). At a more microstructural level, rapid plastic deformation during cold spraying may also result in the formation of nano-sized grain zones at the interfacial regions between particles, which may have further implications on mechanical properties of cold spray deposits (Jahedi et al. 2013). In summary, the low-temperature solid-state condition of the cold spray technique makes this method attractive to process temperature-sensitive materials, such as nanostructured and amorphous materials (Kim et al. 2005; Karthikeyan 2007).

1.3.1.3 No Phase Changes

The properties of all materials are affected by the chemical, microstructural, and phase compositions of the processed material. In most cases, these properties have been engineered through material processing techniques that, at some point, involved high-temperature phase transformations (Melendez and McDonald 2013).

In high-temperature processes, such as plasma spraying, molten species (including ceramic species) can react during their very short fly from the spray gun to the substrate. For example, NiAl plasma-sprayed powder can show all possible phases in the deposit, including Ni, alpha-Ni, NiAl, Ni_2Al_3, $NiAl_3$, and Al; Al_2O_3 and TiO_2 plasma-sprayed powder blends can result in deposition of Al_2O_3 significantly enriched with TiO_2 (Chraska et al. 1992). Even during lower-temperature spray processes, such HVOF, WC–Co powders tend to undergo detrimental decarburization, with by-products of reactions that are fundamentally undesirable, such as W_2C, W, and WO_3 (Kim et al. 2005; Melendez and McDonald 2013).

In the cold spray process, thermally induced phase transformations are avoided. Researchers have used cold spray to consolidate WC-based powders. Using X-ray diffraction (XRD), it has been confirmed that cold spray does not induce changes to the chemistry, phase composition, or grain structure (Kim et al. 2005; Smith 2007; Al-Mangour et al. 2013; Melendez and McDonald 2013). Yet, another classical example refers to the preservation of nanocrystalline microstructure, which yields exceptional mechanical properties (Smith 2007; Fig. 1.3). The cold spray process has been successfully used to consolidate nanostructured powder materials into useful forms without destroying their fine grain size (Karthikeyan 2007).

1.3.1.4 Minimum Thermal Input to the Substrate

Repairing damaged ion vapor deposition (IVD) aluminum coatings on high-strength steel substrates, such as 300M, 4340 or 4130, requires better than 99 wt.% aluminum coatings along with a coating process that does not raise the substrate temperature to

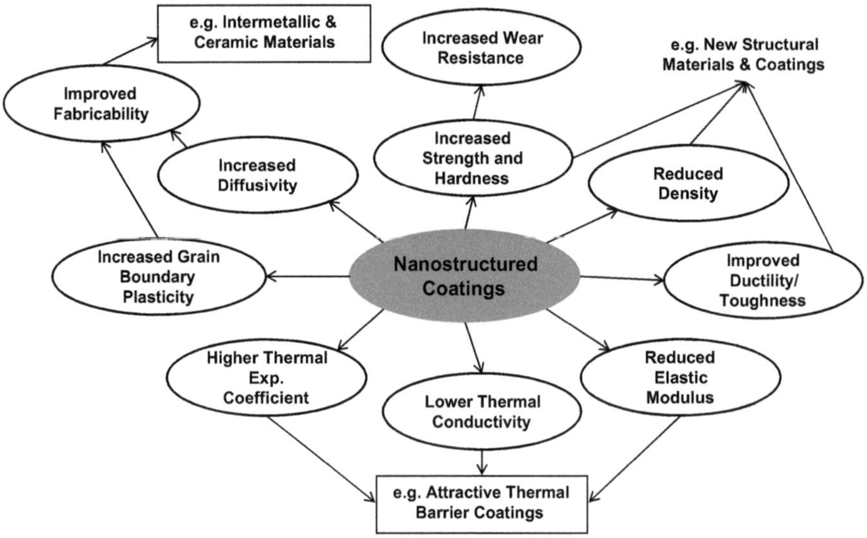

Fig. 1.3 Cold spray nanostructured coating advantages

more than 204 °C (per MIL-DTL-83488D). Because of its low-temperature deposition, cold spray has become the ideal process to repair damaged IVD aluminum on thin plates. Qualification tests have demonstrated that during the cold spray coating process, the temperature at the reverse side of steel sheets (as thin as 1 mm) did not reach more than 120 °C (Gaydos 2011). In addition to IVD, cold spray can also be used to repair damaged alumiplate, sputter aluminum, chemical vapor deposition (CVD) aluminum, and ionic liquid aluminum coatings (Gaydos 2011). Furthermore, cold spray could be used to spray any temperature-sensitive materials such as magnesium, nanostructured materials, amorphous materials, carbide composites, and many polymers. According to experts, cold spray technology could be used for almost 70 % of materials that could be spray coated but that are ruled out because of the high-process temperatures associated with traditional thermal spray processes (Kaye and Thyer 2006).

1.3.1.5 No Oxidation

In-process oxidation constitutes one of the main limitations of most traditional thermal spray processes. In-flight oxidation of particles results in internal oxide inclusions, while postimpact oxidation produces surface oxide layer between splat layers (Gan and Berndt 2013). Low-cost processes such as air plasma spray (APS) and wire arc spray produce coatings with the most oxidation and porosity compared to, for example, HVOF (Gan and Berndt 2013). Yet, the increased particle velocity in HVOF generally correlates well with improved splat deformation and less porosity but unfortunately has no effect on oxidation (Hanson and Settles 2003). Oxidation is particularly critical when spraying oxygen-sensitive materials such as aluminum, copper, magnesium, titanium, and others as small amounts of undesirable oxygen may degrade the physical properties of the deposits (Smith 2007). One example is alloy 600, a nickel-based alloy used in heat exchangers in the nuclear industry, which is prone to stress corrosion cracking (SCC), one of the most challenging modes of material failure. SCC failures have been directly associated with the presence of Fe- and Cr-rich oxide films along grain boundaries (Dugdale et al. 2013).

During particle impact in cold spraying, the brittle oxide skin that covers most metal surfaces shatters, making the oxide swept away by the high-velocity gas jet and readying the bare surface for a clean bond with particles coming behind; in practice, it has been demonstrated that cold spray deposits show same or lower oxygen content than the starting powder material (Karthikeyan 2007). Figure 1.4 illustrates porosity, oxygen content, and thermal properties of Cu, Sn, and Al when sprayed by conventional thermal spray versus cold spray.

The ability of cold spray not to introduce but rather diminish oxide content in the deposit is very appealing for a number of exciting future applications. One example is the deposition of intermetallic compounds such as FeAl-based intermetallic alloys, which exhibit good mechanical properties and excellent corrosion resistance in oxidizing and sulfidizing atmospheres at elevated temperatures. These materials are lighter (5.56 g/cm^3) than steels or Ni-based alloys, have a high melting point, high creep strength, excellent thermal conductivity, and are relatively inexpensive. Because of these attributes, they have been considered as a substitute for stainless steels or Ni-based superalloys for high-temperature service (Wang et al. 2008). Fe–Al intermetallics show limited ductility at low temperatures, and their mechanical strength degrades at temperatures higher than 600 °C. Thermal spray processes, such as plasma spray, HVOF, wire arc, or flame spray, have been used to spray with FeAl-based alloys for corrosion protection of carbon steels. However, such deposits

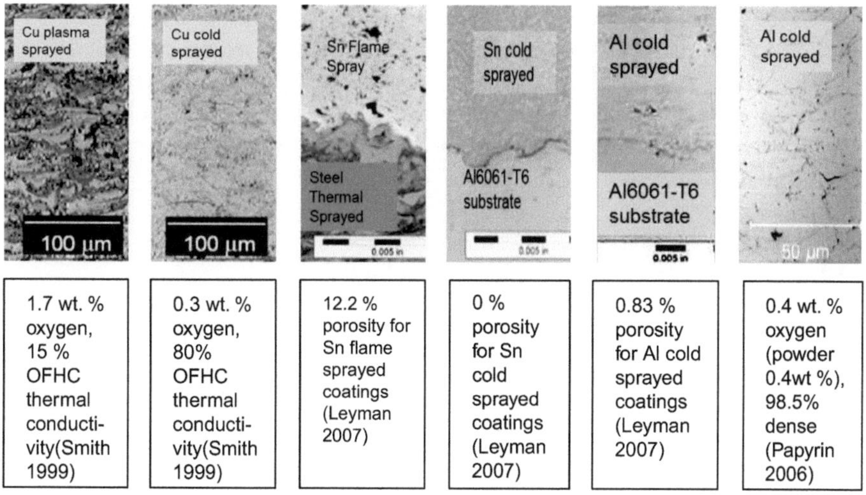

Fig. 1.4 Porosity, oxygen content, and thermal property comparison between thermal spray and cold spray. *OFHC* oxygen-free high conductivity

end up with excessive oxide content which leads to poor corrosion performance and other problems, emphasized by the significant difference in melting points between Al and Fe as well as the exothermic nature of formation of iron aluminides (Lee et al. 2006; Wang et al. 2008, 2012). Cold spraying of Fe–Al-based materials can be achieved by utilizing powder blends, which then, with an appropriate annealing post-spray treatment (Lee et al. 2010), can induce the complete transformation of Fe(Al) solid solution into FeAl intermetallics (Wang et al. 2008). Other powder mixtures, such Al/Ni, Al/Ti, W/Cu, Zn/Al, Ti/Al, and Ni/Al can be cold sprayed and then annealed to form a dense well-dispersed distribution of their intermetallics (Wang et al. 2008; Lee et al. 2010).

1.3.1.6 High Density, Low Porosity

The outer zone of the typical lamellar structure of a thermal spray deposit shows splats that are not well bonded together and that, in their turn, lead to many micron-sized pores (Dong et al. 2013). A high level of porosity—such as 5–15 % for flame and arc spray and 3–8 % for plasma spray—may lead to corrosion (Maev and Leshchynsky 2008).

Cold spraying is a solid-state process with no splashing. When particles impact the substrate at speeds higher than the material's critical velocity, they plastically deform at high strain rates and bonding occurs (as explained later in Chap. 2). High strain rate deformation produces additional thermal energy at interfaces, which may lead to the generation of interfacial metal vapor jet. This jet in effect produces "vapor deposition" of material at the inter-particulate interfaces which fills any pores and cracks that exist. In this respect, it has been suggested that cold spray can be viewed as a combination of a particulate and microscopic vapor deposition processes (Papyrin 2006). On top of that, every subsequent pass of the spray plum effectively "shot peens" the underlying layers, thus increasing their density. The combination of all these phenomena in cold spray produces near-theoretical density coatings (Papyrin 2006). Furthermore, when post-spray heat treatments (such as annealing) are applied, the deposits experience even further consolidation and densification approaching ideal levels due to closure of pores, inter-splat boundaries, and cracks (Chavan et al. 2013; Fig. 1.4).

1.3.1.7 High Thermal and Electrical Conductivity

Electrical conductivity is a good indicator of coating quality in terms of material density and presence of dispersed oxide phases (Koivuluoto et al. 2012). Because of its exceptional electrical and thermal conductivities combined with its commercial availability, copper represents a key material in today's industrialized world (Phani et al. 2007). It has been demonstrated that the presence of oxide inclusions in plasma-sprayed copper coatings lowers the electrical conductivity of the deposits down to about 15 % of the conductivity of oxygen-free high-conductivity (OFHC) copper

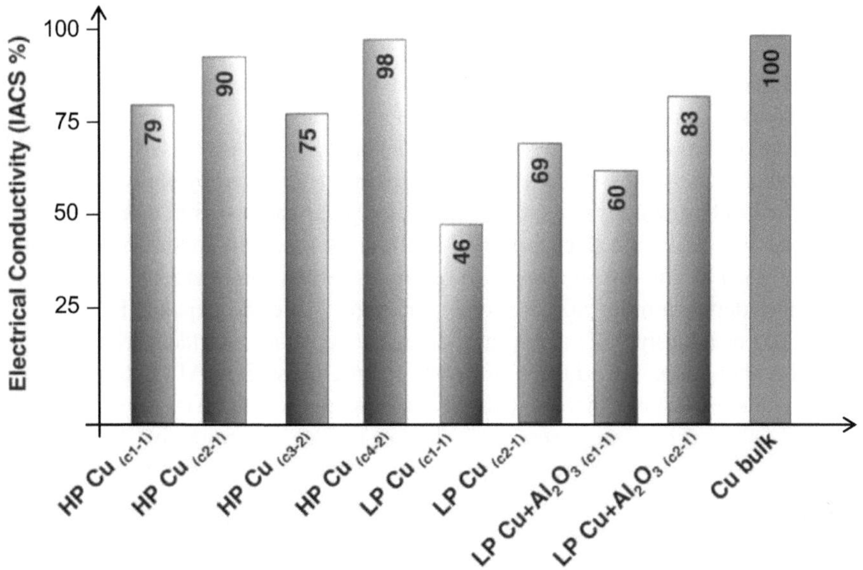

Fig. 1.5 Electrical conductivity for cold-sprayed copper coatings. *IACS* International Annealed Copper Standard, *HP* high gas pressure, *LP* low gas pressure

(Smith 2007). In contrast, dense cold-sprayed copper coatings may display conductivities better than 85 % OFHC copper (Smith 2007; Karthikeyan 2007); similar copper material would typically display conductivities of about 40–63 % OFHC copper when sprayed by conventional thermal spray processes. If the process can afford it, post-annealing after cold spraying will further increase the conductivity of the deposits by densification and recrystallization (Phani et al. 2007; Koivuluoto et al. 2012).

Figure 1.5 compares the "as-sprayed" electrical conductivity of copper coatings using upstream injection (high gas pressures, HP) and downstream injection (lower gas pressures, LP) cold spray systems. International Annealed Copper Standard (IACS) values were given as an average of values measured by using four-point measurements for the following sample conditions: (c_1) on steel as-sprayed, (c_2) on steel heat treated at 400 °C, (c_3) on ceramic heat treated at 280 °C, and (c_4) on ceramic heat treated at 280 °C. The subscript that follows the condition indicates the authors: (1) Koivuluoto et al. (2012) and (2) Donner et al. (2011).

Upstream injection cold spray systems operating at HP would produce copper coatings with higher electrical conductivities than downstream injection systems operating at LP. However, when using downstream injection systems in combination with powder blends (such as $Cu+Al_2O_3$), the deposits can be sufficiently densified as to attain electrical conductivities that are acceptable for most electrical applications. The main function of Al_2O_3 particle additions, in this case, is to activate (cleaning/roughening) the underlying surfaces as well as hammer-deposited particles so that a high-density low-oxygen deposit is built up, become more receptive to adhesion, fresh impact of sprayed particles and better adhere to the surface.

1.3.1.8 Bond Strength

Both adhesive and cohesive strengths of a deposit are key to determine its usefulness within any particular application (Huang and Fukanuma 2012). The mechanical and other properties associated with cold spray deposits will be discussed in Chap. 4. In general, average adhesive/cohesive strengths are determined by spraying the top surface of a cylindrical tensile sample, then gluing the sprayed area to a respective counter-body of the same size, and finally pulling the assembly in tension to failure (ASTM C633). Using this method with certain materials combinations, it has been shown that deposits may fail cohesively (fracture inside the deposit; Koivuluoto et al. 2008) or adhesively (fracture at the deposit/substrate interface; Irissou et al. 2007). In other cases, the addition of ceramic particles, such as Al_2O_3, to pure aluminum can significantly increase adhesive/cohesive strength (Irissou et al. 2007; Lee et al. 2005). There are cases where the test results are limited by the strength of the glue itself. To overcome the latter, others (Huang and Fukanuma 2012) have attempted other testing techniques, which have indicated that cold-sprayed deposits can display high adhesive strengths (as high as 250 MPa for aluminum alloys, Karthikeyan 2007).

1.3.1.9 Compressive Residual Stresses

It is generally accepted that the presence of surface tensile stresses may contribute to the formation and propagation of micro-cracks, which may accelerate fatigue failure. Because of thermal expansion and contraction during melting and solidification, deposits made by thermal spray may develop surface residual tensile stresses (Maev and Leshchynsky 2008). A distinct feature of the cold spray process is the development of superficial compressive residual stresses instead (Spencer et al. 2012). Compressive stresses have the opposite effect on fatigue life. Because of the importance of this subject, Chap. 5 has been entirely dedicated to discuss the topic of the formation of residual stresses during cold spraying.

Through modeling and experimentation, it has been shown that the cold spray process can generate desirable surface compressive stresses which are responsible for improvements in fatigue life of certain materials. For example, cold-sprayed Al7075 over Al5052 substrates, using downstream injection system at low pressure, showed an improvement of close to 30 % in fatigue life (Ghelichi et al. 2012). Others have reported a fatigue life improvement of 10 % for cold-sprayed aluminum over AZ31B magnesium substrates (Shayegan et al. 2014). Compressive residual stresses generated during the cold spray process also contribute to the possibility of producing ultra-thick, well-bonded, and near-room-temperature coatings for near-net-shape manufacturing of components made of metallic, composite, and polymeric materials (Maev and Leshchynsky 2008).

1.3.1.10 Corrosion Resistant

Conventional thermal spray processes such as flame spray, twin-wire arc spray, and air plasma spraying (APS) represent some of the common methods to deposit aluminum over steel and other materials for corrosion protection. Although they are typically less expensive compared to cold spraying, the severe oxidation, phase transformations, and high porosity are conducive to relatively poor corrosion performance compared to fully dense materials (Dong et al. 2013; Chavan et al. 2013), not to mention some of the environmental challenges associated with the processes (Villafuerte and Zheng 2007).

The high density, phase purity, and homogeneous microstructure of cold-sprayed coatings are characteristics that yield exceptional corrosion resistance (Karthikeyan 2007). Consequently, cold spray is increasingly becoming a preferred method to provide localized corrosion repair and protection in a vast number of applications, many without other means available (Villafuerte and Zheng 2007). Some of these applications include aircraft as well as automotive magnesium castings (Villafuerte and Zheng 2011; Suo et al. 2012).

Others (Al-Mangour et al. 2013) have experimented mixing stainless steel 316L particles with particles made of Co–Cr alloy L605; the latter known to display superior corrosion resistance than 316L alone but difficult to manufacture. These studies suggested that cold spray may be used to consolidate these metals as a blend (67 % 316L—33 % L605), then post-heat treated so that both corrosion and mechanical properties of the resulting composite are better than 316L alone, potentially becoming a new class of metallic biomaterial (Al-Mangour et al. 2013).

1.3.2 Manufacturing Advantages

Often, the rate at which new technologies are developed, accepted, and adopted is proportional to the need for new solutions to engineering problems (Dorfman and Sharma 2013). In its current state, cold spray technology is becoming a tool for remanufacturing; an emerging trend born from our urgency to reduce the negative environmental impact of today's manufacturing practices such as overexploitation of resources, waste disposal, contamination, and greenhouse emissions. Some of the advantages of cold spray represent more environmentally friendly alternatives to technologies such as electroplating, soldering, and painting (Grujicic et al. 2003).

1.3.2.1 No Masking

In traditional thermal spray, masking is a necessity as hot particles in the overspray tend to stick very well to surfaces outside the target; often masking must be done manually which significantly adds to the manufacturing costs (Smith 2007). For example, the specifications for repair and assembling of blades in land-based power turbines require that an abradable aluminum seal layer be applied to the base of each blade where the blade is inserted into the main shaft. This minimizes leaks which, otherwise, would lead to significant efficiency losses. Conventional thermal spray processes require extensive masking to protect the rest of the blade from overspray, which represent the most significant portion of the cost (SST 2014).

For other industrial applications, such as electrical circuits and heat conducting surfaces requiring patterned deposits, cold spray represents an ideal process because of its ability to lay down well-defined patterns without masking. The width of the tracks is controlled by the nozzle exit diameter, and many of today's applications require tracks that are narrower than the standard available nozzle diameters (Wielage et al. 2010). In practice, nozzles can be modified by squeezing the cross section into a slim rectangle with the traverse side significantly smaller (1–2 mm) than longitudinal side (Karthikeyan 2007). This geometry change does not significantly affect particle velocity (Sova et al. 2013). As we move forward, with the possibility of using cold spray for additive manufacturing, there is a need to develop smaller nozzles capable of delivering smaller footprints for better shape resolution.

1.3.2.2 Flexibility in Substrate–Coating Selection

Cold spraying has been used for a wide selection of coating–substrate combinations. To name a few: Al on Ni substrate (Lee at al. 2008); Al–10Sn and Al–20Sn substrate—SUS304, Al6061 and Cu (Ning et al. 2008); aluminum alloy Al-5Fe-V-Si onto internal combustion (IC) aluminum engine piston heads (Berube et al. 2012); Cu+Al_2O_3 on steel substrate (Koivuluoto and Vuoristo 2010); Mg powder on stainless steel and aluminum plates (Suo et al. 2012); Al+SiC and Ti+SiC mixtures on aluminum and steel substrates (Sova et al. 2010); or Al and Al/Cu bimetallic coating on carbon fiber-reinforced polymer matrix composite (PMC; Zhou et al. 2011). Because cold spray bonding is a combination of mechanical interlock and metallurgical bonding, the nature of the substrate is of lesser importance and therefore an incredible number of combinations can be created.

1.3.2.3 Coupling Dissimilar Materials

The possibility to integrate dissimilar materials into products with exotic properties is a strong driving force for exploring new fabrication methods. As an example, the wear resistance of aluminum alloy substrates can be enhanced by cladding with iron-based alloys. However, the required heat input in traditional cladding processes may produce undesirable intermetallic phases, such as FeAl, Fe_3Al, Fe_2Al, and Fe_2Al_5 at the interface that induce crack failure upon cooling (Wilden et al. 2008). In other cases, it is the steel substrate that requires corrosion protection, for which a layer of aluminum is typically deposited using twin-wire arc spray. Although the latter is common, it is also known that the resulting porosity promotes the formation of interconnected paths inside the coating that allow corrosive electrolytes to reach the steel substrate (Esfahani et al. 2012). This corrosion behavior has also been observed in coatings made by flame spray, HVOF, and plasma spray (Esfahani et al. 2012).

Because of its much lower process temperature combined with higher particle velocity, the cold spray process can readily form dense deposit of many materials, such as aluminum, copper, nickel, 316L, and Ti64, on dissimilar substrates (Grujicic et al. 2004). Materials, such as aluminum and copper, could be cold sprayed directly onto smooth, unprepared glass surfaces (Dykhuizen and Smith 1998), and some polymeric surfaces (Lupoi and O'Neill 2010). Conversely, polymeric materials (such as polyethylene) can be sprayed onto Al 7075 substrate (Alhulaifi et al. 2012). The ability to mix a diverse range of materials makes cold spray ideal for creating engineered metal matrix composites (MMC) and free forms with custom-graded properties (Karthikeyan 2004).

1.3.2.4 Ultra-Thick Coatings

When building thick deposits by conventional thermal spray, there is a progressive buildup of superficial tensile stresses as the thickness increases; the bond strength decreases progressively. Eventually, the stress buildup overcomes the resistance of the material causing spontaneous spalling or delamination (Karthikeyan 2007). Unlike thermal processes, cold spray coatings tend to be compressively stressed at the surface, therefore minimizing or eliminating the through thickness gradient that is created in the process of producing thick coatings by thermal spray.

1.3.2.5 Deposition Efficiency

As described in subsequent chapters of this book, deposition efficiency (DE) represents the ratio, usually expressed in percentage of the weight of the powder deposited on the substrate to the weight of the powder sprayed (Schmidt et al. 2009). Higher DE is not necessarily associated with better sprayability as some materials can be readily deposited but with poor characteristics of the deposit, such as high porosity and poor bond strength. DE is not only a function of the nature and condition of the surfaces of the particle and/or substrate but also the amount of kinetic energy or impact velocity. In general, high DE values, in excess of 95%, can be attained using very high impact velocities; for example, when depositing copper and/or aluminum alloys. Very high impact velocities can be obtained using helium at high pressures and temperatures, however, at a very high cost tag, as explained in Chap. 11. Conversely, high DE values could also be attained at lower impact velocities by manipulating the characteristics of the spray powder. Depending on the materials to be sprayed, DE can be an important material cost consideration. In most cases, however, it represents a more important consideration from the health and safety standpoint as there may be a significant cost associated with the proper collection and disposal of waste metal powder, as better explained in Chap. 9.

1.3.2.6 Environmental, Health, and Safety

The presence of hexavalent chromium, a well-documented human carcinogen, during processing of traditional chromium plating applications made users of these technologies switch to HVOF thermal spray of tungsten carbide (Dorfman and Sharma 2013), a more benevolent process on this regards. Because of the absence of fumes, combustible gases, sparks, and flames, cold spray represents an even better alternative from the health and safety standpoint (Grujicic et al. 2003). In fact, WC–Co coating has already been applied by cold spray as a replacement of electrolytic hard chrome (EHC) plating (Ang et al. 2012).

In other high operational risk scenarios, downstream injection portable cold spray has been proven acceptable for remote corrosion repair of heavy water aluminum vessels in nuclear reactors (SST 2014). Cold spray technology was also demonstrated to be a reliable process for applying thick copper coatings to cast iron canisters for the disposal of nuclear fuel waste (Irissou et al. 2012).

Medical antibacterial surface applications are also becoming increasingly interesting, as described in Chap. 10. The antibacterial benefits of copper are well documented; the ability of cold spray to deposit dense copper on most surfaces has triggered interest on cold spray for such applications. Some suggest that using inoculants during spraying, such as meticillin-resistant *Staphylococcus aureus* (MRSA), could even yield at least three times the efficiency of eliminating bacteria by other traditional processes making cold spray an exceptional tool for food processing, health care, and air-conditioning applications (Champagne and Helfritch 2013).

However, like any other processes including thermal spray, cold spray technology poses a number of potential environmental, health, and safety risks, mostly associated with the management of waste powder as well as level of noise. The reclamation and disposal of waste powders is extensively reviewed in Chap. 9.

1.3.3 Process Limitations

As with any other materials processing technique, the cold spray process has its own limitations depending on the perspective from which it is evaluated, as discussed below.

1.3.3.1 Near-Zero Ductility

The main disadvantage of cold spray process arises from the necessary plastic deformation of particles, which leads to a loss of ductility of the coating. One study (Ogawa et al. 2008), for example, indicated that the elastic modulus of cold-sprayed aluminum coatings on a cold-rolled plate of pure aluminum alloy A1050 was consistently higher than the corresponding elastic modulus of the cold-rolled aluminum substrate, showing early signs of cracking during tensile loading. When compressive loading was applied, no cracking was observed in the cold-sprayed aluminum, which also displayed higher strength as compared to cold-rolled aluminum. What was interesting is the fact that post-annealing of cold-sprayed specimens at temperatures as low as 270 °C readily restored ductility compared to untreated specimens. In cold-sprayed materials, there is a great deal of stored energy as plastic deformation which can turn into recrystallization and consolidation when the material is heated.

1.3.3.2 Limited Range of Sprayable Materials

Unlike most traditional thermal spray processes that are capable of depositing a wide range of materials from metals to ceramics, at its current state, cold spray is essentially limited to depositing metals or composites possessing a sufficient degree of low-temperature ductility; Examples include metals such as Al, Cu, Ni, Ti, Ag, Zn, Ni (Champagne 2007), as well as blends of these metals with ceramics or other non-ductile species including Al–Al_2O_3 mixtures (Irissou et al. 2007), WC–12Co blended with Ni (Melendez and McDonald 2013), WC–Co blended with Cu or Al (Wang and Villafuerte 2009), and even Al–12Si alloy composites (Yandouzi et al. 2009).

Over these many years, there has also been a great deal of research in an attempt to expand the range of cold sprayable materials. Some examples of exotic attempts include Al–Ni intermetallic compounds (IMC; Lee et al. 2010), SiC without metal matrix binder on Ni–Cr-based superalloys to improve high temperature oxidation resistance (Seo et al. 2012); many of which are impossible to deposit using traditional thermal processes.

1.3.3.3 Substrate Material Must be Hard Enough

In cold spraying, the substrate material must be hard enough (relative to the spray material) to induce sufficient plastic deformation of the incoming particles for acceptable bonding (Karthikeyan 2007). On one side of the spectrum, extremely soft substrates (such as polyethylene) would cause incoming particles to crater well beyond the surface without any possibility of buildup. On the other side of the spectrum, extremely brittle substrates with structures that delaminate easily (such as carbon) would likely experience erosion by the incoming particles. Substrates must be as well resilient or well supported to accept coating.

1.3.3.4 Gas Consumption

Gas consumption in cold spray is much higher than in many thermal spray processes. This is because of the high velocities and flows necessary to propel particles.

Helium, nitrogen, and plain air are the choices for carrier gas. The main difference is the density; helium being the most desirable gas just because of its ability to attain high gas velocities as well as its inertness. High gas velocities often produce better results in terms of quality and DE. Yet, helium is the most expensive and scarce gas, which has limited its usage to very specialized applications; in these cases, recycling of helium may be necessary, which undoubtedly elevates the complexity of the cold spray operation (Champagne 2007).

Therefore, over the past few years, there has been a trend to develop cold spray procedures and specifications with nitrogen and/or air as the gases of choice for all types of cold spray equipment as well as sprayable materials. As explained in Chap. 6, the upper operational range for cold spray includes nitrogen gas temperatures as high as 1000 °C at a pressure of 70 bars. High operational parameters are necessary when spraying materials with less low-temperature ductility than traditional cold sprayable metals; some examples include Ni-based super alloys, titanium alloys, stainless steels, and tantalum. Under these operational conditions, capital and operational costs, including gas consumption, are very high, so the application must very well justify them. The decision on a specific cold spray system should be based on both the actual technical and operational requirements of intended applications.

1.3.3.5 Line of Sight

Just like all other thermal spray processes and unlike processes such as electroplating, physical vapor deposition, and chemical vapor deposition, cold spray is a line-of-sight process (Davis 2004). It is, therefore, difficult to spray materials on inside surfaces, such as the internal diameter of pipes, without special consideration of nozzle design. Unlike traditional thermal spray, the standoff distance from the substrate for cold spraying is in the order of 10 mm, much shorter than for thermal spraying. Therefore, it becomes easier to design nozzle assemblies that can fit within cavities that otherwise would be impossible to spray. At the time of this write-up, manufacturers of commercial cold spray equipment were able to supply 90° nozzle assemblies capable of spraying materials in inside diameters as small as 90 mm.

1.3.3.6 Limited Availability of Standard Specifications

At the time of writing this book, there was only one standard specification, MIL-STD-3021, in the public domain (US ARMY RESEARCH LAB ARL 2008). However, there were numerous companies which already had their own internal cold spray specifications and/or were in the process of developing their specification. Many of these first adopters operate within aerospace and transportation industries.

1.4 Cold Spray Versus Thermal Spray

The previous sections of this chapter presented, in a succinct way, an overview of the advantages and limitations of cold spray, in reference to some practical applications of the process. In the world of thermal spray, there is a wide range of commercial processes to apply coatings for a wide range of coating materials. Surface engineering requirements for specific components vary considerably according to specific service conditions, which generally are a combination of wear, stress, and corrosion. Therefore, not every thermal spray process is suitable for every application; on the other hand, the diversity of available spraying methods helps satisfy the diversity of needs. The role of an applicator is to, as best as possible, identify the processes that make more sense (technical, economical, environmental) to a given application (Sulzer Metco 2014).

In this section, we compare cold spray technology against the more general thermal spray family of processes. In order to visualize their differences and similarities, please refer to Fig. 1.6. This figure shows a selected number of thermal spray and cold spray processes positioned in the flame (or arc) temperature versus particle velocity coordinates as well as substrate temperature, included as an third axis on the right-hand side of the graph.

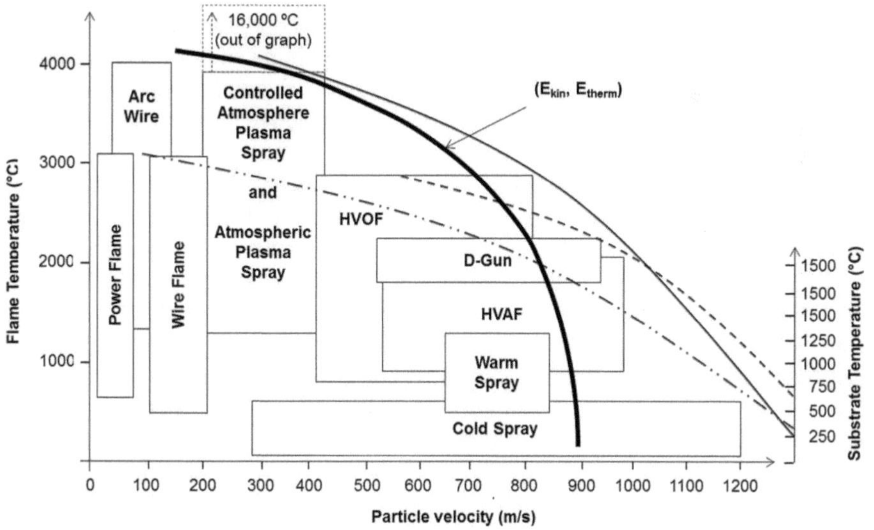

Fig. 1.6 Cold spray versus thermal spray. *HVOF* high-velocity oxy-fuel spray, *HVAF* high-velocity air fuel

1.4.1 Old Versus New

Thermal spray recently celebrated its first century of existence, classifying this family of processes as mature. Over the past few decades, thermal spray has become one of the leading surface technologies alongside physical vapor deposition and weld overlaying, an estimated industry worth approximately US$ 6.5 billion with the majority of revenue generated in coating services (Dorfman and Sharma 2013). Cold spray technology is a relatively new material consolidation technique which has commercially emerged in the past two decades (Smith 2007). The implementation of cold spray technology, probably its most intense development area (Papyrin 2007), generally occurs in situations where conventional hot metal spray technology cannot be successfully used (Champagne 2007). Eliminating the detrimental effects of high process temperatures has been perhaps one of the unique strengths of cold spray for many applications.

1.4.2 Energy Source

In order to produce material consolidation, all spray processes (including cold spray) require a combination of thermal and kinetic energies. Both of these energies are generated through diverse energy sources such as combustion flames, electric arcs, laser beams, and/or simply heating compressed carrier gases (nitrogen, air, helium). In order to achieve quality coatings, certain amounts of both energies ought to be transferred to the spray particles (Klassen 2014); thermal energy is needed to melt or soften the spray material. Kinetic energy, coupled to the particle velocity, is needed to achieve density and bond strength.

The thermal energy released by different spray techniques is transferred into energy in transit or heat. As soon as the transferred energy is absorbed by the coating material, it is no longer heat and once again becomes kinetic energy, which then forms part of the coating material's total internal energy (Nahle 2009). Whether high thermal energy or high kinetic energy of the particle is favorable for material consolidation, it all depends on the nature of the spray material and the desired properties of the deposit. Plasma spraying is common for spraying oxide ceramics; it uses extreme elevated temperatures and relatively low kinetic energy of the particles (or liquid droplets) to achieve ceramic coatings. HVOF covers a wide range of thermal and kinetic energy ranges to produce dense coatings. The temperature and velocity of particles impinging on the substrate is affected by the gas temperature, density, and pressure as well as by the time of exposure to the gas and the particle mass. Cold spray utilizes more of the kinetic rather than the thermal component of the required energy to produce material consolidation (Schmidt et al. 2009). This is quite suitable when spray materials have sufficient low-temperature ductility which promote the bonding mechanisms associated with cold spray, as discussed in detail in Chap. 2. At the same time, low process temperature makes cold spray suitable for material consolidation of temperature-sensitive materials such as magnesium, polymers; nanostructured, amorphous, and phase-sensitive materials including carbide composites.

1.4.3 Oxidation and Porosity

One of the undesirable effects of an elevated temperature process, when carrier gases are oxidative, is the in-flight reaction of hot or molten particles with oxygen leading to the formation of oxide inclusions and porosity into the deposit. For example, the degree of oxidation in combustion processes becomes higher for higher spray temperatures, longer exposure times in the flame, and higher level of free oxygen in the flame (Klassen 2014).

High level of porosity is an important limitation of many thermal spray processes it could significantly degrade corrosion resistance, mechanical properties, electrical conductivity, and thermal conductivity (Smith 2007; Champagne 2007; Maev and Leshchynsky 2008). The highest levels of porosity are produced after flame and arc spray processes, except for HVOF which can produce very dense deposits (>99.5 %). On the other hand, typical plasma coatings have approximately 1–2 % porosity. Controlled atmosphere plasma spray can produce near 100 % dense (Sulzer Metco 2014).

Because of its low process temperature, cold spraying has the benefit of preventing the formation of undesirable oxides to the point that many have demonstrated that trace oxygen levels in cold-sprayed deposits tend to be even lower than oxygen levels in either the spray or the substrate materials. Therefore, cold spray represents a solution to deposit oxygen-sensitive materials like aluminum, copper, and titanium.

1.4.4 Solid State: No Phase Transformations

Cold spraying consolidates materials in the solid state (Smith 2007; Papyrin 2007), which represents another important characteristic of this process. Under these conditions, thermally induced phase transformations are avoided or minimized. Subsequently, after spraying, the original properties of the spray materials are retained, making cold spray attractive to consolidate thermally sensitive materials, such as nanostructured powders (Karthikeyan 2007). On the other hand, conventional thermal spray processes, such as plasma spray, HVOF, or even oxy-fuel flame spraying, often exert undesirable levels of heat input into the coating and/or substrate materials likely resulting in phase changes or warping and distortion. In order to avoid overheating, it is common for thermal spray processes to use some type of cooling media including carbon dioxide, compressed air, or nitrogen mixed into the combustion gases, which are not necessary in cold spray.

It is important to clarify that to achieve higher gas flow velocities in the nozzle, it is necessary to preheat the compressed gas before the diverging section of the nozzle; despite the fact that preheating temperatures can be as high as 1000 °C, due to the fact that the contact time of spray particles with the hot gas is quite short and that the gas rapidly cools as it expands in the diverging section of the nozzle, the temperature of the particles remains substantially below the melting temperature of the feedstock (Grujicic et al. 2003).

As explained in Chap. 11, the economics of thermal spray and cold spray vary widely depending on many factors. Some of the key cost elements include gas consumption and cost of equipment and consumables (Davis 2004). In comparison to conventional thermal spray processes, the cold spray process has the additional advantage of being relatively simple to implement. The process control is carried out mainly through monitoring appropriate adjustment of gas pressures and temperatures. To optimize the process and to achieve best coating qualities, however, these parameters have to be finely tuned with respect to the spray material, powder size range, and the nozzle type (Schmidt et al. 2009).

1.5 Surface Engineering Complexity

Complexity has been considered a notion of inherent difficulty of a problem, a solution, or an approach (Szyperski et al. 2002) with no single formalism, technique, or tool capable of generating useful decisions in all cases (Kusiak 2000). Consequently, the new interdisciplinary field of complex systems cuts across all traditional disciplines and studies how parts of a system give rise to the collective behaviors of the system and how the system interacts with its environment (Bar-Yam 2003). Therefore, with so many complex systems surrounding our modern life, one should be aware that the cold spray is only one technique among a related network of disciplines, theories, technologies, systems, and processes.

Many engineered components operate in aggressive environments characterized by high temperatures, high pressures, large stresses, presence of oxidizing or corroding environments, and presence of particulate materials that induce erosion damage (Bose 2007). In other words, all materials of which these components are made are normally exposed to degradation at some level, either at the surface or internally (bulk).

Surface damage, such as corrosion, oxidation, wear, fretting, or erosion, affects surface finish and dimensional integrity of the components. Surface damage, in the form of cracks, dents, or fretting wear, in the absence of adequate maintenance could lead to fatigue failure affecting the functionality of the component. Internal damage such as aging, creep, and fatigue could affect the microstructure of highly stressed parts that may reduce the strength of the component. The accumulation of internal damage could cause the initiation of flaws which ultimately lead to component failure, as with surface damage, there is a common denominator (NATO 2000).

The increasing abundance and diversity of engineered products from many sources has, at the same time, affected the ability of end users to repair or replace failed components. Reasons for this include, in no particular order, the unavailability of

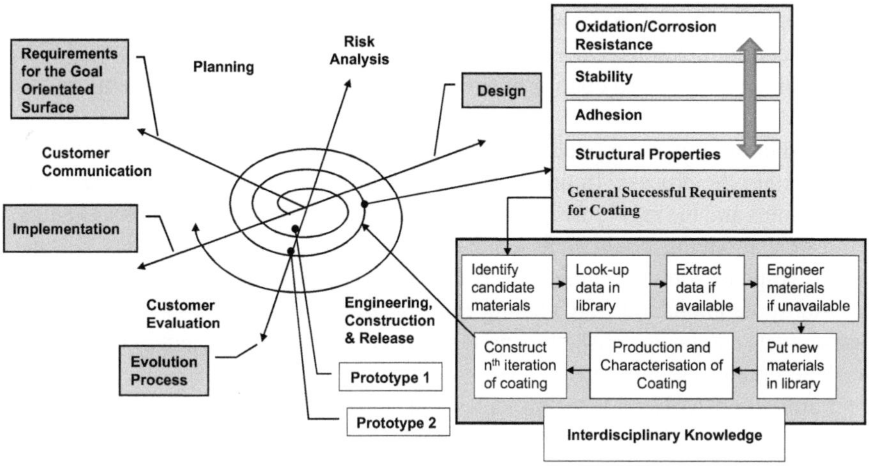

Fig. 1.7 Process model life cycle engine for goal-orientated surface development

replacement parts, decrease in support available from original manufacturers, and rapid technological obsolesce.

In view of the above, there is a need for the concurrent materials design of the surface and substrate together to form a functionally graded system capable of giving a product a cost-effective performance enhancement of which neither of them is capable on its own (Bruzzone et al. 2008). Therefore, there is a need for an interdisciplinary surface engineering approach to solve some of the issues mentioned above and so, provide one of the most important means of engineering product differentiation in terms of quality, performance, and life cycle costs. This approach model is illustrated in Fig. 1.7. The model life cycle engine in this figure represents an evolutionary spiral that starts with customer communication. It is here that the basic problem domain is identified and defined. Then, planning and risk analysis establish a foundation for the project plan. The technical work associated with the design and development of a new goal-orientated surface follows the iterative path. The general successful requirements of a metallic coating must be analyzed and prioritized. For example: (1) a coating required to be oxidation and corrosion resistant should be thermodynamically stable, with protective surface scale of uniform thickness, and show slow growth rate of protective surface scale; (2) a coating required to be stable should have no undesired phase changes within the coating, have a low diffusion rate across interface at use temperature, and show a minimum brittle phase formation; (3) a coating required to have a good adherence to substrate should have a good adhesion to the substrate, have matched coating/substrate properties to reduce thermal stress, and minimum growth stresses; and (4) a coating required to have structural properties must withstand service-related creep, fatigue, and impact loading of surface without failure of function (Bruzzone et al. 2008).

Due to the complexity of the development of a new or improved goal-orientated surface, there is a need for a reiterative process; that is, after solving one challenge another challenge often arises. Based on lessons learnt, this might include new questions about the refinement of the concept, improvement of the design, and new manufacturing methods. Only this integrated interdisciplinary approach could lead to the understanding and control of surface phenomena, particularly at a micro- and nanoscale.

In addition, process modeling is one way to better understand complex interactions within the process. In order to better manage the various parts of the model, sub-models and the relationships between them should be considered. For example, cold spray modeling could include sub-models such as gas flow dynamics, gas–particle two-phase flow dynamics, convergent–divergent nozzle optimization, particle–substrate interaction, and coating characterization. Relationships between these sub-models can be established using real-life cold spray parameters in conjunction with acceptable assumptions.

Furthermore, cold spray technology should have a clear mission, objectives, strategy, and approach (Kestler 2011). For example, a mission could be to provide full life cycle support to applicable industrial operations; the strategy could include actions to implement cold spray technology into these operations, which should lead to better performance and overall cost reduction. To implement the strategy, one could also consider recommendations to: decompose the complex problem into smaller more manageable subproblems (Kochikar and Narendran 1999); create the conditions for learning, adapting, optimizing, and reconfiguring at various levels of the process (Wang et al. 2002); establish a well-equipped cold spray laboratory where the new technology could be demonstrated and its associated knowledge preserved; increase the community's interest in the cold spray process via informed demonstrations; tailor the technology for each specific application, and develop process certification procedures and methods; cooperate with the best in the field from universities, services, and industry; support cold spray development efforts (Botef 2013).

In summary, the cold spray process is much more than spraying particle and achieving coatings. Like any other technology, cold spraying requires sufficient understanding of relevant disciplines such as metallurgy, chemistry, physics, production, and management. Cold spray is a multidisciplinary process that defines the method of transforming a powdered material into a final material consolidation, being that a functional coating or a free form.

1.6 Concluding Remarks

This chapter explored cold spray technology in the context of the much wider thermal spray family, its beginnings as well as its relative advantages and limitations. An attempt was made at discussing cold spray as a single component of a more complex multidisciplinary surface engineering field, aimed at solving engineering problems. It is indeed, both theoretically and practically, possible to use the cold spray process as a method to solve a number of modern surface engineering challenges including recyclability, life cycle, remanufacturing, emissions, energy consumption, and environmental friendliness.

Cold spray was presented as a challenger to traditional problem-solving approaches, drawing attention to the need for new methods that allow us to continue developing and implementing advanced materials for aerospace, electronics, information technology, energy, optics, tribology, and bioengineering applications.

Finally, this chapter served as a brief introduction to the vast amount of current information to be presented in subsequent chapters by well-known professionals in the field of thermal and cold spray.

References

Alhulaifi, Abdulaziz S., Gregory A. Buck, and W. J. Arbegast. 2012. Numerical and experimental investigation of cold spray gas dynamic effects for polymer coating. *Journal of Thermal Spray Technology* 21 (5): 852–862.

Al-Mangour, B., R. Mongrain, E. Irissou, and S. Yue. 2013. Improving the strength and corrosion resistance of 316L stainless steel for biomedical applications using cold spray. *Surface & Coatings Technology* 216:297–307.

Ang, A. S. M., C. C. Berndt, and P. Cheang P. 2012. Deposition effects of WC particle size on cold sprayed WC-Co coatings. *Surface & Coatings Technology* 205:3260–3267.

Atkinson, Joe. 2013. NASA researchers to flying insects: Bug Off!, NASA, http://www.nasa.gov/aero/bug_off.html#.VMqE9tJ4qK8

Bar-Yam, Y. 2003. The dynamics of complex systems. Reading: Westview.

Berube, G., M. Yandouzi, A. Zuniga, L. Ajdelsztajn, J. Villafuerte, and B. Jodoin. 2012. Phase stability of Al-5Fe-V-Si coatings produced by cold gas dynamic spray process using rapidly solidified feedstock materials. *Journal of Thermal Spray Technology* 21:240–254.

Bose, S. 2007. High temperature coatings. Manchester: Elsevier.

Botef, I. 2013. Complex simplicity: A case study. In *Recent advances in mathematical methods & computational techniques in modern science*, eds. H. Fujita, M. Tuba, and J. Sasaki. 1st International conference on complex systems and chaos, Morioka, April 2013. Mathematics and computers in science and engineering series, vol. 11, p. 47. Wisconsin: WSEAS.

Bruzzone, A. A. G., H. L. Costa, P. M. Lonardo, and D. A. Lucca. 2008. Advances in engineered surfaces for functional performance. *CIRP Annals—Manufacturing Technology* 57:750–769.

Champagne, V. K. 2007. Introduction. In *The cold spray materials deposition process: Fundamentals and applications*, 1st ed., ed. V. K. Champagne. Cambridge: Woodhead.

Champagne, V., and D. J. Helfritch. 2013. A demonstration of the antimicrobial effectiveness of various copper surfaces. *Journal of Biological Engineering* 7:8.

Chavan, N. M., B. Kiran, A. Jyothirmayi, P. S. Phani, and G. Sundararajan. 2013. The corrosion behavior of cold sprayed zinc coatings on mild steel substrate. *Journal of Thermal Spray Technology* 22:463–470.

Chraska, P., J. Dubsky, B. Kolman, J. Ilavsky, and J. Forrnan. 1992. Study of phase changes in plasma sprayed deposits. *Journal of Thermal Spray Technology* 1:301–306.

Davis, J. R. 2004. *Handbook of thermal spray technology*. Materials Park: ASM International.

Dong, S-J., B. Song, G. Zhou, C. Li, B. Hansz, H. Liao, and C. Coddet. 2013. Preparation of aluminum coatings by atmospheric plasma spraying and dry-ice blasting and their corrosion behavior. *Journal of Thermal Spray Technology* 22:1222–1228.

Donner, K. R., F. Gaertner, and T. Klassen. 2011. Metallization of thin Al_2O_3 layers in power electronics using cold gas spraying. *Journal of Thermal Spray Technology* 20 (1/2): 299–306.

Dorfman, M. R., and A. Sharma. 2013. Challenges and strategies for growth of thermal spray markets: The six-pillar plan. *Journal of Thermal Spray Technology* 22 (5): 559–563.

Dugdale, H., D. Armstrong, E. Tarleton, S. G. Roberts, and S. L. Perez. 2013. How oxidized grain boundaries fail. *Acta Materialia* 61:4707–4713.

Dusek, V. 2006. Philosophy of technology: An introduction. Malden: Blackwell Publishing.
Dykhuizen, R. C., and M. F. Smith. 1998. Gas dynamic principles of cold spray. *Journal of Thermal Spray Technology* 7 (2): 206–212.
Esfahani, E. A., H. Salimijazi, M. A. Golozar, J. Mostaghimi, and L. Pershin. 2012. Study of corrosion behavior of arc sprayed aluminum coating on mild steel. *Journal of Thermal Spray Technology* 21 (6): 1195–1202.
Gan, J. A., and Christopher C. Berndt. 2013. Review on the Oxidation of Metallic Thermal Sprayed Coatings: A Case Study with Reference to Rare-Earth Permanent Magnetic Coatings. *Journal of Thermal Spray Technology* 22:1069–1091.
Gaydos, S. 2011. Qualification of cold spray for repair of MIL-DTL-83488 aluminum coatings. ASETS defense workshop, Boeing, US.
Ghelichi, R., D. MacDonald, S. Bagherifard, H. Jahed, M. Guagliano, and B. Jodoin. 2012. Microstructure and fatigue behavior of cold spray coated Al5052. *Acta Materialia* 60:6555–6561.
Grujicic, M., C. Tong, W. DeRosset, and D. Helfritch. 2003. Flow analysis and nozzle-shape optimisation for the cold-gas dynamic-spray process. *Proceedings of the Institution of Mechanical Engineers* 217:1603–1613.
Grujicic, M., C. L. Zhao, W. S. DeRosset, and D. Helfritch. 2004. Adiabatic shear instability based mechanism for particles/substrate bonding in the cold-gas dynamic-spray process. *Materials and Design* 25:681–688
Hanson, T. C., and G. S. Settles. 2003. Particle temperature and velocity effects on the porosity and oxidation of an HVOF corrosion-control coating. *Journal of Thermal Spray Technology* 12:403–415.
Hermanek, F. J. 2013. What is thermal spray? *International Thermal Spray Association*. http://www.thermalspray.org. Accessed 2013.
Huang, R., and R. Fukanuma. 2012. Study of the influence of particle velocity on adhesive strength of cold spray deposits. *Journal of Thermal Spray Technology* 21:541–549.
Irissou, E., J. G. Legoux, B. Arsenault, and C. Moreau. 2007. Investigation of Al-Al_2O_3 cold spray coatings formation and properties. *Journal of Thermal Spray Technology* 16:661–668.
Irissou, E., J. G. Legoux, A. N. Ryabinin, B. Jodoin, and C. Moreau. 2008. Review on cold spray process and technology: Part I intellectual property. *Journal of Thermal Spray Technology* 17:495–516.
Irissou, E., P. Vo, D. Poirier, P. Keech, and J. G. Legoux. 2012. Cold sprayed corrosion protection coating for nuclear waste repository canister. North American Cold Spray Conference, Worcester Polytechnic Institute, Worcester, 30 Oct–1 Nov 2012.
Jahedi, M., S. Zahiri, P. King, S. Gulizia, and C. Tang. 2013. "Cold spray of Titanium", ASM International, Aeromat 2013 Conf. Proc., Apr 2–5, Belleview, Washington.
Karthikeyan, J. 2004. Cold spray technology: International status and efforts. Barberton: ASB Industries.
Karthikeyan, J. 2007. The advantages and disadvantages of cold spray coating process. In *The cold spray materials deposition process: fundamentals and applications*, 1st ed., ed. V. K. Champagne. Cambridge: Woodhead.
Kaye, T., and Thyer, R. 2006. Spray coatings: Cold gold—cool moves. Solve Issue 6, CSIRO.
Kestler, R. 2011. NAVAIR cold spray efforts, cold spray action team presentation, Fleet Readiness Center East.
Kim, H. J., C. H. Lee, and S. Y. Hwang. 2005. Fabrication of WC–Co coatings by cold spray deposition. *Surface & Coatings Technology* 19:335–340.
Klassen, T. 2014. HVOF introduction. Institute of Materials Technology http://www.hsu-hh.de. Accessed 09 March 2014.
Knight, P. 2008. *Thermal spray: Past, present and future. A look at Canons and Nanosplats.* Philadelphia: Drexel University.
Kochikar, V. P., and T. T. Narendran. 1999. Logical cell formation in FMS using flexibility-based criteria. *I. J. of Flexible Manufacturing Systems* 10:163–181.

Koivuluoto, H., and P. Vuoristo. 2010. Effect of powder type and composition on structure and mechanical properties of Cu+ Al2O3 coatings prepared by using low-pressure cold spray process. *Journal of Thermal Spray Technology* 19 (5): 1081–1092.
Koivuluoto, H., J. Lagerbom, M. Kylmalahti, and P. Vuoristo. 2008. Microstructure and mechanical properties of low-pressure cold-sprayed (LPCS) coatings. *Journal of Thermal Spray Technology* 17:721–727.
Koivuluoto, H., A. Coleman, K. Murray, M. Kearns, and P. Vuoristo. 2012. High pressure cold sprayed (HPCS) and low pressure cold sprayed (LPCS) coatings prepared from OFHC Cu feedstock: Overview from powder characteristics to coating properties. *Journal of Thermal Spray Technology* 21 (5): 1065–1075.
Kusiak, A. 2000. *Computational intelligence in design and manufacturing*. London: Wiley-Interscience.
Lee, H. Y., S. H. Jung, S. Y. Lee, Y. H. You, K. H. Ko. 2005. Correlation between Al_2O_3 particles and interface of Al-Al_2O_3 coatings by cold spray. *Applied Surface Science* 252:1891–1898.
Lee, H. Y., S. H. Jung, S. Y. Lee, K. H. Ko. 2006. Fabrication of cold sprayed Al-intermetallic compounds coatings by post annealing. *Materials Science and Engineering A* 433:139–143.
Lee, H., H. Shin, S. Lee, and K. Ko. 2008. Effect of gas pressure on Al coatings by cold gas dynamic spray. *Materials Letters* 62:1579–1581.
Lee, H., H. Shin, and K. Ko. 2010. Effects of gas pressure of cold spray on the formation of Al-based intermetallic compound. *Journal of Thermal Spray Technology* 19:102–109.
Li, C. J., H. T. Wang, Q. Zhang, G. J. Yang, W. Y. Li, and H. L. Liao. 2010. Influence of spray materials and their surface oxidation on the critical velocity in cold spraying. *Journal of Thermal Spray Technology* 19:95–101.
Lupoi, R., and W. O'Neill. 2010. Deposition of metallic coatings on polymer surfaces using cold spray. *Surface and Coatings Technology* 205:2167–2173.
Luzin, V., K. Spencer, and M. X. Zhang. 2011. Residual stress and thermo-mechanical properties of cold spray metal coatings. *Acta Materialia* 59:1259–1270.
Maev, R. Gr., and V. Leshchynsky. 2008. *Introduction to low pressure gas dynamic spray, physics and technology*. Weinheim: Wiley-VCH.
Melendez, N. M., and A. G. McDonald. 2013. Development of WC-based metal matrix composite coatings using low-pressure cold gas dynamic spraying. *Surface & Coatings Technology* 214:101–109.
Murrone, A., and P. Villedieu. 2011. Numerical modeling of dispersed two-phase flows. *Journal of Aerospace Lab* AL02–AL04:1–13.
Nahle, N. 2009. Thermal energy and heat (biophysics). http://www.biocab.org. Accessed 09 March 2014.
NATO. 2000. RTO-EN-14 aging engines, avionics, subsystems and helicopters. Research and Technology Organization, North Atlantic Treaty Organization (NATO).
Ning, X-J., J-H. Jang, H-J. Kim, C-J. Li, and C. Lee. 2008. Cold spraying of Al–Sn binary alloy: Coating characteristics and particle bonding features. *Surface and Coatings Technology* 202:1681–1687.
Ogawa, K., K. Ito, K. Ichimura, Y. Ichikawa, S. Ohno, and N. Onda. 2008. Characterization of low-pressure cold-sprayed aluminum coatings. *Journal of Thermal Spray Technology* 17:728–735.
Papyrin, A. 2006. Cold spray: State of the art and applications. European Summer University, St-Etienne, Sept 11–15.
Papyrin, A. 2007. The development of the cold spray process. In *The cold spray materials deposition process: Fundamentals and applications*, 1st ed., ed. V. K. Champagne. Cambridge: Woodhead.
Papyrin, A., V. Kosarev, S. Klinkov, A. Alkhimov, and V. Fomin. 2007. *Cold spray technology*, 1st ed. Oxford: Elsevier.
Phani, P. S., D. S. Rao, S. V. Joshi, and G. Sundararajan. 2007. Effect of process parameters and heat treatments on properties of cold sprayed copper coatings. *Journal of Thermal Spray Technology* 16:425–434.

Rudinger, G. 1976. Flow of solid particle in gases. AGARD-AG-222, Advisory Group For Aerospace Research and Development (AGARD), North Atlantic Treaty Organization (NATO), London.

Sampath, S, U. Schulz, M. O. Jarligo, and S. Kuroda. 2012. Processing Science of advanced thermal-barrier systems. *MRS Bulletin* 37 (12): 903–910.

Schmidt, T., H. Assadi, F. Gartner, H. Richter, T. Stoltenhoff, H. Kreye, and T. Klassen. 2009. From particle acceleration to impact and bonding in cold spraying. *Journal of Thermal Spray Technology* 18:794–809.

Seo, D., M. Sayar, and K. Ogawa. 2012. SiO2 and MoSi2 formation on Inconel 625 surface via SiC coating deposited by cold spray. *Surface & Coatings Technology* 206:2851–2858.

Shayegan, G., H. Mahmoudi, R. Ghelichi, J. Villafuerte, J. Wang, M. Guagliano, H. Jahed. 2014. Residual stress induced by cold spray coating of magnesium AZ31B extrusion. *Materials and Design* 60 (2014): 72–84.

Smith, M. F. 2007. Comparing cold spray with thermal spray coating technologies. In *The cold spray materials deposition process: Fundamentals and applications*, 1st ed., ed. V. K. Champagne. Cambridge: Woodhead.

Sova, A., D. Pervushin, and I. Smurov. 2010. Development of multimaterial coatings by cold spray and gas detonation spraying. *Surface and Coatings Technology* 205:1108–1114.

Sova, A., M. Doubenskaia, S. Grigoriev, A. Okunkova, and I. Smurov. 2013. Parameters of the gas-powder supersonic jet in cold spraying using a mask. *Journal of Thermal Spray Technology* 22 (4): 551–556.

Spencer, K., V. Luzin, N. Matthews, M. X. Zhang. 2012. Residual stresses in cold spray Al coatings: The effect of alloying and of process parameters. *Surface & Coatings Technology* 206:4249–4255.

SST. 2014. Practical cold spray coatings. http://www.supersonicsprat.com. Accessed 10 March 2014.

Sulzer Metco. 2014. An introduction to thermal spray. http://www.sulzer.com. Accessed 05 March 2014.

Suo, X. K., X. P. Guo, W. Y. Li, M. P. Planche, and H. L. Liao. 2012. Investigation of deposition behavior of cold-sprayed magnesium coating. *Journal of Thermal Spray Technology* 21:831–837.

Szyperski, C., D. Grunts, and S. Murer. 2002. *Component software. Beyond object-oriented programming*. London: Addison-Wesley.

US ARMY RESEARCH LAB ARL. 2008. http://www.arl.army.mil/www/pages/375/MIL-STD-3021.pdf. Accessed 2013.

Villafuerte, J., and W. Zheng. 2007. Corrosion protection of magnesium alloys by cold spray. Advanced materials & proceses, September, 53–54.

Villafuerte, J., and Zheng, W. 2011. Corrosion protection of magnesium alloys by cold spray. In *Magnesium alloys corrosion and surface treatments*, 1st ed., ed. F. Czerwinski, 185–194. Croatia: InTech.

Walpole, H. 1754. Letter to Horace Mann dated 28 January 1754, Wikipedia website. Accessed Aug 2013.

Wang, J., and J. Villafuerte. 2009. Low pressure cold spraying of tungsten carbide composite coatings. *Advanced Materials and Process* (ASM International 2009). 167 (2): 54–56.

Wang, C., Y. Zhang, G. Song, C. Yin, and C. Chu. 2002. An integration architecture for process manufacturing systems. *International Journal of Computer Integrated Manufacturing* 15:413–426.

Wang, H. T., C. J. Li, G. C. Ji, G. J. Yang. 2012. Annealing effect on the intermetallic compound formation of cold sprayed Fe/Al composite coating. *Journal of Thermal Spray Technology* 21:571–577.

Wang, H. T., C. J. Li, G. J. Yang, and C. X. Li. 2008. Cold spraying of Fe/Al powder mixture: Coating characteristics and influence of heat treatment on the phase structure. *Applied Surface Science* 255:2538–2544.

Wielage, B., T. Grund, C. Rupprecht, and S. Kuemme. 2010. New method for producing power electronic circuit boards by cold-gas spraying and investigation of adhesion mechanisms. *Surface and Coatings Technology* 205 (4): 1115–1118.

Wilden, J., S. Jahn, S. Reich, and S. Dal-Canton. 2008. Cladding of aluminum substrates with iron based wear resistant materials using controlled short arc technology. *Surface and Coatings Technology* 202:4509–4514.

Xie, L., D. Chen, E. H. Jordan, A. Ozturk, F. Wu, X. Ma, B. M. Cetegen, and M. Gell. 2006. Formation of vertical cracks in solution-precursor plasma-sprayed thermal barrier coatings. *Surface & Coatings Technology* 201:1058–1064.

Yandouzi, M., P. Richer P, and B. Jodoin. 2009. SiC particulate reinforced Al–12Si alloy composite coatings produced by the pulsed gas dynamic spray process: Microstructure and properties. *Surface and Coatings Technology* 203:3260–3270.

Zhou, X. L., A. F. Chen, J. C. Liu, X. K. Wu, and J. S. Zhang. 2011. Preparation of metallic coatings on polymer matrix composites by cold spray. *Surface and Coatings Technology* 206:132–136.

Chapter 2
The Physics of Cold Spray

P. King, M. Yandouzi and B. Jodoin

2.1 Introduction

As introduced in Chap. 1, cold spray (CS) is considered part of the larger family of thermal spray processes. CS is a material deposition technique in which micron-size solid particles are accelerated to high velocities prior to impacting on a substrate and produce a coating through complex deformation and bonding mechanisms. It is also possible to produce near-net-shape parts in a similar way to additive manufacturing processes. The growing interest in CS has been associated with a continual search for optimal spraying conditions, in order to maximize both the coating properties and the process cost-effectiveness. Nowadays, CS can be regarded as an established method for metallic coating/repair production and a promising technique for net-shape rapid manufacturing/additive manufacturing.

Through over two decades of research and development activities, many commercial CS systems have emerged and are now available from equipment manufacturers in various forms, as depicted in Chap. 6. In the CS process, particles are accelerated to high impact velocities by means of a pressurized propellant gas flow that expands in a converging–diverging (de Laval type) nozzle. Two different approaches can be adopted to introduce the particles into the propellant gas stream. In the first approach, the particles are introduced before the converging–diverging nozzle throat, where the gas pressure is higher than the ambient pressure. This is referred to as upstream injection—the powder particles must be pushed at high

P. King (✉)
CSIRO, Clayton, VIC, Australia
e-mail: Peter.King@csiro.au

M. Yandouzi
University of Ottawa, Ottawa, ON, Canada
e-mail: yandouzi@uottawa.ca

B. Jodoin
University of Ottawa, Ottawa, ON, Canada
e-mail: bertrand.jodoin@uottawa.ca

J. Villafuerte (ed.), *Modern Cold Spray*, DOI 10.1007/978-3-319-16772-5_2

pressure into the stream, thus requiring powder feeder systems adapted to high pressure. In the second approach, particles are introduced downstream of the converging–diverging nozzle throat, and this approach is categorized as downstream injection CS. In the CS process, the powder particles are predominantly accelerated in the supersonic portion of the gas jet (in the diverging section), which is at a temperature that is considerably lower than the material's melting point, resulting in coating formation from particles that remain in the solid state throughout their flight. As a result, the harmful effects of high-temperature oxidation, evaporation, melting, crystallization, thermal residual stresses and other common problems for traditional thermal spray processes are minimized or even eliminated. Removing the deleterious effects of high temperature on coatings and substrates offers significant benefits and new prospects and makes the CS process promising for many industrial applications.

It is well known that material deposition in CS results from high-velocity impact and subsequent deformation of particles (Champagne 2007). A most determining factor for bonding is the velocity of the impacting particle. Bonding occurs if the particle velocity upon impact, V_p, becomes greater than a critical value, V_c (Kosarev et al. 2003). The latter is defined as the minimum velocity that any sprayed particle must reach in order to adhere to the substrate, and it is believed to be the main key of the CS process. In general, high particle impact velocity is necessary for optimal deposition efficiency and packing density.

Deposition rate and coating quality depend directly on a large number of process parameters. Furthermore, many process parameters are interrelated, making the investigation of their independent effects a challenging task. Although all the CS process parameters affect the deposition rate and/or the coating quality, some have shown to be more imperative than others. The major parameters can be broadly grouped into four categories:

- *Propellant gas:* The nature of the gas, the gas stagnation pressure and temperature.
- *Feedstock material:* The characteristics of the selected powder, such as the material composition, particles size, particles geometry and size distribution, the particles feeding rate, as well as the particle impact temperature.
- *Substrate:* The nature of the material to be coated and its surface preparation prior to spraying, as well as surface temperature.
- *Spray setup:* This includes the nozzle geometry and material, traverse speed, orientation and distance from the substrate (standoff distance).

2.2 Process Parameters

A typical CS system is schematically illustrated in Fig. 2.1. A high-pressure gas supply (typically helium, nitrogen or air) is used to generate the driving flow that accelerates the powder particles. This flow can be heated by means of an electric gas heater, thus allowing operation at gas temperatures ranging from room temperature

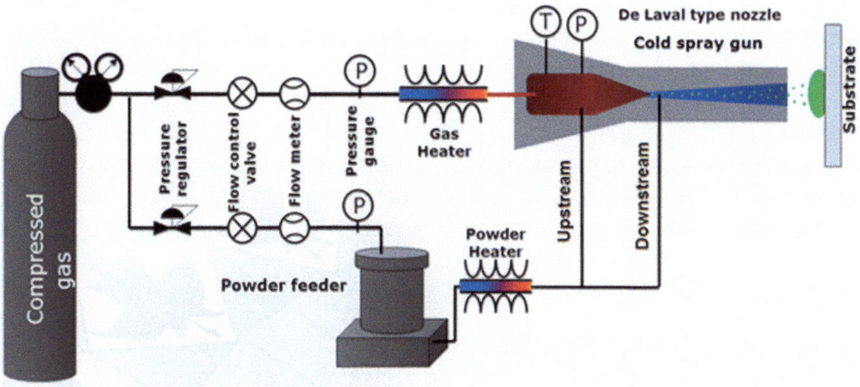

Fig. 2.1 Schematic illustrating the cold spray processes for both: **a** upstream powder injection and **b** downstream powder injection

up to 1200 °C. The primary purpose of heating the gas is to achieve greater gas velocities and consequently higher particle velocities. Softening of the feedstock particles may take place during the CS process and may be beneficial in some cases. The driving flow is fed into a converging/diverging (de Laval type) nozzle where it expands and accelerates to supersonic speeds through the conversion of thermal energy into kinetic energy (Davis 2004; Papyrin 2001). The gas velocity at the nozzle exit is a function of the nozzle geometry, gas type and gas stagnation conditions (pressure and temperature). The feeding flow is directed to a powder feeder where it entrains the powder particles into the nozzle where they are accelerated by the propellant flow. Depending on the choice of process gas and the spray parameters, impact velocities between 200 and 1200 m/s can be obtained. The particle impact temperature depends on various factors such as the gas temperature, nozzle type and design and heat capacity of particles.

The particles having acquired sufficient kinetic energy from the driving gas impact the substrate, plastically deform and adhere to the surface to form a coating. Not all the particles impacting the substrate surface have sufficient kinetic energy to plastically deform and bond to form a coating. Particles having insufficient velocity will bounce off the substrate and potentially erode its surface.

2.2.1 Propellant Gas: Gas Nature

As detailed previously, in the CS process, a high-pressure gas supply is used to accelerate the powder particles while a secondary flow is used to move the powder particles from the powder feeder to the nozzle.

In principle, any pressurized gas can be used. However, it is customary to use air, nitrogen (N_2), or helium (He), with the latter two presenting the advantage of being inert. It can be shown that gases with lower molecular weight, such as helium, are beneficial as they reach a higher velocity for a specific nozzle geometry. Therefore, from a purely thermodynamic point of view, helium might be considered the best candidate. However, the cost of helium can be prohibitive, making it economically unviable for many applications unless recycled using a dedicated (and expensive) helium recovery system. In some applications, a mixture of helium and nitrogen is used as the driving gas. Nitrogen is a diatomic gas, and its addition into helium increases the enthalpy of the carrier gas, yielding improved heat transfer with the spray particles. However, it also comes at the expense of gas velocity due to the heavier atomic mass of the gas mixture, potentially resulting in coatings with reduced density and hardness (Balani et al. 2005).

2.3 Supersonic Gas Flow Phenomena: Nozzle Design, Process Gas Parameters

2.3.1 *Why Cold Spray Requires the Study of High-Velocity Flows*

As stated previously, the CS process requires that the feedstock powder particles be accelerated to high velocity prior to impact with the substrate to be coated. Failure to consistently achieve high impact velocity will result at best in porous coatings with the presence of poorly deformed particles and more often in particles bouncing off the substrate rather than deforming and bonding to it. A more detailed discussion on critical velocity and the nature of particle deformation and bonding follows in Sect. 2.4.

In order to be able to accelerate particles to such large impact velocities, the propellant gases used to impart them their kinetic energy must be flowing at even larger velocities, typically over thousands of metres per second.

2.3.2 *Differences Between Incompressible and Compressible Flows: Pressure, Density and Temperature Evolution*

At these orders of magnitudes (thousands of metres per second), it is typical for gases to be flowing in the supersonic regime, that is, the gas local velocity is larger than the local speed of sound in this gas. The field dealing with this type of flows is a combination of classical fluid mechanics and classical thermodynamics and is known as gas dynamics. The major difference between classical fluid mechanics and gas dynamics relies on the fact that at speeds close to or beyond the local speed of sound, a gas flow will start to experience pressure gradients large enough to induce gas density variations that needs to be accounted for. It is customary in gas dynamics to set the minimal gas velocity limit at which these changes start to be important at a value of local Mach number of 0.3. The local Mach number is defined as the ratio of the local gas velocity to the local speed of sound, and flow regimes

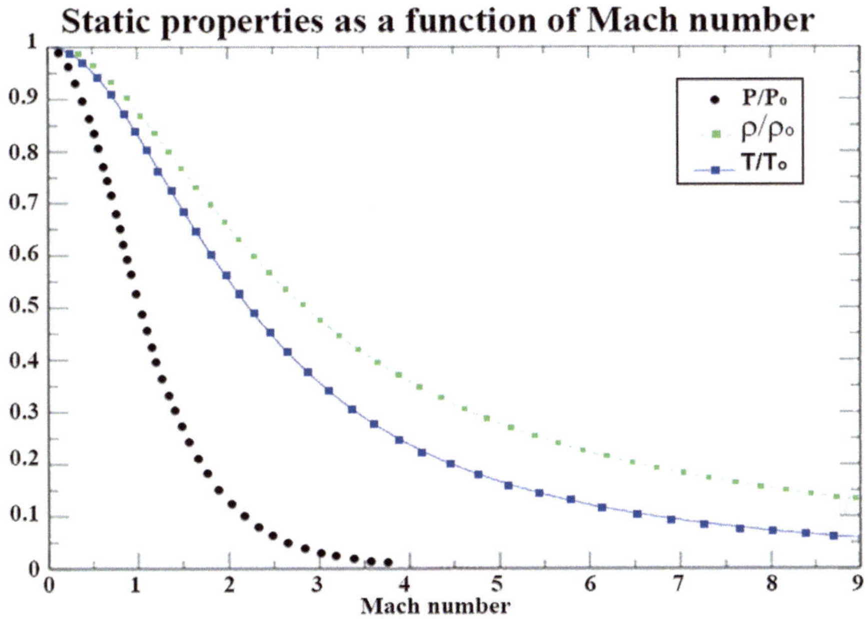

Fig. 2.2 Static/stagnation ratio of pressure, density and temperature as a function of flow Mach number

are usually categorized as: incompressible flow for $M>5$. The gas density is not the only flow variable that can drastically change when the flow approaches or expands beyond the transonic regime. The gas local static temperature and pressure also vary significantly, as shown in Fig. 2.2.

One can observe from Fig. 2.2 that at low flow Mach numbers, in the incompressible regime, there is no noticeable change of pressure and temperature (and thus enthalpy). In the subsonic regime, both pressure and temperature start to experience some changes while these changes become more important in the supersonic regime. This behaviour can be better understood when considering the first law of thermodynamics (energy conservation) applied to an infinitesimal control volume through which a gas is accelerated in an adiabatic (no heat transfer) and reversible way (no friction). Assuming this process to be occurring in a steady state (no change in time of any fluid properties in the control volume), the first law can be expressed as follows:

$$h_1 + \frac{v_1^2}{2} = h_2 + \frac{v_2^2}{2}, \tag{2.1}$$

where h is the gas local enthalpy and v the gas local velocity. This equation, ensuring that energy is conserved, simply states that in a flow accelerating according to the assumptions stated above, the local energy of the flow is conserved but can be

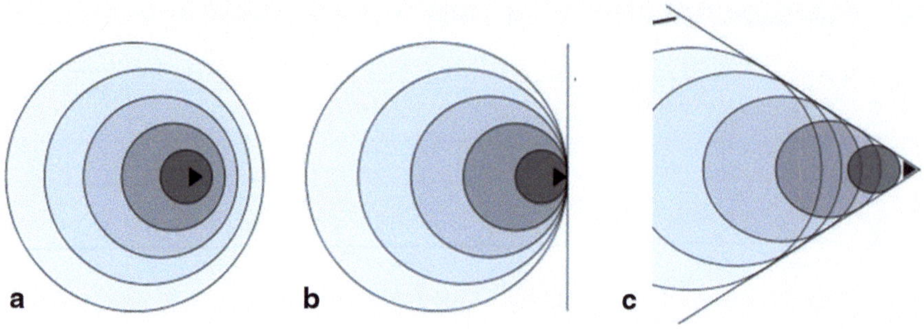

Fig. 2.3 **a** Pressure wave pattern emitted by a point perturbation in a subsonic flow. Flow from right to left. **b** Pressure wave pattern emitted by a point perturbation in a sonic flow. Flow from right to left. **c** Pressure wave pattern emitted by a point perturbation in a supersonic flow. Flow from right to left

expressed in two forms: enthalpy and kinetic energy. An increase in kinetic energy (and thus flow velocity) must be accompanied by a decrease in enthalpy, which for calorically perfect gases such as those used in CS can directly be translated as a decrease in gas temperature since the enthalpy of a calorically perfect gas is given by Eq. 2.2:

$$\mathrm{d}h = C_{\mathrm{P}}\,\mathrm{d}T. \tag{2.2}$$

In this equation, C_{P} is the gas constant pressure specific heat while dh and dT are the changes in enthalpy and temperature, respectively. At low Mach numbers, the increase in gas velocity, and thus increase of kinetic energy, is not sufficient to significantly affect the gas enthalpy, but as the velocity increases (and since the kinetic energy is a function of the square of the gas velocity), the effect on the gas enthalpy (and therefore temperature, and consequently pressure if we assume that the gases in CS do obey the perfect gas law) becomes more and more important, as reflected in Fig. 2.3.

2.3.3 Differences Between Classic Fluid Mechanics and Gas Dynamics: Normal Shock Wave

Apart from the large density changes (as well as pressure and temperature changes) that differentiate compressible flows from incompressible flows, another major difference between them resides in the potential and possible appearance of shock waves in compressible flows. Shock waves are abrupt discontinuities in the flow properties that occur over extremely short distances, typically over a few mean free paths (order of magnitude of a few angstroms). Shock waves are quite common, and one of the most classical examples is the crack of a whip, which is the result of a shock wave that forms at the tip of the whip as it travels faster than the speed of sound. The noise (or crack) emitted by the whip is the result of large changes in flow properties through the shock wave, in particular pressure, that is perceived by the ears.

The appearance of shock waves requires the flow to be in the supersonic regime and is the result of the flow having to accommodate suddenly for a downstream condition or perturbation. The idealized point perturbation model provides a simplified visualization of this effect. When an infinitesimal perturbation exists in a flow, its presence is signalled throughout the fluid by infinitesimal pressure waves that are continuously emitted at the perturbation surface and travel in the fluid at a relative velocity equal to the local speed of sound of the fluid. The pattern of these pressure waves in an incompressible flow can be compared to the perturbation pattern created on the surface of water by a rock that was dropped at one location: The pressure waves travel in a symmetric way throughout the fluid and eventually reach every location of the fluid. If the fluid moves at a certain velocity in the subsonic regime, the pressure waves emitted by the downstream perturbation (triangle in Fig. 2.3) are transmitted according to the pattern shown in Fig. 2.3. The pattern is no longer symmetrical as the incoming fluid velocity must be subtracted from the pressure wave's velocity for those travelling against the flow and added to those travelling in the flow direction.

However, if the fluid is moving exactly at the speed of sound, then the pressure waves emitted at the surface of the infinitesimal point perturbation cannot reach any fluid located ahead of it and rather accumulate exactly at its location, as shown in Fig. 2.3. If the fluid is moving faster than the speed of sound (supersonic regime), then the pressure waves emitted at the surface of the infinitesimal point perturbation are now confined to a limited zone called the Mach cone, as shown in Fig. 2.4.

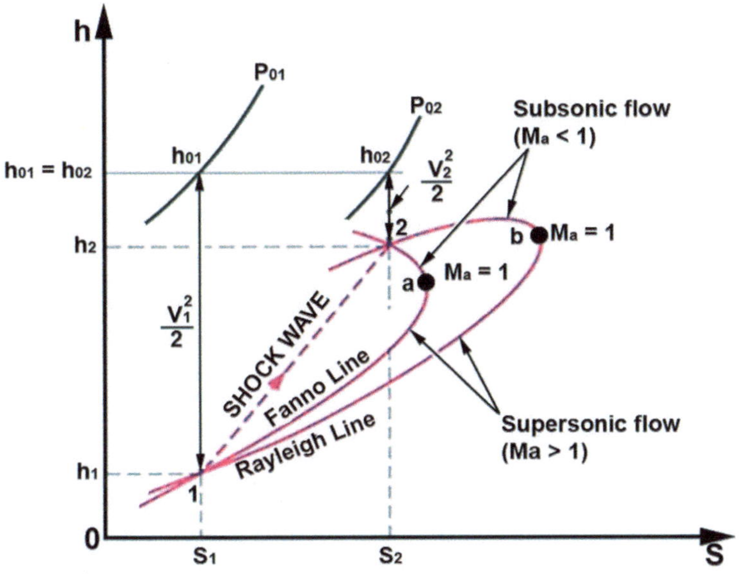

Fig. 2.4 Fanno–Rayleigh line

While simplistic, the infinitesimal point perturbation model allows one to understand that pressure waves will form different patterns based on the flow regime and that they may even coalesce locally to create a large finite pressure wave. When dealing with a complete surface (such as a solid wall), rather than simply an infinitesimal point disturbance, the coalescing waves will become a shock wave through which large flow property changes occur. One important type of shock wave in CS is the one that forms in front of the substrate. Since the jet exiting the CS nozzle is supersonic, the perturbation that represents the substrate cannot be perceived by the jet, and as such, a shock wave is formed ahead of the latter in order to abruptly slow down the jet to subsonic speeds. Once subsonic, the jet now receives the perturbations created by the presence of the substrate allowing the flow to reach stagnation at the substrate and flow around the edges of the substrate. However, shock waves are said to be detrimental to CS, due to the way they affect the flow properties.

2.3.4 Flow Property Changes Across a Normal Shock Wave

An enthalpy–entropy (h–s) diagram presenting the Fanno–Rayleigh lines is useful for visualizing and deducing how normal shock waves affect the properties of a supersonic flow.

Despite the large and sudden variation of properties experienced by a flow going through a normal shock wave, the flow must still obey the fundamental laws of mass, momentum and energy conservation, and since the gas used in CS can be assumed to be a calorically perfect gas, the flow must also obey the perfect gas law. The Fanno and Rayleigh lines represent the locus on an h–s diagram that obey the mass, energy and perfect gas laws for the former and the mass, momentum and perfect gas laws for the latter, for a given initial flow state. The interest in these lines lies in the fact that for an initial flow state, only the loci that are on both lines simultaneously obey all the fundamental conservation equations, that is, mass, momentum and energy equations along with the perfect gas law. Since the Fanno–Rayleigh lines intersect only twice, as shown in Fig. 2.4, this allows one to be able to predict the end state of the flow if the initial state is known.

This jump in flow properties that is observable on the Fanno–Rayleigh line diagram is furthermore dictated by the second law of thermodynamics. Detailed analysis using this law indicates that the jump can only be from the lowest entropy state to the highest entropy state (thus from left to right on the Fanno–Rayleigh diagram). As a result, and since it can be shown that the lower branches of both lines represent supersonic flow properties while the upper branches represent subsonic flow properties, it is then concluded that a normal shock wave can only occur in a supersonic flow and always results in the flow deceleration to subsonic speed. A close examination and analysis of the Fanno–Rayleigh lines allow the evolution of flow properties across the shock wave to be determined.

As stated previously, the flow is suddenly slowed down to subsonic speed, and the entropy increases across a normal shock wave. Furthermore, as can be observed in Fig. 2.4, the flow enthalpy increases across the shock, as well as the pressure.

Consequently, the gas temperature increases. Since the flow obeys the energy conservation equation across the normal shock wave, it can then be deduced that the increase of enthalpy is balanced by an equivalent decrease in kinetic energy and therefore velocity. Following the perfect gas law, the gas density will increase across the shock wave.

The effect of normal shock waves on CS flows now becomes more apparent. CS particles flowing across a normal shock wave will suddenly be exposed to a substantially lower speed flow that will at best reduce their acceleration, but that can even slow down the particles if they are travelling at a higher speed than the new subsonic gas velocity. Furthermore, the increase in gas density will promote the slowdown of the particles by contributing to the drag force acting on them. For all these reasons, it is therefore a good practice in CS to try to prevent the appearance of shock waves in the nozzle, to ensure a smooth and continuous acceleration of the feedstock particles. Nevertheless, since the flow must be supersonic in order to allow proper acceleration of the particles through momentum transfer with the propellant gas, and since the substrate always represents a downstream perturbation to the supersonic flow coming out of the spray nozzle, a shock wave will likely always be present in front of the substrate. It has been shown that the standoff distance can affect the strength of the shock wave and also that the particle size as well as particle materials can influence the effect of the shock wave on particle deceleration (Jodoin 2002).

As the standoff distance increases, the strength of the shock wave decreases, that is, the changes in the flow properties across the shock waves are reduced. This is due to the fact that longer standoff distances will result in a natural jet deceleration due to viscous forces and consequently a reduction of the impacting jet velocity (thus Mach number). It has also been demonstrated that since the distance between the shock wave and the substrate is small, only the small/light particles are truly affected by the sudden flow deceleration, as the larger/heavier particles have sufficient momentum to overcome the enhanced drag force created by the subsonic flow behind the shock wave over the short shock–substrate distance (Jodoin 2002).

2.3.5 Nozzle Design

As mentioned previously, it is considered good practice in CS to try to prevent the appearance of shock waves in the nozzle, to ensure a smooth and continuous acceleration of the feedstock particles. Furthermore, gas dynamics analysis reveals that in order to accelerate a flow to supersonic speeds, the use of a converging–diverging (also known as a de Laval nozzle, after Carl. G.P. de Laval) is required. One of the most important results of early gas dynamics development is Eq. 2.3 that correlates the local change of area with the local flow Mach number:

$$\left(\frac{dV}{V}\right)(1-M^2) = -\left(\frac{dA}{A}\right) \tag{2.3}$$

The $(1-M^2)$ term results in a difference of the required change of flow area depending on the flow regime. For a subsonic flow, an increase in flow velocity requires a decrease of flow passage area as the term $(1-M^2)$ is positive. However, for a supersonic flow, an increase in flow velocity requires an increase of the flow passage area as the same term becomes negative. Consequently, the acceleration of a gas from zero velocity to supersonic regime requires first a converging section, up to it reaching $M=1$, then subsequently a diverging section to allow the acceleration process to go on. This is realized through the use of a converging–diverging nozzle.

However, the use of a converging–diverging nozzle does not guarantee achievement of a supersonic flow in the latter part of the nozzle. For that to occur, the driving force acting on the gas must be sufficient for it to reach $M=1$ at the junction of the two sections, the throat. Therefore, the pressure at which the gas enters the nozzle is a crucial parameter for achieving supersonic speeds. Assuming the gas velocity to be negligible at the nozzle inlet and that the pressure at the location where the nozzle is discharging can be varied from the pressure level found at the inlet all the way to zero, various cases are possible with a few illustrated in Fig. 2.5.

Since it is assumed that the gas velocity is negligible at the nozzle inlet, the properties at that location are usually referred to as stagnation properties, with the subscript 0. For a nozzle exposed to a discharging pressure equal to the inlet stagnation pressure, no flow occurs. As the discharging pressure is reduced, the gas starts to accelerate in the converging section of the nozzle, but decelerates in the diverging section as it remains subsonic at the throat. For discharging pressures below a specific value, the gas reaches the sonic velocity at the throat ($M=1$). At that point, no further discharging pressure affect the flow in the converging section as the perturbations brought about the new discharging pressure cannot be transmitted upstream of the nozzle throat. The nozzle is said to be choked. Furthermore, at this point, any reduction in the discharging pressure will lead to a supersonic flow, at least in a portion of the diverging section. Only two shockless flow regimes are

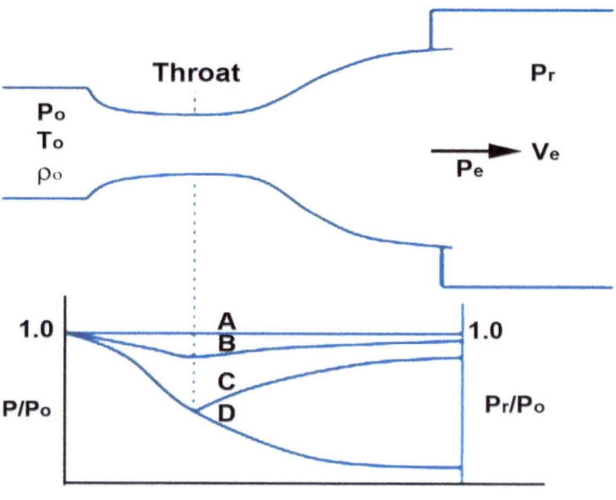

Fig. 2.5 Pressure distribution in converging–diverging nozzle, assuming one-dimensional isentropic flow

possible once the nozzle is choked. In the first, the discharging pressure is too high to sustain a supersonic flow in the diverging part of the nozzle. In that case, the flow decelerates without the presence of any shockwaves in the nozzle. For the second case, the discharging pressure is low enough to allow the gas to accelerate without the presence of any shockwave in the diverging section. For any discharging pressure between these two limiting cases, the flow will experience a shockwave in the diverging section of the nozzle, which appears to allow the flow to adjust to the discharge pressure. The ideal CS nozzle set-up is the one which generates a supersonic flow at the nozzle exit, without the presence of any shockwaves inside. Given the design nozzle exit Mach number M_e and the nozzle discharge pressure P_e (which is usually the ambient pressure unless the system operates in a vacuum or partial vacuum), the required stagnation pressure can be found using Eq. 2.4:

$$\frac{P_0}{P_e} = \left[\left(1+\frac{\gamma-1}{2}\right)M_e^{\,2}\right]^{\frac{\gamma}{\gamma-1}}. \tag{2.4}$$

The pressure decrease through a CS nozzle that creates an increase of gas velocity is also accompanied by a temperature decrease, which is a function of the local Mach number and given by Eq. 2.5:

$$\frac{T_0}{T} = \left[\left(1+\frac{\gamma-1}{2}\right)M^2\right]. \tag{2.5}$$

As mentioned previously, the increase of kinetic energy is accomplished at the expense of gas enthalpy, and therefore pressure and temperature. The drop in gas temperature can be large and typically results in gas temperatures in the diverging section of CS nozzles that are well below the particles' melting point and potentially lower than room temperature. This explains how melting, oxidation and other chemical reactions at the surface of the particles can be avoided in their flight through the nozzle. It also explains why particle preheating may be beneficial as otherwise the particles may cool down to temperatures that could affect their ductility, thus ability to deform upon impact with the substrate.

As the gas expands in the nozzle, its density also decreases. This affects negatively the ability of the flow to accelerate CS particles as the drag force is proportional to the gas density and the square of the relative particle/gas velocity. While one might be tempted to design CS nozzles that achieve extremely high Mach numbers (thus velocity), the associated drop in gas density can considerably reduce the benefits of higher gas velocity. Furthermore, the strength of the shock wave standing in front of the substrate will also increase with the gas velocity. For all these reasons, the CS nozzle design Mach number are typically kept in the realm of supersonic speeds rather than hypersonic.

Further use of the simple one-dimensional isentropic gas dynamics theory has established that the local gas velocity inside a de Laval nozzle is solely dictated by the nozzle geometry, and in particular by the local-to-throat area ratio, expressed by Eq. 2.6:

$$\frac{A}{A^*} = \frac{1}{M}\left\{\frac{1+\frac{\gamma-1}{2}M^2}{\frac{\gamma+1}{2}}\right\}^{\frac{\gamma+1}{2(\gamma-1)}} \tag{2.6}$$

This equation allows the prediction of the local flow Mach number if the ratio of the local (A)-to-throat area (A^*) is known, or to determine the nozzle local area if the nozzle design Mach number is set as well as the throat area. The latter typically varies between 1 and 3 mm in existing commercial CS systems and is usually set by the maximum gas flow rate the system is designed to handle for a specific range of operating temperature, based on the gas heater system capacity. The gas flow rate can be expressed by Eq. 2.7:

$$\dot{m} = \frac{A^* P_0}{\sqrt{T_0}}\sqrt{\frac{\gamma}{R}\left(\frac{\gamma+1}{2}\right)^{\frac{-(\gamma+1)}{2(\gamma-1)}}} \tag{2.7}$$

2.3.6 *Nozzle Operating Conditions: Spray Parameters*

To partially overcome the effect of reduced gas density in the nozzle, it is customary to operate CS nozzles at high gas stagnation pressures, up to 5 MPa. As the stagnation gas pressure is increased, the gas density increases inside the nozzle, thus providing higher drag forces for particle acceleration. Nevertheless, the benefits of increased gas stagnation pressure eventually level off, with the specific optimal gas stagnation pressure being dependant on the particle nature and size (Schmidt et al. 2009). Figure 2.6 presents this effect.

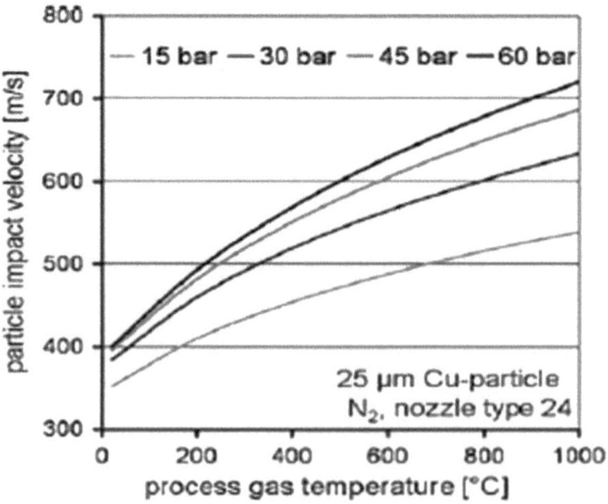

Fig. 2.6 Effect of process gas pressure on particle impact velocity. (Schmidt et al. 2009. Reproduced with kind permission from Springer Science and Business Media)

Furthermore, the increase of gas stagnation pressure leads to an increase in gas flow rate, which in turn affects the potential maximum stagnation temperature achievable by the CS system, as set by the gas heater power availability and heat transfer performance.

A higher gas stagnation temperature is beneficial to the CS process. As indicated in the previous sections, the local flow Mach number in a de Laval nozzle is solely determined by the nozzle geometry. However, the local gas velocity (expressed as $V = M \times C$, where V is the gas velocity, M the local Mach number, and C the local speed of sound) is proportional to both the local flow Mach number and the local speed of sound. The latter is a function of the local flow temperature. As such, an increase in gas stagnation temperature, which results in an increase of the local gas temperature throughout the nozzle, leads to an increase of the local value of the speed of sound throughout the nozzle, and therefore an increase in the local gas velocity throughout the nozzle. The latter directly affects (positively) the drag force exerted on the CS particles travelling inside the nozzle and increases the acceleration of the particles as seen previously in Fig. 2.7. Furthermore, the increased local gas temperature also presents the benefit of reducing the cooling effect of the gas on the particles, although this benefit can be minor in many cases.

The previous sections have outlined the fundamentals of gas dynamics that have been used as tools and applied throughout the early years of CS development. A strong understanding of gas dynamics principles is beneficial to anyone who wants to design, build and operate a CS system. The conclusions that can be drawn from these principles have been verified by numerous researchers active in the CS field. Verification through particle velocity measurements as well as by observation of the coatings microstructure (Schmidt et al. 2009) has confirmed that these principles do apply well to the CS process.

Despite the benefits of the simplified one-dimensional isentropic approach classically used in gas dynamics, many researchers have developed enhanced nozzle designs using computational fluid dynamics (CFD). These tools have allowed fine tuning of nozzle designs and have brought the performance of CS systems to a higher level, expanding the number of materials that can be sprayed by CS.

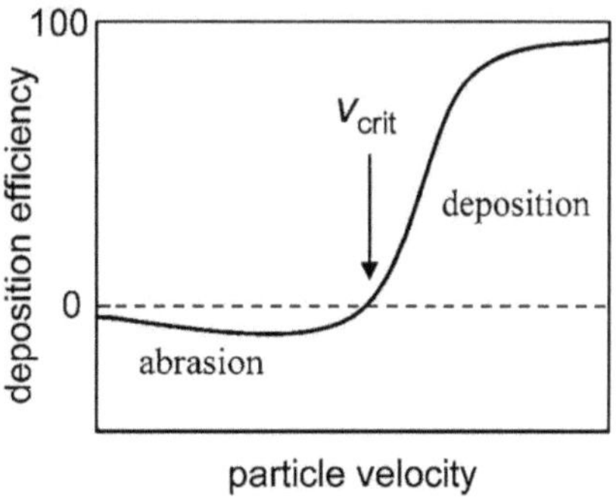

Fig. 2.7 Idealized deposition efficiency curve. (From Gartner et al. 2006. Reproduced with kind permission from Springer Science and Business Media)

2.4 CS Bonding

When a metal specimen is fractured and the two surfaces brought back into contact, the whole process being performed under high vacuum, a cohesive bond may be formed with strength equal to that of the parent metal (Conrad and Rice 1970; Ham 1963). However, to achieve a strong bond between particles of a metal powder under atmospheric conditions is an entirely different matter. Metal surfaces that have been exposed to air are covered by a native oxide film and other organic contaminants. In order for strong, metallurgical bonds to form between two metal bodies, the metal lattices must be brought into intimate contact and the intervening layers removed.

Mankind has invented numerous ways to overcome this problem over the centuries. In 1724, John Theophilus Desagulier, in his studies into friction, demonstrated that if two lead balls were cut to expose fresh surfaces, then pressed together and twisted, they could be joined together. In cold pressure welding, the ends of ductile metal wires or rods are forced together causing lateral extrusion of the interfacial material. The whole operation is done at room temperature. Friction welding employs a combination of applied compressive force and heat generated by relative movement between two surfaces to facilitate plastic deformation (Crossland 1971b). The two common elements in these and other solid-state joining processes are (a) the removal of contaminant layers by plastic flow of the interface material and (b) the use of pressure to force the surfaces into intimate contact.

In CS, particles strike the substrate or already deposited layer at high velocity, resulting in deformation at extreme strain rates—up to $10^9\ s^{-1}$ at the interface (Assadi et al. 2003; Lemiale et al. 2014). This creates the necessary conditions at the contact surface between the impacting particles and the receiving surface for strong bonding to take place. Establishing exactly what these conditions are is no trivial task. The following difficulties are encountered:

- The length scale over which bonding occurs is extremely small, demanding high-resolution characterization techniques. Oxide films, amorphous and intermetallic layers, if present, are only nanometres thick.
- Due to the wide variety of materials which can be deposited by CS, process conditions vary widely: Material purity and particle morphology are not always consistent from one powder batch to another, and many results reported in the literature do not fully state the experimental conditions used. Thus, it is difficult to compare results from different sources. Real powders comprise a distribution of particle sizes, and particles leave the CS nozzle with a range of velocities and angles. As will be discussed, the geometry of particle flattening produces a broad variety of deformation states within each individual particle. As a result, the physical conditions at any point of the interface are not 'known' in the same way that they might be in processes which involve the atom-by-atom or molecule-by-molecule deposition of material (e.g. vapour deposition, electroplating).

2.4.1 Critical Velocity

Since the discovery of CS in Russia in the 1980s by Papyrin et al., it has been recognized that deposition of solid particles only occurs over a limited range of particle sizes and velocities. Thus, bodies exposed to repeated particle impact at low velocities (of the order 10^1–10^2 m/s) generally experience erosion. Submicron-sized particles impinging at < 100 m/s are able to adhere to the surface by van der Waals or electrostatic forces (Klinkov et al. 2005). However, that does not fall within CS regime, which is characterized by layered build-up of material and strong) bonding between particles (Klinkov et al. 2005).

The most important physical limit to be overcome is the critical velocity, V_c. Below V_c, material loss occurs by erosion of the substrate. Above V_c, particles deposit and there is material gain. The use of high-pressure gas, gas preheating, and the particular attention paid to nozzle design (Sect. 2.3) are all directed towards providing sufficient particle acceleration to exceed V_c.

Three methods of determining critical velocity, V_c, have been used in the literature:

- Spraying a sufficiently large number of particles onto a sample to allow the accurate measurement of deposition efficiency, usually by weight (Sect. 2.4.2)
- Spraying a limited number of particles and examination of individual impact events (bonded particles, rebounds; Sect. 2.4.5)
- Numerical modelling of particle impact (Sect. 2.4.7)

At the high-velocity end of the spectrum (>3000 m/s), we begin to enter the hypervelocity field, where impact stresses greatly exceed the strength of the materials involved, strong shock waves develop, and solid materials behave like liquids (Murr et al. 1998).

In CS, there is a transition from deposition to strong erosion at high velocity. According to one analysis, the transition occurs at around 2 V_c (Schmidt et al. 2006). For many materials with high V_c, the strong erosion threshold is unattainable, apart from when helium is used, and the focus from a development point of view is to maximize particle acceleration so that as many particles as possible exceed V_c.

2.4.2 The Transition from Erosion to Deposition: Deposition Efficiency Curves

In many thermal spray processes, a certain proportion of the sprayed material will rebound or splash off the surface rather than being 'captured' in the coating, resulting in a process efficiency lower than 100 %. Thus, the term deposition efficiency (DE) has come into use and is defined according to Eq. 2.8.

$$\mathrm{DE} = \frac{\Delta m}{M_0}, \tag{2.8}$$

where Δm is the increase in weight of the sample during spray, and M_0 is the total mass of material fired at the sample.

In low-temperature, solid-state particle impingement, an identical definition of DE may be adopted. At low velocities, such as may be encountered in micro-grit blasting and peening operations (<200 m/s), the vast majority of particles rebound elastically, while only a few may embed or remain loosely adhered, and so **DE** is close to zero. In fact, a net mass loss is normally registered due to erosion of the substrate. If the sprayed particles present some ductility level and the spray system is capable of accelerating the particles to higher velocities, **DE** will begin to rise steeply. With sufficiently high impact velocity, **DE** may eventually approach 100 %, depending on the powder material. Figure 2.7 shows an idealized representation of the response of DE to increasing particle velocity.

Figure 2.8 shows some early DE measurements for the CS process using different metal powders (Alkhimov et al. 1990).

The transition from DE ~0 % to a high DE is relatively steep, often occurring within a few hundred metres per second. Different-sized particles are accelerated to varying degrees depending on their level of inertia. Furthermore, critical velocity itself is found to be particle-size dependent (Sect. 2.4.4). The sharpness of the DE–V_p curve is therefore limited in practice since in the transition region, only a fraction of the particles (the fastest travelling, i.e. smallest ones) exceed V_c. Accurate determination of V_c is achieved by careful correlation of DE data against the particle size distribution, combined with in situ measurement or calculation of particle velocities (Schmidt et al. 2006).

Gas heating is a standard practice in CS not only in order to increase the velocity of the gas and hence particle acceleration (Dykhuizen and Smith 1998) but also because thermal softening of the particles improves their deformation, lowers residual coating porosity, and reduces V_c. Figure 2.8 shows the shift in the DE curves to lower velocities that occurs when particles are accelerated by heated gas. Lee et al. measured the effect of process gas temperature on CS of Cu–20Sn bronze and found that V_c decreased by 50 m/s for every 100 °C increase in gas stagnation temperature (Lee et al. 2007).

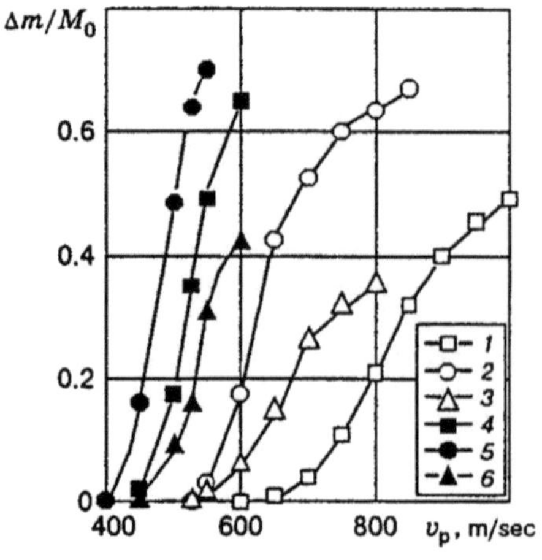

Fig. 2.8 Deposition efficiency versus particle velocity for *1*—aluminium, *2*—copper and *3*—nickel accelerated by an air–helium mixture at room stagnation temperature and *4*—aluminium, *5*—copper and *6*—nickel accelerated by heated air. (From Alkhimov et al. 1998. Reproduced with kind permission from Springer Science and Business Media)

2.4.3 The Effect of Material Properties on Critical Velocity

The coating industry has evolved towards spraying new materials for increasingly varied applications. In the early development stage, CS was mostly confined to copper, nickel and aluminium. This repertoire has expanded and nowadays a wide range of metals, alloys and metal–matrix composites have been successfully deposited by CS.

A low melting point and low mechanical strength (low resistance to deformation) were common attributes of the materials that initially showed promise for CS. These included pure aluminium, copper, zinc, silver, tin and their alloys. They all have low yield strength and exhibit significant softening at elevated temperatures. There are a large number of studies available in the literature on depositing Al and its alloys (Spencer et al. 2009; Ajdelsztajn et al. 2006; DeForce et al. 2011) as well as Cu and its alloys (Fukumoto et al. 2009; Karthikeyan et al. 2005).

Figure 2.8 shows that aluminium, copper and nickel powders with particle size <50 μm have V_c in the range 500–600 m/s. Assadi et al. found 570 m/s for 5–22-μm copper and 660 m/s for <45-μm aluminium (Assadi et al. 2003). These metals belong to the same isomechanical group according to the classification by Frost and Ashby (Frost and Ashby 1982). They share the face-centred cubic (FCC) crystal structure and exhibit similar bonding. In the FCC lattice, slip occurs along the close packed plane, and there are 12 slip systems overall. Thus, FCC metals have good deformability.

In contrast, there are fewer sliding planes available to hexagonal close-packed (HCP) metals, and so they are considerably less deformable. An example is titanium, whose V_c was measured at ~750 m/s in (Schmidt et al. 2009). Several studies have been published on the deposition of Ti and its alloys (Price 2006; Wong 2010; Cinca 2010). In the body-centred cubic (BCC) lattice, there are no close-packed planes, the coordination number is lower than that of FCC or HCP and deformability is poorest of all (Vlcek et al. 2005).

Apart from these major crystal groups, alloying also generally lowers deformability, and as a result, alloys have higher V_c than their pure metal counterparts. Following the trend further, oxides, nitrides, carbides and other ceramics which are inherently brittle ought not to be cold sprayable at all. Using conventional CS equipment, thin layers of ZrO_2 (Vlcek et al. 2005) or TiO_2 (Kliemann et al. 2011) have been deposited onto ductile, metallic substrates, but they require deformation of the substrate itself, and once this is covered, coating build-up ceases. Xu and Hutchings have demonstrated that polyolefins may be deposited by CS at velocities of ~135 m/s and DE <0.5 % (Xu and Hutchings 2006).

CS of powder blends, where one component is a metal and another is an intermetallic, a ceramic (e.g. oxide, carbide) or a different alloy, is a broadly applied technique. The resulting coating is a metal–matrix composite (MMC). Bonding and build-up of a dense deposit occur largely by plastic deformation of the metallic component, while the hard component embeds upon impact or is prevented from rebounding by deformation of the metal particles. In the literature, MMCs with compositions varying from 10:1 to 1:1 wt.% have been successfully deposited (Lee et al. 2004; Irissou et al. 2007; Sova et al. 2010; Bu et al. 2012). Compared to spraying the pure metal by itself, inclusion of hard particles creates a peening effect, further compacts and densifies the coating and improves the bond strength of the coating to the substrate (Yandouzi et al. 2007; Wolfe et al. 2006).

For CS bonding of metals and alloys, deformability is the prime consideration. However, other factors have been found to influence V_c, particularly melting point, T_m and density, ρ. Thus, from Fig. 2.8, it is apparent that aluminium is more difficult to deposit than Ni or Cu, even though all three metals lie within the one isomechanical group. The low density of aluminium (2.70 g/cm^3) reduces the impact energy compared to nickel (8.91 g/cm^3) or copper (8.96 g/cm^3). Furthermore, its high specific heat capacity, C_p, makes it more difficult to reach shear instability (more on shear instability in Sect. 2.4.4; Moridi et al. 2014).

Assadi et al. (2003) derived Eq. 2.9, which relates V_c to three material properties and the particle temperature:

$$V_c = 667 - 14\rho + 0.08T_m + 0.1\sigma_u - 0.4T_i, \tag{2.9}$$

where ρ is the density in g/cm^3, T_m is the melting temperature in °C, σ_u is the ultimate tensile strength (UTS) in MPa at room temperature, and T_i is the particle impact temperature in °C. The latter is dependent on the particle material and size as well as the spray process parameters that affect the in-flight heat transfer between the propellant gas and the sprayed particles, as stated in Sect. 2.3.6. The authors stress that Eq. 2.9 is valid over only a limited range of material properties and process conditions. Nevertheless, it serves as a useful demonstration of the relative influence of material parameters on V_c.

This kind of analysis that effectively ranks the relative influence of different material properties and allows V_c to be plotted out for a range of metals or alloys beyond what might be otherwise practicable experimentally. In Klassen et al. (2010) and Schmidt et al. (2006), critical velocities were first measured empirically for a limited set of particle types, then based on a theoretical working of the effects of material properties, size and temperature, the results extended to other materials (Fig. 2.9). The top of each bar in Fig. 2.9 represents the upper erosion limit for each material, determined by mass loss measurements of impacts of single millimetre-sized spheres and extrapolated into the CS size range. Although there have now been reported spray trials of most of the materials in Fig. 2.9, a systematic comparison of V_c across such a wide range of materials is lacking. The contrast between low-V_c materials such as tin, lead and gold and the high-V_c titanium and nickel alloys is quite surprising and highlights the need for the CS application developer to tune the CS conditions based on materials knowledge.

2.4.4 Effects of Particle Size and Purity

The reader will appreciate from the earlier section on particle acceleration (Sect. 2.3) that a particle size considerably smaller than those used for thermal spray (e.g. plasma or high-velocity oxy-fuel spraying) is needed in order to achieve the high velocities required for CS. It is worth noting that in the development of new CS applications, researchers have often had to settle for significantly coarser, conventional thermal spray powders. Because the deformation and bonding mechanisms in CS are quite different from thermal spray processes, a serious investigation needs to be taken, by researchers and industrial powder developers and producers, to develop dedicated feedstock powders for CS.

An excellent discussion of the effect of particle size on V_c has been given by Schmidt et al. (2006, 2009). Critical velocity increases with decreasing particle size due to a number of factors. Small particles have a higher surface area-to-volume ratio and contain greater levels of impurities such as oxygen. With powders produced by cooling from the melt, such as atomization, higher quench rates will occur for smaller particles. The resulting smaller grain size gives rise to Hall–Petch hardening. During the CS process itself, strain-rate hardening will be greater for particles with smaller dimensions. Thermal gradients are steeper, and so thermal diffusion has a more limiting effect on peak interfacial temperatures.

The following empirical relationships have been determined for copper (Eq. 2.10) and 316L stainless steel (Eq. 2.11) at a particle impact temperature of 20 °C and over a particle size range of 5–200 m/s

$$V_c^{Cu} = 900.d_p^{-0.19}, \tag{2.10}$$

$$V_c^{316L} = 950.d_p^{-0.14}, \tag{2.11}$$

where d_p is the diameter of a spherical particle in micrometers (Schmidt et al. 2006).

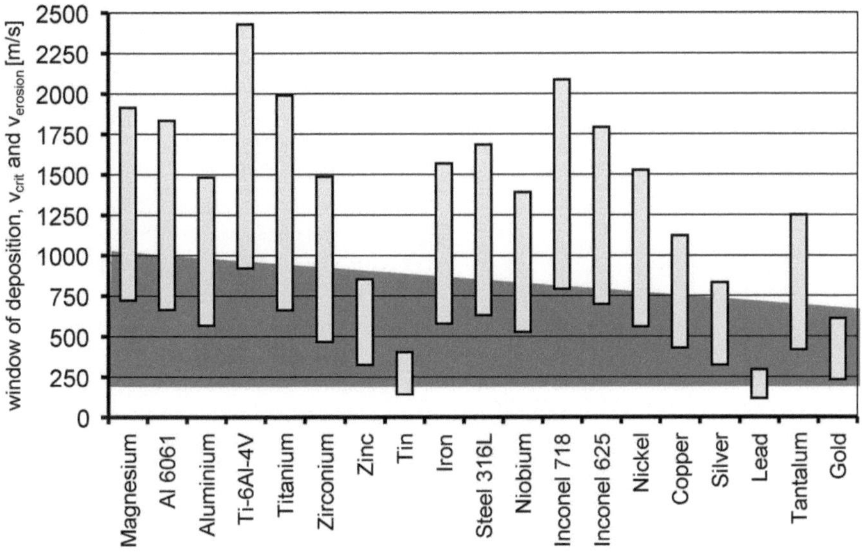

Fig. 2.9 Calculated critical velocities and windows of deposition for impact temperature of 20 °C, for various materials. (From Klassen et al. 2010. Reproduced with kind permission from Wiley)

Of course, larger particles have greater inertia and are more difficult to accelerate to high velocity. Thus, there exists an optimum particle size range, where particles are not so small that the above effects dominate and not so large that the amount of energy required to accelerate them to V_c is prohibitive. Typically, this range might be somewhere in the region of 10–45 μm for a ductile material such as copper (Schmidt et al. 2006).

Apart from particle size distribution, material purity is an important consideration for CS powders. It has been shown by artificial oxidation of aluminium (Kang et al. 2008), copper, 316L stainless steel and Monel powders (Li et al. 2009) that increased oxygen content increases V_c. Oxidation has a twofold effect on metal powders. Firstly, it causes a loss in ductility due to oxide dispersion and interstitial hardening, reducing the extent of deformation on impact. Secondly, it alters the structure and/or thickness of the oxide film on the surface, impeding bonding (Kang et al. 2008; Li et al. 2009).

When comparing values of V_c from the literature, the effects of particle size distribution and purity must be taken into account. Furthermore, the lack of any standardized method for performing the measurement causes some variation amongst different researchers. Table 2.1 reports seven values for copper critical velocity. Apart from the one measurement of 327 m/s by Li et al. (2006), they range from 520 to 610 m/s. At first glance, 327 m/s appears to be at odds with the rest. However, it can be noticed that particle sizes are smaller. Secondly, Li et al. used a higher-purity copper, which was shown in oxidation tests to be a considerable factor. For

Table 2.1 Measured critical velocities of copper

Reference	Particle size (μm)	Powder properties	Measurement technique	Critical velocity (m/s)
Assadi et al. (2003)	5–22	Inert gas atomized, 99.8 % purity	Deposition efficiency, combined with particle velocity distribution by CFD, verified by laser Doppler anemometry	570
Gartner et al. (2006)	5–25	Spherical morphology, oxygen content < 0.2 %	Deposition efficiency, particle velocity determined by CFD	550
Raletz et al. (2006)	10–33	–	Deposition efficiency, particle imaging	538
			Particle imaging, velocity calculated by 1-D isentropic model	520
Li et al. (2006)	Average 56	Gas atomized, oxygen content ~0.01 %	Deposition efficiency, particle velocity determined by CFD	327
		As above, oxidized to 0.14 % oxygen		550
		As above, oxidized to 0.38 % oxygen		610

CFD computational fluid dynamics

industrial application, however, an oxygen concentration of 0.1–0.2 wt.% is a more realistic specification in a <25 μm powder.

From the various factors mentioned in Sects. 2.4.3 and 2.4.4—material purity, crystal structure, alloying and temperature at impact—it is clear that bonding of cold-sprayed particles is intimately related to their deformation. In order to understand why this relationship exists, requires a more detailed look at the physical processes that occur during flattening and deformation of CS particles.

2.4.5 Splat Studies: Variety of CS Impacts

Two types of bonding situations are brought about in CS; particle-onto-substrate bonding and bonding of particles onto already deposited particles (particle-onto-particle bonding). Particle-on-substrate bonding is necessary for effective adhesion of the coating to the workpiece surface whereas particle-on-particle bonding determines the cohesive strength within the deposit itself.

One of the most useful features of the CS process is its flexibility regarding the different coating/substrate combinations which may be sprayed. Often the particle and substrate materials have contrasting physical properties. The velocity required for adhesion of particles to a dissimilar surface is not the same as the intrinsic critical velocity of the particles found by bulk measurement (Sect. 2.4.2). Once a complete covering layer of particles has formed and only particle-onto-particle impacts occur, the bulk value of V_c comes to define the velocity for particle adhesion.

Some researchers have sought to measure a particle-to-substrate critical velocity, that is, the threshold for particle adhesion onto a dissimilar, freshly prepared substrate surface. It may be obtained by inspection of adhered particles under high magnification (otherwise known as splat study). From the deformed particle volume, and with suitably accurate modelling or measurement of in-flight velocity, V_c is found from the largest (slowest moving) particle found to adhere. Lee et al. (2007) measured a 160 m/s increase in V_c when bronze particles were sprayed onto aluminium compared to bronze onto bronze. On the other hand, Raletz et al. using an in situ particle imaging technique (Raletz et al. 2005) measured only a small difference in V_c in spraying copper onto copper (520 m/s) versus copper onto 316L stainless steel (523 m/s).

Splat studies also reveal useful information about the mechanics of particle/substrate deformation. Several scenarios may occur:

- When the particle and substrate are of an identical or similar ductile material then both will deform. The total amount of deformation will depend on the material properties (Bae et al. 2008). With softer materials, for example, copper impact onto copper (Fig. 2.10), there is more particle flattening and deeper embedment into the substrate surface. With harder materials, for example, titanium onto titanium (Fig. 2.11), the particle retains more of its original shape and sits prouder on the surface. It should always be borne in mind that spray conditions and particle size also have a large bearing on the overall amount of deformation.

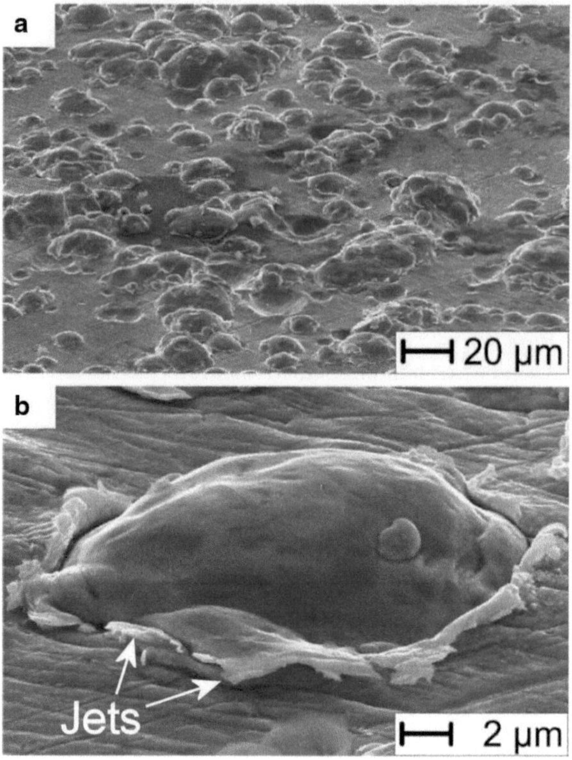

Fig. 2.10 Copper surface, cold sprayed with copper particles. (From Assadi et al. 2003. Reproduced with kind permission from Elsevier)

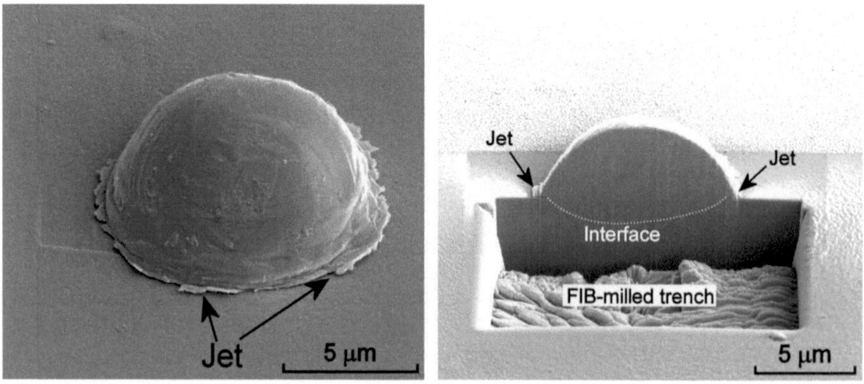

Fig. 2.11 Gas atomized, pure titanium particle bonded to a pure titanium surface. The image on the left is a secondary electron image taken before focused ion beam (FIB) milling. The image on the right shows the same particle dissected by FIB. Despite some deformation of the substrate, jetting was entirely restricted to the particle

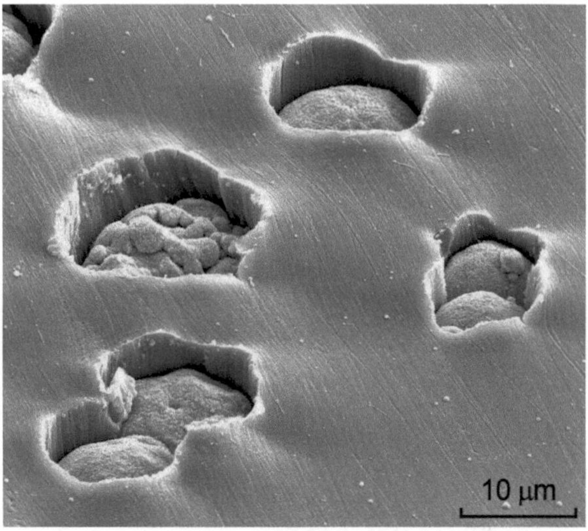

Fig. 2.12 Copper particles embedded into a pure aluminium surface. Molten splashes from the impacts were shown by EDS to consist of aluminium only

- If the particle is considerably harder than the substrate, then it will physically embed into the surface. Depending on the resistance to embedment offered by the substrate, the particle may plastically deform to a small extent or not at all. Penetration is also aided if the particles are denser than the substrate. Deep embedment is often seen in CS onto aluminium. Figure 2.12 shows how embedded copper particles are prevented from rebounding from a pure aluminium surface by deformation of the aluminium around them. Metal or ceramic particles may also be embedded into polymer surfaces even when build-up of a coating is difficult (Poole et al. 2012; Vucko et al. 2012; King et al. 2013).
- If the particle material is softer than that of the substrate, then deformation will mainly occur in the particle. In the extreme case, the substrate may be a material with negligible ductility such as a ceramic. If the surface is smooth (i.e. microscopically flat), then particle adhesion must occur by chemical interaction between the two materials. On real surfaces, there is usually some degree of roughness or in the case of ceramics, granularity or porosity. Furthermore, the initial impact of particles onto brittle materials can cause micro-fracture, grain removal and other defects which provide anchor points for adhesion (King et al. 2008; Zhang et al. 2005).

The manner in which the final cold-sprayed coating adheres to the substrate will also depend on the extent of substrate deformability. On many surfaces, CS has a roughening effect akin to a light grit blast. For this reason, blasting with coarse ceramic grit or sand prior to coating, a common pretreatment for plasma spray and other coating techniques, is often considered unnecessary prior to CS. Blasting media can become embedded in the substrate, which then promotes crack initiation and propagation, resulting in lower adhesion strength. For this reason, it is often observed with CS that polished substrates lead to better coating adhesion compared with grit-blasted ones.

With embedment of cold-sprayed particles, there is an interlocking or 'keying in' of the materials, providing a mechanical component to the total coating adhesive strength, additional to any chemical component. The relative importance of mechanical anchoring versus metallic bonding depends on the materials involved and spray conditions. For example, anchoring plays a comparatively larger role in the adhesion of coatings made using low gas pressures, such as in the case of downstream injection CS systems.

This also highlights the changing nature of the substrate surface from the moment that the first particles strike. In practice, there may be a noticeable delay in the initiation of deposition. For example, when aluminium was sprayed onto an aluminium alloy by Zhang et al. (2005), deposition was hampered with the presence of a tenacious oxide layer. A single spray pass produced a thinner-than-expected deposit. On other metallic (e.g. copper-based or ferrous) substrates, increasing substrate hardness promoted particle deformation, allowing deposition to commence earlier and resulting in a comparatively thicker first pass. On non-metallic substrates, aluminium deposition was comparatively slow due to a lack of metallic bonding (Zhang et al. 2005).

If the number of rebounding impacts onto a fresh surface is initially high, they may cause activation of the surface. Activation is a change in chemical state due to removal of passive films and increased dislocation concentration at the surface. Klinkov and Kosarev analysed the activation phenomenon for CS of aluminium particles with 30.2 μm mean diameter onto polished copper (Klinkov and Kosarev 2006). They defined two critical velocities, $V_{c-1} = 550$ m/s and $V_{c-2} = 850$ m/s. Below V_{c-1}, there is only erosion of the surface by the particles. Between V_{c-1} and V_{c-2}, there is an induction or delay time before the beginning of particle attachment. At a powder mass flow of 0.06 kg/m^2 s, the delay time is over 1 min at 550 m/s, and the surface is covered with empty craters before large numbers of particles begin to adhere. The delay period decreases with increasing velocity until above V_{c-2} there is immediate particle adhesion and no delay. Although the study (Klinkov and Kosarev 2006) was not extended to other powder sizes and material combinations, it can be expected from the descriptions above that V_{c-1}, V_{c-2} and the appearance of a delay time at all would depend on the materials involved. For instance, embedment into a sufficiently soft substrate (Fig. 2.12) initially occurs at a negligible rebound rate while further particle-to-particle adhesion then requires higher impact velocities.

2.4.6 *Adiabatic Shear-Based Mechanism of Bonding*

A common feature of all CS impacts that result in the formation of strong bonds, as is necessary for the build-up of cohesive deposits, is that there is intense localization of strain at the interface. This is manifested as a fast, outward-moving layer of material which protrudes radially from the edge of the particle—substrate contact zone. The layer is often referred to as a material jet, not to be confused with the CS gas jet. In Figs. 2.10 and 2.11, the interfacial jets have been labelled.

Interfacial jets form due to the development of an adiabatic shear instability. High-strain-rate deformation at the interface releases heat, which causes thermal softening, which further intensifies localized flow, and so on. There is a breakdown in the material's resistance to shear flow at the interface. The criterion for instability is given by Eq. 2.12:

$$\mathrm{d}\sigma = \frac{\partial \sigma}{\partial T}\mathrm{d}T + \frac{\partial \sigma}{\partial \varepsilon}\mathrm{d}\varepsilon + \frac{\partial \sigma}{\partial \dot{\varepsilon}}\mathrm{d}\dot{\varepsilon} \leq 0. \tag{2.12}$$

Equation 2.12 states that at some point, a maximum in-shear stress σ occurs, depending on the variables temperature T, strain ε, and strain rate $\dot{\varepsilon}$. If thermal softening is dominant over strain hardening and strain-rate hardening, then strain localization occurs, that is, $\mathrm{d}\sigma \leq 0$.

Jetting is a phenomenon which has been well documented prior to CS in at least two established industrial processes: explosive welding and dynamic powder compaction (shock consolidation). In explosive welding, metal plates are bonded by explosive-driven oblique impact (Crossland 1971a; Crossland and Williams 1970). The pressures that develop at the collision point between opposing metal surfaces are many times greater than the shear strength of the material. The material begins to act as a viscous fluid when strain rates exceed about 10^4 s^{-1} or as an inviscid fluid above 10^7 s^{-1} (Robinson 1977). Ultimately, the material behaviour is best described using the laws of hydrodynamics (Walsh et al. 1953). Under certain collision angles, a wavy interface forms, which can develop into turbulent vortices and intermixing zones.

Explanations for bonding in explosive welding fall into two categories. The first can be described as a cold welding—extreme deformation in the jet disrupts or removes the native oxide films and other surface contaminants so that the metal lattices of the two opposing surfaces come into direct contact under high pressure. The mechanism is analogous to cold pressure welding mentioned at the start of Sect. 2.4 except that it occurs at much higher strain rates. An alternative is fusion welding, whereby the heat evolved by deformation gives rise to local melting and transient atomic diffusion across the interface while the material is in the liquid state. Signs of melting are clearly present in the microstructures of many explosive welded joints. It may occur in isolated pockets within vortex regions, or combine to form a continuous molten layer around 50 μm thick if the kinetic energy is high enough (Crossland and Williams 1970). The latter is generally not considered desirable for bond strength particularly if brittle intermetallic compounds are produced (Crossland and Williams 1970).

Dynamic powder compaction is the shock-wave compression of metal or ceramic powders to form solid components (Raybould 1980). There are many common features with explosive welding, including jet-like flow of material into the voids between particles, and local melting which often occurs at the tips of the jets (Mamalis et al. 2001).

While jetting is a common characteristic of CS impacts above V_c, fluid-like mixing of materials has only been observed in some cases. Champagne et al. (2005)

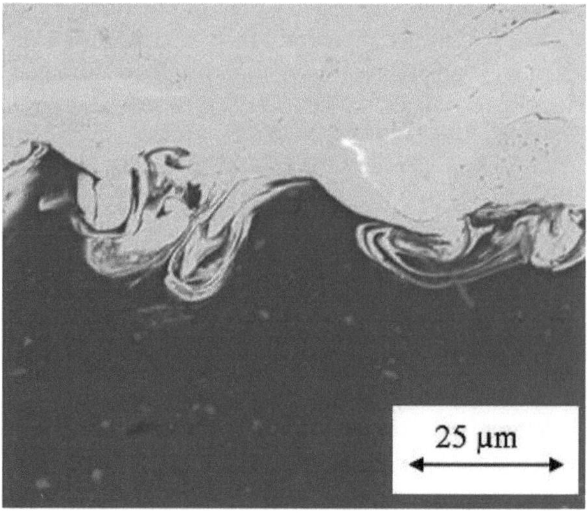

Fig. 2.13 EDS map showing interfacial mixing between a cold-sprayed copper coating and 6061 aluminium substrate. (From Champagne et al. 2005. Reproduced with kind permission from Springer Science and Business Media)

found rollups and vortices from the penetration of copper particles into aluminium alloy AA6061 (Fig. 2.13). Similar features have been seen in CS of nickel onto aluminium (Ajdelsztajn et al. 2005). Barradas et al. (2007) observed 'small waves' (<1 μm) in a pure aluminium substrate after copper impact, which however they noted were not widespread. Vortices have been found at the interface between zinc-based coatings and aluminium alloy or magnesium substrates (Wank et al. 2006). One interpretation of wavy interfaces is that they are a form of Kelvin–Helmholtz instability, which occurs between two flows having different tangential velocities. Analysis of the viscosity ratios of copper and aluminium and their jet thicknesses has been used to explain why this instability might arise in the Cu-on-Al case, but not for instance with the reverse, Al-on-Cu (Grujicic et al. 2003).

2.4.7 Modelling of CS Impact

Finite element modelling (FEM) has played an integral part in the development of a theoretical understanding of CS bonding. Mostly, this is due to the very small scale of particle impact, and the dynamic nature of the deformation process, which makes it impossible to perform accurate in situ observations of individual bonding events or analytical calculations without oversimplifying the problem. The following is not a comprehensive discussion of CS modelling, but is meant to provide the reader with an overview of the contribution of FEM towards an adiabatic shear-based explanation of bonding.

Modelling studies have focused on a spherical particle impinging normally against a flat surface (the numerical approximation of a polished substrate). This geometry has a high degree of symmetry. Real CS deposit surfaces are highly deformed and rough, but that brings additional complexities when it comes to deciding upon a representative model.

Mostly, FEM models rely on the Johnson–Cook plasticity model, given by Eq. 2.13.

$$\sigma = \left(A + B\ \bar{\varepsilon}^{n}\right)\left(1 + C\ \log\frac{\dot{\bar{\varepsilon}}}{\dot{\varepsilon}_0}\right)\left(1 - \left(\frac{T - T_\mathrm{r}}{T_\mathrm{m} - T_\mathrm{r}}\right)^{\mathrm{m}}\right), \tag{2.13}$$

where σ is the flow stress; $\bar{\varepsilon}$ the plastic strain; $\dot{\bar{\varepsilon}}$ the plastic strain rate; T_m the melting temperature; T_r a reference temperature; and *A, B, C* and *n* are material constants (Johnson and Cook 1983). It should be noted that *A, B, C* and *n* are obtained from low-strain-rate experiments ($< 10^4\ \mathrm{s}^{-1}$), and do not take into account the change in strain-rate sensitivity at extremely high strain rates. In comparison, in CS particle impact most of the particle strain rate is of the order to $10^7\ \mathrm{s}^{-1}$ (King et al. 2009), while in the jetting region it is around $10^9\ \mathrm{s}^{-1}$ (Assadi et al. 2003).

We consider now the case of identical particle and substrate materials; this being the most important because it relates to the build-up of coatings and is free of the complex dynamics that evolve with dissimilar materials (Sect. 2.4.5). Copper is perhaps the most popular choice of material due to an availability of empirical CS data (Table 2.1), known Johnson–Cook parameters, and its inherent 'cold sprayability'. Of course, aluminium, stainless steel, titanium and other commonly sprayed materials have also been modelled. However, copper-on-copper impact appears in several key studies on bonding mechanisms (Grujicic et al. 2004; Bae et al. 2008; Lemiale et al. 2011; Assadi et al. 2003; Schmidt et al. 2006; Li et al. 2006).

Figure 2.14 shows the final deformed state of a 20-μm copper particle following 200, 400 and 600 m/s impact onto a copper surface. The simulation times were 38, 35 and 38 ns for Fig. 2.14a, b and c, respectively. Depending on the size of the particle, velocity and dominant deformation mode (elastic vs plastic), the deformation period may be shorter or longer; however, in general for most <25 μm impacts, it is shorter than 100 ns.

Coloration denotes equivalent plastic strain, which is clearly highly concentrated at the interface. At 600 m/s, a prominent jet has formed, involving both particle and substrate. During the impact process, the peak pressures at the central contact point (sometimes referred to as the south pole) are of the order of 1–10 Gpa (Grujicic

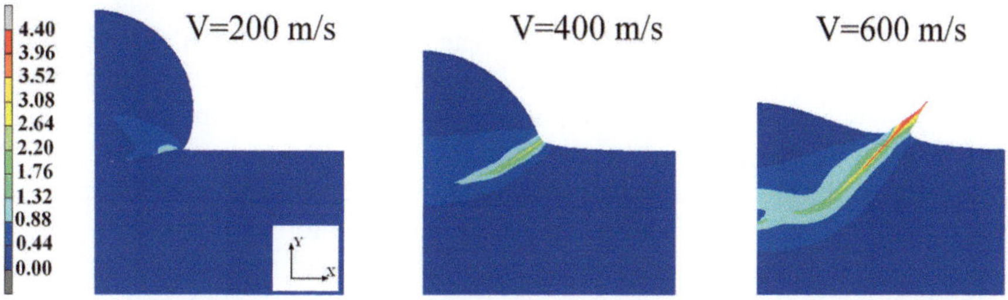

Fig. 2.14 Two-dimensional axisymmetric simulations of a 20-μm copper particle impact onto a copper surface using a model which includes heat transfer. The initial impact velocities were (a) 200 m/s, (b) 400 m/s and (c) 600 m/s. (From Lemiale et al. 2011. Reproduced with kind permission from Springer Science and Business Media)

et al. 2003; Dykhuizen et al. 1999). Deformation, however, is greater towards the periphery, within the shear jet zone. Plastic strains are of the order of 10, causing mesh elements to become highly distorted.

Part of the mechanical energy expended during the plastic deformation process in metals is converted into heat, while the remainder is stored in the material microstructure. This stored energy produces the cold-worked state, modifies the internal energy of the deformed metal, and it remains in the material after the deformation process (impact) is completed. It is generally assumed that most of the mechanical energy expended during plastic deformation is dissipated as heat. The fraction of the rate of plastic work dissipated as heat is typically assumed to be a constant parameter of 0.9, for most metals. Consequently, there is also a sharp increase in temperature within the first 10–20 ns of impact (Lemiale et al. 2011).

At low impact velocity, plastic strain, temperature and stress at the interface change monotonically with time. However, at higher velocities in adiabatic FEM models (Bae et al. 2008; Assadi et al. 2003; Grujicic et al. 2004; Li et al. 2006), there is an additional, sudden increase in plastic strain and temperature in the second half of the deformation period. Maximum interfacial temperatures approach or reach melting point, and concurrently, von Mises stress falls to zero as the jetting material loses strength. These events have been interpreted as manifestations of shear instability, consistent with Eq. 2.12. Assadi et al. (2003) also reported that the drop in flow stress was accompanied by large fluctuations in stress due to the competing effects of strain-rate hardening (second term in Eq. 2.12) and thermal softening (third term in Eq. 2.12). For Cu-on-Cu impact, shear instability in the model was shown to occur only at impact velocities of 580 m/s and higher (Assadi et al. 2003), an almost identical value to V_c obtained by DE measurements (570 m/s, Table 2.1; Assadi et al. 2003).

The adiabatic assumption is often justified with reference to the dimensionless quantity $x^2/D_{th}t$, where x is a characteristic system dimension, D_{th} is the thermal diffusivity and t is the process time. Assadi et al. (2003) assumed a thermal diffusivity of 10^{-6} m²/s for copper, and 10^{-6} m and 10^{-8} s for x and t, respectively, with the result that $x^2/D_{th}t \gg 1$. However, other researchers who have incorporated heat transfer into their simulations have found that the adiabatic assumption does cause an overestimation of interface temperatures. The deformed shape of the particle is largely unaffected, and jetting does occur. However, for Cu-on-Cu impact at 600 m/s, the interface does not approach melting point, and a complete loss of shear strength does not eventuate (Lemiale et al. 2011; Wang et al. 2014). Interestingly, Schmidt et al. assumed the thermal conductivity of the highly deformed material to be 60 % of annealed bulk copper, and at this limited level of heat transfer, the shear instability behaviour was similar to fully adiabatic simulations (Schmidt et al. 2006).

Adiabatic FEM models do not achieve full mesh size independence (Assadi et al. 2003). Extreme mesh distortion in the jetting zone is a further cause for concern. Remeshing schemes have been attempted, but are found to result in unrealistically distorted jets (Lemiale et al. 2011). Wang et al. (2014) and Yu et al. (2012) questioned the validity of a Lagrangian approach to CS impact. When an Eulerian method was employed, shear instability did not develop for a 20-μm copper impact up to 700 m/s.

2.4.8 Differences in Bonding Conditions over the Contact Area

FEM analysis of the normal impact of a spherical particle (Sect. 2.4.7) has shown that different interfacial conditions are produced at the base (south pole), compared with the edges (peripheral region). Intense shearing occurs at the periphery (Fig. 2.14).

Figure 2.15 shows an FEM simulation of a 15-μm copper particle impact onto an aluminium alloy AA7075 surface. A thin layer of aluminium has reached melting point. On the copper side, temperatures are lower. The discontinuity is allowed by the absence of any thermal diffusion in this particular model. On both sides, there has been considerably greater deformation and therefore heat evolution towards the periphery (King et al. 2010a).

If bonding is associated only with adiabatic shearing, then it follows that not all of the contact surface will have bonded. The question of what fraction of the interface remains unbonded and furthermore how that may be reduced is important from the point of view of coating adhesion, deposit mechanical properties and electrical conductivity (Stoltenhoff et al. 2006). The bond strength of copper coatings on copper substrates has found to range from 30 to 40 MPa when the particles are sprayed to velocities near V_c. These measurements are ~20% of the UTS of copper (Assadi et al. 2003). Comparison with modelling of Cu-on-Cu impact showed 15–25% of the interfacial area was subject to shear instability (Assadi et al. 2003). Thus, there appears to be agreement between the two values, simplifications assumptions of the FEM method notwithstanding.

Increasing particle velocity, powder and substrate preheating are all techniques known to raise bond strength (Marrocco et al. 2006). From etched micrographs of copper coatings, use of helium gas (which roughly doubles the gas velocity compared to nitrogen) increased the bonded fraction of particle boundaries to ~75% (Stoltenhoff et al. 2006). Price et al. used a heat-treating method to determine the extent of inter-particle bonding between co-sprayed copper and aluminium and found an improvement with increasing gas pressure (Price et al. 2006). Similarly, preheating of nickel particles has been shown in FEM modelling to increase the interfacial area subject to adiabatic shear instability, while in experiments corresponding enhancements in DE and coating bond strength (from ~15 to 20% of Ni UTS) were obtained (Bae et al. 2012).

Direct experimental confirmation that bonding is favoured at the periphery of the contact zone has come from a few sources. In the CS of Al–Si particles onto mild steel (Wu et al. 2006) and of aluminium particles onto lead zirconate titanate (King et al. 2008), there is little or no permanent deformation of the substrate. However, particles that elastically rebound have a momentary adhesive interaction, and they leave behind a mark consisting of submicron fragments of powder material. In both cases, the pattern is clearly ring shaped with fewer or no fragments in the middle of each mark.

Careful dissection of particles by milling with a focused ion beam (FIB) may reveal a crack between particle and substrate, usually around the south pole region (Fig. 2.16). This part of the interface has not survived the tensile rebound forces

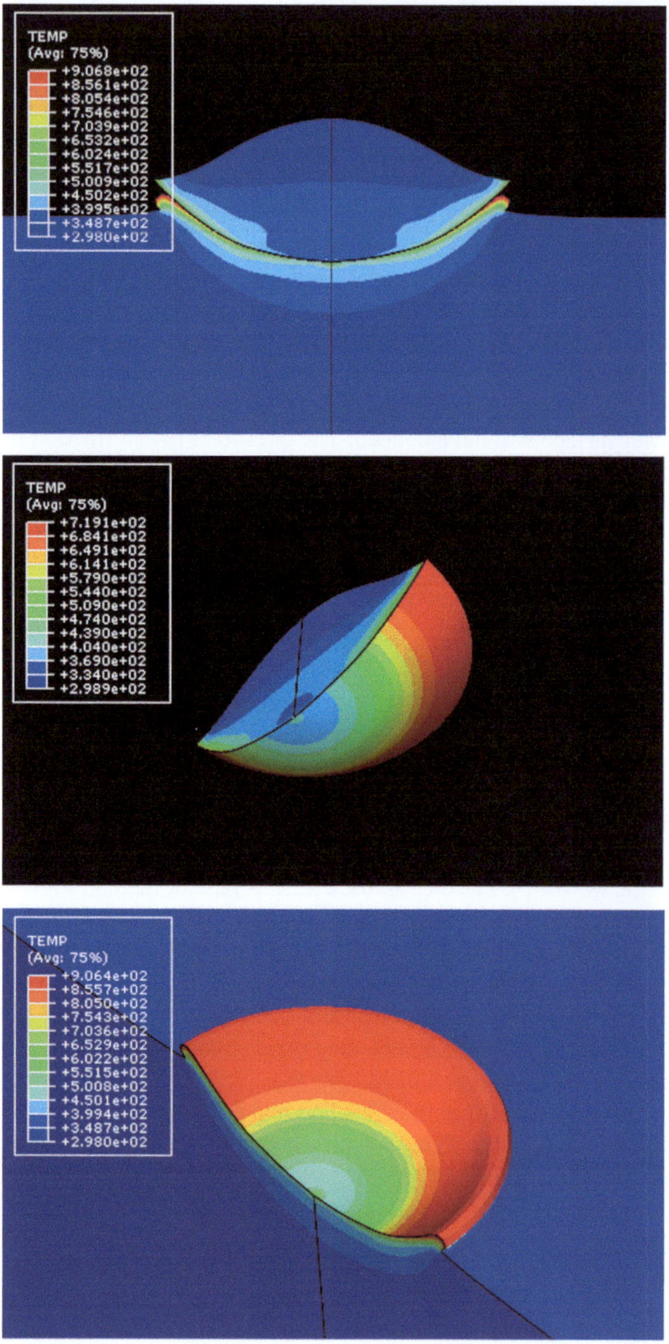

Fig. 2.15 FEM simulation of a 15-μm copper particle impact onto aluminium AA7075 at 430 m/s. (a) cross-sectional view of particle and substrate, (b) view of particle bottom surface without substrate, (c) view of substrate surface with particle removed. (Taken from a three-dimensional, axisymmetric simulation by G. Bae at Hanyang University, Republic of Korea using the commercial finite-element package ABAQUS 6.7–2. Details of the modelling procedure are given in King et al. (2010a). Figure 2.15a, c is reproduced with kind permission from Springer Science and Business Media)

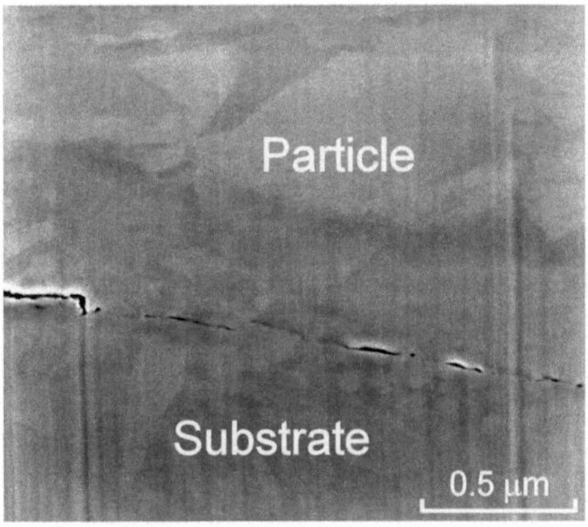

Fig. 2.16 Interface between a titanium particle and titanium substrate, revealed by focused ion beam (FIB) milling. Defects due to voids and oxide layers can be seen

although the particle as a whole remains adhered due to bonding closer to the edges. Study of the impact of gas-atomized titanium particles onto a polished Ti surface has shown that even when no obvious crack is present, higher magnifications reveal micro-voids, that is, sections of the interface which are not perfectly co-conforming (Fig. 2.16). At the peripheries, the interface appears 'melded'—there is no distinguishing feature at the boundary that can be observed by FIB-SEM (focused ion beam scanning electron microscopy). From particle to particle the linear melded fraction of the interface varied from 26 to 77 % (King et al. 2014).

According to the cold welding model of bonding such as has been proposed for explosive welding (Sect. 2.4.6), removal of the oxide layers by violent jetting and conformal contact of the particle and substrate under pressure are the two necessary ingredients for metal-on-metal bonding. To simulate oxide film behaviour in FEM modelling, a thin Al_2O_3 shell has been added to an AA6061-T6 particle and its properties described using the Johnson–Holmquist plasticity damage model (Yin et al. 2012). The oxide film is immediately crushed upon contact with the substrate, with fracture spreading as the contact area increases. At 6 ns, the outward moving jet begins extruding the cracked oxides in the peripheral region outwards, whereas at the south pole the cracked oxides tend to remain. Substrate hardness has a large bearing on jetting and in turn on the amount of oxide removal (Yin et al. 2012).

Of course, in metal-on-metal impact, oxide films exist on both sides of the interface. The contribution to interfacial shear instability by the particle and substrate is not necessarily equal, even when both are the same material. On close inspection of Fig. 2.10, jets can be seen to originate from both particle and substrate. However, particle jetting is often favourable over substrate jetting, particularly in less penetrating impacts like titanium on titanium (King et al. 2014). Goldbaum et al. (2012) have defined three regimes for titanium bonding:

- Below V_c, there is weak conformal adhesion and limited metallurgical bonding.
- Above V_c, adiabatic shear instability becomes pronounced. Jetting increases the interfacial area, leading to a greater fraction of metallurgical bonding.
- At velocities greatly in excess of V_c, there is extensive deformation in both particle and substrate. A combined adiabatic shear process leads to a more

continuous, void or oxide-free bond. The particle adhesion strength approaches bulk shear strength.

2.4.9 Melting and Other Interfacial Features

FEM modelling shows that the majority of the particle experiences only a small rise in temperature due to plastic work, typically less than 100 K (Fig. 2.15). However, the temperature in the jetting zone approaches melting point.

Over the past 10 years of research into CS, a growing body of evidence has accumulated that shows that interfacial melting can occur during CS. Any molten layer that forms is on the nanometre scale. However, there is also a considerable amount of contra-evidence of successful bonding without any sign of melting. The reason for this apparent contradiction may lie in the sheer range of spray temperatures, pressures, particle sizes, shapes and materials that are possible. While basic set-point conditions are usually reported accurately, the most practical aspects of spray technique are sometimes not. By way of example, the temperature at the substrate surface is a function of its size (i.e. thermal mass) and the speed and number of passes of the CS gun. Since detailed characterization is required to observe nanoscale features, any single study can naturally only be focused on a limited set of spray conditions. Obtaining an overall picture then becomes a difficult task of collating information from a variety of disparate sources—see Table 2.2.

When CS particles impact a surface, a number of features are often left on the surface that may be associated with melting. Most readily identifiable are ejecta—fine particulates that become detached from the jets and are thrown outwards. Figure 2.17 shows ejecta in the form of long splashes and droplets which have spheroidized while molten due to surface tension. They can be seen to radiate from bonded particles and rebound sites, or they are attached to crater walls (King et al. 2010a). Nanometre-scale, spheroidized droplets have also been found buried within the cold-sprayed microstructure at particle–particle interfaces (Li et al. 2005, 2007).

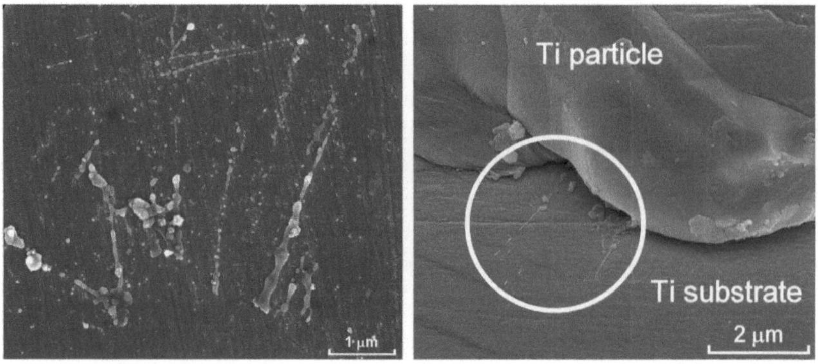

Fig. 2.17 Ejecta from **a** aluminium impact onto lead zirconate titanate at 350 °C, and **b** from titanium impact onto titanium at 600 °C

Table 2.2 Empirical evidence for interfacial melting during cold spray

Reference	Material system	Stagnation temperature (°C)	Type of investigation	Evidence of melting
Li et al. (2005)	Zn	320, 410	TEM of deposit cross section, SEM of deposit surface	Spheroidized particulates, recrystallized grains, amorphous regions
Wank et al. (2006)	Zn and Zn5Al onto Al 7022 and Mg AZ91	250	TEM studies of coating/substrate interface	Submicron and nanosized precipitates at interface: $MgZn_2$ and $Mg_5Zn_2Al_2$ and $Mg_{11}Zn_{11}Al_6$
Zhang et al. (2005)	Al onto Sn	Room temperature	SEM	Melting of substrate—details not given
Xiong et al. (2008)	Al onto Al/ Ni onto Al6061-T6	300, 400	TEM	Formation of amorphous layer, 3 nm thick
Ning et al. (2008)	Al–10 %Sn onto 304 stainless steel	~ 300	Splat studies	Ejecta
Bolesta et al. (2001)	Al onto Ni	Not given	Grazing incidence XRD analysis of coating–substrate interface	Detection of Ni_3Al. Estimated interfacial layer thickness 20–50 nm
King et al. (2008)	Al onto lead zirconate titanate	350	Splat studies	Spheroidized particulate ejecta and signs of adhesive interaction
Song et al. (2013)	Al onto soda-lime glass	350–400	TEM	Amorphous/nanocrystalline reaction layer up to 80 nm thickness, containing elements from both particle and substrate
Li et al. (2007)	Various; Al2319, Al12Si, Ti, Ti–6Al–4V, NiCoCrAlTaY	520–620	SEM observation of deposit fracture surfaces	Ductile failure, dimples
Barradas et al. (2007)	Cu onto Al	550	TEM of coating–substrate interface	α_2-$AlCu_4$, η_2-AlCu and θ-$CuAl_2$ phases found, in some places layers were 20 nm thick, in others, $CuAl_2$ grains were 400 nm in size

Table 2.2 (continued)

Reference	Material system	Stagnation temperature (°C)	Type of investigation	Evidence of melting
Guetta et al. (2008)	Cu onto Al	600	FIB/ TEM study of single particle interface with substrate	> 500-nm-thick diffusion layer containing intermetallic phases θ-$CuAl_2$ and Cu_9Al_4. Also, Al-$CuAl_2$ eutectic region
King et al. (2008)	Cu onto Al	600	Splat studies	Spheroidized Al particulate ejecta
King et al. (2010a)	Cu onto Al, Cu onto Al7050	200–600	FIB/SEM, TEM of particle/ substrate interface	Spheroidized Al ejecta, dimpled, lumpy crater walls, intermittent θ-$CuAl_2$ intermetallic at interface
Xiong et al. (2011)	Ni onto Cu	400	TEM	Amorphous layer with intermixed Cu, Ni and some oxygen
King et al. (2014)	Ti onto Ti	600	FIB/SEM	Ejecta; detached chips and spheroidized droplets
Bae et al. (2009)	Ti onto Ti	600	SEM	Spheroidized particulates
Vlcek et al. (2005)	Ti-6Al-4V	Not given	SEM, TEM	Spheroidized particulates

FIB focused ion beam, *SEM* scanning electron microscope, *TEM* transmission electron microscope

Rebound craters sometimes have a dimpled morphology due to intense thermal softening of the surface layer, although that on its own does not necessarily imply melting (King et al. 2008).

At the bonded interface between two different metals, reaction layers consisting of intermetallic compounds have been found. For example, at the boundary of cold-sprayed copper coatings and aluminium substrates, intermittent precipitates (King et al. 2010a) and 'grains' (Barradas et al. 2007; Guetta et al. 2009) of θ-$CuAl_2$ and layers tens of nanometres thick (Barradas et al. 2007; Guetta et al. 2009) comprising CuAl, $CuAl_2$ and Cu_9Al_4 are reported. Similar reaction zones have been observed in the Ni–Al (Bolesta et al. 2001) and Zn–Al–Mg (Wank et al. 2006) systems.

By the following argument, we see that thick intermetallic regions such as these could not have formed by solid-state diffusion, but rather require that the interface was molten for a period of time. It is known that the particle deformation period is less than 100 ns (FEM modelling, Sect. 2.4.7). Furthermore, since the high-temperature shear layer is of submicron thickness, it could be expected to be cooled rapidly by the surrounding metal. Now, referring to the example of copper and aluminium again, the diffusivity of copper in solid solution in aluminium at the Al–$CuAl_2$ eutectic temperature (812 K) is 1.1×10^{-13} m^2/s (Anand et al. 1965). The diffusion distance $\sqrt{Dt}$ is thus 0.1 nm—less than the lattice parameter of aluminium (0.4 nm), and certainly insufficient to be the mechanism through which tens-of-nanometers-thick layers are produced. By contrast, the diffusivity of Cu in liquid aluminium by conduction alone is 42.2×10^{-9} m^2/s (Isono et al. 1996)—four orders of magnitude higher than in the solid (giving a diffusion distance $\sqrt{Dt}$ two orders of magnitude greater). When the effect of liquid convection is included, even longer mixing distances could be expected (King et al. 2010a).

Interfacial amorphous layers are also frequently observed. Here, the interpretation is complicated by the probable retention of native oxide films from the metal particle or substrate surfaces. As mentioned earlier (Sect. 2.5.8), it has been proposed that native oxides are fractured and removed by high impact pressures, scrubbing jets, etc.; however, that process could not be expected to be entirely effective at all points along the interface, and so any unremoved, amorphous oxide might nevertheless be mistaken for a reaction layer. However, in bimetallic systems, energy dispersive X-ray analysis has proven the presence of atoms from both substrate and particle within the amorphous layer.

Thus, if these layers do indeed incorporate atoms from opposing sides of the interface, then they must also be thought of as mixing zones. The presence of differently sized atoms may improve the glass-forming ability of the melt. The amorphous structure is retained down to room temperature due to rapid cooling by the steep thermal gradient acting normal to the interface.

It is also possible that the extremely high strain rates ($\sim 10^9$ s^{-1}) known to exist within the jets may in themselves be a contributing factor towards amorphization. In CS, as in other 'real-world' processes, high-strain-rate deformation is always accompanied by a temperature rise as the vast majority of plastic work is dissipated as heat. However, over the past ~15 years, molecular dynamics simulations of metal nanowires have allowed the effects of loading and loading rate to be studied independently of any thermal effect. At sufficiently high strain rates at 300 K, a perfect metal or alloy crystal may undergo a transformation to a homogeneous disordered state in a process known as strain-rate-induced amorphization (Branício and Rino 2000; Ikeda et al. 1999). The enthalpy of fusion is overcome purely by shock-induced kinetic energy rather than thermal energy, such as occurs in heat-induced melting.

Reaction layers also result from the impact of metal particles onto non-metallic surfaces, such as aluminium on oxide ceramics (King et al. 2010b) or soda-lime glass (Song et al. 2013). Characterization of an aluminium–glass interfacial layer by transmission electron microscopy showed it to be partly amorphous and partly nanocrystalline, with elements from both materials and also significant Na enrichment (Song et al. 2013). The low thermal conductivity of a ceramic substrate raises interfacial temperatures, making conditions more favourable for melting. Alkhimov et al. have calculated the interface temperature for 800 m/s impact of a 25-μm aluminium particle to be 970 K against Al_2O_3, compared with 630 K against copper. Furthermore, some contribution of frictional heating must be expected as the particle flattens and flows outwards against a nondeforming substrate (Alkhimov et al. 2000). However, Drehmann et al. found the interface between an aluminium coating on Al_2O_3 to be sharp, without any intermixing or amorphous intermediate layer (Drehmann et al. 2013).

Apart from the possibility of melting, other microstructural effects resulting from the intense shearing within the jetting region may influence bond strength. Typically, large microstructural inhomogeneities are seen throughout the bulk of a cold-sprayed particle. This is because the deformation history varies locally within the particle, depending on the geometry of the initial particle impact and of subsequent impact by later-arriving particles. It will also depend on local crystal orientation, the availability of slip systems and the occurrence of deformation twinning in some metals. Furthermore, static recovery or recrystallization can alter the deformed structure. Within the shear jets, however, it is known that the strain rate is sufficiently high as to cause (at least) near-melting-point softening while deformation is still in progress. Under these conditions, dynamic recrystallization may occur.

In various publications, elongated subgrains, dense dislocation walls and highly misoriented, equiaxed grains have been reported at the interface. In copper, fine, equiaxed grains about 100 nm in diameter (Borchers et al. 2004) and down to 30–50 nm (King et al. 2009) have been observed. In nickel, electron backscatter diffraction (EBSD) analysis has shown poorer pattern quality at particle boundaries due to higher defect densities and/or lattice strain, and in these regions 100–200 nm sub-grains were identified (Zou et al. 2009). Aluminium is a high stacking fault energy (SFE) material, and its as-deformed grain sizes are larger relative to Cu or even Ni (Borchers et al. 2004). At Al interfaces, reports of grain sizes vary from around 500 nm with some smaller grains (Borchers et al. 2004) to a few nanometres at Al–Al_2O_3 interfaces (Drehmann et al. 2013). Wang et al. found a mixture of low-angle ($\leq 15°$) and high-angle ($\geq 15°$) boundaries at Al interfaces (Wang et al. 2011). Titanium is known to have a high 'adiabaticity' owing to its low thermal conductivity (21.9 $Wm^{-1}K^{-1}$) and is prone to shear localization during impact or other high-strain-rate deformation. In this material, broad zones containing large concentrations of defects and 20–50 nm subgrains have been reported (King and Jahedi 2011), while in another study, nanocrystals as small as 10 nm (Rafaja et al. 2009) were seen.

It has been proposed that the formation of large numbers of fine grains at CS interfaces by dynamic recrystallization may improve bonding by allowing epitaxial growth of certain crystals that have a favourable orientation with the substrate (Drehmann et al. 2013; Rafaja et al. 2009). For example, titanium coatings on smooth, (001) oriented sapphire (Al_2O_3) single crystals proved strong bonding was possible between these two materials. High-resolution transmission electron microscopy (HRTEM) study showed only a small degree of lattice misfit ($<7\,\%$) between some recrystallized Ti nano-grains and the adjacent sapphire (Rafaja et al. 2009).

2.5 Concluding Remarks

In this chapter, the fundamentals that govern the CS process and their correlation with the quality of CS deposits were presented. An extensive overview was made of how the impact velocity of particles influences microscopic deformation phenomena that are at the root of the physics of bonding, including the adiabatic shear instability phenomena. In order to better understand the process of accelerating particles, a detailed description of compressible fluid dynamics theory, which is at the heart of the CS process, was also presented. These fundamentals supported a direct correlation to nozzle geometry, gas pressure and gas temperature as the most crucial parameters in CS.

References

Ajdelsztajn, L., B. Jodoin, G. E. Kim, and J. M. Schoenung. 2005. Cold spray deposition of nanocrystalline aluminum alloys. *Metallurgical and Materials Transactions A* 36A (3): 657–666.

Ajdelsztajn, L., A. Zúñiga, B. Jodoin, and E. J. Lavernia. 2006. Cold gas dynamic spraying of a high temperature Al alloy. *Surface and Coatings Technology* 201 (6): 2109–2116.

Alkhimov, A. P., V. F. Kosarev, and A. N. Papyrin. 1990. A method of Cold gas-dynamic deposition. *Soviet Physics Doklady* 35 (12): 1047–1049.

Alkhimov, A., V. Kosarev, and A. Papyrin. 1998. Gas-dynamic spraying. An experimental study of the spraying process. *Journal of Applied Mechanics and Technical Physics* 39 (2): 318–323.

Alkhimov, A., A. Gudilov, V. Kosarev, and N. Nesterovich. 2000. Specific features of microparticle deformation upon impact on a rigid barrier. *Journal of Applied Mechanics and Technical Physics* 41 (1): 188–192.

Anand, M. S., S. P. Murarka, and R. P. Agarwala. 1965. Diffusion of copper in nickel and aluminum. *Journal of Applied Physics* 36 (12): 3860–3862.

Assadi, H., F. Gartner, T. Stoltenhoff, and H. Kreye. 2003. Bonding mechanism in cold gas spraying. *Acta Materialia* 51 (15): 4379–4394.

Bae, G., Y. Xiong, S. Kumar, K. Kang, and C. Lee. 2008. General aspects of interface bonding in kinetic sprayed coatings. *Acta Materialia* 56 (17): 4858–4868.

Bae, G., S. Kumar, S. Yoon, K. Kang, H. Na, H.-J. Kim, and C. Lee. 2009. Bonding features and associated mechanisms in kinetic sprayed titanium coatings. *Acta Materialia* 57 (19): 5654–5666.

Bae, G., J. I. Jang, and C. Lee. 2012. Correlation of particle impact conditions with bonding, nanocrystal formation and mechanical properties in kinetic sprayed nickel. *Acta Materialia* 60 (8): 3524–3535. doi:10.1016/j.actamat.2012.03.001.

Balani, K., A. Agarwal, S. Seal, and J. Karthikeyan. 2005. Transmission electron microscopy of cold sprayed 1100 aluminum coating. *Scripta Materialia* 53 (7): 845–850.

Barradas, S., V. Guipont, R. Molins, M. Jeandin, M. Arrigoni, M. Boustie, C. Bolis, L. Berthe, and M. Ducos. 2007. Laser shock flier impact simulation of particle-substrate interactions in cold spray. *Journal of Thermal Spray Technology* 16 (4): 475–479.

Bolesta, A. V., V. M. Fomin, M. R. Sharafutdinov, and B. P. Tolochko. 2001. Investigation of interface boundary occurring during cold gas-dynamic spraying of metallic particles. *Nuclear Instruments and Methods in Physics Research Section A* 470 (1–2): 249–252.

Borchers, C., F. Gartner, T. Stoltenhoff, and H. Kreye. 2004. Microstructural bonding features of cold sprayed face centered cubic metals. *Journal of Applied Physics* 96 (8): 4288–4292.

Branício, P. S., and J.-P. Rino. 2000. Large deformation and amorphization of Ni nanowires under uniaxial strain: a molecular dynamics study. *Physical Review B* 62 (24): 16950–16955.

Bu, H., M. Yandouzi, C. Lu, D. MacDonald, and B. Jodoin. 2012. Cold spray blended Al + Mg17Al12 coating for corrosion protection of AZ91D magnesium alloy. *Surface and Coatings Technology* 207 (0): 155–162. doi:http://dx.doi.org/10.1016/j.surfcoat.2012.06.050.

Champagne, V. K. 2007. *The cold spray materials deposition process: fundamentals and applications*. Cambridge: Woodhead.

Champagne, V. K., D. Helfritch, P. Leyman, S. Grendahl, and B. Klotz. 2005. Interface material mixing formed by the deposition of copper on aluminium by means of the cold spray process. *Journal of Thermal Spray Technology* 14 (3): 330–334.

Cinca, N., J. M. Rebled, S. Estradé, F. Peiró, J. Fernández, and J. M. Guilemany. 2013. Influence of the particle morphology on the Cold Gas Spray deposition behaviour of titanium on aluminum light alloys. *Journal of Alloys and Compounds* 554:89–96.

Conrad, H., and L. Rice. 1970. The cohesion of previously fractured Fcc metals in ultrahigh vacuum. *Metallurgical Transactions* 1 (11): 3019–3029 doi:10.1007/BF03038415.

Crossland, B. 1971a. The development of explosive welding and its application in engineering. *Metals and Materials* 5 (12): 401–413.

Crossland, B. 1971b. Friction welding. *Contemporary Physics* 12 (6): 559–574. doi:10.1080/00107517108205660.

Crossland, B., and J. D. Williams. 1970. Explosive welding. *Metals and Materials* 4: 79–100.

Davis, J.R. ed. 2004. *Handbook of thermal spray technology*. Materials Park: ASM International.

DeForce, B. S., T. J. Eden, and J. K. Potter. 2011. Cold spray Al-5% Mg coatings for the corrosion protection of magnesium alloys. *Journal of Thermal Spray Technology* 20 (6):1352–1358.

Drehmann, R., T. Grund, T. Lampke, B. Wielage, K. Manygoats, T. Schucknecht, and D. Rafaja. 2013. Splat formation and adhesion mechanisms of cold gas-sprayed Al coatings on Al2O3 substrates. *Journal of Thermal Spray Technology* 23 (1–2): 68–75. doi:10.1007/s11666–013-9966-z.

Dykhuizen, R. C., and M. F. Smith. 1998. Gas dynamic principles of cold spray. *Journal of Thermal Spray Technology* 7 (2): 205–212.

Dykhuizen, R. C., M. F. Smith, D. L. Gilmore, R. A. Neiser, X. Jiang, and S. Sampath. 1999. Impact of high velocity cold spray particles. *Journal of Thermal Spray Technology* 8 (4): 559–564.

Frost, H. J., and M. F. Ashby. 1982. *Deformation mechanism maps: the plasticity and creep of metals and ceramics*. Oxford: Pergamon.

Fukumoto, M., H. Terada, M. Mashiko, K. Sato, M. Yamada, and E. Yamaguchi. 2009. Deposition of copper fine particle by cold spray process. *Materials Transactions* 50 (6): 1482–1488.

Gartner, F., T. Stoltenhoff, T. Schmidt, and H. Kreye. 2006. The cold spray process and its potential for industrial applications. *Journal of Thermal Spray Technology* 15 (2): 223–232.

Goldbaum, D., J. Shockley, R. Chromik, A. Rezaeian, S. Yue, J.-G. Legoux, and E. Irissou. 2012. The effect of deposition conditions on adhesion strength of Ti and Ti6Al4V cold spray splats. *Journal of Thermal Spray Technology* 21 (2): 288–303. doi:10.1007/s11666–011-9720–3.

Grujicic, M., J. R. Saylor, D. E. Beasley, W. S. DeRosset, and D. Helfritch. 2003. Computational analysis of the interfacial bonding between feed-powder particles and the substrate in the cold-gas dynamic-spray process. *Applied Surface Science* 219 (3–4): 211–227.

Grujicic, M., C. L. Zhao, W. S. DeRosset, and D. Helfritch. 2004. Adiabatic shear instability based mechanism for particles/substrate bonding in the cold-gas dynamic-spray process. *Materials and Design* 25 (8): 681–688.

Guetta, S., M. H. Berger, F. Borit, V. Guipont, M. Jeandin, M. Boustie, F. Poitiers, Y. Ichikawa, and K. Ogawa. 2008. Influence of particle velocity on adhesion of cold-sprayed splats. In *Thermal spray: crossing borders, Maastricht, Netherlands 2008*, ed. E. Lugscheider. Dusseldorf: DVS-Verlag.

Guetta, S., M. Berger, F. Borit, V. Guipont, M. Jeandin, M. Boustie, Y. Ichikawa, K. Sakaguchi, and K. Ogawa. 2009. Influence of particle velocity on adhesion of cold-sprayed splats. *Journal of Thermal Spray Technology* 18 (3): 331–342.

Ham, J. L. 1963. Metallic cohesion in high vacuum. *American Society of Lubrication Engineers Transactions* 6 (1): 20–28. doi:10.1080/05698196308971995.

Ikeda, H., Y. Qi, T. Çagin, K. Samwer, W. L. Johnson, and W. A. Goddard. 1999. Strain rate induced amorphization in metallic nanowires. *Physical Review Letters* 82 (14): 2900–2903.

Irissou, E., J.-G. Legoux, B. Arsenault, and C. Moreau. 2007. Investigation of Al-Al2O3 cold spray coating formation and properties. *Journal of Thermal Spray Technology* 16 (5): 661–668.

Isono, N., P. Smith, D. Turnbull, and M. Aziz. 1996. Anomalous diffusion of Fe in liquid Al measured by the pulsed laser technique. *Metallurgical and Materials Transactions A* 27 (3): 725–730.

Jodoin, B. 2002. Cold spray nozzle mach number limitation. *Journal of Thermal Spray Technology* 11 (4): 496–507.

Johnson, G. R., and W. H. Cook. 1983. A constitutive model and data for metals subjected to large strains, high strain rates and high temperatures. Paper presented at the Proceedings of the 7th International Ballistics Symposium, Hague, The Netherlands.

Kang, K., S. Yoon, Y. Ji, and C. Lee. 2008. Oxidation dependency of critical velocity for aluminum feedstock deposition in kinetic spraying process. *Materials Science and Engineeing A* 486 (1–2): 300–307.

Karthikeyan, J., and A. Kay. 2005. Cold spray processing of copper and copper alloys. *Advanced Materials and Processes* 163 (8): 49–55.

King, P. C., and M. Jahedi. 2011. Transmission electron microscopy of cold sprayed titanium. In Thermal Spray: *global solutions for future application, Singapore 2010*, eds. B. Marple, A. Agarwal, M. Hyland, Y.-C. Lau, C.-J. Li, R. S. Lima, and G. Montavon. ASM International, Materials Park, USA.

King, P. C., S. H. Zahiri, and M. H. Jahedi. 2008. Focussed ion beam micro-dissection of cold sprayed particles. *Acta Materialia* 56 (19): 5617–5626.

King, P. C., S. H. Zahiri, and M. Jahedi. 2009. Microstructural refinement within a cold sprayed copper particle. *Metallurgical and Materials Transactions A* 40 (9): 2115–2123.

King, P. C., G. Bae, S. Zahiri, M. Jahedi, and C. Lee. 2010a. An experimental and finite element study of cold spray copper impact onto two aluminum substrates. *Journal of Thermal Spray Technology* 19 (3): 620–634.

King, P. C., S. Zahiri, M. Jahedi, and J. Friend. 2010b. Aluminium coating of lead zirconate titanate–A study of cold spray variables. *Surface and Coatings Technology* 205 (7): 2016–2022.

King, P. C., A. J. Poole, S. Horne, R. de Nys, S. Gulizia, and M. Z. Jahedi. 2013. Embedment of copper particles into polymers by cold spray. *Surface and Coatings Technology* 216 (0): 60–67. doi:http://dx.doi.org/10.1016/j.surfcoat.2012.11.023.

King, P. C., C. Busch, T. Kittel-Sherri, M. Jahedi, and S. Gulizia. 2014. Interface melding in cold spray titanium particle impact. *Surface and Coatings Technology* 239 (0): 191–199. doi:http://dx.doi.org/10.1016/j.surfcoat.2013.11.039.

Klassen, T., F. Gärtner, T. Schmidt, J. O. Kliemann, K. Onizawa, K. R. Donner, H. Gutzmann, K. Binder, and H. Kreye. 2010. Basic principles and application potentials of cold gas spraying.

Bindemechanismen und potenzielle Anwendungen des Kaltgasspritzens. *Materialwissenschaft und Werkstofftechnik* 41 (7): 575–584. doi:10.1002/mawe.201000645.

Kliemann, J. O., H. Gutzmann, F. Gärtner, H. Hübner, C. Borchers, and T. Klassen. 2011. Formation of cold-sprayed ceramic titanium dioxide layers on metal surfaces. *Journal of Thermal Spray Technology* 20 (1–2): 292–298. doi:10.1007/s11666–010-9563–3.

Klinkov, S. V., and V. F. Kosarev. 2006. Measurements of cold spray deposition efficiency. *Journal of Thermal Spray Technology* 15 (3): 364–371.

Klinkov, S. V., V. F. Kosarev, and M. Rein. 2005. Cold spray deposition: Significance of particle impact phenomena. *Aerospace Science and Technology* 9 (7): 582–591.

Kosarev, V. F., S. V. Klinkov, A. P. Alkhimov, and A. N. Papyrin. 2003. On some aspects of gas dynamics of the cold spray process. *Journal of Thermal Spray Technology* 12 (2): 265–281.

Lee, H. Y., Y. H. Yu, Y. C. Lee, Y. P. Hong, and K. H. Ko. 2004. Cold spray of SiC and Al2O3 with soft metal incorporation: a technical contribution. *Journal of Thermal Spray Technology* 13 (2): 184–189.

Lee, J., S. Shin, H. Kim, and C. Lee. 2007. Effect of gas temperature on critical velocity and deposition characteristics in kinetic spraying. *Applied Surface Science* 253 (7): 3512–3520.

Lemiale, V., Y. Estrin, H. S. Kim, and R. O'Donnell. 2011. Forming nanocrystalline structures in metal particle impact. *Metallurgical and Materials Transactions A* 42A (10): 3006–3012. doi:10.1007/s11661–010-0588–5.

Lemiale, V., P. C. King, M. Rudman, M. Prakash, P. W. Cleary, M. Z. Jahedi, and S. Gulizia. 2014. Temperature and strain rate effects in cold spray investigated by smoothed particle hydrodynamics. *Surface and Coatings Technology* 254: 121–130. doi:10.1016/j.surfcoat.2014.05.071.

Li, C.-J., W.-Y. Li, and Y.-Y. Wang. 2005. Formation of metastable phases in cold-sprayed soft metallic deposit. *Surface and Coatings Technology* 198 (1–3): 469–473.

Li, C.-J., W.-Y. Li, and H. Liao. 2006. Examination of the critical velocity for deposition of particles in cold spraying. *Journal of Thermal Spray Technology* 15 (2): 212–222. doi:10.1361/105996306 × 108093.

Li, W.-Y., C. Zhang, X. Guo, C.-J. Li, H. Liao, and C. Coddet. 2007. Study on impact fusion at particle interfaces and its effect on coating microstructure in cold spraying. *Applied Surface Science* 254 (2): 517–526.

Li, C.J., H. T. Wang, Q. Zhang, G. J. Yang, W. Y. Li, and H. L. Liao. 2009. Influence of spray materials and their surface oxidation on the critical velocity in cold spraying. *Journal of Thermal Spray Technology* 19 (1–2): 95–101. doi:10.1007/s11666–009-9427-x.

Mamalis, A. G., I. N. Vottea, and D. E. Manolakos. 2001. On the modelling of the compaction mechanism of shock compacted powders. *Journal of Materials Processing Technology* 108 (2): 165–178.

Marrocco, T., D. G. McCartney, P. H. Shipway, and A. J. Sturgeon. 2006. Production of titanium deposits by cold-gas dynamic spray: numerical modeling and experimental characterization. *Journal of Thermal Spray Technology* 15 (2): 263–272.

Moridi, A., S. M. Hassani-Gangaraj, M. Guagliano, and M. Dao. 2014. Cold spray coating: Review of material systems and future perspectives. *Surface Engineering* 30 (6): 369–395.

Murr, L. E., S. A. Quinones, T. E. Ferreyra, A. Ayala, L. O. Valerio, F. Horz, and R. P. Bernhard. 1998. The low-velocity-to-hypervelocity penetration transition for impact craters in metal targets. *Materials Science and Engineering: A* 256 (1–2): 166–182.

Ning, X.-J., J.-H. Jang, H.-J. Kim, C.-J. Li, and C. Lee. 2008. Cold spraying of Al-Sn binary alloy: coating characteristics and particle bonding features. *Surface and Coatings Technology* 202: 1681–1687.

Papyrin, A. 2001. Cold spray technology. *Advanced Materials and Processes* 159 (9): 49–51.

Poole, A.J., R. de Nys, P. C. King, S. Gulizia, and M. Jahedi. 2012. Protecting polymer surface against fouling in boat hulls involves embedding in polymer surface particles with antifouling properties by spray mechanism, where particles are accelerated and sprayed onto polymer surface with suitable velocity. WO2012006687–A1.

Price, T., P. Shipway, and D. McCartney. 2006. Effect of cold spray deposition of a titanium coating on fatigue behavior of a titanium alloy. *Journal of Thermal Spray Technology* 15 (4): 507–512.

Rafaja, D., T. Schucknecht, V. Klemm, A. Paul, and H. Berek. 2009. Microstructural characterisation of titanium coatings deposited using cold gas spraying on Al2O3 substrates. *Surface and Coatings Technology* 203 (20–21): 3206–3213. doi:10.1016/j.surfcoat.2009.03.054.

Raletz, F., M. Vardelle, and G. Ezo'o. 2005. Fast determination of particle critical velocity in cold spraying. In Thermal spray 2005: Thermal Spray connects: Explore its surfacing potential! 2005, ed. E. Lugscheider. ASM International Materials Park, USA.

Raletz, F., M. Vardelle, and G. Ezo'o. 2006. Critical particle velocity under cold spray conditions. *Surface and Coatings Technology* 201 (5): 1942–1947.

Raybould, D. 1980. The cold welding of powders by dynamic compaction. International *Journal of Powder Metallurgy and Powder Technology* 16 (1): 9–19.

Robinson, J. L. 1977. Fluid mechanics of copper: viscous energy dissipation in impact welding. *Journal of Applied Physics* 48 (6): 2202–2207.

Schmidt, T., F. Gartner, H. Assadi, and H. Kreye. 2006. Development of a generalized parameter window for cold spray deposition. *Acta Materialia* 54 (3): 729–742.

Schmidt, T., H. Assadi, F. Gärtner, H. Richter, T. Stoltenhoff, H. Kreye, and T. Klassen. 2009. From particle acceleration to impact and bonding in cold spraying. *Journal of Thermal Spray Technology* 18 (5): 794–808.

Song, M., H. Araki, S. Kuroda, and K. Sakaki. 2013. Reaction layer at the interface between aluminium particles and a glass substrate formed by cold spray. *Journal of Physics D: Applied Physics* 46 (19): 195301.

Sova, A., V. F. Kosarev, A. Papyrin, and I. Smurov. 2010. Effect of ceramic particle velocity on cold spray deposition of metal-ceramic coatings. *Journal of Thermal Spray Technology* 20 (1–2): 285–291. doi:10.1007/s11666–010-9571–3.

Spencer, K.,and M.-X. Zhang. 2009. Heat treatment of cold spray coatings to form protective intermetallic layers. Scripta Materialia 61 (1): 44–47.

Stoltenhoff, T., C. Borchers, F. Gartner, and H. Kreye. 2006. Microstructures and key properties of cold-sprayed and thermally sprayed copper coatings. *Surface and Coatings Technology* 200 (16–17): 4947–4960.

Vlcek, J., L. Gimeno, H. Huber, and E. Lugscheider. 2005. A systematic approach to material eligibility for the cold-spray process. *Journal of Thermal Spray Technology* 14 (1): 125–133.

Vucko, M. J., P. C. King, A. J. Poole, C. Carl, M. Z. Jahedi, and R. de Nys. 2012. Cold spray metal embedment: an innovative antifouling technology. *Biofouling* 28 (3): 239–248.

Walsh, J.M., R. G. Shreffler, and F. J. Willig. 1953. Limiting conditions for jet formation in high velocity collisions. *Journal of Applied Physics* 24 (3): 349–359.

Wang, Q., N. Birbilis, and M.-X. Zhang. 2011. Interfacial structure between particles in an aluminum deposit produced by cold spray. *Materials Letters* 65 (11): 1576–1578. doi:10.1016/j.matlet.2011.03.035.

Wang, F. F., W. Y. Li, M. Yu, and H. L. Liao. 2014. Prediction of critical velocity during cold spraying based on a coupled Thermomechanical Eulerian Model. *Journal of Thermal Spray Technology* 23 (1–2): 60–67. doi:10.1007/s11666–013-0009–6.

Wank, A., B. Wielage, H. Podlesak, and T. Grund. 2006. High-resolution microstructural investigations of interfaces between light metal alloy substrates and cold gas-sprayed coatings. *Journal of Thermal Spray Technology* 15 (2): 280–283.

Wolfe, D. E., T. J. Eden, J. K. Potter, and A. P. Jaroh. 2006. Investigation and characterization of Cr3C2-based wear-resistance coatings applied by the cold spray process. *Journal of Thermal Spray Technology* 15 (3): 400–412.

Wong, W., A. Rezaeian, E. Irissou, J.-G. Legoux, and S. Yue. 2010. Cold spray characteristics of commercially pure Ti and Ti-6Al-4V. *Advanced Materials Research* 89–91:639–644.

Wu, J., H. Fang, H. Kim, and C. Lee. 2006. High speed impact behaviors of Al alloy particle onto mild steel substrate during kinetic deposition. *Materials Science and Engineering: A* 417 (1–2): 114–119.

Xiong, Y., K. Kang, G. Bae, S. Yoon, and C. Lee. 2008. Dynamic amorphization and recrystallization of metals in kinetic spray process. *Applied Physics Letters* 92: 194101.

Xiong, Y. M., X. Xiong, S. Yoon, G. Bae, and C. Lee. 2011. Dependence of bonding mechanisms of cold sprayed coatings on strain-rate-induced non-equilibrium phase transformation. *Journal of Thermal Spray Technology* 20 (4): 860–865. doi:10.1007/s11666–011-9634–0.

Xu, Y., and I. M. Hutchings. 2006. Cold spray deposition of thermoplastic powder. *Surface and Coatings Technology* 201 (6): 3044–3050.

Yandouzi, M., E. Sansoucy., L. Ajdelsztajn, and B. Jodoin. 2007. WC-based cermet coatings produced by cold gas dynamic and pulsed gas dynamic spraying processes. *Surface and Coatings Technology* 202 (2): 382–390. doi:http://dx.doi.org/10.1016/j.surfcoat.2007.05.095.

Yin, S., X. Wang, W. Li, H. Liao, and H. Jie. 2012. Deformation behavior of the oxide film on the surface of cold sprayed powder particle. *Applied Surface Science* 259: 294–300. doi:10.1016/j.apsusc.2012.07.036.

Yu, M., W. Y. Li, F. F. Wang, and H. L. Liao. 2012. Finite element simulation of impacting behavior of particles in cold spraying by eulerian approach. *Journal of Thermal Spray Technology* 21 (3–4): 745–752. doi:10.1007/s11666–011-9717-y.

Zhang, D., P. H. Shipway, and D. G. McCartney. 2005. Cold gas dynamic spraying of aluminium: the role of substrate characteristics in deposit formation. *Journal of Thermal Spray Technology* 14 (1): 109–116.

Zou, Y., W. Qin, E. Irissou, J. G. Legoux, S. Yue, and J. A. Szpunar. 2009. Dynamic recrystallization in the particle/particle interfacial region of cold-sprayed nickel coating: electron backscatter diffraction characterization. *Scripta Materialia* 61 (9): 899–902. doi:10.1016/j.scriptamat.2009.07.020.

Chapter 3
Characteristics of Feedstock Materials

T. Hussain, S. Yue and C.-J. LI

3.1 Introduction

Feedstock in any spray processing is very important as the quality of the coating depends on the quality of the feedstock materials. A cold-sprayed coating is characterised by a set of parameters related to its application, such as electrical conductivity, heat isolation, corrosion resistance, oxidation resistance, wear resistance, etc. All these parameters are intrinsically related to the starting powder and the cold spray process parameters used to deposit the coatings. Unlike thermal spraying where the feedstock material changes chemically to a large extent during spraying, cold spraying tends to retain the characteristics of the feedstock powder in the coating process. This makes the selection of feedstock materials a critical factor in depositing cold-sprayed coatings.

The powder manufacturing industry is increasingly moving towards higher purity, narrower cut of powder particle sizes and powder morphologies suitable for cold spraying. At the beginning of the twenty-first century, during the early stage of development of the cold spray, the powder feedstock used in cold spray was very limited, as most of the powders were manufactured for thermal spraying and powder

T. Hussain (✉)
Division of Materials, Mechanics and Structure, Faculty of Engineering,
University of Nottingham, Nottingham NG7 2RD, UK
e-mail: tanvir.hussain@nottingham.ac.uk

S. Yue
Department of Mining and Materials Engineering, McGill University,
Montreal, QC H3A 0C5, Canada
e-mail: steve.yue@mcgill.ca

C.-J. Li
State Key Laboratory for Mechanical Behavior of Materials, School of Materials Science
& Engineering, Xi'an Jiaotong University, Xi'an 710049, Shaanxi, PRC
e-mail: licj@mail.xjtu.edu.cn

J. Villafuerte (ed.), *Modern Cold Spray,* DOI 10.1007/978-3-319-16772-5_3

metallurgy industries. With the introduction of commercial cold spray equipment, the powder manufacturing industry migrated towards producing powder size ranges and compositions suitable for cold spraying. Typically, cold spraying requires a feedstock which has a narrow size distribution and finer than the typical thermal spraying feedstock, the reasons behind these are explained later in this chapter. The required particle-size fractions are specific for each of the different cold spray equipment design (Champagne 2007). The most widely used particle-size fractions are in the ranges of 5–25 and 15–45 μm. This is typically the smaller end of the powder size ranges manufactured by the commercial powder manufacturers. The price of the powders is a function of the particle size and morphology specifications. Needless to say, the costs of the powders go up with decreasing particle-size range and the increasing purity.

Feedstock material is considered as an essential part of the coating design process which includes the properties of the substrates and the intermediate layer (in case of polymers, composites, etc.). The properties of the feedstock materials vary depending on the manufacturing routes. A number of well-established techniques such as atomisation, mechanical alloying, sintering and spray drying are currently used to produce feedstock for cold spraying. This chapter is divided into three major sections: the first section describes the properties of the feedstock materials followed by the powder manufacturing routes, effect of feedstock properties on the cold sprayability and finally the deposition of composite powders using a cold spray.

3.2 Feedstock Properties and Characterisations

There are several characteristics of powders which can make them suitable or unsuitable for feeding and cold spraying. The properties of the feedstock material considerably affect the quality of the cold-sprayed coatings. A fundamental requirement of the powder feeder in the cold spray process is to transport the powder to the upstream or the downstream of the convergent–divergent nozzle where the powder particles are accelerated towards the substrates. Powder flowability, or as it is known in the industry as the 'powder feeding problem', is an aspect related to the sprayability of the powder (Davis 2004). Powder flowability is the first criterion in selecting a feedstock material for cold spray. Powder feeding problems or clogging of nozzles at the throat (the smallest cross-section in a convergent–divergent nozzle) can result in interrupted spray runs and the coatings of poor quality. A quantitative test method for measuring the flowability of the powder is ASTM B213 using a Hall flowmeter funnel (Davis 2004). A Hall flowmeter is a simple device which determines the flow rate of the metallic and fine free-flow powders using a pre-calibrated funnel. Hall flowmeter also determines the apparent density of the test powder according to ASTM B212 (Pawlowski 2008). The flow rate of the powder is an essential characteristic, which determines the feasibility of the powder for the cold spray process. A flow rate that is too high will build up at the convergent section of the nozzle and block the throat; alternatively, a poor flowability will result in intermitted feeding and inconsistent coating properties.

The properties of the feedstock powder can be broadly divided into two categories: physical and chemical characteristics. The physical characteristics include:

- Particle sizes and grain sizes
- External morphology (spherical, globular, angular, with/without satellites) as well as internal morphology (porosity)
- Flowability and apparent density
- Thermal properties
- Electrical conductivity/resistance

And, the typical chemical properties of the powders include:

- Chemical composition (purity, level of unwanted species such as oxygen, nitrogen levels)
- Distribution of precipitates and phases
- Crystallographic information (solid solutions)

It is desirable to have powders with good flowability and free from satellite particles. Satellite particles are fine particles which stick to the larger particles and cause agglomeration and feeding problems. This section summarises the characteristics of the powders and their measurement techniques. The effect of various powder characteristics on the cold sprayability will be discussed in the next section.

3.2.1 Particle Size Distribution

Powder particle size in a batch of powder is defined by the upper and lower limit of the powder size distribution. The particle sizes are measured using a laser diffraction particle-size analyser (such as Malvern 3000, Malvern, UK) which produces a size distribution of all the measured sizes. A good powder size distribution is normally distributed (Gaussian) and are not skewed meaning the difference between the 50th percentile and the mean particle size is not significant (Crawmer 2004). In reality, some skew is unavoidable in the powder size distribution. The powder sizes are typically established by the 90th and the 10th percentiles of the distribution. A typical Cu powder for cold spray has a size range of $+5-25$ μm, which means 10 % of the powder is below 5 μm and 10 % of the powder is above 25 μm. Classification is the technique to separate powders in different size ranges after the production stages. Air classification is commonly utilised for size ranges below 45 μm as the screens/sieves are not practical.

3.2.2 Particle Morphology

Particle morphology or shape is a function of the manufacturing route of the powder production. Powder morphology affects the flowability, apparent density, cold sprayability and ultimately the coating porosity and deposition efficiency (DE). In general, irregular shaped particles do not feed as easily as the spherical particles as the irregular particles pack together closely (Berndt 2004a). Gas-atomised particles are spherical and hence have a better flowability.

Powder morphology needs to be investigated to determine the manufacturing route of the feedstock. In addition to, external features such as satellite particles powders can also have internal porosities. External morphology of the powder can be examined directly using a scanning electron microscope, and internal morphology and porosity can be examined through a metallographic preparation of powder cross-sections. Powders can be encapsulated in the resin and prepared using grinding and polishing for microstructural observations.

3.2.3 Chemical Purity

Powders used in cold spraying are of higher purity grades than those typically used for thermal spraying. Materials are graded according to various levels of impurities (O_2, N_2) content in them, which also affects the mechanical properties of the material. The bonding in cold spraying is considered to be a result of breaking up of the oxide shells surrounding the materials, and hence higher level of oxygen can make it difficult for bonding. The role of oxides in cold sprayability is discussed later in this chapter. The impurity levels in the powder depend on the manufacturing routes and powder storage.

Absorbed moisture on the powder surface can create significant powder feeding problem and degrade the powder. To remove the moisture, it is recommended to dry the powder in a heated oven above 100 °C (boiling point of water) before spraying. Finer powders due to their high surface area to volume ratios can hold a large amount of moisture than coarser powders (Crawmer 2004). A sphere has the lowest surface area to a volume ratio of any geometry and hence a spherical shape absorbs less surface moisture than irregularly shaped particles.

The chemical composition of a powder can be analysed using X-ray fluorescence (XRF) spectroscopy and the powder surface can be analysed using X-ray photoelectron spectroscopy (XPS). XPS is a useful technique to measure the surface contamination of the powders and the resulting coatings. X-ray diffraction (XRD; Cullity and Stock 2001) is widely used to identify the phases present in the powder. Comparing the initial phases in the powder and the phases in the cold-sprayed coating can provide valuable information about the phase changes during spraying. Elemental distribution of the composite or the agglomerated powders can be obtained from the energy-dispersive X-ray (EDX) analysis on the metallographic cross-section of the powder particles.

3.2.4 Flowability

Flowability is a major factor in powder feeding in the cold spraying industry. As previously mentioned, flowability of a powder and apparent density is measured using a Hall flowmeter. The term density in relation to powder is used loosely in the thermal/cold spray industry. Apparent density is the specific gravity of the powder including the spaces between the powder particles and any porosity within the particles (Crawmer 2004). Apparent density is related to the packing of the powders and affected by the powder particle-size distribution. Apparent density of the feedstock affects the spray feedrate during cold spraying.

3.3 Powder Production Methods

The materials used in cold spraying are principally metallic. During the infancy of the cold spray process in the 1990s, commercially pure copper feedstock was extensively used to optimise nozzle dimensions and process parameters. The widespread use of copper was due to its ductility and the availability of the powder in desired size ranges. The research and development involving copper was closely followed by aluminium and nickel. Most of these metallic powders were produced using atomisation techniques and had a spherical morphology. With the introduction of commercial cold spray equipment during the early twenty-first century, significant research and development effort were focussed on developing commercially pure titanium coatings, which was followed by angular/ sponge titanium and titanium alloy powders (such as Ti-6Al-4V). The current interest in the cold spraying community is in the spraying of alloy powders such as In 718, MCrAlY and composite powders from blends.

Novel hard-metal coatings with ceramics, which were not possible to deposit using cold spraying in the past, can now be successfully deposited using novel powder design and manufacturing routes. Moreover, the development in the cold spraying equipment with higher gas pressure and temperature means difficult-to-spray materials can now be deposited. The following types of powders have been successfully deposited using cold spray:

- Metals: Cu, Al, Ni, Zn, Ti
- Refractory metals: Zr, Ta
- Alloys: Stainless steel 316, Al alloys, MCrAlY
- Oxides: TiO_2
- Cermet (combination of a ceramic and a metal): WC–Co/Ni
- Intermetallics: Fe/Al

It is not the authors' intention to provide a comprehensive review of all the feedstock powders, which have so far been successfully sprayed in cold spraying, but to give an overview of the materials which could be used for cold spraying. A summary of the common powder production methods is described below.

3.3.1 Atomisation

Atomisation is a process of breaking up of a melt into droplets which can be achieved using a number of ways, such as by spraying through a nozzle, by pouring over a rotating disc, electrostatically and ultrasonically (Yule and Dunkley 1994).

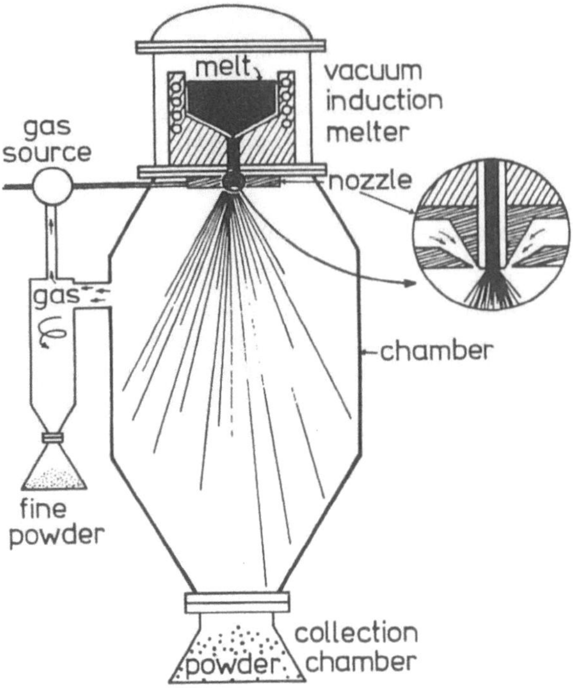

Fig. 3.1 Schematic of an inert gas atomisation. (Yule and Dunkley 1994)

The science and engineering of atomisation is rather well understood and documented. The readers are recommended to the textbook by Yule and Dunkley for a detailed understanding of the atomisation process. Metal and alloy powders are mostly manufactured using the atomisation route as the metals melt at well-defined high temperature to produce low-viscosity liquid. Atomisation is suitable for most metals such as tin, lead, zinc, aluminium, magnesium, silver, copper, gold, palladium, cobalt, nickel, iron and steel and even refractory metals such as tungsten. Metals melting over 2000 K are difficult to atomise due to naturally present difficulties. Although not widely used, atomisation of glass, ceramic and polymers has been studied to some limited extent. Figure 3.1 shows the schematic of an inert gas atomisation process.

In atomisation, a bulk liquid is formed by melting a substance, and the end product is a solid which is a powder. Only the gas and water atomisation are widely used for cold spray powders. In an atomiser, the metal or alloy is melted in an inductive heater creating a melt which is then poured into a crucible with calibrated outlet where the melt drops down as a liquid stream into a nozzle. The continuous stream of liquid metal is broken down into droplets by impingement of a gas or water stream inside the nozzle (Pawlowski 2008). The metal solidifies during the free fall inside the chamber forming small powder particles.

There are ranges of variables which affect the powder quality produced by atomisation technique such as nozzle geometry, velocity and pressure of the atomising media, gas purity, atomising jet geometry, melt superheat (i.e. temperature above the melting point), composition of the melt, viscosity and surface tension of the melt, height of the cooling tower, etc.

3.3.2 *Gas Atomisation*

Gas atomisation is carried out using air or inert gases such as argon and helium which breaks up the flowing stream of liquid metal. There are several models to describe the liquid break-up during atomisation process; the idealised modes include the formation of waves in sheets, then ligament formation and finally the ligament breakdown into droplets resulting in spheroidisation (Berndt 2004b). There are two variations of the gas atomisation process which are commonly used in the powder manufacturing industry: (a) close confined nozzles, (b) open nozzles (Klar and Shafer 1972). In a closed nozzle, the melt passes through a feed tube before the gas strikes the melt as it emerges from the nozzle. On the other hand, in open nozzles, the metal stream falls some distance from the nozzle into the confluence of gas stream that has been created by the gas nozzle.

Cold spray powder requires very high purity that is virtually oxygen free. This is the reason inert gas atomisation and vacuum atomisation are frequently used to produce cold-spray grade powder. There are systematic studies published in literature with Cu, which showed increasing the oxygen content of the powder increased the critical velocity, the minimum velocity required for deposition (Li et al. 2006). Typical morphology of gas-atomised copper, aluminium and titanium powder are shown in Fig. 3.2.

In gas atomisation, powder particles generally experience rapid cooling, subsequent rapid solidification rate and exhibit quenched microstructure with fine grains (even ultrafine grain structure), high density dislocation and saturated solution of alloy elements (Rokni et al. 2014). This quenching effect increases the hardness of spray powder particles, and thus it reduces its deformability. Higher magnification images of the gas-atomised titanium particles showing the rapidly quenched morphology is shown in Fig. 3.3.

3.3.2.1 Water Atomisation

Waterjets can be used in the open nozzle configuration previously described as the water does not tend to lose the kinetic energy nearly as rapidly as gas jets. Waterjets also not tend to spread as much as gas jets. The designs typically allow the melt stream to fall 100–500 mm before being hit by the water. The waterjet tends to produce slightly more irregular particles than gas atomisation (Berndt 2004b), but industry favours this process due to its high production rates. Dewatering or reduction stages are necessary to bring the moisture and oxygen content of the powder within the specific limit.

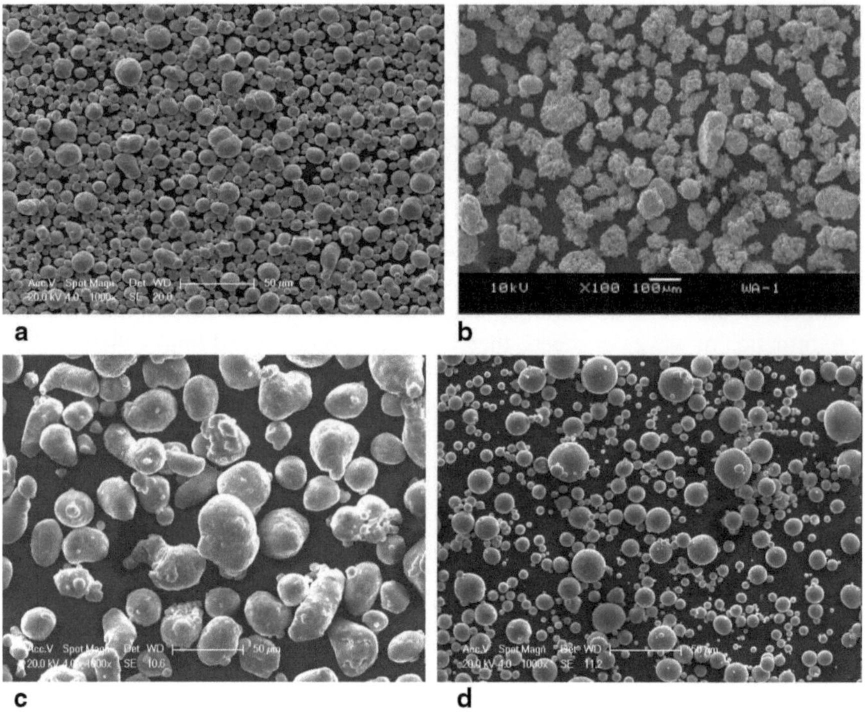

Fig. 3.2 Scanning electron micrograph (secondary electron) of **a** gas-atomised copper (Hussain et al. 2009), **b** water-atomised copper (Chiu et al. 2007), **c** gas-atomised aluminium (Hussain et al. 2012) and **d** gas-atomised titanium used in cold spray process. (Hussain et al. 2011a)

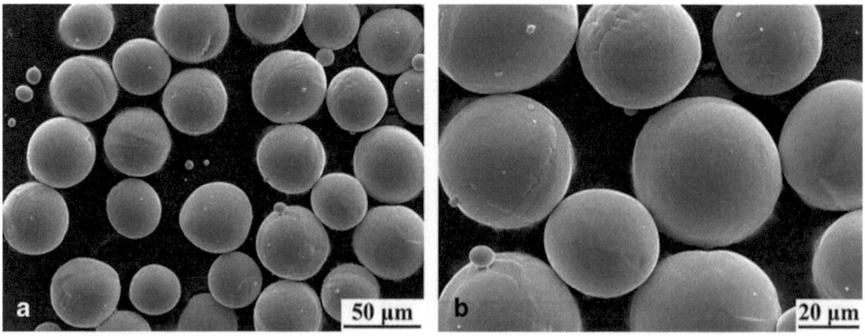

Fig. 3.3 Gas-atomised titanium powder showing the surface morphology **a** low magnification and **b** high magnification

The primary advantage of gas atomisation over water atomisation is the enhancement of the spherical morphology. The powders produced via gas atomisation are relatively cleaner and have low oxide contents. The level of oxygen in the feedstock powder is crucial as the bonding in cold spraying requires oxide-free clean metallic interfaces. Both process have very high cooling rates, for example, in water atomisation the particles are cooled at 10^4–10^6 °C/s, whereas the cooling rate in gas atomisation is 10^3–10^5°C/s—an order of magnitude lower (Berndt 2004b).

In general, atomisation in metal and alloy powder production can handle a wide range of outputs, typically 1–100,000 t/year, and can produce a wide range of particle sizes from 10 μm to 10 mm. The main advantage of atomisation is the ability to control the size distribution and the shape of the particles rather well, within very tight limit. One can produce near-spheroidal particles using gas atomisation which has better flow properties than any milled particles.

3.3.3 Spray Drying (Agglomeration)

Spray drying is a powder manufacturing route which allows various types of materials to be agglomerated. A typical spray-drying process consists of several stages: preparation of slurries, atomisation of slurries, drying of the spray and densification of the particles. A slurry of finely dispersed precursors, organic binder and water are injected into a large empty chamber using a slurry pump. A centrifugal or a nozzle can be used to atomise the slurry, which is subsequently dried using a gas. The solid particles are collected in the powder collector. The powder produced via this technique is usually porous and requires a densification stage such as sintering. Agglomeration and sintering is a preferable route for powders such as oxides, nitrides and cermets. The powders produced via the spray-drying process have yet to find a commercial cold spray application, but several experimental studies reported the cold sprayability of such powders (Fig. 3.4).

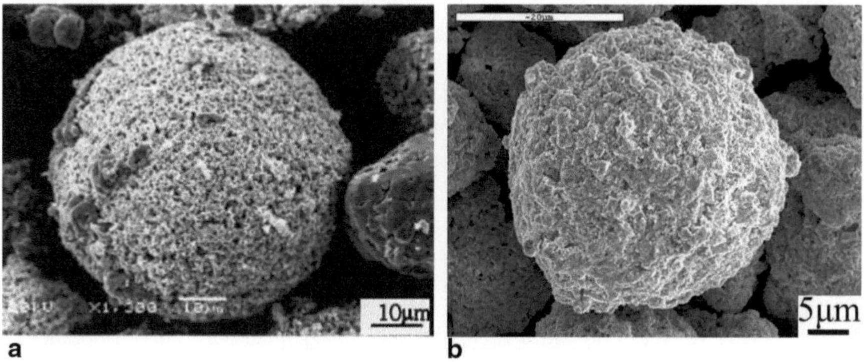

Fig. 3.4 **a** Agglomerated W–Cu powder (Moridi et al. 2014) and **b** agglomerated and sintered WC–Co powder used in cold spraying. (Li et al. 2007b)

3.3.4 Sintering and Crushing

Typically, oxides and cermets are examples of powders produced via sintering and crushing. Sintered powders are irregular in shape which has poor flowability and sprayability in cold spraying. Sintering usually takes place at temperatures around 0.7 × melting point. The powders produced from sintering are typically dense and blocky. WC is produced by carburising tungsten followed by crushing and screening to get the desired size range. The crushed powder is then mixed with Co with an organic binder and sintered under a reducing temperature.

3.3.5 Hydride–Dehydride

Commercially pure titanium powder and titanium alloys are of significant interest to the cold spraying community as the lack of particle melting and oxidation during spraying results in maintaining the initial powder properties. Hydride–dehydride (HDH) is a chemical process which is widely used to produce titanium and titanium alloy feedstock. The manufacturing cost of this powder is much less than the gas-atomised titanium feedstock. The typical morphology of the powders produced using HDH process is irregular. The manufacturing process relies on the brittle nature of some metal hydrides which can be milled and screened (Hussain 2013).

The level of impurities in titanium is crucial as this can affect the mechanical properties of the finished product. For finer powders, the oxygen content is sensitive to the particle surface area to volume ratio. As the particle diameter decreases, the particle surface area to volume ratio increases significantly. The significance of this is the fact that as the particle size decreases the oxygen content increases.

In the HDH process, the raw material is initially loaded into a hydride unit, and the material is heated under a hydrogen atmosphere. The reaction results in the formation of titanium hydride (TiH_2). This brittle TiH_2 is crushed into finer particle size. The fine particles are returned to the hydride unit for the dehydride process. The particles are placed under a high vacuum and heated up to release the hydrogen from the particles during a reversible reaction. The powder is then screened to remove any sintered particles. The powder morphology is impacted by the starting raw materials in this process. Currently, there are three sources of raw titanium: wrought titanium, sodium- or magnesium-reduced titanium. A flow diagram of the HDH process is shown in Fig. 3.5. Typical morphology of the HDH commercially pure titanium and Ti6Al4 alloy are shown in Fig. 3.6.

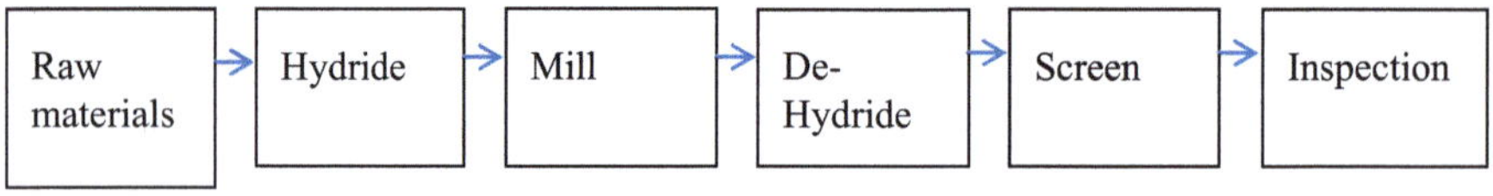

Fig. 3.5 Flow diagram of an HDH process to make titanium powder

Fig. 3.6 Scanning electron morphology (SEM) images of the hydride–dehydride (a–b) commercially pure titanium (c–d) Ti–6Al–4V powder

3.3.6 Others

Mechanical alloying is another technique to produce small batches of experimental powder for cold spraying. The technique has so far been utilised to manufacture intermetallic and nanostructured coatings from composite materials such as Fe–Si and Fe–Al (Li et al. 2007a) (Wang et al. 2007). In a mechanical alloying process, the friction energy from the milling process induces plastic deformation of the particles.

There are other processes such as hydroxy manufacturing which is also used to manufacture powder for cold spraying. Typically, nickel and iron powders are manufactured via this route. The powders produced by this method are not spherical, although they may exhibit spherical-like shape.

3.4 Effect of Powder Characteristics in Cold Sprayability

Two important characteristics in cold spraying are DE and porosity of the coatings. DE is the ratio of the quantity of powder sprayed to powder deposited. For a given set of powder characteristics (i.e. composition, microstructure, morphology, size and size distribution) and substrate combination, increasing the spray gas temperature and pressure will increase the DE and most likely decrease the porosity, primarily because the particle velocity is increased. However, it is important to note that increasing the velocity by increasing the gas temperature alone will not necessarily lead to exactly the same increments in DE and porosity as would be seen by reaching the same increment of velocity by increasing gas pressure alone (Wong et al. 2009). Nevertheless, velocity remains the key process variable in producing an acceptable coating. In turn, for a given gas, gas temperature and pressure, the velocity of the powder is affected by powder morphology, size and size distribution.

Many of the conclusions concerning the effect of powder characteristics on cold sprayability are based on mathematical modelling. In terms of qualitative and quantitative verification, the experimentalists are constrained mainly by the limited availability of powders of the same alloy with a wide range of particle characteristics. For example, to quantify the optimum particle size and associated maximum powder velocity, there are very few published investigations that include a sufficient number of powders with different average particle sizes to be able to even approximately determine an optimum. A common approach to obtaining different powder sizes is by sieving, but to produce enough powder to perform a campaign using off-the-shelf cold spray guns is very time consuming, especially at the fine end of the powder size spectrum. Even with the availability of powders, measuring the impact velocity is extremely difficult with a substrate present, compounding the experimental problems. Finally, it is not easy to vary each powder characteristic independently of the others, complicating the analysis of the experimental results. Thus, unfortunately, there are very few studies that comprehensively investigate, experimentally, the effects of powder characteristics on cold sprayability. This section provides a comprehensive overview of the mathematical and experimental investigation of powder particle size, morphology and oxidation states on the particle velocity, DE and porosity of the coatings.

3.4.1 Effect of Powder Size on Particle Velocity

Powder particle size significantly influences the accelerating behaviour and subsequently particle velocity. Figure 3.7 is a typical example for the effect of particle size on its velocity for a spherical Cu powder (Li and Li 2004) where generally particle velocity is increased with the decrease of powder particle size. The smaller particles are easily accelerated to a higher velocity, which is favourable for cold spraying. Therefore, using the powders with a small particle size can easily fulfil the particle velocity requirement for deposition. Spray powder particles generally possess a certain particle-size distribution, and with cold spraying, the particle size of the spray powders is usually less than 50 μm.

Fluid dynamics studies have revealed that increasing the particle coarseness decreases the velocity of the particle. However, in the presence of a substrate, there is the so-called bow shock immediately in front of the substrate, which tends to

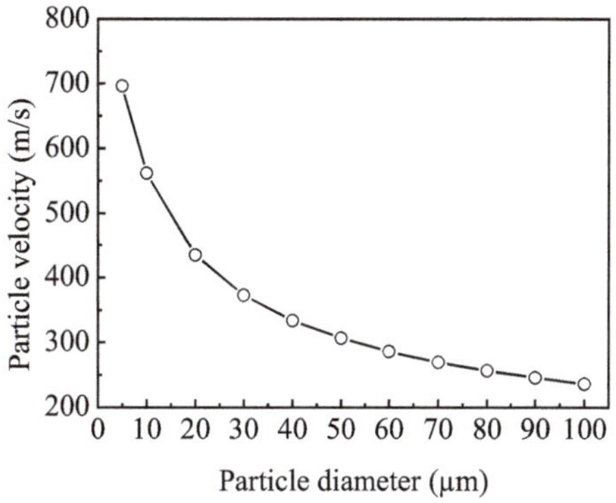

Fig. 3.7 Effect of the particle size on particle impact velocity using N_2. (Li and Li 2004)

decelerate all the particles. The level of deceleration depends on the particle size, with very small particles being strongly affected by the bow shock.

Besides particle size, the density of particle also influences the accelerating behaviour. Generally, the lower the particle density, the higher the accelerating rate of spray particles. Therefore, the particle of lower density may reach a higher velocity than the high-density particle. According to calculations by Helfritch and Champagne (2006), the optimum particle size is a relatively weak function of density, with values of a few micrometers for a range of densities that encompass Al and Fe, amongst other metals, as shown in Fig. 3.8. These calculations are based on nitrogen gas with pressure 2.76 MPa and temperature 673 K.

With regard to experimental verification of the effect of particle size on velocity, Ning et al. (2007) have performed a comprehensive study with five different sizes of spherical copper powder covering a range from about 12 to 60 MPa. Low-pressure cold spray was used with nitrogen and helium at 0.7 MPa and 573 K. The effect of Cu particle size on velocity is shown in Fig. 3.9. The mean particle velocity decreases with increasing mean particle size, and the particles sprayed with helium generated higher velocities than nitrogen.

It is argued that there is an optimum particle size that will lead to a maximum impact velocity. However, this value seems to be of the order of a few micrometers,

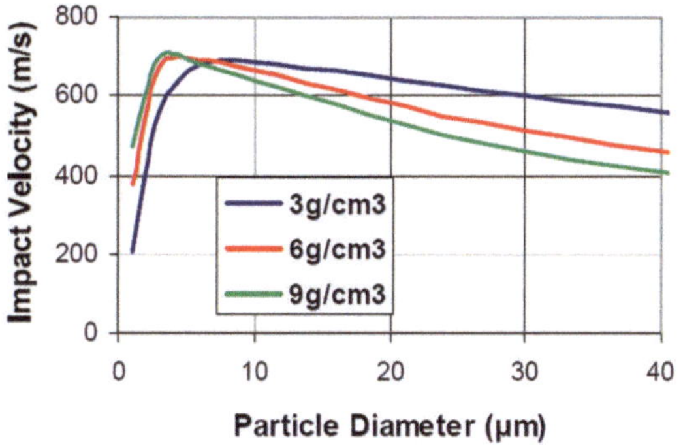

Fig. 3.8 Calculated effect of particle size on impact velocity for three densities (nitrogen gas, pressure 2.76 MPa and temperature 673 K; Helfritch and Champagne 2006)

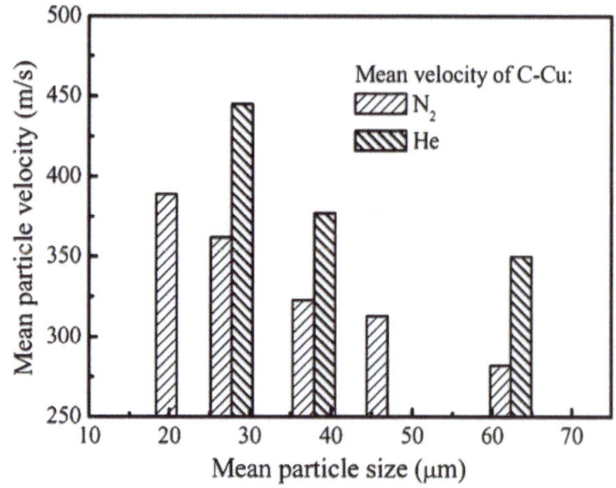

Fig. 3.9 Experimental results showing the effect of particle size on velocity for low-pressure cold spray; pressure 0.7 MPa, temperature 573 K

and spraying with such fine powders can lead to other problems. The generally accepted particle-size range is of the order of 20–30 μm, which is possibly a compromise between the impact velocity and the constraints of powder manufacturing. Within this range of sizes, the velocity does not change drastically. However, as mentioned previously and as illustrated in Fig. 3.10, the DE increases rapidly over a small range of velocities between V_{crit} and a velocity (here termed V_{max}) at which the rapid increase in DE transforms to an asymptotic increase to the maximum DE. Thus, to maximize the DE, it is important that the velocity comfortably exceeds V_{max} (as opposed to V_{crit}). Operating above V_{max} will facilitate process control, minimising any sensitivity of cold spray to particle size.

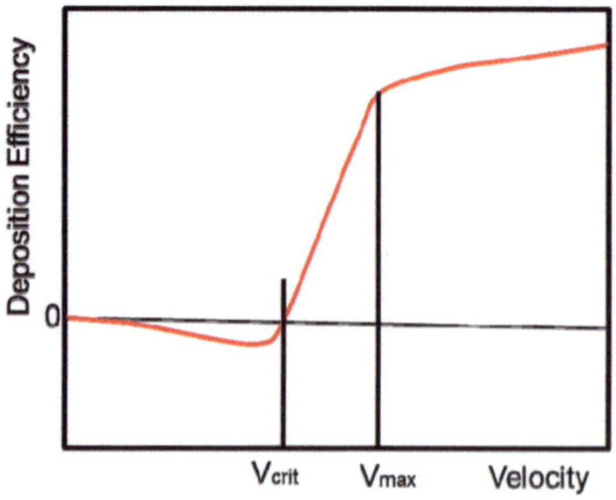

Fig. 3.10 Deposition efficiency versus velocity; the particle velocity must be above V_{max} rather than V_{crit} to avoid any strong influence of particle size on deposition efficiency

3.4.2 Effect of Powder Morphology on Particle Velocity

The morphology of a spray powder depends on its manufacturing process. The gas atomisation is usually applied to produce metal alloy powders which have a spherical shape. Besides gas atomisation process, hydriding–dehydriding and mechanical alloying are employed to produce metal alloy powders for cold spraying. However, the powders produced by those methods are not spherical, although they may exhibit spherical-like shape. The particle velocity is influenced by the morphology of spray powder. The morphology effect is usually characterised by the shape factor, which is defined as the surface area ratio of the equivalent spherical particle with the same volume as the current particle to the surface area of the current particle. Therefore, the shape factor is less than one. The simulation exhibits that the particles with irregular shape have a higher velocity than spherical particle (Fig. 3.11). Therefore, the powder particles with irregular shape could be deposited with higher DE than spherical ones. However, it should be noted that due to high specific surface area of irregularly shaped powders, the oxidation content of the coating deposited may be high. High surface areas will also be retained into resultant coating, which affects the mechanical properties of cold spray coatings by post-spray annealing.

It is conceptually unequivocal that irregular powders will exhibit a higher drag coefficient, thereby achieving higher particle velocities, all else being equal. However, mathematically modelling the velocity is difficult because of the irregular nature of the particles; scanning electron microscopy of irregular powders is shown in Fig. 3.12 to give an idea of the wide range of irregular powders. Thus, with regard to quantifying the effect of real irregularities on particle velocity, measurement is required. Wong et al. (2013) compared the commercially pure Ti feedstock in five forms, three spherical powders and two 'irregular' powders. (Figs. 3.12a and b are the irregular powders used in this work).

The volume-weighted powder size distribution for all powders was determined experimentally with a laser diffraction particle-size analyser and is shown in

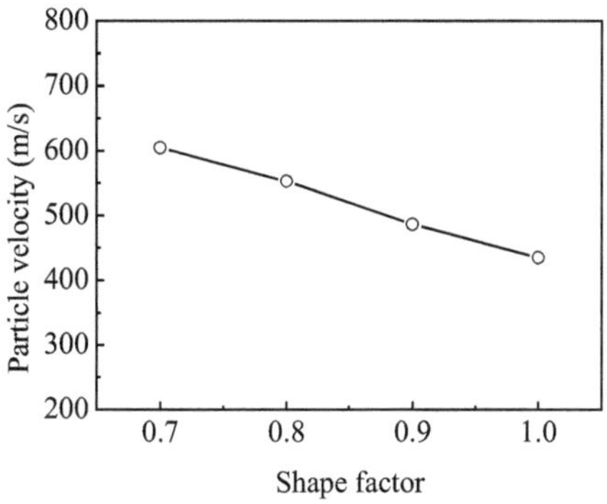

Fig. 3.11 Effect of particle shape factor on its velocity for Cu with an equivalent diameter of 20 μm using N_2. (Li and Li 2004)

Fig. 3.12 Examples of irregular powders. **a** commercial purity (CP) Ti 'sponge'. **b** CP Ti 'irregular'. **c** 'Pure' Fe showing a rough surface which will also increase the drag coefficient compared to a smooth surfaced spherical powder

Fig. 3.13a. The velocities were measured for a range of nitrogen cold spray conditions, condition 1 being 3 MPa pressure and 573 K, condition 6 being 4 MPa and 1073 K and the other four conditions being combinations of temperature and pressure within these extremes. Based on the average sizes, the best comparison to quantify the effect of the 'irregular' powder is to compare it with the 'small' spherical powder. Comparing the average velocities, it seems that the increase in velocity due to morphology decreases with increasing spray intensity; for spray condition 1, the increase is about 12 %, whereas for spray condition 6, the increase falls to about 8 %. A similar trend is observed when comparing the small and large spherical powders, the smaller particle size imparting a velocity increase of about 8 % for condition 1 and 6 % for condition 2.

The results of a comparison between angular and spherical stainless steel powders by Fukanuma et al. (2006) are shown in Fig. 3.14, where the Micro-Melt powder is the spherical one. At 50 % cumulative volume fraction, and for the more intense spray condition, there is approximately a 20 % increase in velocity. In the same study, perhaps a better way to illustrate the potential of irregular powders in

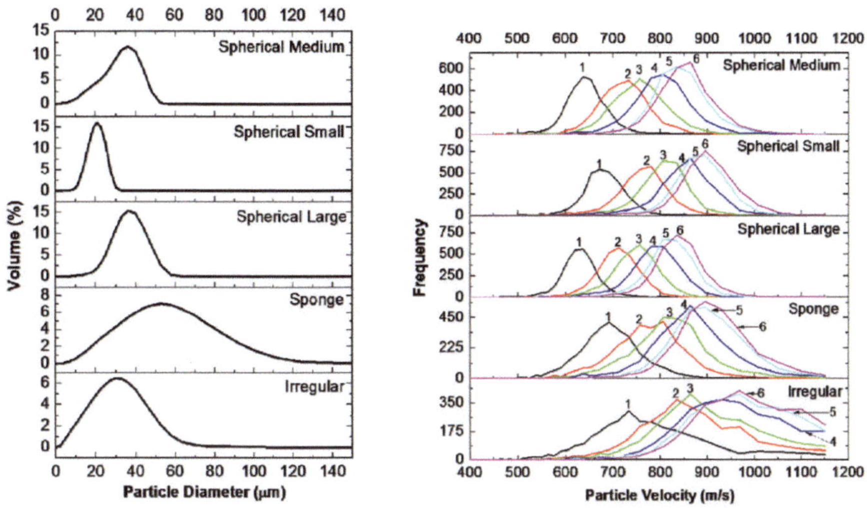

Fig. 3.13 Size distributions (**a**) and the velocity distribution (**b**) of powders for six spray conditions used in the study of Wong et al. (2013). The sponge and irregular powders were shown in Fig. 3.12

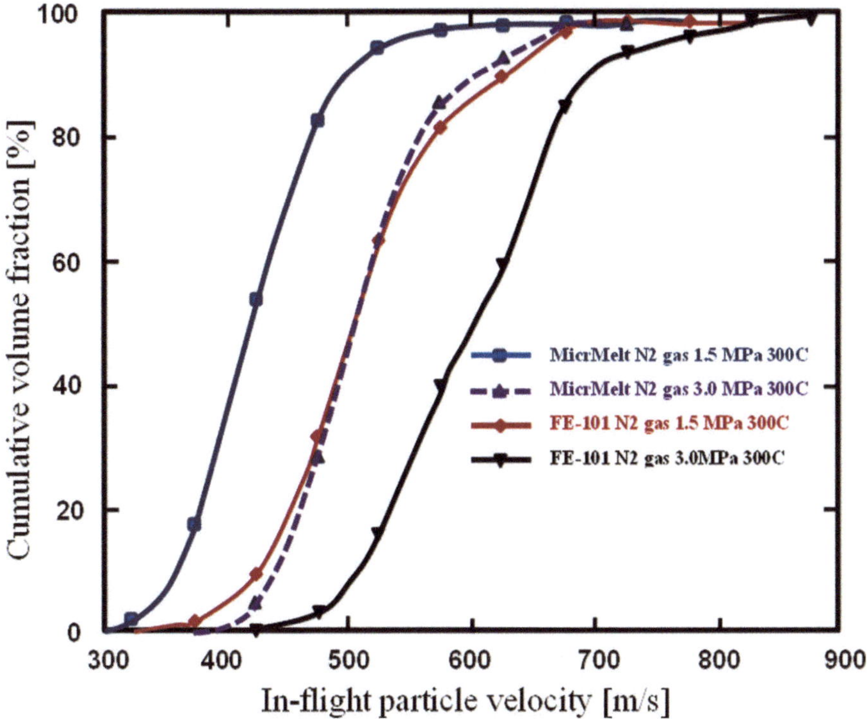

Fig. 3.14 Comparison of the cumulative volume fraction velocities of spherical and angular stainless steel powders as a function of particle velocity. Micro-melt is the spherical powder

improving cold sprayability is the result that the irregular stainless steel powder sprayed with nitrogen achieved more or less the same velocity as spherical powders sprayed with He (3 MPa and 573 K for both gases).

3.4.3 *Critical Velocity*

As noted earlier, there are two problems with cold spraying very fine particle sizes, the decelerating effect of bow shock and the increasing V_{crit} with decreasing particle size. As schematically illustrated in Fig. 3.15 (adapted from Schmidt et al. 2006), for a given cold spray condition, there is a particle-size window where the particle velocity exceeds the V_{crit} and deposition takes place. The optimum particle size is one which *maximises* the increment of the particle velocity above V_{crit}. In Fig. 3.15, there is a range of particle sizes which give approximately the same (particle velocity–V_{crit}) increment, which is useful for process control in desensitising the influence of particle size on this metric. At sizes above this range, the cold sprayability decreases because the powder velocity decreases faster than V_{crit}; at particle sizes lower than this range, V_{crit} increases faster than the particle velocity.

Early work on modelling V_{crit} did not include any particle-size effects on V_{crit}; however, experimental evidence (Schmidt et al. 2009) suggested that V_{crit} is dependent on particle size. As was shown in Fig. 3.15, V_{crit} decreases with increasing size, until a plateau is reached. Assuming a particle-bonding mechanism based on breaking oxide films to create reactive surfaces and the adiabatic shear mechanism (Assadi et al. 2003), a decreasing particle size leads to increasing oxide film surface area per unit volume and increasing rates of thermal diffusivity, the latter making it difficult to reach the temperatures required for the adiabatic shear mechanism to be activated.

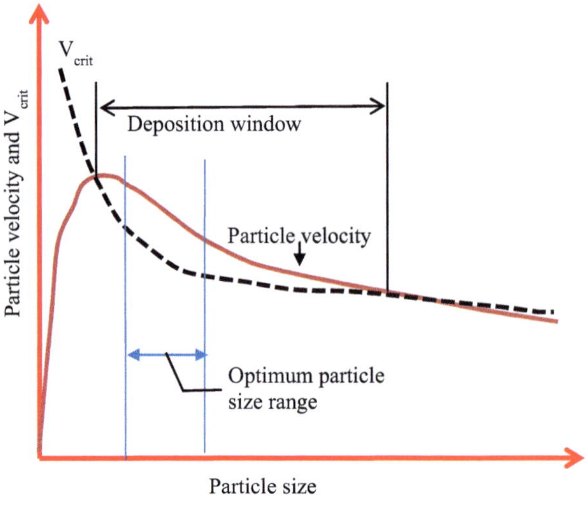

Fig. 3.15 Schematic diagram showing the effect of particle size on critical velocity and particle velocity, which defines the optimum particle-size range for cold sprayability. (Adapted from Schmidt et al. 2006)

Table 3.1 Effect of powder size on morphology on critical velocity. (Wong et al. 2013)

Cold spray condition	v_{Cr} (m/s)	v_{Cr} (m/s)	v_{Cr} (m/s)	v_{Cr} (m/s)	v_{Cr} (m/s)
	Spherical medium	Spherical small	Spherical large	Sponge	Irregular
1	610	619	605	604	618
2	582	594	576	574	593
3	568	582	560	559	581
4	546	563	537	536	563
5	546	563	537	534	560
6	539	557	529	526	554

The effect of powder morphology can also affect the critical velocity V_{crit}. For example, for the powders investigated by Wong et al. and characterised in Fig. 3.13, the V_{crit} values are tabulated in Table 3.1. Note that V_{crit} varies with spray condition, which is explained by the effect of gas temperature and particle time at that temperature; increasing gas temperature and residence time tends to increase the softness of the particles, leading to lower V_{crit} values (Wong et al. 2009). Comparing the 'irregular' powder with the small spherical powder values, there is little effect of morphology. Similarly, comparing the 'sponge' and the large spherical V_{crit} values show no significant differences.

3.4.4 Effect of Particle Velocity on DE and Porosity

So far, the effect of powder characteristics on powder velocity and V_{crit} has been discussed. As shown in Fig. 3.16, increasing velocity leads to increasing DE for a given powder. This suggests that anything that increases the velocity will tend to increase DE. Thus, with larger powders generally decreasing velocity, it is possible that the consequence is a decreased DE. However, the effect of velocity is relative to V_{crit} and V_{max} shown in Fig. 3.10; if the particle velocity is below V_{crit}, there is no deposition; if it is above V_{max}, there is no effect of particle size. In fact, the findings of Schmidt et al. (2006), Fig. 3.16, show that there is no effect of particle size on DE when plotted as a function of (particle velocity–V_{crit}). Thus, there is no additional effect of particle size beyond that of V_{crit} and particle velocity. In fact, the schematic diagram of Fig. 3.15 suggests that there is a particle-size range, the optimum particle-size range, over which the DE is more or less constant. This range will vary from alloy to alloy and may also vary with cold spray processing parameters.

Regarding the effect of particle characteristics on porosity, in general, the findings are similar to that of DE, in that the key metric is the degree to which the particle velocity exceeds the critical velocity. Wong et al. (2013) have noted a power–law relationship between porosity and the ratio of the powder velocity to V_{crit}, for spherical powders, as shown in Fig. 3.17. Over this particular size range (Fig. 3.13),

Fig. 3.16 Effect of the increment of particle velocity above the critical velocity on deposition efficiency showing no effect of powder size. *DE* deposition efficiency. (Adapted from Schmidt et al. 2006)

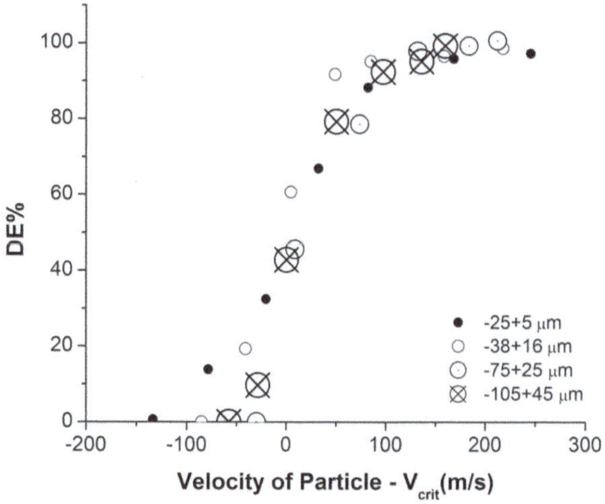

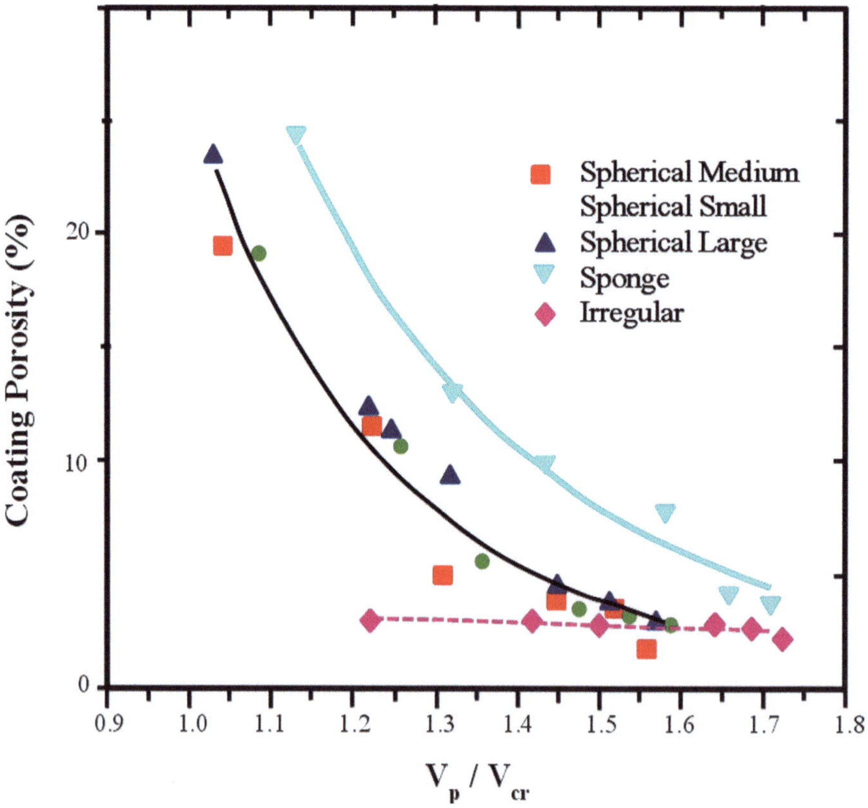

Fig. 3.17 Effect of ratio of particle velocity, V_p, to V_{crit} on porosity for CP Ti of particle sizes and morphologies illustrated in Figs. 3.12 and 3.13. (Wong et al. 2013)

the spherical powders seem to belong to the same population. For the 'irregular' powders, the range of V_p/V_{crit} values is too narrow to examine the effect on porosity, but the powders of the 'sponge' morphology (Fig. 3.12a) follow a behaviour similar to the spherical powders, but exhibit a somewhat higher porosity for a given V_p/V_{crit} value compared to the spherical morphology. This suggests that irregular powders are more difficult to pack than spherical powders. In fact, the irregular powders in this study exhibited a pronounced porosity gradient with increasing porosity at increasing distances from the substrate, whereas the porosity is more or less constant throughout the coating thickness for the spherical powders. This suggests that peening is very important for irregular particles to decrease porosity and again points to a 'packing' issue.

Conflicting results have been reported for CP-Ti feedstock powders having an irregular morphology. In one study by Zahiri et al. (2009) revealed that coating porosity had decreased (9.5–8.0 %) with decreasing average particle size (22–16 µm). The authors believed that this was due to the smaller particles having a higher particle impact velocity. In another study by Marrocco et al. (2006) showed that coating porosity decreases (22–14 %) with increasing average particle size (28–47 µm). The explanation for this behaviour was an enhanced peening effect caused by the larger impinging particles. It was believed that these larger particles would travel at velocities lower than the critical velocity and, hence, they would only provide an enhanced peening effect and would not deposit onto the substrate. Consequently, this higher peening intensity would result in a lower coating porosity, but could lead to a lower DE. This latter hypothesis suggests that particle-size distribution should also influence the DE and porosity. Blose (2005) examined the cold spray characteristics of three different Ti–6Al–4V feedstocks with similar particle sizes but made by different processes culminating in different powder characteristics, as shown in Table 3.2.

The Ti–6Al–4V hydride–dehydride angular feedstock exhibited the lowest porosity but the lowest DE. The low DE is likely due to the high hardness but the low porosity was attributed to the relatively large number of fines available to fill the voids. These latter results not only illustrate that DE and porosity are not necessarily related but also demonstrate that powder characteristics can be coupled. In this case, the powder processing technique not only changed the morphology but also

Table 3.2 Powder characteristics of reference (Blose 2005)

Powder type	Manufacturing technique	Morphology	Manufacturer	Mean size (mm)	Hardness (VHN)
Ti–6Al–4V	Gas atomized	Spherical	Crucible Research, USA	29	291
Ti–6Al–4V	Plasma atomized	Spherical	Pyrogenesis Canada	27	280
Ti–6Al–4V	Hydride–dehydride	Angular	Affinity China	30.7	351
Ti-CP	Hydride–deHydride	Angular	Affinity China	21	153

VHN Vickers hardness number

changed the hardness of the powder. In fact, the correlation of powder characteristics to hardness is a fairly common theme. For example, ball milling is a commonly used technique to produce nanograin sizes and to change the morphology. The morphology changes increase the velocity of the powder but the work hardening reduces the cold sprayability, greatly offsetting the benefit of increasing velocity. In fact, refining the particle size can also change mechanical properties by increasing the bulk oxygen levels, possibly leading to solid solution strengthening.

3.4.5 Surface Oxidation Effect of Powders

With metallic alloy powder particles, it is inevitable that powder particles contain a certain level of oxygen since metal alloy surface is usually covered by oxide scale of several tens of nanometres in thickness (Temples et al. 1993). The oxide scale on the particle surface prevents the impacting particle from direct contact with the underlying deposited particles at the contact interface. The formation of an effective bonding requires sufficiently breaking and dispersion of oxide scale at the contact interface. Thus, a higher particle velocity is naturally required for the powder particles of thicker surface oxide scale. Moreover, the thicker the oxide scale on the particle surface, the higher the critical velocity becomes. With copper powders, a wide range of the critical velocity values were reported by different investigators as summarized with available oxygen content in Table 3.3 (Li and Li 2009). The critical velocity varied from 290 to 640 m/s for normal spray powders. By using copper powder particles of low oxygen content and subsequently oxidised to different high oxide levels, Li et al found out that the oxidation of copper powders significantly influences the critical velocity and consequently DE (Li et al. 2006). With the copper powder of oxygen content less than 0.02 wt% the critical velocity was about 300 m/s. When the same powder was oxidised in air to the oxygen contents of 0.13 and 0.38 wt%, the critical velocity increased to 550 and 610 m/s, respectively. Since the oxidation of most metal alloy powders proceeds naturally in

Table 3.3 Typical critical velocities for Cu reported in literature. (Li and Li 2009)

Investigators	Critical velocity (m/s)	Oxygen content (wt%)	Particle size	Ref.
Alkimov et al.	500	–	10 μm	Alkimov et al. (1990)
Stoltenhoff et al.	550–570	0.1–0.2	5–25 μm	Stoltenhoff et al. (2002)
Gilmore et al.	640	0.336	19 μm, 22 μm	Gilmore et al. (1999)
Li et al.	290–360	0.02	64.1 μm	Li et al. (2006)
Li et al.	550	0.13	20.5 μm	Li et al. (2006)
Li et al.	610	0.38	20.5 μm	Li et al. (2006)
Schmidt et al.	250–280	–	20 mm	Schmidt et al. (2006)
Schmidt et al.	500	–	25 μm	Schmidt et al. (2006)
Raletz et al.	422–437	–	10–33 μm	Raletz et al. (2005)

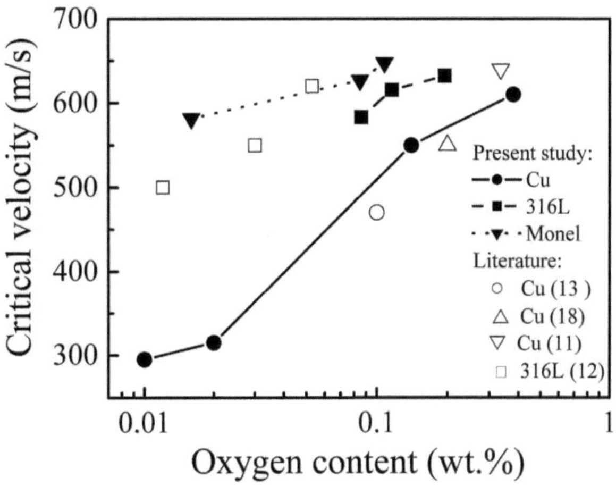

Fig. 3.18 Effect of oxygen content of the feedstock on the critical velocity. (Li et al. 2010)

air, the storage of metallic powder becomes very important. The DE of spray powder particles will gradually decrease with the increase of the storage duration. Thus, with the increased storage time, the data on the coatings obtained with the powder might not correspond to the coatings obtained by the virgin powder. A study by Li et al. also reported that the critical velocity dependence on the surface oxidation is more significant on the soft ductile materials (Li et al. 2010) (Fig. 3.18). With the increase of alloy hardness, the effect of oxidation may become less significant.

3.5 Composite Materials in Cold Spraying

3.5.1 Blended Powder

Cold spraying is a promising process to deposit composite materials or composite coatings. Since the composition of individual powder particles is generally retained in the deposit without compositional change, the coatings can be manufactured by powder composition design. The deposition of composite coatings can be realised by different types of powders, for example, a blend of different starting powders, pre-alloyed powder and quasi-alloyed composite powders such as mechanically alloyed powders. Using a powder mixture is a much simpler and a cheaper approach to deposit composite coatings. Therefore, many studies have been conducted for different materials combinations via such route (Wang et al. 2008, 2013; Feng et al. 2012; Melendez and McDonald 2013; Koivuluoto et al. 2012; Shockley et al. 2013; Irissou et al. 2007; Sansoucy et al. 2008; Spencer et al. 2012; Luo et al. 2011; Luo and Li 2012). However, since the deposition behaviour of the individual particle depends on its deformation ability, this usually leads to a large deviation of deposit composition from its starting powder. In Fig. 3.19, typical results are summarised

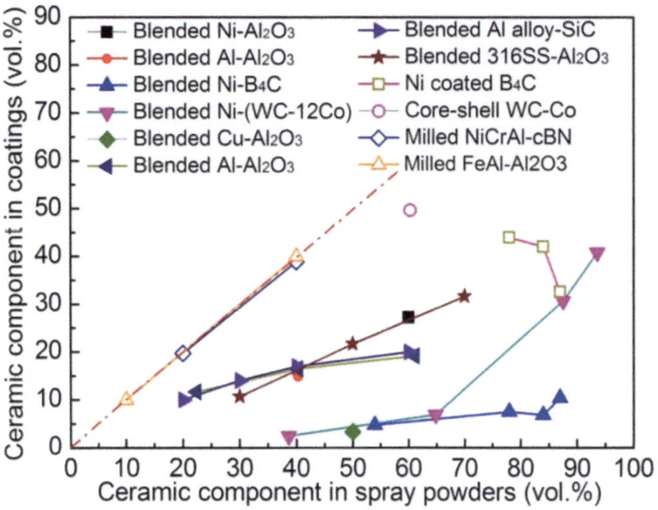

Fig. 3.19 Relationship between the hard reinforcement content in the starting powder and the content in the coating for different composite powders produced by different methods

for ceramic particulates reinforced metal matrix composites deposited by cold spraying using blended powders. One fact is that when blend composite powders of two phases are used, the content of hard ceramic phase in the deposit is significantly lower than the initial powder mixture. It is difficult to deposit a composite containing hard reinforcements at a concentration higher than 40 vol%. For example, to deposit Ni–(WC–12Co) composite with about 30 vol% WC–12Co content, a powder mixture with about 90 vol% WC–12Co needs to be used. This means that at least over two thirds of WC–12Co in the starting powder will rebound off the substrate during deposition. Therefore, an experimental correlation between the deposit composition and starting powder composition should be established for the determination of the coating composition.

3.5.2 Pre-alloyed Powder and Intermetallic

The solution to retain powder compositions in the deposits is to use a pre-alloyed powder or quasi-composite powder. When mechanically alloyed powders are used, since there is no possibility for specific constituent in a powder to preferably rebound off, the deposit of the same composition as that of the powder can be deposited (such as NiCrAl–cBN, Luo et al. 2011, 2012b and FeAl–Al_2O_3 powders, Luo and Li 2012). With the composite powders consisting of metal constituents, the compositions of the powders can be altered based on design requirements. However, with the metal–ceramics composite due to significant difference in deformation ability of metal from ceramic phases, during mechanical alloying, ceramic constituents are primarily included in the metal matrix. There are limitations during preparation of

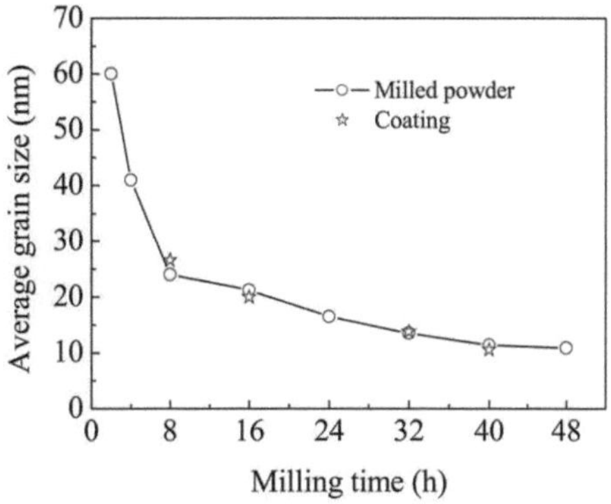

Fig. 3.20 Effect of milling time on the grain size of Fe–12Si alloy powder. (Li et al. 2010)

metal–ceramic composite powders by mechanically alloying. One is the ceramic particle content in the composite powder and the other is to develop a homogenous microstructure without any ceramic–ceramic contact. By using a step-fashioned process, the ceramic constituent with a content of up to 40 vol% can be uniformly included in metal–alloy matrix (Luo and Li 2012). In order to prepare uniform composite powder for cold spraying with a ceramic content higher than 40 vol%, the sintering followed by crushing or spray-sintering process can be employed. In the literature (Gao et al. 2008, 2010; Yang et al. 2012), WC–12Co porous powder was produced by sintering and crushing for cold spraying.

Mechanical alloying is also a suitable route to produce nanostructured powder and intermetallic coatings. By controlling the alloying process and milling duration, the uniformity of milled alloy can be altered remarkably and grain size can be significantly reduced. Figure 3.20 illustrates the grain size change with the milling time for 90%Fe–10%Si powder and the corresponding coating (Li et al. 2010). After several hours of milling, the grain size of the powder changed to several tens of nanometres. Generally, mechanically alloyed powders present a spherical-like morphology with a rough surface as shown in Fig. 3.21 (Wang et al. 2007). Following milling, the powders with a suitable size range for cold spraying are sieved.

Since nanostructured materials have many unique physical, chemical and mechanical properties, the effective deposition of nanostructured coating is of great interest. Most importantly, cold spraying can deposit bulk-like thick nanostructured coatings. Due to a lack of melting in cold spraying, these nanostructured powders can be used as feedstock. Many investigations (Li and Li 2010; Wang et al. 2007; Ajdelsztajn et al. 2006; Lima et al. 2002; Kim et al. 2005; Li et al. 2007a; Zhang et al. 2008) on the cold spraying of nanostructured powders have been carried out, and the general conclusion is that the nanostructure of the feedstock is retained in the coatings. An example of nanostructured NiCrAlY coating produced via cold spraying is shown in Fig. 3.22. This illustrates the XRD patterns for (a) starting NiCrAlY powder, (b) mechanically alloyed NiCrAlY and (c) cold-sprayed coating and a transmission electron microscope (TEM) image of nanostructured NiCrAlY coating deposited by mechanically alloyed nanostructured NiCrAlY coating. The XRD patterns of the mechanically alloyed powder and the resultant coating show

Fig. 3.21 Morphology (a) and cross-sectional microstructure (b) of the as-milled Fe–40Al powder. (Wang et al. 2007)

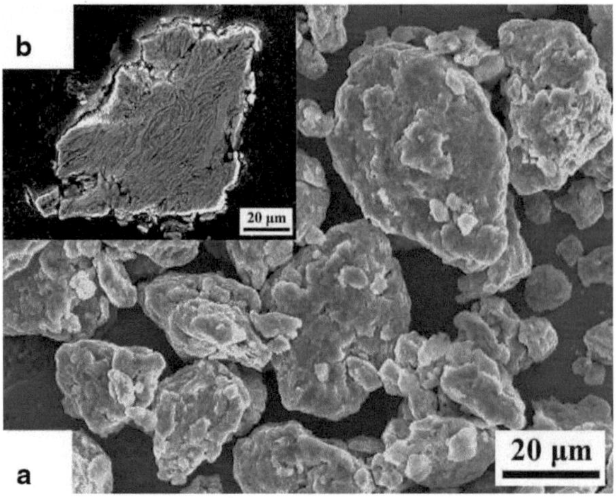

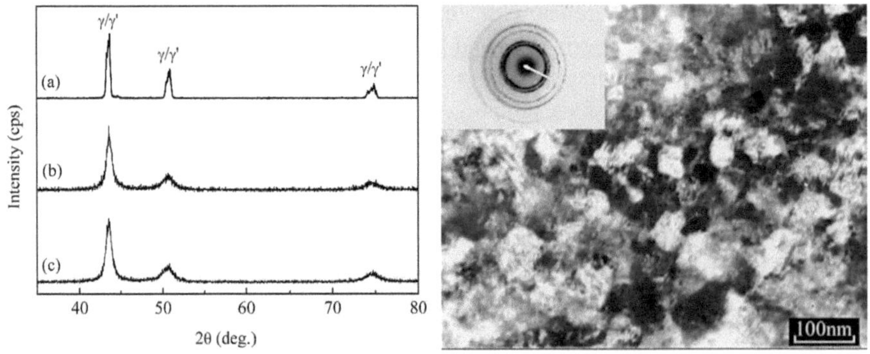

Fig. 3.22 **a** XRD patterns of the NiCrAlY as-received powder *(a),* milled powder *(b)* and cold-sprayed nanostructured coating *(c)*. **b** TEM image of nanostructured NiCrAl coating. (Zhang et al. 2008)

significant broadening of diffraction peaks suggesting nanostructured features in the microstructure. The TEM image of the deposit clearly reveals that the grain size is less than 100 nm.

Intermetallic compounds such as iron and nickel aluminides are attractive materials for many industrial applications in medium to high temperatures owing to their excellent corrosion resistance in oxidising and sulfidising atmospheres (Stoloff et al. 2000). Compared to steels and other commercial Fe-based alloy, FeAl alloy exhibits improved oxidation resistance and has lower density. Moreover, the abnormal strengthening effect also makes intermetallics promising high-temperature wear-resistant materials. The oxidation during thermal spraying makes those promising properties of the materials ineffective (Cinca and Guilemany 2012). Since direct deposition of intermetallics by cold spraying is difficult due to their low ductility and brittleness at low temperature, a route using mechanical alloying for Fe/Al and Ni/Al was proposed to fabricate intermetallic coatings (Li et al. 2007a, 2011). With this route, the mechanically alloyed powder is produced with sufficient deformation ability upon high velocity impact. Through post-spray annealing, the mechanically alloyed deposit is transformed to intermetallic phases. With Fe/Al or Ni/Al alloys, the phase transformation to ordered intermetallics occurs at a temperature higher than 500 °C (Yang et al. 2013).

3.5.3 Cermet and Porous Powder Design for Hard Coatings

WC–Co hard metals and Cr_3C_2–NiCr are widely used as hard coating materials applied by thermal spray processes (Li and Yang 2013). The former is employed at oxidation environment at temperatures less than about 500 °C, while the latter is used at temperatures higher than 500 °C. The coatings of these materials are usually deposited by high-velocity oxy-fuel (HVOF) process. Although HVOF became an effective process to deposit WC–Co with superior wear performance, it is still difficult to deposit a dense nanostructured WC–Co coating with limited decarburisation. However, cold spray process makes it possible to achieve nanostructured WC–Co coating without any decarburisation as mentioned previously.

To build up a deposit through cold spraying, both the approaching particles and the substrate need to experience necessary plastic deformation. Therefore, a thick cermet coating with high content of ceramic, especially hard WC–Co coating, cannot be easily deposited by cold spraying due to low deformability of the hard cermets (Lima et al. 2002). In some cold-sprayed coatings (such as titanium), a porous top layer and a dense bottom layer are observed in the coating, which is due to the tamping effect of the particles (Li and Li 2003; Hussain et al. 2011b). This tamping effect can be utilised to build-up the hard cermet coatings through porous powder design, which can contribute to the pseudo-deformability of both the impacting and deposited hard particles (Gao et al. 2010).

The deposition of individual porous WC–Co particles have been studied to identify the deformation mechanisms. When a porous cermet particle impacts on a substrate, deformation occurs to the region around the particle impact zone resulting in densification of the lower region of the particle, while the porous top part of particle remains unaffected (Li et al. 2007b) (Fig. 3.23). The top region of the porous particle still maintains the deformability for the successive particle impact. Thereafter, when a porous cermet particle impacts on the already deposited cermet layer, the deformation occurs at the top porous layer of the deposited particles. This makes it possible to fulfil the requirements of the deformation of both the deposited layer and the impacting particles. Thus, a dense WC–12Co coating can be obtained by using porous powder design in cold spraying. The coating microhardness measurements indicated that the cold-sprayed nanostructured WC–12Co coating yielded hardness values from 1870 Hv to about 2000 Hv (Kim et al. 2005; Li et al. 2007b), which is comparable to that of the sintered bulk.

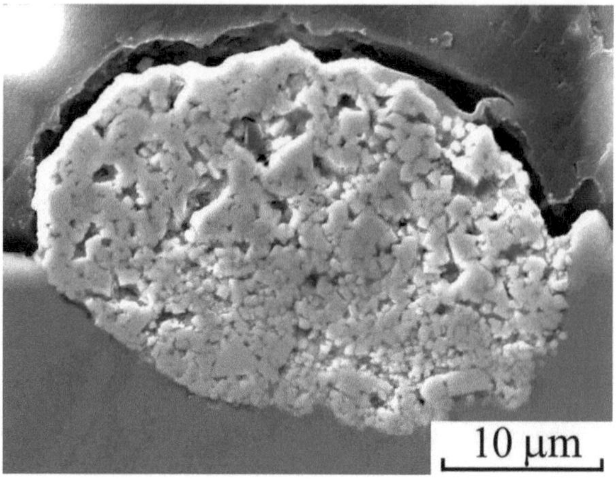

Fig. 3.23 Cross-section of cold-sprayed WC–Co particle produced by porous powder design. The nonuniform deformation leads to the densification at the lower part and retains the porous structure at the upper part. (Li et al. 2007b)

3.5.4 Metal-Coated Ceramic Particles for Hard Coatings

As an alternative to simply mechanically mixing powders to form a blend for cold spray, single tungsten carbide particles were instead encapsulated with pure aluminium via a proprietary chemical vapour deposition (CVD) method; the high activity of aluminium makes it difficult to encapsulate WC particles using other alternative wet chemical methods. Although it was generally difficult to deposit aluminium homogeneously over the surfaces of individual WC particles, the encapsulation technique used was sufficient to stimulate solid-state particle bonding (Wang and Villafuerte 2009). This was ascribed to the ability of small amounts of aluminium around brittle WC particles to provide sufficient ductility to stimulate particle deposition by cold spray, even at relatively low gas pressures. The same CVD technique was also attempted to deposit pure copper on similar tungsten carbide particles without much success; instead, electroplating was used to encapsulate tungsten carbide particles with pure copper. The microstructures of Al-coated WC and Cu-coated WC feedstock powder following cold spraying are shown in Fig. 3.24. The metal matrix

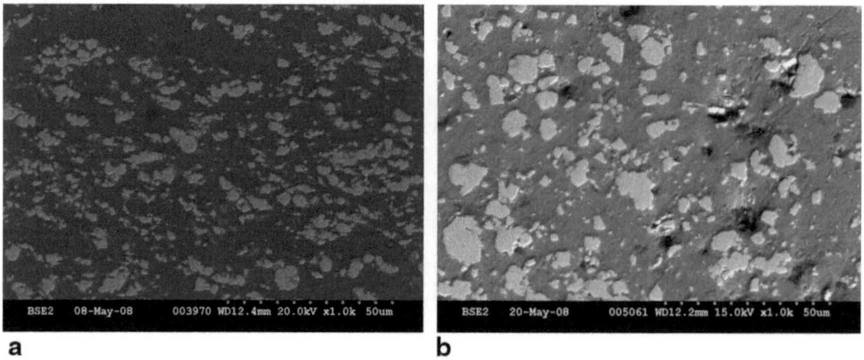

Fig. 3.24 Microstructure of cold-sprayed **a** Al-coated WC and **b** Cu-coated WC. The metal matrix is made of **a** aluminium and **b** copper. (Wang et al. 2008)

(aluminium or copper) worked as a ductile binder for cold spraying. The coatings were characterised by a high percentage of well-dispersed, retained carbide phase and low porosity. Hardness measurements indicated higher hardness values for cold spray deposits produced from encapsulated carbide feedstock compared to cold spray deposits produced from metal–carbide blends.

3.6 Concluding Remarks

Feedstock selection in cold spraying requires a thorough understanding of the cold spray process, specific spray equipment design and the characteristics of the powders. This chapter summarised the characteristics of the feedstock material and their associated measurement techniques. A brief section on powder manufacturing has been presented which included both commercial and experimental powders for cold spraying. Powders required for cold spraying need to have good flowability, free from satellite particles, high purity and a narrow size distribution. Metal and alloy powders for cold spray applications are generally manufactured using an inert gas atomisation technique, which produces powders with spherical morphology and tight specifications on size ranges. Irregular powders such as angular titanium powder, manufactured using hydride–dehydride process, is now widely used in cold spray applications due to lower cost and availability. Moreover, agglomerated cermet powders and mechanically alloyed intermetallic have been used to deposit coatings from cold spraying.

The cold sprayability of the feedstock materials has been discussed in detail in light of particle velocity (critical velocity), DE and porosity of the coatings. The following general conclusions can be drawn from the discussion:

- In general, for a given gas temperature and pressure, increasing the particle size decreases the particle velocity; however, the very fine particles are strongly decelerated by the bow-shock effect due to the presence of a substrate in the spray stream. Also, particles with lower density reach higher particle velocities compared to the higher-density particles. The generally accepted optimum particle size range is of the order of 20–30 μm, which is possibly a compromise between the impact velocity and the constraints of powder manufacturing.
- The effect of particle morphology (spherical vs angular) on the coating microstructure is rather complex. Theoretically, an irregular powder reaches a higher impact velocity than a spherical one because of the higher drag coefficient in the case of the irregular powder. Compared to spherical powders, this velocity tends to increase the DE of irregular powders, but may not necessarily lead to a lower porosity because of differences in particle deformation, bonding mechanism and packing. Irregular powders are more difficult to pack than spherical powders which can contribute to higher porosity in the coating.
- Critical velocity of the powders largely depends on the oxidation state of the particles. Typically, increasing the oxygen content of the powders results in higher critical velocities as more energy is required to break-up the oxide shells upon impact.

Cold spraying is increasingly gaining popularity for the deposition of composite coatings (such as metal–ceramic coatings) for various high-end engineering applications. Using a blended powder mixture (metals and ceramics) as a feedstock is a simpler and a cheaper way to deposit composite coatings in cold spraying as the composition of the powders do not change during spraying. However, the ceramic particles may rebound during spraying thus reducing the fraction of ceramic phases in the coating. Pre-alloying and mechanical alloying of powders are alternatives to depositing a blended feedstock which can eliminate the ceramic rebounding effect. Both WC–Co cermet and intermetallics (Fe–Al and Fe–Si) have been deposited using cold spraying using pre-alloyed powders. A novel approach of porous powder design has also shown promising results with WC–Co cermet. In the porous particles, the deformation primarily takes place at the bottom of the particle which leaves the top layer unaffected and suitable for successive particle impact and coating build-up.

References

Ajdelsztajn, L., A. Zuniga, B. Jodoin, and E. J. Lavernia. 2006. Cold-spray processing of a nanocrystalline Al-Cu-Mg-Fe-Ni Alloy with Sc. *Journal of Thermal Spray Technology* 15:184–190.

Alkimov, A. P., V. F. Kosarev, and A. N. Papyrin. 1990. A method of cold gas dynamic deposition. *Doklady Akademii Nauk SSSR* 315 (5): 1062–1065.

Assadi, H., F. Gartner, T. Stoltenhoff, and H. Kreye. 2003. Bonding mechanism in cold gas spraying. *Acta Materialia* 51: 4379–4394.

Berndt, C. 2004a. Feedstock material considerations. In *Handbook of thermal spray technology*, ed. J. R. Davis, 137–141. Materials Park: TSS/ASM International.

Berndt, C. 2004b. Material production process. In *Handbook of thermal spray technology*, ed. J. R. Davis, 147–158. Materials Park: TSS/ASM International.

Blose, R. E. 2005. Spray forming titanium alloys using the cold spray process. Proceedings of the international thermal spray conference, Basel/DVS-Verlag, Dusseldorf.

Champagne, V. 2007. *Introduction. Cold spray materials deposition process—fundamentals and applications*. Woodhead: CRC Press, 1–7.

Chiu, L.-H., C.-H. Wu, and P.-Y. Lee. 2007. Comparison between oxide-reduced and water-atomized copper powders used in making sintered wicks of heat pipe. *China Particuology* 5 (3): 220–224.

Cinca, N., and J. M. Guilemany. 2012. Thermal spraying of transient metal aluminides: an overview. *Intermetallics* 24:6.

Crawmer, D. 2004. Process control equipment. In *Handbook of thermal spray technology*, ed. J. R. Davis, 85–98. Materials Park: TSS/ASM International.

Cullity, B. D., and S. R. Stock. 2001. *Elements of X-ray diffraction*. New York: Prentice Hall.

Davis, J. R. 2004. *Handbook of thermal spray technology*. Ohio: TSS/ASM International.

Feng, C., V. Guipont, M. Jeandin, O. Amsellem, F. Pauchet, R. Saenger, S. Bucher, and C. Iacob. 2012. B4 C/Ni composite coatings prepared by cold spray of blended or CVD-coated powders. *Journal of Thermal Spray Technology* 21 (3/4): 561–570.

Fukanuma, H., N. Ohno, and R. Huang. 2006. The Influence of Particle Morphology on In-flight Particle Velocity in Cold Spray. Proceedings of the 2006 international thermal spray conference May 15–18. Seattle: ASM International.

Gao, P.-H., Y.-G. Li, C.-J. Li, G.-J. Yang, and C.-X. Li. 2008. Influence of powder porous structure on the deposition behavior of cold-sprayed WC-12Co coatings. *Journal of Thermal Spray Technology*. 17 (5/6): 742–749.

Gao, P.-H., C.-J. Li, G.-J. Yang, Y.-G. Li, and C.-X. Li. 2010. Influence of substrate hardness transition on built-up of nanostructured WC-12Co by cold spraying. *Applied Surface Science* 256:2163–2168.

Gilmore, D. L., R. C. Dykhuizen, R. A. Neiser, T. J. Roemer, and M. F. Smith. 1999. Particle velocity and deposition efficiency in the cold spray process. *Journal of Thermal Spray Technology* 8 (4): 576–582.

Helfritch, D., and V. Champagne. 2006. Optimal particle size for the cold spray process. Proceedings of the international thermal spray conference. 15–18 May 2006, Seattle, Washington.

Hussain, T. 2013. Cold spraying of titanium: a review of bonding mechanisms, microstructure and properties. *Key Engineering Materials* 533:53–90.

Hussain, T., D. G. McCartney, P. H. Shipway, and D. Zhang. 2009. Bonding mechanisms in cold spraying: the contributions of metallurgical and mechanical components. *Journal of Thermal Spray Technology* 18 (3): 364–379.

Hussain, T., D. G. McCartney, and P. H. Shipway. 2011a. Impact phenomena in cold-spraying of titanium onto various ferrous alloys. *Surface and Coatings Technology* 205 (21/22): 5021–5027.

Hussain T, D. G. McCartney, P. H. Shipway, and T. Marrocco. 2011b. Corrosion behavior of cold sprayed titanium coatings and free standing deposits. *Journal of Thermal Spray Technology* 20 (1/2): 260–274.

Hussain, T., D. G. McCartney, and P. H. Shipway. 2012. Bonding between aluminium and copper in cold spraying: story of asymmetry. *Materials Science and Technology* 28 (12): 1371–1378.

Irissou, E., J.G. Legoux, B. Arsenault, and C. Moreau. 2007. Investigation of Al-Al_2O_3 cold spray coating formation and properties. *Journal of Thermal Spray Technology* 16 (5/6): 661–668.

Kim, H.-J., C.-H. Lee, and S.-Y. Hwang. 2005. Superhard Nano WC-12 %Co coating by cold spray deposition. *Materials Science and Engineering A* 391:243–248.

Klar, E., and W. M. Shafer. 1972. In *Powder metallurgy for high-performance applications*, eds. J.J. Burke, and V. Weiss, 57. New York: Syracuse University Press.

Koivuluoto, H., A. Coleman, K. Murray, M. Kearns, and P. Vuoristo. 2012. High pressure cold sprayed (HPCS) and low pressure cold sprayed (LPCS) coatings prepared from OFHC Cu feedstock: Overview from powder characteristics to coating properties. *Journal of Thermal Spray Technology* 21 (5): 1065–1075.

Li, C.-J., and W.-Y. Li. 2003. Deposition characteristics of titanium coating in cold spraying. *Surface and Coating Technology* 167 (2/3): 278–283.

Li, C.-J., and W. Y. Li. 2009. Recent advances in coating development by cold spraying. In *Surface modification of materials by coatings and films research signpost Kerala, India*, ed. X.Y. Liu, P.K. Zhu, and C.X. Ding.

Li, C.-J., and G.-J. Yang. 2013. Relationships between feedstock structure, particle parameter, coating deposition, microstructure and properties for thermally sprayed conventional and nanostructured WC-Co. *International Journal of Refractory Metals & Hard Material* 39:2–17.

Li, C.-J., W.-Y. Li, and H. Liao. 2006. Examination of the critical velocity for deposition of particles in cold spraying. *Journal of Thermal Spray Technology* 15 (2): 212–222.

Li, C.-J., G.-J. Yang, and H.T. Wang. 2007a. Manufacturing method of intermetallics coatings. ZL 2007 10017976.X Chinese patent.

Li, C.-J., G.-J. Yang, P.-H. Gao, J. Ma, Y.-Y. Wang, and C.-X. Li. 2007b. Characterization of nanostructured WC-Co deposited by cold spraying. *Journal of Thermal Spray Technology* 16 (5/6): 1011–1020.

Li, C.-J., H.-T. Wang, Q. Zhang, G.-J. Yang, W.-Y. Li, and H. L. Liao. 2010. Influence of spray materials and their surface oxidation on the critical velocity in cold spraying. *Journal of Thermal Spray Technology* 19 (1/2): 95–101.

Li, C.-J., H.-T. Wang, G.-J. Yang, and C.-G. Bao. 2011. Characterization of high-temperature abrasive wear of cold-sprayed FeAl intermetallic compound coating. *Journal of Thermal Spray Technology* 20:227–231.

Li, W.Y., and C.-J. Li. 2004. Optimization of spray conditions in cold spraying based on the numerical analysis of particle velocity. *Transactions–Nonferrous Metals Society of China* 14 (S2): 43–48.

Li, W.Y, and C.-J. Li. 2010. Characterization of cold-sprayed nanostructured Fe-based alloy. *Applied Surface Science* 256:2193–2198.

Lima, R. S., J. Karthikeyan, C. M. Kay, J. Lindemann, and C. C. Berndt. 2002. Microstructural characteristics of cold-sprayed nanostructured WC-Co coating. *Thin Solid Films* 416: 129–135.

Luo, X.T., and C.-J. Li. 2012. Dual-scale oxide dispersoids reinforcement of Fe–40 at.% Al intermetallic coating for both high hardness and high fracture toughness. *Materials Science and Engineering A* 555:85–92.

Luo, X.T., G.J. Yang, and C.J. Li. 2011. Multiple strengthening mechanisms of cold-sprayed cBNp/NiCrAl composite coating. *Surface and Coatings Technology* 205 (20): 4808–4813.

Luo, X.-T., G.-J. Yang, and C.-J. Li. 2012b. Preparation of cBNp/NiCrAl nanostructured composite powders by a step-fashion mechanical alloying process. *Powder Technology* 217:591–598.

Marrocco, T., D. G. McCartney, P. H. Shipway, and A. J. Sturgeon. 2006. Production of titanium deposits by cold-gas dynamic spray: numerical modeling and experimental characterization. *Journal of Thermal Spray Technology* 15 (2): 263–272.

Melendez, N. M., and A. G. McDonald. 2013. Development of WC-based metal matrix composite coatings using low-pressure cold gas dynamic spraying. *Surface and Coatings Technology* 214:101–109.

Moridi, A., S. M. Hassani-Gangaraj, M. Guagliano, and M. Dao. 2014. Cold spray coating: Review of material systems and future perspectives. *Surface Engineering* 30 (6): 369–395.

Ning, J., J. -H. Jang, and H.-J. Kim. 2007. The effects of powder properties on in-flight particle velocity and deposition process during low pressure cold spray process. *Applied Surface Science* 253:7449–7455.

Pawlowski, L. 2008. *The science and engineering of thermal spray coatings*, 2nd edn. Wiley.

Raletz, F., M. Vardelle, and G. Ezo'o. 2005. Fast determination of particle critical velocity in cold spraying. In thermal spray connects: Explore its surface potential, 2005 international thermal spray conference Basel, DVS, Germany Welding Institute Dusseldolf in CD–ROM.

Rokni, M. R., C. A. Widener, and V. R. Champagne. 2014. Microstructural evolution of 6061 aluminum gas-atomized powder and high-pressure cold-sprayed deposition. *Journal of Thermal Spray Technology* 23 (3): 514–524.

Sansoucy, E., P. Marcoux, L. Ajdelsztajn, and B. Jodoin. 2008. Properties of SiC-reinforced aluminum alloy coatings produced by the cold gas dynamic spraying process. *Surface and Coatings Technology* 202 (6): 3988–3996.

Schmidt, T., F. Gärtner, H. Assadi, and H. Kreye. 2006. Development of a generalized parameter window for cold spray deposition. *Acta Materialia* 54: 729–742.

Schmidt, T., H. Assadi, F. Gartner, H. Richter, T. Stoltenhoff, H. Kreye, and T. Klassen. 2009. From particle acceleration to impact and bonding in cold spraying. *Journal of Thermal Spray Technology* 18 (5/6): 794–808.

Shockley, J. M., H. W. Strauss, R. R. Chromik, N. Brodusch, R. Gauvin, E. Irissou, and J. G. Legoux. 2013. In situ tribometry of cold-sprayed Al-Al_2O_3 composite coatings. *Surface and Coatings Technology* 215: 350–356.

Spencer, K., D.M. Fabijanic, and M. X. Zhang. 2012. The influence of Al_2O_3 reinforcement on the properties of stainless steel cold spray coatings. *Surface and Coatings Technology* 206 (14): 3275–3282.

Stoloff, N. S., C. T. Liu, and S. C. Deevi. 2000. Emerging applications of intermetallics. *Intermetallics* 8: 1313.

Stoltenhoff, T., H. Kreye, and H. J. Richter. 2002. An analysis of the cold spray process and its coatings. *Journal of Thermal Technology* 11 (4): 542–555.

Temples, L. B., M. F. Grununger, and C. H. Londry. 1993. Influences of oxygen content on MCrAlY's. In *Thermal spray: Research design and applications*, ed. C.C. Berndt, 359–363. Materials Park: ASM International.

Wang, J., and J. Villafuerte. 2009. Low pressure cold spraying of tungsten carbide composite coatings. *Advanced Materials & Processess* (ASM International 2009). 167 (2): 54–56.

Wang, H.-T., C.-J. Li, G.-J. Yang, C.-X. Li, Q. Zhang, and W.-Y. Li. 2007. Microstructural characterization of cold-sprayed nanostructured FeAl intermetallic compound coating and its ball-milled feedstock powders. *Journal of Thermal Spray Technology* 16 (5/6): 669–676.

Wang, H. T., C. J. Li, G. J. Yang, and C. X. Li. 2008. Effect of heat treatment on the microstructure and property of cold-sprayed nanostructured FeAl/Al_2O_3 intermetallic composite coating. *Vacuum* 83 (1): 146–152.

Wang, Q., N. Birbilis, H. Huang, and M. X. Zhang. 2013. Microstructure characterization and nanomechanics of cold-sprayed pure Al and Al-Al_2O_3 composite coatings. *Surface and Coatings Technology* 232 (15): 216–223.

Wong, W., A. Rezaeian, S. Yue, E. Irissou, and J. G. Legoux. 2009. Effects of gas temperature, gas pressure, and particle characteristics on cold sprayed pure titanium coatings. In Proceedings of the international thermal spray conference Las Vegas, NV.

Wong, W., P. Vo, E. Irissou, A. N. Ryabinin, J.-G. Legoux, and S. Yue 2013. Effect of particle morphology and size distribution on cold-sprayed pure titanium coating. *Journal of Thermal Spray Technology* 22 (7): 1140.

Yang, G.-J., P.-H. Gao, C.-X. Li, and C.-J. Li. 2012. Simultaneous strengthening and toughening effects in WC-(nanoWC-Co). *Scripta Materialia* 66: 777–780.

Yang, G.-J., S.-N. Zhao, C.-X. Li, and C.-J. Li. 2013. Effect of phase transformation mechanism on the microstructure of cold-sprayed Ni/Al-Al_2O_3 composite coatings during post-spray annealing treatment. *Journal of Thermal Spray Technology* 22 (2): 398–405.

Yule, A. J., and J. J. Dunkley. 1994. *Atomization of melts: for powder production and spray deposition*. New York: Oxford University Press.

Zahiri, S. H., C. I. Antonio, and M. Jahedi. 2009. Elimination of porosity in directly fabricated titanium via cold gas dynamic spraying. *Journal of Materials Processing Technology* 209 (2): 922–929.

Zhang, Q., C.-J Li, C.-X. Li, G.-J. Yang, H.-T. Wang, and S.-C. Lui. 2008. Study of oxidation behavior of cold-sprayed nanostructured NiCrAlY bond coatings. *Surface and Coatings Technology* 202 (14): 3378–3384.

Chapter 4
Coating Properties

M. Jeandin, H. Koivuluoto and S. Vezzu

4.1 Microstructure

Due to the ever-spreading range of applications for materials, it is more and more relevant to consider a given microstructure as made of two components, that is, on the one hand, that made of plain material and on the other hand, that made of voids. The latter is commonly termed as porosity. This division was kept for this subchapter.

4.1.1 Particulate and Grain Microstructure

A typical cold spray (CS) microstructure is dual due to the mere fact that the starting material is powder. At the particle scale, the microstructure is made of splats, that is, particles which were deformed at the impact. At a lower scale, the microstructure is of a metallurgical type due to the intra-particle grains. At both scales, the driving force is deformation which results in two types of phenomena, that is, material flow and grain transformation. These can lead to either a rather general description of the microstructure or a more local description which involves crystallography and interface considerations.

M. Jeandin (✉)
MINES ParisTech, Paris, France
e-mail: michel.jeandin@mines-paristech.fr

H. Koivuluoto
Department of Materials Science, Tampere University of Technology, Tampere, Finland
e-mail: heli.koivuluoto@tut.fi

S. Vezzu
Veneto Nanotech, Venezia, Italy
e-mail: simone.vezzu@venetonanotech.it

J. Villafuerte (ed.), *Modern Cold Spray*, DOI 10.1007/978-3-319-16772-5_4

4.1.1.1 General

Deformation due to particle impacts results in a splat-typed microstructure which is not always easy to reveal through conventional metallography. Depending on the nature of the cold-sprayed material, the use of specific etching and/or image analysis may be required. When successful, metallography reveals a typical microstructure of splats (Fig. 4.1), the shape of which looks like that of a blobfish (Fig. 4.2). This is due to the fact that plastic deformation of a particle at the impact can be compared to the adaptation of the blobfish body to the high pressure in deep water.

Material flow behaviour is shown not only by the outline of the deformed particles but also by the former grain (in the broadest sense of the term) boundaries as a sign of particle heredity (Fig. 4.1). Depending on the feedstock particle production, the starting particles showed a more or less marked fine-grained microstructure (see Chap. 3). For example, the microstructure of a given particle from an atomized powder can range from a dendritic microstructure to a finely cellular microstructure as a function of the cooling rate when atomized (Fig. 4.3). Since rather fine pow-

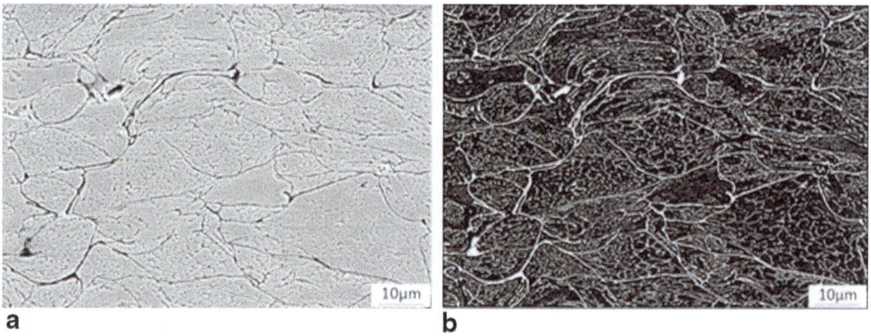

Fig. 4.1 Cross-sectional scanning electron microscopy (SEM) image of cold-sprayed Al **a** as slightly "Keller's" etched and **b** after image processing. (Courtesy of Quentin Blochet, MINES ParisTech 2014)

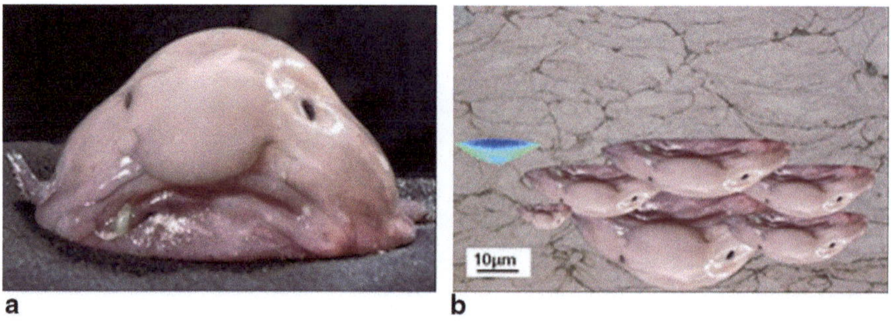

Fig. 4.2 Blobfish. **a** General. **b** Upside down inserted in cold-sprayed Cu (cross-sectional SEM image); in comparison with finite element (FE)-simulated splat (in *blue*)

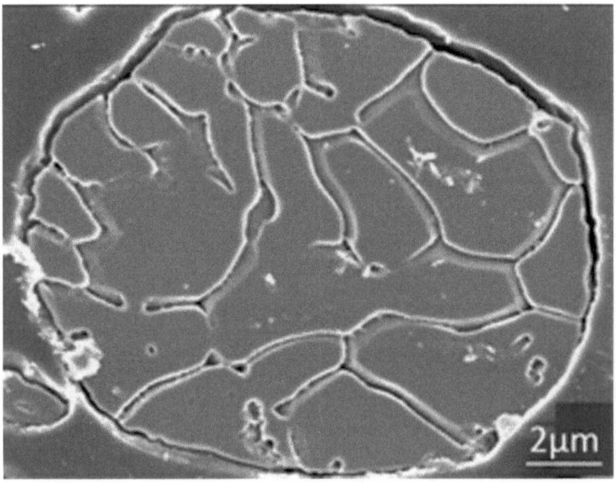

Fig. 4.3 Cross-sectional SEM image of a nitrogen-atomized Al particle, "Keller's reagent". (Courtesy of Quentin Blochet, MINES ParisTech 2014)

ders, that is, below 30 µm in size, are generally used for CS, dendritic particles are not frequent.

Coating observation at this scale gives an idea of the homogeneity of the material, which reflects the degree of uniformity in the coating build-up process (Chap. 2) as a function of powder characteristics. To go into this aspect, with a quantitative assessment in particular, one has to develop an approach to morphological parameters using specific tools, on which Sect. 4.2 elaborates. From the observation of the overall microstructure, that is, when considered at the scale of the splat/particle, as described in this section, one may suspect what could happen at a lower scale during the build-up process. The corresponding phenomena are actually those which govern the final (mechanical and physical) properties of the coating.

4.1.1.2 Crystallographic and Interface Characteristics

When colliding, a given particle can undergo extreme conditions for plastic deformation, as described in Chap. 2. For further details, one may refer to various comprehensive descriptions such as recently those by Moridi et al. (2014a, b), Jeandin et al. (2014), and Cinca et al. 2013a, b). As a reminder, one may say that, at particle impact, strain and heating rates can, respectively, reach $10^9\ s^{-1}$ and $10^9\ K\ s^{-1}$ typically. In these conditions, three paramount phenomena can occur, that is, grain refinement, strain accommodation and phase/interface transformations and can be distinguished even though these are not entirely independent. Each of them involves various basic mechanisms which result in various microstructure characteristics.

Grain Refinement

Grain refinement results from dynamic recrystallization as a result of high plastic deformation at particle impact. The basic phenomenon can be well exhibited using

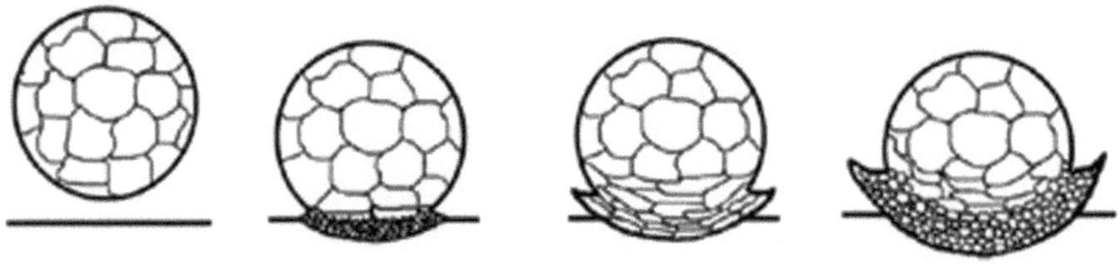

Fig. 4.4 Schematic illustration of dynamic recrystallization at impact between a given particle and the substrate. (After Kim et al. 2008)

transmission electron microscopy (TEM) of a whole elementary slat after CS in nominal conditions. Specific splat-collecting experiments, namely "splat experiments", were carried out on this for different materials (e.g. the "historical" paper by Dykhuizen et al. (1999), and, more recently, that by Descurninges et al. (2011)). For example, a thin foil of a Ti splat ascertains the popular schematic which was already proposed in 2009 by Kim et al. (2008; Fig. 4.4).

Depending on the type of materials, the degree of recrystallization is more or less pronounced. When spraying Ti onto Ti–6Al–4V, grain refinement through recrystallization can involve about half of the volume of the splat (Fig. 4.5).

In the actual coating, during the build-up stage, dynamic recrystallization occurs at the particle interfaces. The process is in keeping with the general basic mechanism which was proposed by Meyers et al. (2007) but applied to the particle–particle interface (Fig. 4.6). Recrystallized grain size and misorientation depend on the particle melting temperature and stacking fault energy (SFE) of the sprayed material (Borchers et al. 2005). This was particularly evidenced in the cold spraying of face-centred cubic (fcc) materials such as Cu, Al or Ni. The latter, for example, due to a rather high melting temperature coupled with a rather low SFE, recrystallizes dynamically in rather small numbers of ultra-fined grains. Electron backscatter diffraction (EBSD) analysis consists of a powerful tool to show this, as successfully applied to Ni in an early work by Zou et al. (2009; Fig. 4.6).

At the prior particle boundaries (ppbs), grains can grow till a size in the micron range typically, due to temperature increase at impact in adiabatic shearing conditions (Assadi et al. 2003; Guetta et al. 2009).

Fig. 4.5 Dark-field TEM image of a thin foil of a Ti cold-sprayed splat onto a Ti–6Al–4V substrate. (After Giraud et al. 2015)

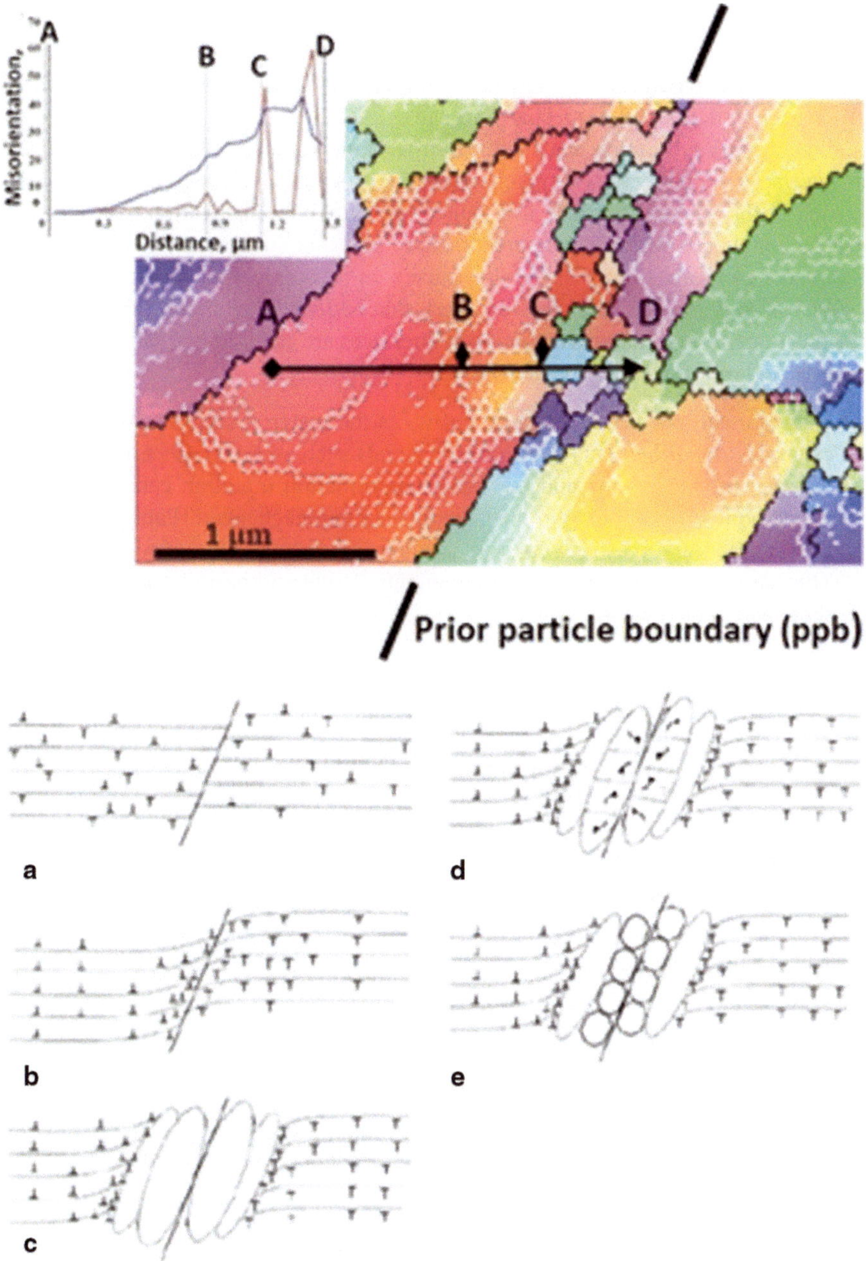

Fig. 4.6 EBSD inverse pole figure (IPF) map of cold-sprayed Ni with, inserted, a misorientation profile crossing a particle–particle boundary (after and below a schematic illustration of the corresponding recrystallization process at this same boundary. (After Zou et al. 2009)

Solid-State Strain Accommodation Mechanisms

In addition to dynamic recrystallization which can also be considered as due to strain accomodation, other mechanisms can occur due to severe plastic deformation at impact. This subsection deals with solid-state phenomena, which are not located at interfaces exclusively. The mechanism of dislocation rearrangement is not discussed in this chapter because it is rather conventional. Dislocation rearranges in conventional cells which can be precursors of grains or subgrains. For illustrations, the reader can refer to one of the first papers on this topic, that is, Mc Cune et al. (2000), or to one of the most recent papers, that is, Jeandin et al. (2014). Moreover, the phenomena which involve melting at interfaces are covered in a subsequent section since they show a prominent role.

- *Twinning* can occur, all the more easily, as the material shows a low SFE, for example, along the (111) planes in fcc metals such as Ag. A 40° misorientation between the slip bands and the plastic deformation direction can thus be obtained typically according to the general shearing mechanism (Paul et al. 2007). Deformation involves regions from a rather large scale to the nanometric scale. This can lead to slip bands crossing an entire splat and/or nanotwinning within shearing bands (Fig. 4.7).

- *Solid-state phase transformation* due to strain accommodation can result in amorphization and disordered structures with randomly oriented nanocrystallites, which can be partly stabilized to some extent due to the presence of impurities (Xiong et al. 2011; Fig. 4.8).

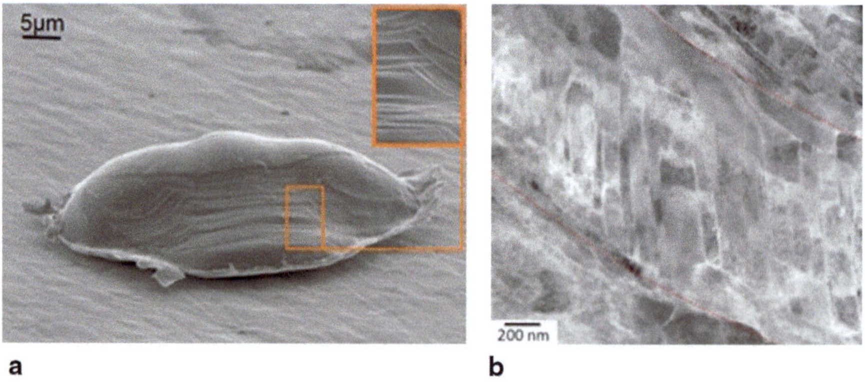

Fig. 4.7 Stain accommodation evidences. **a** Slip planes in a cold-sprayed Dart-Vadered Manta-typed Ti splat (inserted, magnification; courtesy of Damien Giraud/MINES ParisTech 2014). **b** Nanotwinned shear band in cold-sprayed Ag. (Courtesy of Gilles Rolland/MINES ParisTech 2010)

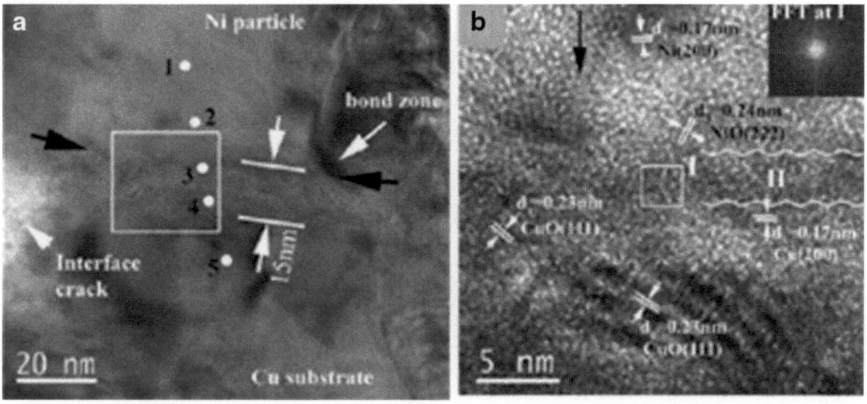

Fig. 4.8 High-resolution transmission electron microscopy (HRTEM) images of cold-sprayed Ni onto Cu. **a** General view. **b** Magnification on the box region, showing the disordered and amorphous-like structure. *FFT* fast Fourier transform. (After Xiong et al. 2011)

Phase/Interface Transformations

Due to high-energy and short-term material interactions at impact, as already seen, CS microstructure exhibits submicronic, not to say nano-sized, features, the knowledge of which is crucial to understand, therefore possibly control, coating adhesion and cohesion. These features are located at splat–splat or splat–substrate interfaces. The approach to the corresponding phenomena is rather complex due to nonequilibrium conditions. One may, however, put them into two classes depending on whether they contribute to melting or/and cleaning the interface. Solid-state transformations were discussed in the previous sections.

- *Melting* can be considered as the culmination of material interaction due to the impact. Evidences of melting are rather difficult to find out due to the small size of the interaction areas which, moreover, cannot be described through—powerless—modelling. TEM analysis is therefore the best tool for investigation, especially when involving materials which can react with each other or when using a low-melting temperature spray material. For example, in an early study, Barradas et al. (2007) thoroughly described the formation mechanisms of intermetallic phases when cold spraying Cu onto Al and proposed an interface phenomenological diagram. These phases revealed transient melting at the coating–substrate interface through eutectic or peritectic zones in particular (Fig. 4.9a). This is all the easier as the melting point can decrease with increasing strain, for example, for fcc metals (Lynden-Bel 1995). When considering the CS coating itself, using a low-temperature material such as zinc can promote a liquid phase (Fig. 4.9b, Li et al. 2010).

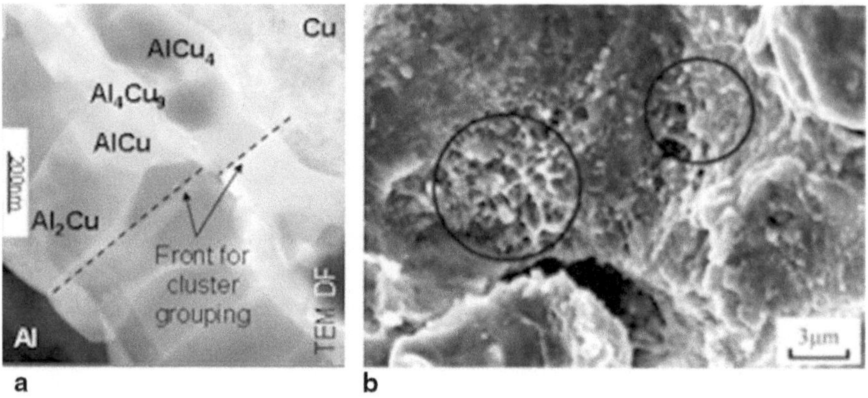

Fig. 4.9 Evidences of melting at cold spray interfaces. **a** Transmission electron microscopy *(TEM)* image of (Al, Cu) intermetallic (Al, Cu) phases for cold-sprayed Cu onto Al (after Jeandin 2011). **b** SEM image of molten zones *(circled)* in cold-sprayed Zn. (Li et al. 2010)

- An advanced stage of the melting process can result in a rather extended interface layer of molten material, possibly amorphous, provided the cooling rate is high enough for the involved material. Amorphization can outline a great part of the splat–substrate interface (Fig. 4.10).
- An intermediate interaction state between purely solid-state transformation as described in a subsequent section (Fig. 4.8) and interface fusion consists in a so-called viscous forced mixing of the two interacting materials. These can remain either partly at the solid state or not, depending on the nature of the starting materials and on processing conditions. The involved mechanism can be compared

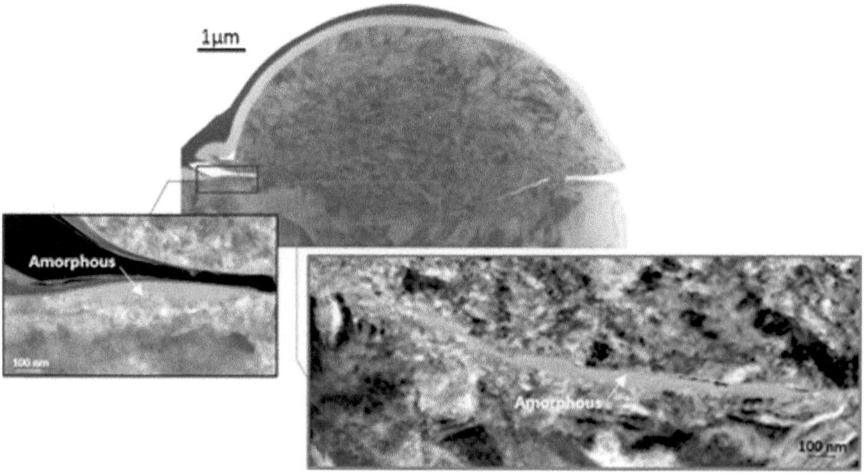

Fig. 4.10 Bright-field TEM image of a Ti cold spray splat onto Ti–6Al–4V above two dark-field magnified TEM images of the amorphous layer at the interface. (Courtesy of Damien Giraud, MINES ParisTech 2014)

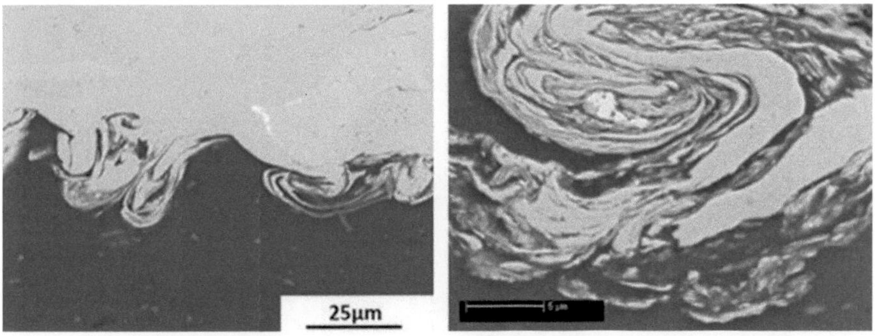

Fig. 4.11 Cross-sectional SEM images of vortices at the interface with an Al substrate *(dark)* of cold-sprayed **a** Cu (after Cha et al. 2005) and **b** Ni *(bright)*. (After Ajdelsztajn et al. 2005)

to that encountered in mechanical alloying, explosive processes or in a two-body contact area under wear–friction conditions. Typical vortices can form at the interface as a result of adiabatic shearing instabilities which relate to the Kelvin–Helmotz instability phenomenon (Fig. 4.11; Champagne et al. 2005; Ajdelsztajn et al. 2005).

- *Cleaning,* to use a general term, means removing, at least partly, contaminants, inclusions and/or external phases such as oxides, nitrides, etc. The most common effect rests on fragmentation and/or partial removal of the oxide layer which exists at the surface of the starting powder (Fig. 4.12). This occurs due to particle impact at the coating build-up stage. The effect is especially marked and beneficial for reactive and oxygen-sensitive materials such as Ti or Ta (Giraud et al. 2015; Jeandin et al. 2014; Descurninges et al. 2011).
- The oxygen content varies along the particle impact interface due to temperature and strain differences at impact. Here again, basic phenomena could be better elucidated from splat experiments, knowing that these can be transposed to the coating build-up level. Oxygen generally decreases from the centre to

Fig. 4.12 High-resolution (HR) TEM image of a commercial feedstock Ta powder. (After Jeandin et al. 2014)

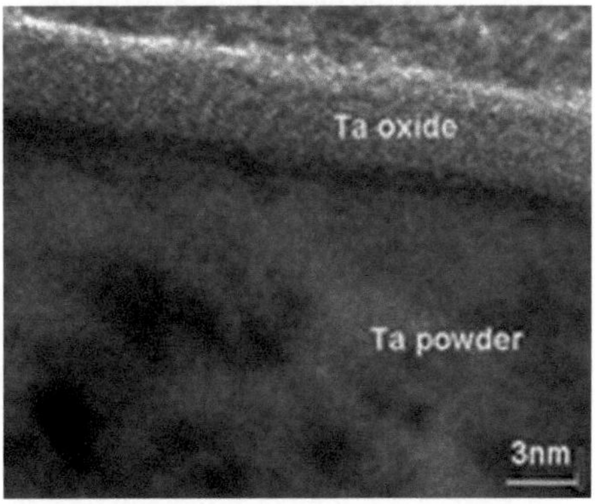

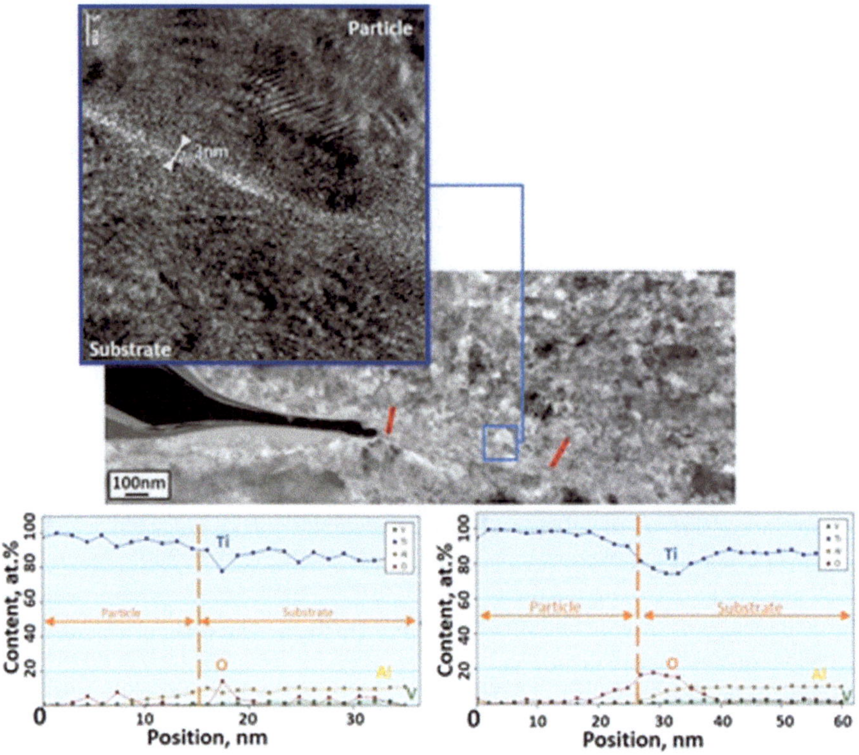

Fig. 4.13. Dark-field TEM image and *(top left)* HRTEM image of a Ti splat–Ti–6Al–4V interface with EDX linescan profiles across two lines (drawn in *red* in the image). (After Giraud et al. 2015)

the periphery of the splat because of temperature increase along the outline, as shown by modelling of the particle impact (e.g. Guetta et al. 2009; Schmidt et al. 2009). Temperature increase promotes oxygen diffusion at both the solid and liquid state. For the latter, it can be assumed that part of the superficial oxide layer at the surface of the sprayed particle could break and leave a purely metal-to-metal contact at impact, as already shown for alumina in the cold spraying of Cu onto Al (Barradas et al. 2007). Solid-state diffusion of oxygen from the splat–substrate/splat interface could be exhibited in several studies of CS microstructure, using Energy-dispersive X-ray spectroscopy (EDX) analysis in TEM (Giraud et al. 2015; Jeandin et al. 2014; Fig. 4.13).

- Oxide fragmentation can also be assumed to contribute to oxygen variation at the interface. However, this has not yet been shown specifically in research work despite a great deal of presumptions. Oxide layer fragmentation can play a specific role in the context of that of the oxygen content because of consequences on the mechanical behaviour of the involved interfaces, for example, for the coating–substrate bond strength and coating cohesion which will be developed in Sect. 4.3.6. Oxide fragmentation should give fine oxide fragments the role they show in oxide dispersion-strengthened alloys. An oxide dispersion at CS interfaces can exist as could be encountered in dynamically compacted powder metallurgical (P/M) superalloys some time ago (Morris et al. 1987), which can be beneficial for resistance due to composite reinforcing effect at interfaces.

More generally, all of the above-mentioned interface microstructural CS features can influence coating mechanical properties, that is, vortices as pegging sites for adhesion, intermetallics as pegging sites also or (more detrimental) as embrittling phases, amorphous interlayer as a protecting barrier, oxide dispersion as a local composite, etc. There is therefore a strong demand for assessing mechanical properties which correspond to these typical CS microstructural characteristics. To meet this demand, a local approach to these properties is required since all these characteristics involve the nanometric range, which may lead to say that cold-sprayed coating systems are nano-length scale governed. A local investigation into mechanical (interface) properties would be used as an input for micro-to-macro modelling provided that a significant development could be done in the future to involve very fine microstructures and ultra-rapid phenomena. First steps in this scope are discussed in Sect. 4.3.6.

4.1.2 Porosity

Even though (or because) CS was formerly developed to achieve fully dense coatings due to high-kinetic processing conditions which were suitable for that, coating porosity assessment is of high concern. This is all the more true because subsequent development also showed that CS can be used to obtain porous coatings deliberately, for example, for biomedical applications (Sun et al. 2008; Cinca et al. 2010).

Porosity strongly depends on the coating build-up process. Porosity forms due to insufficient particle deformation at the impact and/or an insufficient particle velocity, which are not independent parameters. One cannot be more precise since these required parameters, that is, deformation and velocity, are local depending on the size and morphology of the particles and the roughness of the substrate. In the coating formation process, "substrate" means, first, the actual bulk substrate for the first layer to be deposited and, second, that made of the already-deposited particles. Porosity creation therefore consists of a random process governed by the particle impinging. One may give only general trends on its evolution within the coating to integrate the random variation over a high number of splats, typically above a few hundred. The major trend results from the peening effect, that is, tamping due to successive impact from the succeeding particles, which is cumulative till a certain coating thickness. In the upper part of the coating, porosity is therefore higher due to a lower number of impacts the material had to undergo. Consequently, a cold-sprayed coating exhibits a gradient from the coating–substrate interface to the coating surface. The gradient profile depends on the materials and spraying conditions, primarily powder grain size and distribution, therefore the particle velocity field, powder flow rate, number of passes and the nature of the substrate. This can be well exhibited when cold spraying a hard-to-densify material, for example, a Ti-based alloy, using two passes (Fig. 4.14).

However, this peening effect needs a certain time prior to be established, which corresponds to the time from which the underlayer (made of the already-deposited

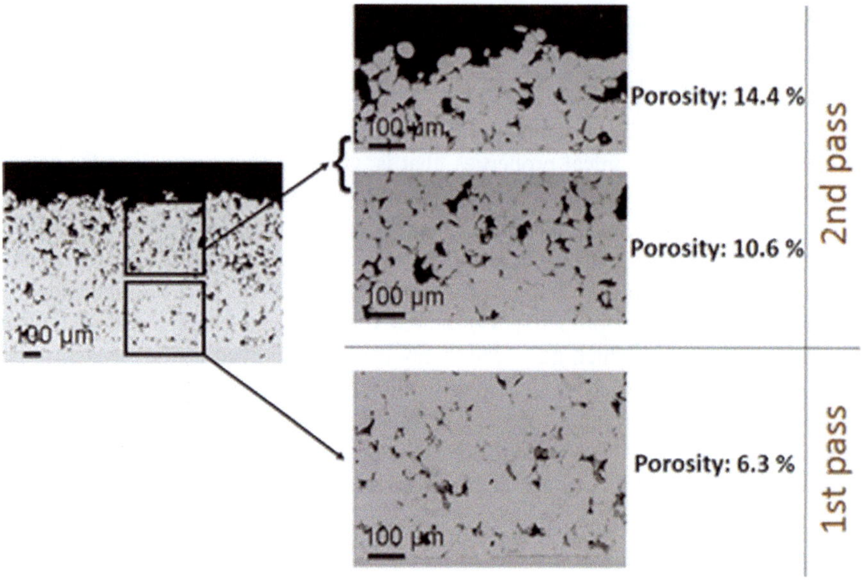

Fig. 4.14 Cross-sectional SEM images of cold-sprayed Ti–6Al–4V onto Ti–6Al–4V. (After Christoulis et al. 2011)

particles) is stable. This time is all the longer the substrate material shows a high shock-absorbing capacity such as a polymer (Fig. 4.15). To shorten this time and promote adhesion and densification, a metallic bond coat can be used, for example, using tin (Ganesan et al. 2012).

To determine the void content, that is, the porosity level as commonly said, several, not to say many, methods exist actually (reviewed in Adreola et al. (2000), for example). Common methods are physical methods such as Archimedean porosimetry, mercury intrusion porosimetry (MIP), gas permeation and pycnometry. However, these are not quite often convenient due to characteristics typical of CS, that is, generally a very low and/or heterogeneous porosity. The best way to proceed consists in using conventional two-dimensional (2D) metallography or three-dimensional (3D) techniques.

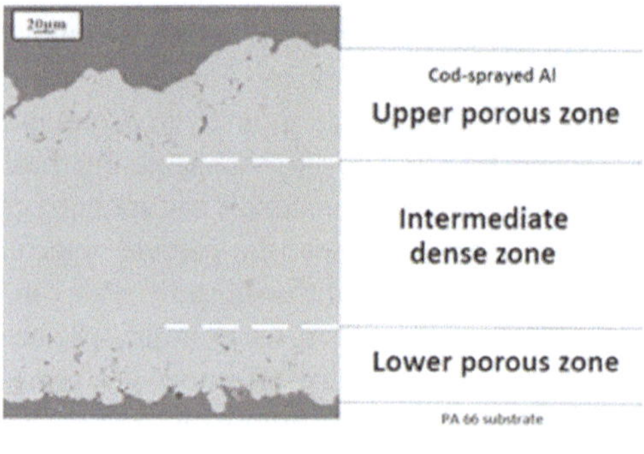

Fig. 4.15 SEM micrographs of cold-sprayed aluminum onto PA66 using three passes at 2.5 MPa–250 °C. (Courtesy of Damien Giraud/MINES ParisTech 2014)

4.1.2.1 Two-Dimensional Metallography

Conventional 2D metallography can be suitable provided that smearing and undeliberate material removal can be prevented when polishing. Smearing is all the more frequent that ductile materials are often employed for CS coating. In contrast, undeliberate material removal is promoted by the presence of hard phases or due to local differences in hardness as can be seen at the coating–substrate interface or at the edge of pores. Even though it is a very common issue in the preparation of materials prior to observation, special care is required when characterizing cold-sprayed materials. If not, this can result in over- or underestimating porosity (Fig. 4.16).

Polishing quality is therefore the main source of potential errors for subsequent porosity measurements, which is now conventionally carried out using quantitative image analysis (QIA; Fig. 4.17). For a given polishing state, the degree of uncertainty to the result is rather low at this stage and can be said to be limited to a maximum of ± 5 % (relative value).

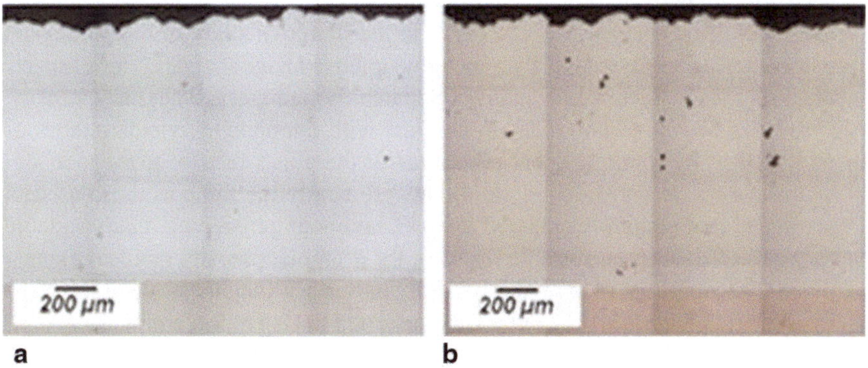

Fig. 4.16 Cross-sectional optical images of a specimen (the same for the two pictures) of cold-sprayed Ag onto Cu. **a** After a rather mediocre polishing. **b** After careful polishing. (Courtesy of Gilles Rolland, MINES ParisTech 2010)

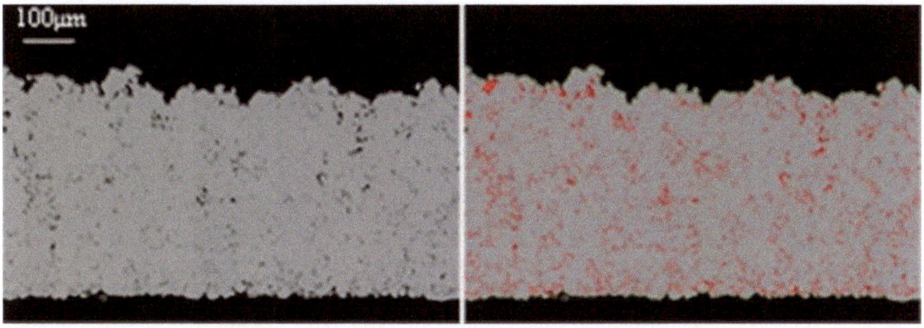

Fig. 4.17 Cross-sectional image of a cold-spray Al coating of PA66 before *(left)* and after (*right*) image processing. (Courtesy of Damien Giraud, MINES ParisTech 2014)

4.1.2.2 Three-Dimensional Metallography

Compared to 2D methods, the strength of 3D methods for porosity assessment results from the bypassing of the materials preparation stage due to the direct observation within the material. Incidentally, pseudo-3D methods based on the use of serial cross sections are not satisfactory due to the need of polishing again, even though there were attempts in the thermal spray field (Ctibor et al. 2006). The most popular techniques are based on X-ray microtomography (XMT) or laminography, despite the development of ultrasmall-angle X-Ray or small-angle neutrons scattering. However, the latter remain rather marginal and mainly restricted to ceramic materials. XMT and variants (primarily laminography) can now show the required high resolution. The application of these techniques to cold-sprayed coatings was particularly developed successfully in the past 5 years (for the most recent, Delloro et al. 2014a, b).

- *XMT* is a powerful tool for investigating into porosity in thermally sprayed coatings which could reveal characteristics which had not yet been exhibited or even suspected (Amsellem et al. 2012). XMT can show the influence of powder characteristics on porosity (Fig. 4.18). Beyond the determination of the mere porosity level and distribution, XMT can result in the thorough study of morphological parameters using stereological protocols coupled to image analysis. These aspects are discussed in Sect. 4.2, including CS materials parameters other than porosity.
- *Computed laminography (CL),* in contrast with tomography, yields images of object slices by a simple linear translation of the object relative to the tube–detector system. Reconstruction algorithms are nearly the same as those used in computed tomography. Compared to XMT, laminography is particularly suitable for characterizating anisotropic features, for example, porosity gradient along a given direction, that of spraying typically or surface roughness along the coating–substrate interface. The latter is discussed in Sect. 4.2. Regarding the assessment of in-depth evolution of porosity within a CS coating, a striking example results from the already-described shock absorbing combined to tamping effects in CS metallization of a polymer (Fig. 4.19).

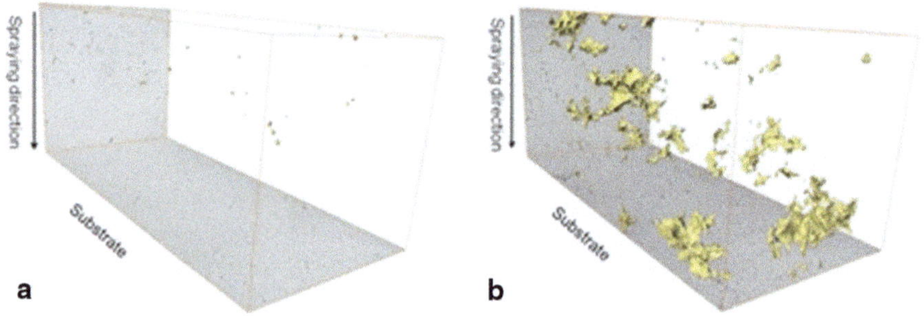

Fig. 4.18 3D XMT images (reconstructed volumes of 280 × 280 × 573 mm^3) of porosity in cold-sprayed Ag. **a** For a fine powder. **b** For a coarser powder. (After Rolland et al. 2008)

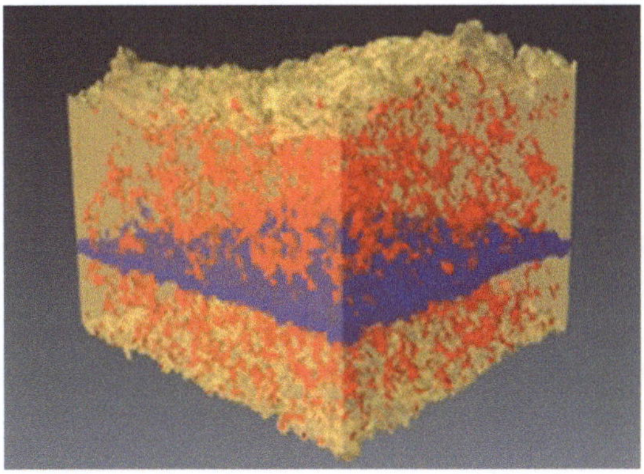

Fig. 4.19 3D CL image of porous cold-sprayed Al. Porosity in *red,* Al in *beige* and reconstructing/analysis plane (of 175 × 183 μm² in size) in *purple*. (Courtesy of Damien Giraud/MINES ParisTech 2014)

4.1.2.3 Comparison between 2D and 3D Techniques

3D techniques for porosity assessment (and more generally for microstructure analysis) can remove all doubts on porosity assessment from the prevention of any artifact due to sample preparation (see Sect. 4.1.2.1). Differences between 3D and 2D assessments are significant, whatever be the analysed region within the coating (Fig. 4.20). Differences can result in either underestimation or overestimation depending on the type of involved materials which would promote either smearing or material removal effects. In Fig. 4.20, the error bars correspond to a given preparation method (2D or 3D) actually. They do not therefore overlap.

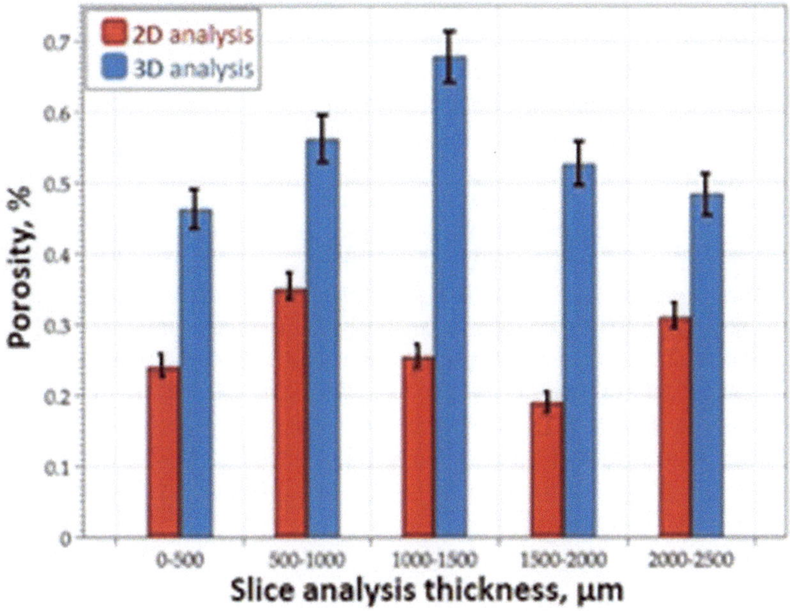

Fig. 4.20 Comparison between the global void content (i.e. porosity) obtained from 2D and 3D techniques. (After Rolland et al. 2008)

CS microstructure is multifaceted. Features range from the nano- (as formerly stated by Grujicic et al. 2004) to the macroscale. The latter is well illustrated by the web video illustration of additive manufacturing of freestanding components by CS (Halterman 2013). In addition to the mere dimensional aspect, an essential part of the CS coating properties relates to the morphology of these same features. This results from specific processing and metallurgical characteristics involved in CS, as previously described. A good knowledge of coating properties therefore requires morphological studies. These result in the material of the subsequent section in which morphological concerns are discussed. The whole will help in developing more powerful and realistic modelling of coating microstructures therefore properties.

4.2 Morphological and Physical Properties

The so-called morphological properties (see the definition in Sect. 4.2.1) are discussed in the same section as that for physical properties, even though these should have been discussed in a separate section. This was not done due to the fact that they result in a constantly changing domain in which a high amount of research work is still in progress with many advances still to be made. The chapter, the scope of which is to deal with these aspects, gives, however, a flavour of these due to the associated promising outlook.

4.2.1 Morphology

To give a definition, morphological properties relate to the shape and size of the various parts which result in the coating, that is, from the particle to the coating itself through all microstructure-relevant features. In addition, some relevant morphological features can relate to the substrate, primarily surface roughness and coating–substrate interface.

As previously shown, the coating microstructure and therefore coating properties strongly depend on local parameters such as particle velocity, temperature, consequently strain rate, which directly result from mainly local morphological characteristics (Cinca and Guilemany 2013; Cinca et al. 2013a, b). A description of these, which can be a 3D description in the most advanced development, is therefore required. The description is carried out prior and after CS deposition, to go into the process and final properties of the products, including the feeding of models. This should also help in enriching approaches to the process, including well-established approaches such as that based on the so-called particle critical velocity. This description is the material of this subchapter. The latter will not revert to porosity, which was already discussed in Sect. 4.1.2.

4.2.1.1 Particle Morphology

A view shared by a larger and larger number of people is that powder is the key factor in the development of CS (Jeandin et al. 2014; see the above introduction of Sect. 4.2.1 and Chap. 2). Spherical powders are no more considered as the best powders for the process, which is inconsistent with what the doxa said in the early stages of CS (and still say from time to time). There is therefore a high demand for developing tailored powders, which corresponds to a major economic issue. This development requires thorough characterization of particle morphology for better understanding of consequences on coating properties consequently for powder optimizing.

Conventional characterization methods such as optical and SEM methods are rather limited even when coupled with image analysis. Moreover, laser-based imaging diagnostics and granulometers are not suitable for morphology assessment. Numerical 3D classification of particles therefore consists of a paramount step, especially for use as a data supplier for modelling input. Advanced classification (Delloro et al. 2014a) from XMT can show three stages typically, that is, (1) image processing (e.g. using segmentation), (2) shape criteria application (using various measuring operations), and (3) cluster analysis (e.g. using the K-means method). The method can involve several thousands of particles, the shape distribution of which can be given (Fig. 4.21).

This morphological approach can be extended to agglomerates the use of which is expected to be promoted in CS, as could be done for WC–Co (Li et al. 2013),

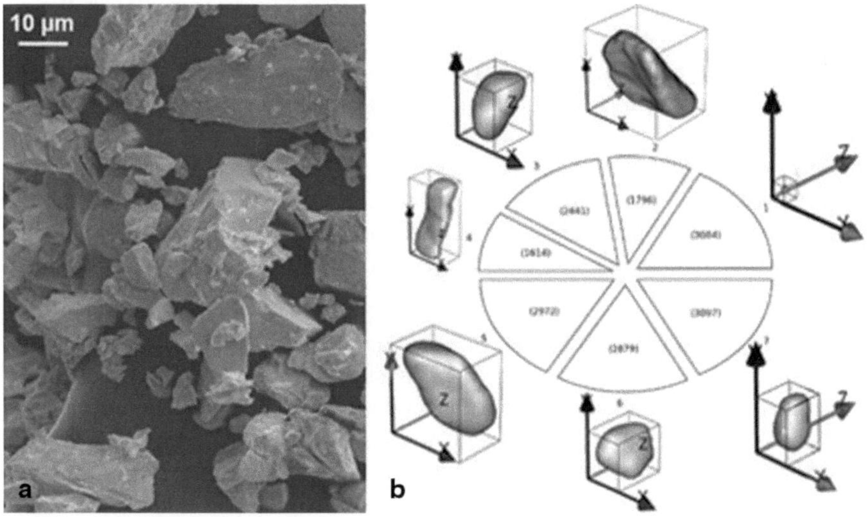

Fig. 4.21 Tantalum irregular powder for cold spray. **a** SEM image of the loose particles. **b** Shape distribution obtained by XMT (each sector of the pie diagram shows the number of analysed particles and the corresponding XMT image of the representative shape with *x*, *y*, *z* axis vectors of 15 μm in length). (After Delloro et al. 2014a)

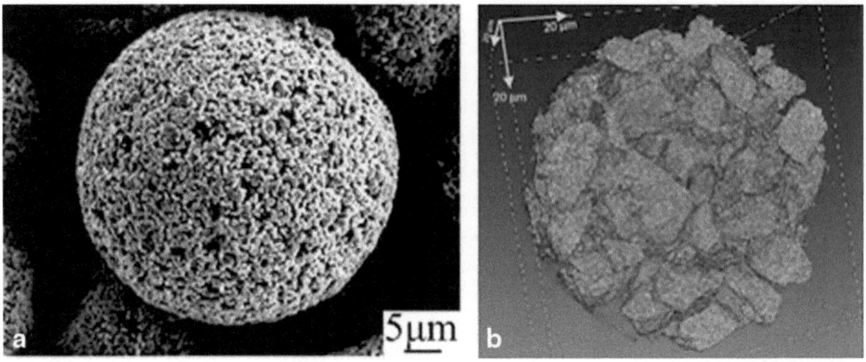

Fig. 4.22 Example of agglomerated powders for cold spray. **a** SEM image of WC–Co powder (after Li et al. 2013). **b** XMT image of Ag–SnO_2 powder. (Courtesy of Yassine Zeralli, MINES ParisTech 2013)

Ag-based composites (Zeralli et al. 2014; Rolland et al. 2012) or ceramics (Yamada et al. 2009). Here again, 3D microtomography is a powerful tool for characterization (Fig. 4.22).

4.2.1.2 Splat Morphology

A prominent interest in the study of splat morphology is to set correlations with that of particle morphology through the deformation behaviour. Proceeding particle by particle, that is, to study it for a given particle, remains an experimental challenge, which could be successful in the near future using advanced techniques such as laser shock-based techniques (Barradas et al. 2007; Jeandin 2011; see Sect. 4.3.6.3). Currently, the approach is statistical and applied to either a single splat or the whole coating. Two- and three-dimensional techniques can be employed. The subsequent subsections will not elaborate on 2D techniques, of which Sect. 4.1 already gave an illustration indirectly. One may say only that 2D imaging is generally combined to image analysis and measuring of the splat deformation ratio, for example. In contrast, in 3D techniques, the principal stage is that of image asquiring, as discussed below exclusively:

- *Single splat* morphology can be studied from linescan-typed collecting experiments. Dimensional and morphology assessment of the emerging part of the splat is fairly easy to achieve by conventional 3D optical or SEM profilometry (Sect. 4.1). Complete characterization, that is, including the part which is embedded into the substrate, requires an additional study of cross sections to establish a shape typology from the knowledge of the emerging part. Otherwise, direct complete characterization can be obtained from XMT or better laminography (CL; Delloro et al. 2014a). This is all the easier to apply as the splat differs from the substrate material due to X-ray absorption contrast. If not, the sample has to be prepared specifically (Delloro et al. 2014b). Selective etching, infiltration

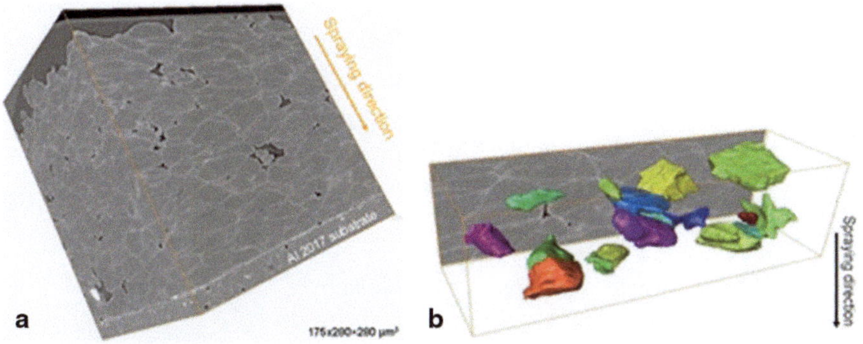

Fig. 4.23 XMT images of cold-sprayed Al onto Al 2017. **a** General. **b** With numerically extracted splats. (After Rolland et al. 2008)

or coating (of the splat, provided that this does not influence the impact) can be used for that purpose (Fig. 4.23). In addition, XMT and laminography could be used to go into inner splat morphological features such as deformation contours due to porosity within a splat aggregate (Li et al. 2013).

- *In-coating splat morphology* can be obtained using XMT or CL provided that the just above-mentioned preparation of the sample could be used to extract a given splat from the surrounding splats (Fig. 4.23; Rolland et al. 2008).
- The use of these 3D techniques ascertained that deformation is not uniform actually at the splat scale, which highlights the role of local parameters such as particle shape therefore local velocity and temperature.

4.2.1.3 Roughness

Interface roughness is the relevant parameter to be considered due to its influence on splat–splat and splat–substrate adhesion properties, consequently on coating cohesion and bond strength (Sect. 4.3.6). Interface roughness results from particle morphology (discussed in Sect. 4.2.1.1) and substrate surface roughness prior to CS.

- *Surface roughness* of the substrate can result from pretreatment, including grit blasting and/or the first CS pass—with heating and cleaning effects. Once again, conventional methods such as those mentioned in the introduction of Sect. 4.2.1.2 are suitable (Gan and Berndt 2014; Blochet et al. 2014). As already mentioned, a more thorough investigation can be based on the use of X-ray laminography (Fig. 4.24).
- *Interface roughness* can be more or less accurately assessed depending on the selected characterization method. For example, the latter can go as far as describing vortex-like features (back to Fig. 4.11) which govern pegging effects. 2D or 3D techniques in addition to quantitative image analysis (Blochet et al. 2014) can be applied (Fig. 4.24).

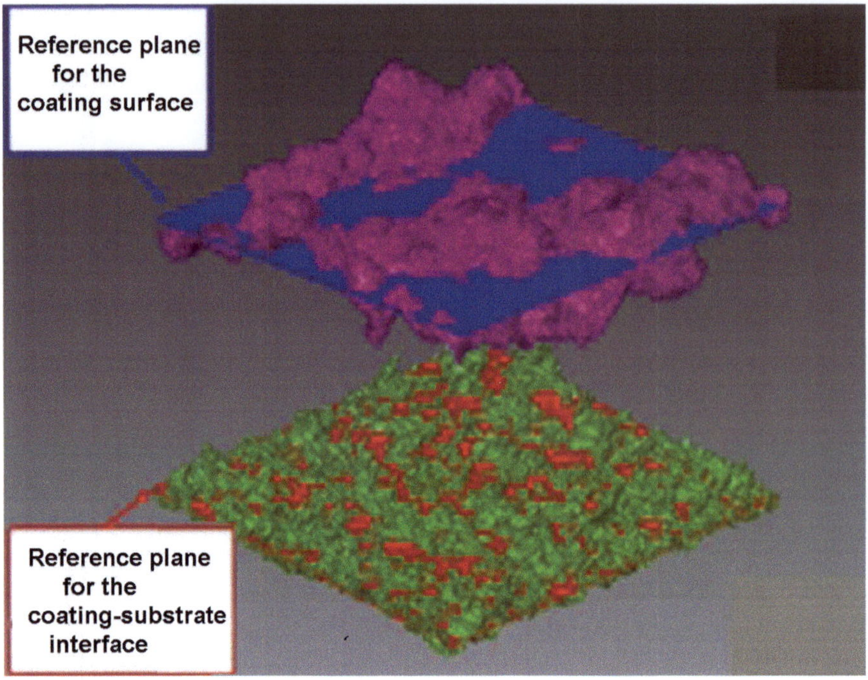

Fig. 4.24 3D CL image of coating surface and coating–substrate interface roughness for cold-sprayed Al onto PA66 (reference planes—in *blue* and *red*—of 175 × 183 μm² in size). (Courtesy of Damien Giraud/MINES ParisTech 2014)

4.2.1.4 Macroscopic Shape

"Macroscopic shape" is understood as a shape related to the substrate or coating geometry. Involved dimensions are therefore commensurable with coating thickness typically. In the first case, the relevant geometry is that of the substrate. A corresponding issue might be that of repair, for which the filling of cavities is of high concern (Blochet et al. 2014; Jones et al. 2011). Second, the relevant geometry is that of the shape of the coating. This can relate to the control of coating thickness for conventional coating applications for CS. However, beyond this, this can relate to application of CS to additive/direct manufacturing of parts, namely freedom fabrication. The latter was already claimed to be very promising at the beginning of CS development, even though at this time the control of the coating build-up could be rather difficult (Pattison et al. 2007). Today, the achievement of rather complex shapes, for example, using micronozzles (Sova et al. 2013a, b), and controlled deposition without masking, for example, for electrode circuits (Kim et al. 2013), can be envisaged (Fig. 4.25).

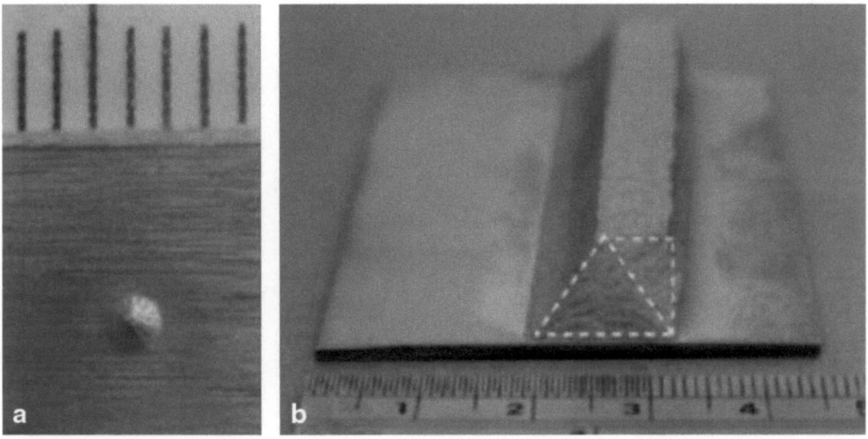

Fig. 4.25 Shaped deposits for application to freedom fabrication by cold spray. **a** Optical top view of an Al deposition cone using a micronozzle (after Sova et al. 2013a, b). **b** Optical view of Al–Cu vertical wall using a triangular tessellation scheme. The scale graduation of the rule is 1 mm. (After Pattison et al. 2007)

4.2.1.5 Approach to Morphology Through Multiscale Modelling

All the points which were discussed in the four previous sections strongly depend on the coating build-up process. With regard to CS, their understanding, validation and prediction impose the obligation of developing modelling and numerical simulation since any empirical approach is basically limited. The main difficulties to overcome result from the multiscale nature of the process. Build-up involves phenomena which occur at the particle scale and the coating scale, knowing that the submicronic scale cannot yet be described through computational modelling, as already mentioned. This section aims only to give a few key elements on modelling from the morphological standpoint to go into CS process and applications. Incidentally, this area is booming, which justifies a snapshot of it only.

- *Computational fluid dynamics (CFD)* must involve particle morphology due to its influence on spraying gas velocity. Since this type of modelling is continuously under development, for example, Lupoi and O'Neill (2011), morphological aspects will play a greater and greater role. The approach will be refined from the former use of drag coefficients of irregularly shaped particles (Tran-Cong et al. 2004) (Fig. 4.26).
- *Finite element (FE) simulation* is very popular to simulate particle deformation at impact. Numerous publications, for example, Gu (2013) and Xie et al. (2013), including impressive videos on the web, are available. However, for the vast majority, they deal with spherical particles and/or in a 2D approach. Recent advances relate to irregular powders (Assadi et al. 2014; Yin et al. 2014), even though the irregular character remains still rather limited and the number

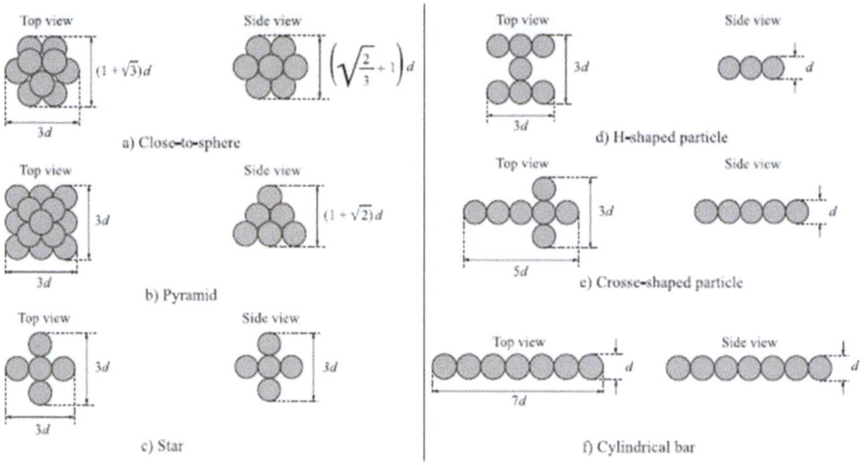

Fig. 4.26 Various typical shapes of particles, based on agglomerated spheres, used for the calculation of drag coefficients. (After Tran-Cong et al. 2004)

of involved particles is low. However, great expectations exist actually from improvements in FE 3D calculations applied to real particles (i.e. obtained from XMT, see Sect. 4.2.1.1) in addition to the involvement of the so-called morphological models based on the use of statistics (Delloro et al. 2014a; Fig. 4.27). The latter permits a significant increase of the number of particles to be simulated in the coating build-up process.

- *Morphological models* consist of a class of models which put ahead the morphology of the particles. Their development in thermal spray date back to about one and a half decades ago when applied to particles which were plasma-sprayed onto rough materials, for example, fibers (Cochelin et al. 1999). At this time, they were based on a lattice-gas automaton that reproduces the hydrodynamical behaviour of fluids. They could be then developed for the modelling of CS

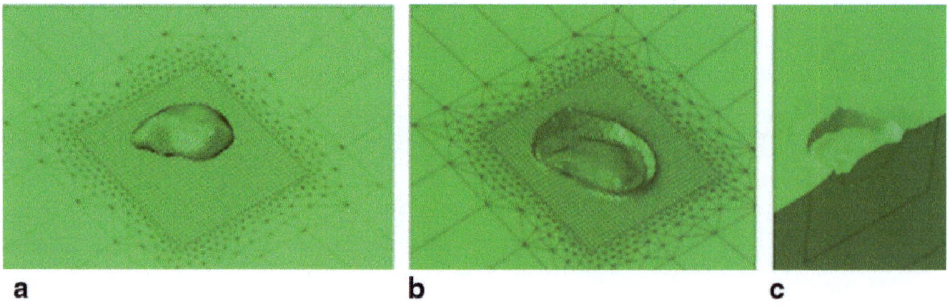

Fig. 4.27 FE 3D simulation of the impact of a real irregular particle. **a** Top view at the initial stage prior to impact. **b** and **c** At the end of the impact with a cross-sectional view (**c**). (After Delloro et al. 2014a)

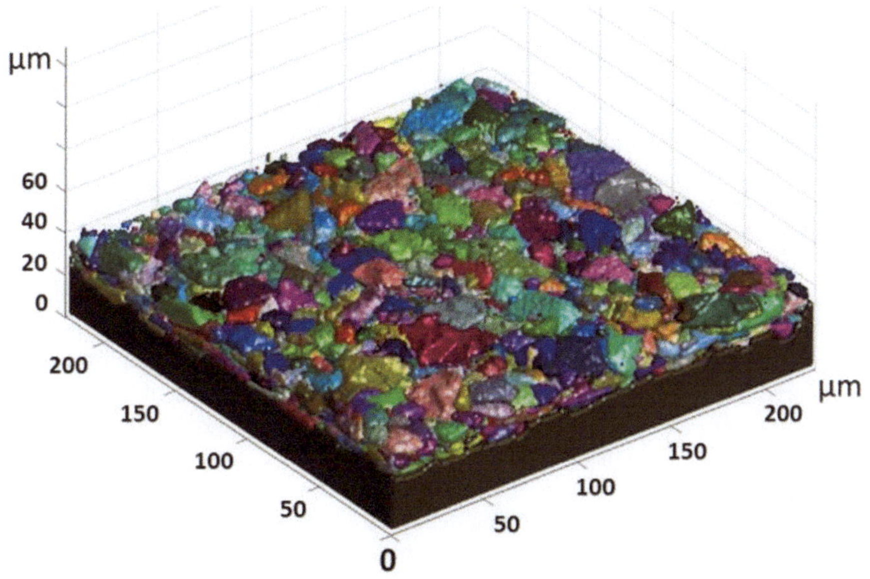

Fig. 4.28 Simulation from 3D morphological modelling of the coating build-up for cold spray of Ta. (Courtesy of Laure-Line Descurninges, MINES ParisTech 2013)

coating build-up (Delloro et al. 2014a; Jeandin et al. 2014; Fig. 4.28) with the final objective to be coupled with FE modelling to result in a general powerful model using real images (see Sect. 4.2.1.1) as numerical inputs.

- A striking advantage of this class of morphological model rests on the high number of particles which can be involved, that is, several thousand.

4.2.2 Electrical and Thermal Conductivity

Conduction is the transfer of energy, either thermal or electrical, through molecular communication within a medium or between mediums in physical contact. The transfer of thermal energy, for example, could be dominated by elastic impact as in fluids, or by free electron diffusion as in metals or phonon vibration as in insulators (Seo et al. 2012a). The thermal and electrical conductivities of metal and composite coatings deposited by CS are reported to be strictly related and directly result from the feedstock powders characteristics, the deposition process (and post-process) conditions and especially from the influence of these factors on the average coating quality in terms of microstructure and morphology (Koivuluoto et al. 2012; Stoltenhoff et al. 2006); in this sense, the behaviour and discussion of both thermal and electrical conductivities can be associated being dominated by the nature of chemical and physical bonding among atoms and grains. Regarding the materials of interest in the world of CS, up to now, the study of thermal and electrical conductivities is still restricted to copper- and aluminum-based coatings as reported, for example, in Seo et al. (2012a, b), Koivuluoto et al. (2012) and Sudharshan et al. (2007) and the reason is essentially the excellent intrinsic high conductivities of these metals which lead to their extensive use for the electrical and thermal management in the majority of industrial applications.

4.2.2.1 Conduction Properties on CS Coatings

The quality of the boundary between grain and particles is the key factor to understand and describe the behaviour of conduction properties in CS coatings. Unlike in bulk annealed materials, the microstructure consists of large grains with low-defect boundaries in metallurgical contact between each other; in the case of CS coatings, the microstructure is very different: the presence of pores, oxides, highly plastically deformed zones and high dislocation density due to the cold working is typically encountered, as well as the presence of extended nonhomogeneous particle–particle boundaries which can depend on deposited materials and process conditions. Indeed, during the coating growth, the impinging particles plastically deform and stick together with the substrate, thanks to several mechanisms such as adiabatic shear instabilities, mechanical anchorage and local micro-welding processes, so that the description and characteristics of the particle–particle boundaries network is a really complex issue and the conditions of things are that only a qualitative evaluation based on micrographic investigation can be performed. Unfortunately, the contact resistance provided by these boundaries is precisely the key factor determining conductivity properties of the deposited coatings and as a consequence the lack of tools to analytically describe these boundaries lead to the impossibility to predict and control the conduction properties of CS coatings. Some attempts have been made during the years, for example, in Sudharshan et al. (2007) a general formula according to the Matthiessen rule has been considered to describe the electrical resistivity of Al and Al–Al_2O_3 composite coatings:

$$\rho = \rho_0 + \Delta\rho_{gb} + \Delta\rho_{disl} + \Delta\rho_{por} + \Delta\rho_{fil}, \tag{4.1}$$

where ρ is the electrical resistivity of the material (coating), ρ_0 is the temperature-dependent contribution caused by the thermal vibration, $\Delta\rho_{gb}$ is the contribution of the grain boundaries, $\Delta\rho_{disl}$ is the contribution of dislocations, $\Delta\rho_{por}$ is the contribution of porosity and $\Delta\rho_{fil}$ is the contribution of the (eventual) ceramic or other filler embedded in the metal matrix. However, after an accurate evaluation of each term of Eq. (4.1), the conclusion was that the only significant contribution is the ρ_0 term or rather the intrinsic properties of the deposited material that is strongly in disagreement with the reported experimental data for cold-sprayed Al and Al–Al_2O_3 composite coatings (Sudharshan et al. 2007). At the same time, in Litovski et al. (2014), an empirical relation for apparent thermal conductivity, λ_{app}, has been formulated to describe the thermal conductivity of Al and Al–Al_2O_3 composite coatings:

$$\lambda_{app} = M\lambda_{solid} f(\text{porosity}), \tag{4.2}$$

where λ_{solid} is the thermal conductivity of the solid phase in the range of interest, *f*(porosity) is a function of total porosity with a numerical value of about 0.5 and *M* is a microstructural parameter accounting for the size of the relative contact area between particles. By using this approach, an *M* value of 0.02 has been back calculated by using the experimental results of thermal conductivities and taking into account that the original particle size was in the range of 20–30 μm. A contact area between particles of about 300 nm was estimated and confirmed experimentally by the authors, thanks to cross-sectional SEM investigations (Litovski et al. 2014). However, the variation of the defined *M* value as a function of coating material and deposition conditions are hard to determine, and the consequent validity of Eq. (4.2) must be checked each time, limiting the powerfulness of the formula.

Anyway, considering that up to now it is hard to give an analytic description about the influence of specific coating microstructural characteristics on conduction properties of cold-sprayed coatings, it is still true that some details and trends based on experimental results are evident and can help the understanding of these phenomena.

4.2.2.2 Effect of Coating Microstructure and Post-deposition Annealing

The effect of coating porosity and crystallite size on thermal conductivity of pure copper CS coatings, for example, is extensively discussed in Seo et al. (2012a, b) as a function of powder manufacturing process, spray parameters and equipment as well as post-deposition annealing conditions as summarized in Fig. 4.29. Certainly higher coating porosity leads to poorer thermal conductivities; however, the quality of the original feedstock is important, and, for example, as-sprayed coatings (indicated with mark 1 in Fig. 4.29) obtained with electrolytic powders (A) or water

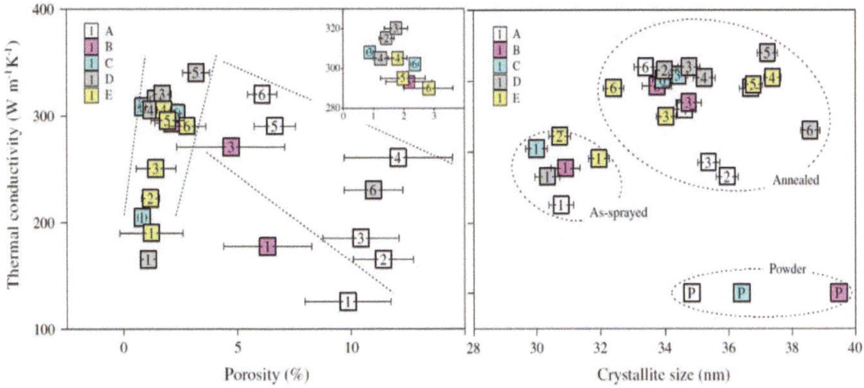

Fig. 4.29 Thermal conductivity of cold spray pure copper coatings as a function of coating porosity and crystallite size. Coatings are obtained by using feedstock powders produced by electrolysis *(A)* water atomization *(B)* and gas atomization *(C, D, E)*. Coatings have been sprayed with air at 400 °C and 0.6 MPa as carrier gas *(A, B, C)* or helium at room temperature and 0.62 MPa *(D)* and 3.0 MPa *(E)*. Samples have been heat-treated isothermally in high vacuum up to 600 °C (as-sprayed, 200, 300, 400, 500 and 600 °C) for 1 h as referred to by marks 1, 2, 3, 4, 5 and 6, respectively. (Seo et al. 2012b)

atomized (B) even if characterized by low porosity exhibit poor thermal conductivity probably due to the larger quantity of oxygen in respect to gas atomized. Moreover, certainly a post-deposition thermal annealing is effective in reducing porosity and heat-treated specimens always exhibit higher thermal conductivities in respect to corresponding as-sprayed coatings, even if the annealing must be properly tuned as a function of coating material and also deposition parameters. In this sense, higher temperature (condition 6, 600 °C) can be either detrimental, for example, in the case of helium-sprayed coatings because of a significant increase in porosity and crystallite size (D), either beneficial in the case of electrolytic powders (A) or leading to no significant effects with respect to a treatment at lower temperature (B, C, E).

Similar considerations are reported in Coddet et al. (2014) to explain the evolution of the electrical conductivity of Cu–0.5Cr–0.05Zr cold-sprayed coatings as a function of post-deposition annealing temperature. The conductivity ranged between 15.5 mS/m (i.e. 25 % International annealed copper standard (IACS)) obtained in as-sprayed coating up to 49 mS/m (i.e. 84.5 IACS) with a properly tuned post-deposition annealing. Same situation is observed for Cu–Al_2O_3 nanocomposite cold-sprayed coatings obtained by using mechanically milled nanocrystalline copper alumina powders where the coatings exhibited electrical conductivity lower than 20 mS/m in as-sprayed conditions, while up to about 50 mS/m resulted after annealing (950 °C) as reported in Sudharshan et al. (2007). The correlation between microstructure evolution and electrical conductivity is further highlighted in Koivuluoto et al. (2012), wherein the performances of high-pressure CS (HPCS) and low-pressure CS (LPCS) are compared in the deposition of Cu coatings starting from Oxygen-free high conductivity (OFHC) Cu feedstock. Figure 4.30 showed

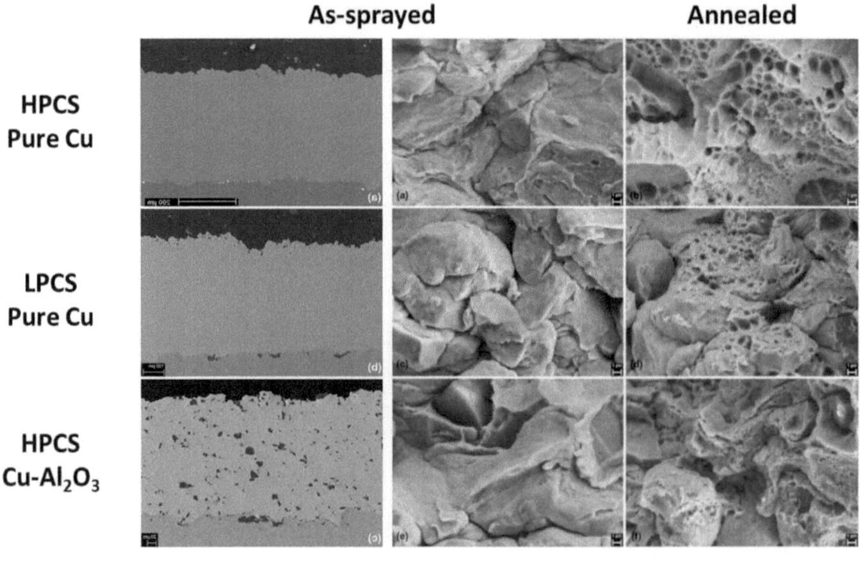

Fig. 4.30 Different cold spray 6-hydroxy-5-flucytosine Cu (OFHC) coatings morphology and fractographies (before and after annealing) obtained with different spray equipment and feedstock. *HPCS* high-pressure cold spray, *LPCS* low-pressure cold spray. (Koivuluoto et al. 2012)

the cross-sectional micrography and fractography observations of as-deposited and annealed coatings obtained with HPCS and LPCS; a further comparison with composite Cu–Al_2O_3 LPCS coating is reported. The high plastic deformation upon impact, particle flattening and related average quality of the microstructure obtained with HPCS enabled the obtainment of electrical conductivity up to 79 IACS in as-deposited conditions compared with only 46 obtained with low pressure. Even after a post-deposition annealing (400 °C—2 h) the gap is still unchanged having 90 IACS versus 69 IACS obtained, respectively, starting by high- and low-pressure cold-sprayed coatings. After annealing, the presence of dimples is a proof of the effectiveness of the treatment for the promotion of atomic diffusion and microstructure consolidation and as a consequence for the enhancement of conduction performances.

When the coating microstructure exhibited a poor particle–particle cohesion as in the case of LPCS pure copper coatings, the addition of a small percent of ceramic alumina particles in the feedstock can enhance the peening effect leading to a more significant particle deformation, porosity reduction and subsequent enhancement of the average compactness of the microstructure. This variation, despite the strongly insulating characteristics of alumina particles embedded in the coating as shown in Fig. 4.30, resulted in an average increase of the electrical conductivity of the LPCS coating (60 and 83 IACS, respectively, before and after thermal annealing) confirming once again the crucial role of coating microstructure in respect to all other parameters, still including material intrinsic characteristics, in determining the conduction properties.

As for electrical conductivity, similar considerations are reported in the evolution of thermal conductivity, for example, in the case of Cu–Cr cold-sprayed composite coatings as reported in Kikuchi et al. (2013). In particular, vacuum heat treatment at 1093 K is reported to promote a beneficial evolution of coating microstructure with the vanishing of particle–particle flattened boundaries typical of as-sprayed morphology and with the obtainment of a full recrystallization of Cu particles and crystal grain growth up to 10 μm. As a consequence of this microstructural change, an increase of thermal conductivity of about 10 % resulted.

Electrical resistivity is reported in Choi et al. (2007) to show anisotropy among in-plane and through-thickness properties in the case of as-sprayed pure Al coatings with through-thickness resistivity value typically higher in respect to in plane.

Summarizing, the coating microstructure and in particular porosity and morphology of particle–particle boundaries are reported to be the major factors influencing the conduction properties of cold-sprayed coatings even if up to now no detailed models are available to analytically describe the mechanisms. As a matter of fact, the key to control these microstructural features is fundamental to enable a significant enhancement in conduction performances of the as-sprayed coatings; in this sense, a post-deposition thermal annealing is certainly the more available, quick and, for these reasons, also investigated approach. A convincing as well as a qualitative mechanism proposed to describe the influence of thermal annealing is reported in Seo et al. (2012a) regarding the evolution of thermal conduction in cold-sprayed pure copper coatings: in the case of as-sprayed material (Fig. 4.31a), interfaces with voids and oxides between splats are formed as well-distributed porosity depending on deposition

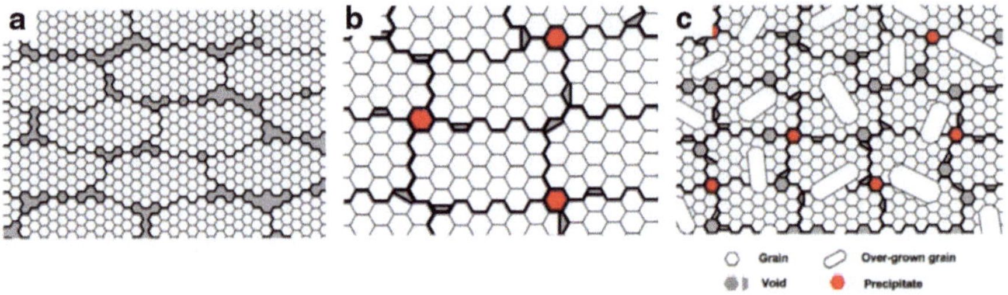

Fig. 4.31 Schematic description on optimization of annealing process. **a** Grains in as-sprayed copper coating. **b** Uniformly grown grains after optimal annealing. **c** Abnormally grown grains over optimal annealing temperature. (Seo et al. 2012a)

conditions and material characteristics (strength, critical velocity, etc.). These interfaces act as an interceptive gap in the conduction, hindering the energy transfer and being responsible for typical poor conduction properties of the as-deposited coatings. Thanks to a properly tuned annealing treatment (Fig. 4.31b), porosity coalesced, and particle–particle interfaces can progressively vanish thanks to the atomic diffusion and deep contact establishment between splats and grains. These are the optimal performances achievable; typically, they provide conduction properties still lower in respect to correspondent bulk material due to the residual presence of porosity and thicker interfaces. Annealing over the optimal conditions (as shown in Fig. 4.31c) lead to abnormal grain growth further increasing the interfaces between grains and rearrangement of voids. This is detrimental by the point of view of conduction properties as well as mechanical cohesion promoting a strong deterioration of the coating properties.

4.2.2.3 Conduction Properties: CS Versus Other Thermal Spray Techniques

Finally, it is further interesting to highlight the conduction properties of CS coatings in respect to other thermal spray technologies. The behaviour of electrical conductivity for copper coatings deposited by CS, high-velocity oxygen fuel (HVOF) and arc spraying in the as-sprayed conditions and after different post-deposition annealing is shown in Fig. 4.32. The low processing temperature of CS in respect to the other thermal spray technology is the key of its suitability to obtain denser coatings with a lower oxygen content and as a consequence with improved conduction performances. As expected and according to the previous discussion, a post-deposition annealing is beneficial for all coatings and considering the specific conditions explored in this study, a higher annealing temperature leads to a progressively significant improvement of the electrical conductivity. The gap between CS coatings and other thermal spray coatings is still preserved confirming the excellence of CS in the deposition of pure metal coatings. On the other hand, all coatings, even after post-deposition annealing, exhibit conduction properties lower in respect to bulk Cu even if it must be noticed that a properly tuned CS process and post-deposition treatment allow the achievement of performances very close to the correspondent bulk material.

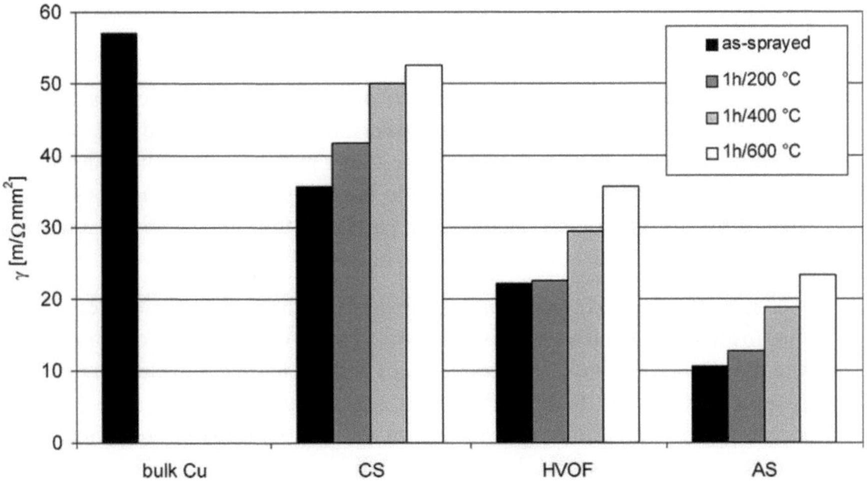

Fig. 4.32 Electrical conductivity of copper coatings deposited by cold spray *(CS)*, high-velocity oxygen fuel *(HVOF)* and arc spraying *(AS)* in the as-sprayed conditions and after different annealing. Annealed bulk Cu data are reported as reference material. (Stoltenhoff et al. 2006)

4.3 Mechanical Properties

The mechanical resistance and structural integrity of thermal spray and CS coatings are generally one of the main issues for several industrial applications. For example, CS is often applied in maintenance, repair and overhaul (MRO) in aeronautic, military and automotive industries (Champagne and Helfritch 2014; Jones et al. 2011) for both refurbishment and structural recovery, where mandatory restrictions on coating adhesion to the base materials and cohesion strength are present. Furthermore, hard metals and cermet are employed as wear-resistant coatings where specific characteristics in terms of surface properties, such as hardness, scratch resistance and specific wear resistance, are required. In the panorama depicted by thermal spray, the coatings obtained by CS offer a significantly different picture regarding coating morphology, microstructure and mechanical properties, due to the low temperature and unique solid-state growth mechanism. The severe plastic deformation during impact and growth and the consequent cold working of deposited coatings, as previously discussed in this book, lead, for example, to a significant compressive residual stresses state along the through thickness of the coatings up to the first layer of the substrate (Shayegan et al. 2014) opening, for example, new opportunities to control and enhance the fatigue behaviour. On the other hand, the typically high stiffness and low elongation properties exhibited by cold-sprayed coatings still represent a restriction for many structural applications (Jones et al. 2011).

This section aims to give a survey on mechanical properties of CS coatings; it has the ambition to represent a useful summary for a CS expert as well as a tool to guide students and industrial end users to a rapid understanding of the process characteristics and potentiality for specific industrial applications.

4.3.1 Hardness

Hardness is defined as the resistance of a material to indentation (Rösler et al. 2007); in a coherent manner, indentation is the universally most employed and diffused experimental method to determine the hardness of a material (Revankar 2000). According to ASM international definitions, the hardness (indentation) tests may be classified using various criteria, including in particular the (1) type of measurement and (2) magnitude of indentation load (Kuhn and Medlin 2004).

1. Concerning the *Type of Measurement,* the more diffused classification criterium distinguishes between the hardness evaluated by the measurement of dimensions of the indentation (Brinell, Vickers, Knoop) and the hardness evaluated by measuring the depth of indentation (Rockwell, nanoindentation).
2. Concerning the *Magnitude of Indentation Load,* it is possible to define three different classes: macrohardness, microhardness, and nanohardness tests. For macrohardness tests, indentation loads are 1 kgf or greater: Rockwell test (max 150 kgf) and Brinell (max 3000 kgf) tests are generally the most diffused and employed. The microhardness tests (Vickers and Knoop in particular) use smaller loads ranging from 1 gf to 1 kgf, the most common being 25–500 gf. The nanoindentation test, also called the instrumented indentation test, depends on the simultaneous measurement of the load and depth of indentation produced by loads that may be as small as 0.1 mN, with depth measurements in the 20 nm range. Berkovich penetrators are used in these tests (Revankar 2000).

Since the nineteenth century, indentation is carried out on minerals and bulk materials to determine macro hardness (DIN 50359-1 1997) In this case, the volume of material interested by plastic deformation upon indentation and the related indented area are so significant with respect to the material microstructure and (eventual) phase distribution that commonly it is allowed to consider the resulting hardness as a representative average behaviour of the indented material. On the contrary, the hardness determination in thermal spray (TS) and CS coatings is a slightly more slipping field: the low coating thickness avoids the possibility to perform indentation on the surface at high load or penetration depth without including the influence of the substrate (a general rule of thumb suggests that the penetration depth should be no more than the 10 or 20% of the whole coating thickness in the case of hard coating on soft substrate and soft coating on hard substrate, respectively; Fischer-Cripps 2000); indeed, the tests are generally performed on cross-sectioned and polished coatings and *micro*indentation loads typically range between 25 and 500 gf depending on whole coating thickness and specific characteristics following, for example, the guidelines reported in American Society for Testing and Materials (ASTM) B933-04. The indented area is reduced progressively with indentation load enhancing the hardness reliance on local microstructure, phase distribution and composition. In addition, due to a coating build-up process significantly out of the thermodynamic equilibrium, the use of composite, agglomerated powder feedstock (i.e. agglomerated carbides such as WC–Co, WC–Ni; Ortner et al. 2014) powder blends (Sevillano et al. 2013) or coated powders as well to the presence of specific microstructural features as splats (especially in plasma spray; Pawloski 2008), voids, cold working and strained and fine grains (especially in CS; Papyrin et al. 2007) quite complex and nonhomogenous microstructures are generally exhibited by TS and CS coatings with respect to traditional bulk materials further enhancing local variation of the hardness results and data scattering.

4.3.1.1 Hardness of CS Coatings

In this scenario, *micro*-hardness of CS coatings is extensively investigated and reviewed in the literature for a wide range of materials (Luo et al. 2014a, b). The microhardness of a metal-based CS coating typically balanced a positive contribution coming from the high particle deformation upon impact and the related cold working that induces an enhancement of coating microhardness in respect to correspondent bulk materials; with a negative contribution coming from the presence of pores and defects that induce a reduction of coating microhardness due to the lack of cohesive strength at the particle–particle boundaries. For these reasons, the ductile materials able to achieve high plastic deformation at low temperature exhibit the highest hardness enhancement with respect to the property of corresponding bulk material. Moreover, the process parameters, able to induce an increase of particle plastic deformation during the coating growth (i.e. carrier gas pressure) and an increase in the final coating compactness, are the ones allowing the achievement of the higher coating microhardness. This behaviour can be evidenced for the deposition of a ductile metal such as pure silver as reported in Chavan et al. (2013) and shown in Fig. 4.33; the experimental trends of microhardness evolution (Vickers penetrator, 100 gf indentation load) as a function of carrier gas temperature and pressure in the range of 250–450 °C and 1.0–2.0 MPa, respectively, (related particle velocity up to 480 m/s) are reported highlighting the beneficial effect of both gas temperature and pressure as beneficial contributions to the increase of particle velocity. Moreover,

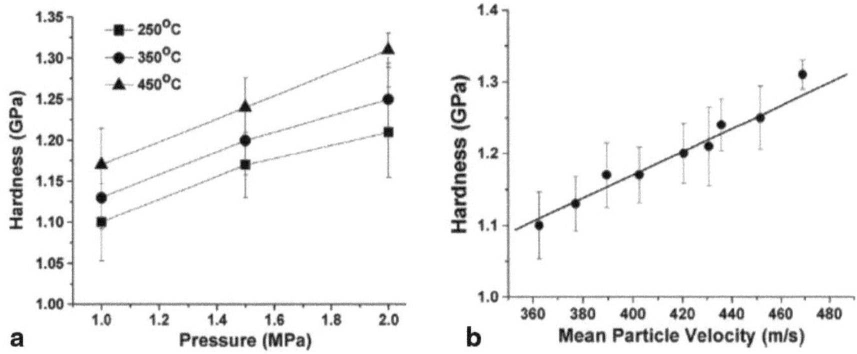

Fig. 4.33 Behaviour of cold spray silver coating microhardness as a function of **a** carrier gas temperature and pressure and **b** corresponding mean particle velocity according to Chavan et al. (2013)

the mean coating hardness is more than triple the hardness of annealed bulk silver (25 Vickers), thanks to the severe cold working of the ductile particle upon impact.

A similar trend is observed in pure copper coatings: The deposition of pure copper coating has also a historical role considering that the whole CS process understanding is essentially based on the deposition of this material; coating microhardness as a function of process parameters are extensively discussed in the case of both high pressure (Stoltenhoff et al. 2001; Schmidt et al. 2009) and low pressure (Papyrin et al. 2007). Luo et al. (2014a, b) recently reviewed the hardness of metal coatings, mainly pure metals, emphasizing the differences among corresponding annealed bulk material, spray powders and as-sprayed coatings as reported in Table 4.1. According to this review, the hardness of copper coatings can be increased by a factor of about 3 as compared to the annealed coarse-grain bulk. A hardness increase by a factor of 2 and 3 was observed for Ni coating compared with initial powder in micrometric grains and annealed Ni bulk, respectively (Ajdelsztajn et al. 2006). Moreover, different materials exhibit different degrees of in situ hardening in comparison to their corresponding starting powders. For pure Ti coatings, a relative reduced hardness increase of ~13 % is reported for irregular-shaped powders and 39 % for spherical powders (Goldbaum et al. 2011), while a remarkable hardness increase of around 140 % for Ta coatings was reported (Koivuluoto and Vuoristo 2010a; Koivuluoto et al. 2010b).

Vickers microhardness of commercially pure titanium coatings deposited by using different spray conditions and equipment have been recently reviewed by Hussain (2013). Typical microhardness of a commercially pure grade 1 bulk titanium is about 145 kgf/mm^2; that of a gas atomized spherical titanium powder is about 141 kgf/mm^2 (Wong et al. 2010), while cold-sprayed coating hardness has been reported ranging between 150 and 320 kgf/mm^2 emphasizing how feedstock powders, cold working, coating microstructure and porosity can influence significantly the coating microhardness.

The effect of gas pressure on microhardness of cold-sprayed CP–Al coatings is reported in Lee et al. (2008) who sprayed pure Al with nitrogen as carrier gas and reported coating microhardness ranging from 42 to 55 Vickers for coatings deposited, respectively, at 0.7 and 2.5 MPa carrier gas pressure. Similarly, to other ductile pure metal coatings, the hardness is more than triple of the corresponding annealed bulk material (15 Vickers) confirming the important contribution of particle plastic deformation upon impact. The mechanical characteristics and microhardness of different aluminum alloys coatings such as, for example, A2024, A7075, A6082, A6061, A5083 deposited with both high- and low-pressure CS are extensively investigated and reported by many authors (Stoltenhoff and Zimmermann (2009); Ghelichi et al. 2012; Rech et al. 2011; Ziemann et al. 2014). The behaviour of hardness in alloys and especially precipitation hardened alloys strongly depend on the thermal history of the coating: Generally, gas-atomized powders are employed in CS, thanks to their spherical shape and size homogeneity; gas atomization involves a fast cooling process of the processed material leading to a not-controlled precipitate distribution. For this reason, typical hardness of gas-atomized powders of precipitation hardened alloys are lower than the corresponding thermally treated

Table 4.1 Summary of cold spray coating microhardness for several pure metals and alloys

Material	Hardness (bulk)	Hardness (as- deposited coating)	Reference
Pure metals			
Ti	97 HV	4.0±0.3 GPa	Li et al. (2003)
		2.76±0.13 GPa	Ajaja et al. (2011)
Ta	87 HV	2.73±0.21 GPa	Koivuluoto and Vuoristo (2010a), Koivuluoto et al. (2010b)
		230 HV	Koivuluoto and Vuoristo (2010a), Koivuluoto et al. (2010b)
Cu	40 HV	150 HV	Borchers et al. (2005)
		105–145 HV	Koivuluoto et al. (2012)
		73–118 HV	Venkatesh et al. (2011)
Ag	0.2 GPa	1.3 GPa	Chavan et al. (2013)
Ni	80 HV	197±21 $HV_{0.3}$	Bae et al. (2010)
Zn	20 HV	50–75 $HV_{0.2}$	Li et al. (2010)
Al		45–55 HV	Rech et al. (2009)
Alloys			
A1100	80 $HV_{0.05}$	115–257 $HV_{0.05}$	Balani et al. (2005a, b)
A2024			
A2224		140–150 HV	Stoltenhoff and Zimmermann (2009)
A2618		3.75 MPa	Jodoin et al. (2006)
Nc-A2618		4.41 MPa	Jodoin et al. (2006)
Nc-A5083	–	261 $HV_{0.3}$	Ajdelsztajn et al. 2005
A6061		90–110 $HV_{0.01}$	Rech et al. (2014)
A6082		70 HV	Moridi et al. (2014a, b)
A7075		142 HV	Stoltenhoff and Zimmermann (2009)
A7075		120–140 HV	Ghelichi et al. (2014a, b)
Nc-A7075		130–170 HV	Ghelichi et al. (2014a, (b)
Cu–4Cr–2Nb		157 $HV_{0.2}$	Yu et al. (2011)
Cu–1Cr–0.1Zr		165 $HV_{0.5}$	Vezzu et al. (2015)
Cu–8Sn		167 $HV_{0.2}$	Guo et al. (2007)
AISI304 SS	200 $HV_{0.2}$	345±18 $HV_{0.2}$	Meng et al. (2011a)
AISI316 SS	2.11 GPa	2.92 GPa	Sundararajan et al. (2009)
Stellite 6		682 $HV_{0.1}$	Cinca and Guilemany (2013) and Cinca et al. (2013a, b)
In 625		5.7 MPa	Poza et al. (2014)
In 718		423–516 HV	Levasseur et al. (2012)
Waspaloy		538–579 $HV_{0.025}$	Vezzu et al. (2014)

materials (Ashgriz et al. 2011). During deposition, the cold work hardening due to the high-velocity impact is opposed to the hardness reduction due to a not-optimized coating microstructure. This is true in the case of A2024 as-deposited coatings with respect to A2024-T4 bulk material or A7075 as-deposited coatings with respect to A7075-T6 bulk material (Stoltenhoff and Zimmermann (2009)) just to have two representative examples. It must be highlighted that the mechanical behaviour of CS coatings and heat-treated bulk materials is very different; even if the average hardness can be similar, its origin is completely diverse, while in thermal-treated materials the hardness is due to a precipitation hardening process, and in as-deposited CS coatings the hardness is only the effect of cold working and strain hardening. In this sense, pointing the attention on microhardness, the CS deposition is able to induce a significant increase in microhardness with respect to the initial powder value even if this enhancement is often not enough to balance the values achieved in thermally treated alloys.

The effect of using pure helium or helium/nitrogen mixtures, rather than nitrogen as a gas carrier, on coating microhardness has been studied in Balani et al. (2005b) on as-deposited A1100 alloy coatings revealing, as expected, the beneficial effect of helium on obtaining more compact and hard coatings, thanks to its higher sonic velocity with respect to nitrogen (or air), leading to a better performing CS process as a whole, increasing process efficiency, general coating quality, microstructure and mechanical properties. However, the extreme cost/benefit balance has led to a continuous replacement of helium with nitrogen as discussed in this book.

Hardness has been reported to increase with the particle velocity or rather with carrier gas temperature and pressure also in the case of AISI304 stainless steel coatings in Meng et al. (2011a) the coatings have been sprayed with Kinetik-3000 deposition using nitrogen at 3.0 MPa in the temperature range 450–550 °C. Starting from gas-atomized powders with hardness of 171 Vickers (50-g indentation load), a coating microhardness (200-g indentation load) up to 267 Vickers is obtained. Villa et al. (2013) deeply investigate the microhardness of AISI316 stainless steel as a function of spray parameters confirming that the optimized coating hardness, up to 358 Vickers, is roughly the double of the initial particle hardness. This study further emphasizes the effect of local coating microstructure on hardness by performing several hardness maps by using nanoindentation and confirming the significant detrimental influence of porosity, defects and particle borders on local hardness. The Vickers microhardness (300-g indentation load) of Cu, Ni and Zn coatings deposited by low-pressure CS on both Cu and steel substrates is reported in Koivuluoto et al. (2008a, b) resulting, respectively, 105 (Cu), 120 (Ni) and 57 (Zn) Vickers. The microhardness of CS coatings of Ni and Co superalloys is also reported by many authors, for example, in the case of Waspaloy (Vezzu et al. 2014), Stellite (Cinca and Guilemany 2013), Inconel625 (Poza et al. 2014) and Inconel718 (Levasseur et al. 2012).

The effect of standoff distance on coating microhardness is studied in Li et al. (2006) in the case of pure metals, Cu, Al and Ti, resulting that despite the decrease of the deposition efficiency (DE), the coating microhardness is essentially not influenced by the standoff distance in the range of 10–110 mm.

Post-deposition thermal treatments are often performed on CS coatings in order to compact the microstructure, promote the metallurgical bonding at the particle–particle interface hence increasing tensile properties. The annealing is always reported to induce a reduction in coating microhardness due to the relaxation of peening stress and cold working (Meng et al. 2011b; Levasseur et al. 2012; Coddet et al. 2014; Bu et al. 2012a).

Despite the large amount of CS parameters and other process conditions, such as, for example, feedstock characteristics, realization of pre- and/or post-deposition treatments on both powders and coatings; some trends can be pointed out to fix some useful even if approximate rule of thumbs. For example, generally the coating microhardness increases with the extent of particle plastic deformation upon impact. For this reason higher is the particle velocity higher is the resulting coating microhardness. In this sense, particle morphology can play a significant role in determining coating microhardness; indeed, irregular particles can reach higher in-flight velocity, thanks to the more effective drag coefficient with respect to spherical particles. At the same time, the use of low hardness powder feedstock, for example, obtained by realizing a thermal annealing on feedstock powders (Li et al. 2013; Ko et al. 2014) or by using dendritic feedstock produced by electrochemical processes can generally enhance the plastic deformation capability of particles upon impact leading to higher strain rate during the coating growth and as a consequence higher enhancement of microhardness in respect to particle hardness. Wong et al. (2013) investigated these effects in the case of titanium CS coatings by using different powders feedstock and spray parameters and summarize their results in the behaviour of coating microhardness as a function of the particle velocity/critical velocity ratio as shown in Fig. 4.34. A slight progressive increase of coating microhardness is reported, increasing the ratio between particle velocity and critical velocity. Moreover, higher coating microhardness can be obtained starting from softer and

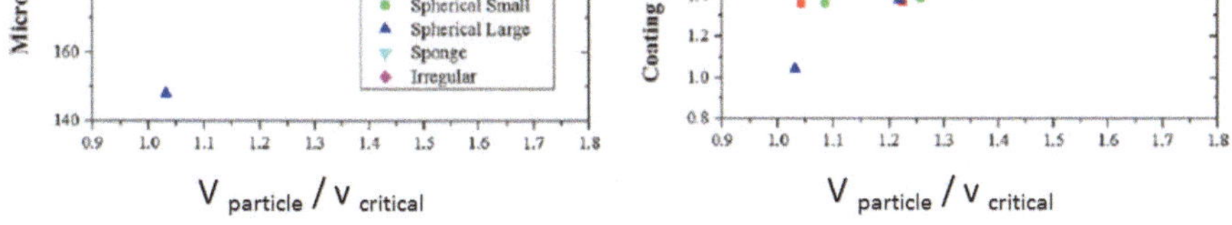

Fig. 4.34 Behaviour of coating microhardness and coating microhardness/powder microhardness ratio for cold spray Ti coatings deposited by using different feedstock and process parameters according to Wong et al. (2013)

irregular feedstock as in the case of sponge or irregular titanium with respect to spherical gas atomized.

The low temperature and solid-state coating growth mechanism allows to take advantage of using temperature-sensitive feedstock. For example, fine composite or agglomerated powders or even nanostructured powders are able to exhibit superior hardness with respect to traditional powder feedstock; however, this hardness is strongly affected by a temperature rise leading to the impossibility to profit of this property when processed by traditional thermal spray techniques. In this field, CS is reported to be really effective (Kim et al. 2005; Jodoin et al. 2006). For example, ball milling of alloys A5083 powder under liquid nitrogen is reported to achieve a nanocrystalline grain size in the range of 20–30 nm and the nanocrystalline grain structure of the cryomilled feedstock powder was proved to be retained after the CS process (Ajdelsztajn et al. 2005). The resulting microhardness enhancement is significant, from 104–261 HV(300 g), comparing the nanocrystalline coating with cast, cold worked, A5083. Generally, the superior hardness and stiffness of nanostructured powders lead to a lower particle plastic deformation upon impact, and for these reasons, the as-sprayed CS coatings have more porosity in comparison to the ones obtained with traditional gas-atomized powder. In this sense, the presence of porosity and microstructural defects is detrimental for both structural properties and microhardness so that the final coating behaviour is the result of a beneficial effect of superior properties of initial powder and a negative effect of reduced compactness of the microstructure and the effectiveness of using nanostructured powders need to be evaluated case by case.

Depth-Sensing Indentation

Depth-sensing indentation or instrumented indentation consists of a traditional indentation test in which the applied normal load and the displacement are continuously detected and collected during the test, resulting in a loading and unloading indentation curves. For both curves, the instrumented indentation can produce an accurate and complete sampling of the load (L) versus penetration depth (h; Fischer-Cripps 2005, 2011). The current main application of depth-sensing indentation is upon low load condition when the size of the indent is too small to be observed and detected by optical microscopy as in microindentation and when the mechanical behaviour of the indented material cannot be considered fully plastic due to the significant contribution of the elastic recovery. This is the case of nanoindentation where the indentation load typically ranges between 0.1 mN up to 0.5 N and the penetration depth ranges between a few tens of nanometers up to some microns. Nanoindentation is widely used in coating technology and surface engineering, and up to now, it is increasingly being considered due to the powerful new instrumentations and its wide diffusion. The measured indentation curve is a function of the mechanical properties of the tested specimen; therefore, if an inverse analysis method can be found, the mechanical properties of the tested specimen can be predicted from the measured indentation curve. Today, this is generally performed by using

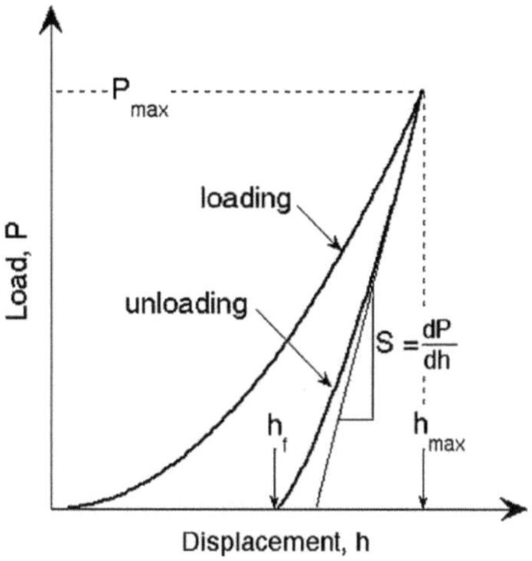

Fig. 4.35 Schematic illustration of indentation load-displacement curve showing important measured paramters. (Oliver and Pharr 2004)

the Oliver and Pharr theoretical model and method that has been developed in 1994 (Oliver and Pharr); a schematic load-displacement indentation curve is shown in Fig. 4.35 for an elastic–plastic specimen. Upon loading, there is an initial elastic response followed by elastic–plastic deformation. Load is increased up to his maximum value, P_{max}, with the correspondent depth, h_{max}. The test can be either a load-controlled or depth-controlled setting, respectively, a maximum indentation load (depth will be determined as a consequence) or maximum penetration depth (load will be determined as a consequence). Once the maximum load (or depth) has been reached, the load is optionally kept constant for a dwell time, and after that removed progressively leading to the unloading curve. Upon unloading, there is first the elastic recovery wherein the *dP/dh* behaviour, *S,* is rather constant, followed by elastic–plastic deformation as schematically shown in Fig. 4.31. Finally, upon complete unload, there is a residual impression of depth h_r employed to estimate the material's hardness. An estimation of reduced elastic modulus, E_r, defined as $E_r = E/1-v$ where *E* is the elastic modulus of the indented material and *v* its Poisson coefficient, is obtained from the slope *S* (approximation of linear behaviour) or fitting with a quadratic function the first part of the unloading curve.

In the practice, some types of discontinuities can be encountered in load-displacement curves, and the more observed are pop-in and pop-out events which are sudden displacement excursions into the target materials during load-controlled experiments. Pop in is observed in the loading curve while pop out in the unloading as shown schematically in Fig. (4.36a, b). Both pop-in and pop-out events are generally associated with dislocation nucleation and movement, phase transformations and crack nucleation and propagation (pop in especially) in bulk defect-free materials (Fischer-Cripps 2011). However, talking about materials coming from P/M and thermal spray coatings pop-in events can be considered a qualitative index of coating cohesion and particle–particle bond strength; indeed, in presence of porosity or no compact microstructure, the indentation can often induce a collapse of the material highlighted as a sudden displacement excursion in the loading curve; as these

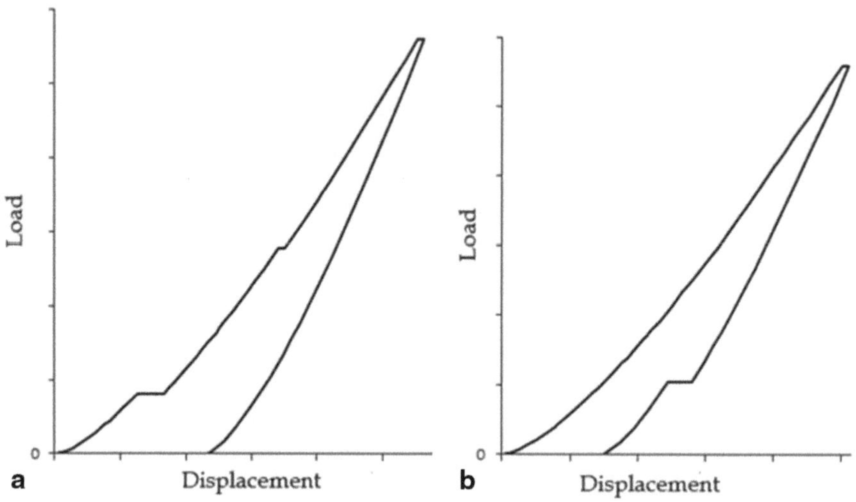

Fig. 4.36 Most encountered discontinuities in load-displacement curves **a** pop-in event and **b** pop-out event

events are more frequently observed and appears more pronounced, they reveal a lack in microstructure compactness and particle–particle cohesion.

The possibility to vary the loading rate during the indentation can be used experimentally to induce different amounts of work hardening upon indentation. Generally, this phenomenon can be avoided or eventually must be considered to properly fit the curve and produce indentation results and for this reason is normal procedure to set the loading rate as high as necessary to avoid the development of work hardening; however, on the other hand, this can be also used to deeply investigate the work-hardening effect as reported, for example, in Kim et al. (2010) where a prediction of the work-hardening exponents of metallic materials has been performed. This has been obtained by means of atomic force microscopy observations of residual indentation impressions in sharp indentation (Kim et al. 2010). These methods are not yet exported and applied in the characterization of thermal spray coatings, even if the amount of work hardening plays a crucial role, especially in CS deposition both by giving information about the particle deformation upon impact and by having important correlation with coating properties, so that a progressive development of these procedures in the near future can be expected.

Depth-Sensing Indentation on CS Coatings

Instrumented indentation and nanoindentation is also useful and fruitful in mechanical characterization of local features in thermal spray and CS coatings. The low size of the indents enhances the spatial resolution of this mechanical investigation opening the opportunity to emphasize local differences, highlight nanostructuring effect or look for correlation between coating microstructure and mechanical properties, especially hardness. Furthermore, the method can be useful for the investigation of composite and multimaterial coatings. For example in Yan et al. (2012), the conditions of applying the Olver–Pharr method to the nanoindentation of particles in composites is investigated, and the limits in which the accuracy of the measure is acceptable in the case of both soft particles in stiff matrix and stiff particles in soft matrix are defined. Berkovich nanoindentation test have been performed in Bae et al. (2012) and Zou et al. (2010) on nickel cold-sprayed coatings, and both highlighted inhomogeneity of the nanohardness values within the particles. The hardness in the vicinity of Ni particle interfaces is higher than that in the particle interior, and this difference is attributed to the CS-induced grain boundaries and dislocation densities. Wang et al. (2013) carried out nanoindentation in order to investigate the effect of localized deformation on mechanical properties of aluminum particles in composite Al–Al_2O_3 CS coatings. The effect of the indentation load on the hardness measurements on CS Ti coatings, referring to the Nix–Gao model is reported in Goldbaum et al. (2011) and Ajaja et al. (2011). This model accounted the effect of the penetration depth on the hardness measurement, and it is based on a consideration of strain-gradient plasticity. The true hardness, H_0, or rather the hardness at infinite depth, is related to the measured hardness H and the penetration depth, h, and a characteristic length scale, h^*, following $H = H_0(1 + h^*/h)^{1/2}$. The true hardness can be obtained by fitting a set of indentation data obtained at different depths as reported in the case of CS Ti coatings in Ajaja et al. (2011). True hardness is reported to be higher with respect to bulk hardness and strictly related to the porosity and presence of defects in the coating. Tantalum cold-sprayed coatings were studied by depth-sensing Berkovich indentation testing in Bolelli et al. (2010). The mechanical properties of the coatings were found to be free of any scale dependence, insensitive to the presence of a lamellar structure, indicating strong, tight bonding between cold-sprayed Ta particles. Again, due to the low size of the indent, depth-sensing indentation represents a useful tool for the realization of high-resolution depth profiles. For example, Poza et al. (2014) reported the evolution of hardness and elastic modulus across a laser remelted track of an Inconel625 coating deposited by high-pressure CS, while in Liang et al. (2011) nanoindentation investigation is performed to distinguish hardness of different areas of Co-based cold-sprayed coatings.

Depth-Sensing Indentation on Feedstock Powders

Due to the reduced size of the indent, nanoindentation can also be performed on feedstock powders. Different techniques of sample preparation for magnetic and nonmagnetic abrasive particles, taking into account the deformation of the embedding medium, are reported in Shorey et al. (2001). It is important in the presence of an embedding matrix to consider that a particle could be pushed into the embedding medium under the influence of the indenting load rather than plastically deform itself. In that case, there is an overestimation of the penetration and residual depth and as a consequence an underestimation of the hardness. Hryha et al. (2009) also discussed the influence of the stiffness of the embedding resin on the indentation hardness and modulus with a special focus on metal powders.

Depth-Sensing Indentation with Spherical Indenter

Depth-sensing indentation performed with spherical indenter can also be employed to have an estimation of the strength of materials. Assuming plastic hardening of metals as described by the Hollomon stress–strain curve in the form:

$$\sigma = \sigma_0 + k\varepsilon_{\mathrm{p}}^{m}, \tag{4.3}$$

where σ_0, k and m are the material parameters and can be identified from spherical indentation tests by measuring compliance moduli in loading and unloading of the load–penetration curve. Several empirical relationships have been proposed for directly correlating hardness with yield and tensile strengths as reported, for example, in Tabor (1951), Shabel et al. (1987), and Fischer-Cripps (2000) and in this sense, the introduction of the instrumented indentation testing machines made many details of the indentation process to be available and stimulated a great effort to develop more refined procedures (Au et al. 1980; Nayebi et al. 2001; Taljat et al. 1998) for getting more accurate estimates of the elastic–plastic properties. The load-displacement curve depends on several physical properties of the tests, but it is mainly affected by the uniaxial stress–strain (σ–ε) curve of the sample material. Some authors as reported, for example, in Fischer-Cripps (1997) used the slope of the load-displacement curve produced during loading to estimate plastic flow properties and deduced the Young modulus by the slope during unloading (Huber et al. 1997; Nayebi et al. 2002). Beghini et al. (2002, 2006), performed an extensive parametrical finite element analysis of the spherical indentation in order to study the dependence of the crater shape to the yield stress and strain hardening and proposed a direct method for deducing the σ–ε curve of a material from load-displacement curves. An approach based on the direct correlation between the load-displacement curve and σ–ε curve was also applied in Nayebi et al. (2001) for characterizing surface structurally graded materials.

Concerning the application of these concepts and models on coatings and CS coatings in particular, it must be stated preliminarily that all estimations are accurate as much as the coating properties and mechanical behaviour approach the properties of the correspondent bulk annealed material. This is generally false due to the presence of porosity and defects, splats, interparticle debonding and other specific microstructural features of CS coatings; however, in the case of pure metal coatings with ductile behaviour, these differences are almost levelled leading to the opportunity to obtain quite accurate results. For example, in Bolelli et al. (2010), the spherical indentation tests are applied according to the multiple partial unloading method by Field and Swain ((1995) on a cold-sprayed tantalum coating behaving as a bulk material, in order to obtain stress–strain curve and elastic modulus with very promising results. One limitation of the analysis is its inability to provide a direct quantification of the yield strength (YS) of the material, because the experimental data points obtained by spherical indentation depart significantly from the elastic regime. So that, these models and procedures can be very useful to predict tensile properties with a direct, quick and nondestructive technique; however, the necessary assumptions of *ideal* bulk material behaviour is mainly too radical in the case of thermal spray and CS coatings hiding that it is risky to obtain poor accuracy results.

4.4 Strength and Elastic Properties

The assessment of strength and elastic properties represent a fundamental issue to describe the mechanical behaviour of a coating material. Generally, tensile tests are used to obtain a stress–strain curve from which these properties can be deduced. A schematic stress–strain curve is shown in Fig. 4.37 in the case of ductile and brittle material; some critical parameters are defined to describe quantitatively the mechanical behaviour of the material: The critical stress to produce appreciable (0.2 % in most case is considered) plastic deformation (YSσ_y); the fracture stress (σ_f) or the maximum stress beard by the material until breaking (ultimate tensile strength, UTS), respectively, for brittle and ductile materials. Further, information concerns the elastic behaviour (Young or Elastic Modulus, E) according to Hooke's law as well as plastic behaviour in terms of ductility (how a material can deform before fracture), resilience (the capacity of a material to absorb energy when it is deformed elastically) and toughness (the energy required to cause fracture) (Rösler 2007).

Tensile tests are widely used to select material for structural and engineering applications, also the test may be used to compare different materials under loads characterizing the main mechanical properties and the quality (Davis 2004). Typical tensile specimen (Fig. 4.38) has enlarged extremity for the locking to the tensile machine grips and a gage region with more restricted section where the deformation and break can take place. The two regions are properly connected and dimensioned to avoid any loads effects outside the gage length. Several standard tests, ASTM and DIN EN, describing specimen type and procedures are available with specimen shape ranging from cylindrical to flat and size from few millimetres up to many centimetres (Fig. 4.34).

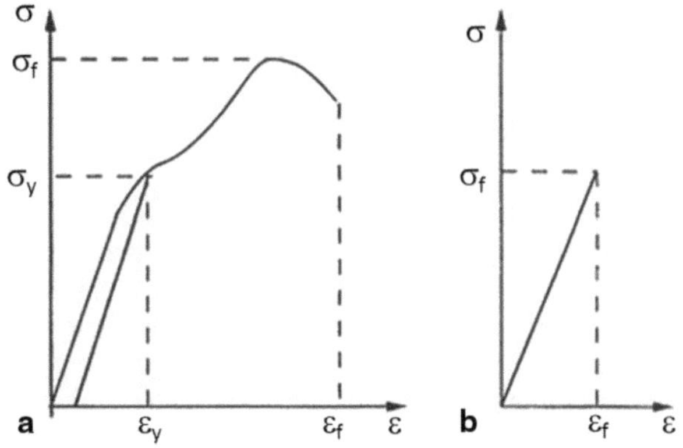

Fig. 4.37 Typical stress–strain curve in the case of **a** ductile material and **b** brittle material

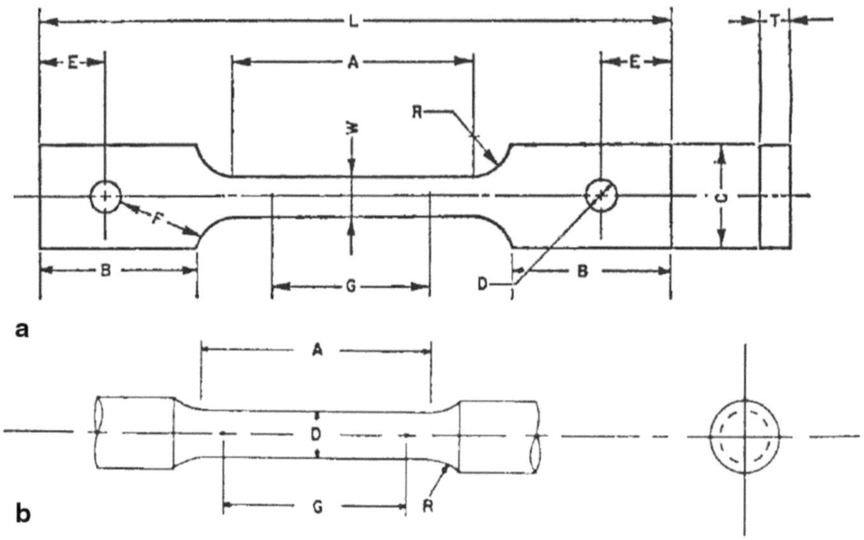

Fig. 4.38 Tensile test specimens specification according to ASTM E8-04. **a** Flat and **b** cylindrical specimens

4.4.1 Tensile Testing on Thick Coatings

Among the procedures defined across the centuries to mechanically characterize materials and metallic materials in particular, there are a few that are extended and customized for coating and coated specimens as well. First, it must be distinguished if the objective of the investigation is (1) the mechanical behaviour of the coating material itself or (2) the mechanical behaviour of the coated system. (1) The more employed and accurate way to assess tensile test and obtain stress–strain curve is by the micro-flat tensile (MFT) test. (2) Some specific procedures have been developed to investigate the mechanical behaviour of the whole-coated system by either obtaining a full stress–strain curve or only have ultimate strength such as 3- and 4-point bending test, tubular coating tensile (TCT) or the use of notched dog-bone specimens.

4.4.1.1 MFT Test

MFT test performed following the guideline of ASTM E8-04 is commonly used in powder metallurgy products to obtain an estimation of the intrinsic properties of deposited (compacted) material. The preparation of the specimen can be a problematic issue requiring the production of a several millimetre thick coating, removal of the coating from the original substrate and properly machining to obtain a freestanding specimen totally composed of the deposited material with shape and size according to the ASTM procedure. A CS-deposited coating can behave like a brittle material in particular when depositing high-strength and low-ductility materials such as hard steel, Ni and Co superalloys. In that case, diffused porosity is located at the particle–particle boundaries leading to weak cohesive strength and plastic deformation capabilities promoting crack propagation and brittle fracture (Vezzu et al. 2014; Levasseur et al. 2012). For these reasons, the MFT test is generally limited to investigate the behaviour of as-deposited coating of ductile materials, or on the other hand, it is devoted to investigate the evolution of cohesive strength when post-deposition annealing treatments are performed on as-deposited coatings as reported, for example, in Yu et al. (2011). This is of particular interest when CS has the ambition to be employed in structural application (Jones et al. 2014) or as a rapid manufacturing technique (Ajdelsztajn et al. 2005; Sova et al. 2013b).

4.4.1.2 TCT Test

Schmidt et al. developed a customized tensile test specifically designed for thermal spray and CS coatings: the TCT (Schmidt et al 2006a, b). The TCT test can be used to determine cohesive strength and have an estimation on UTS of the coating material. The test is not yet classified with an official norm even if its use is rather diffused and consolidated in thermal spray and CS community. The specimen preparation and testing procedure is schematically shown in Fig. 4.39: A pair of cylinders is coupled and joined by an inner screw; the coating is deposited along the external surface of the specimen so that once the inner screw is removed, the coating is the only support of the two-parts specimen. The cylinders are then gripped by screws to the universal test machine and subjected to tensile load until the coating failure occurs. The main advantage of using TCT test is certainly the quickness with respect to MFT; however, it has to be mentioned that the geometrical design of the two-coated substrates leads to a stress concentration in the pulled coating. This stress concentration increases the Mises stress at the gap between the substrates to a factor of 1.5–1.7 of the average Mises stress in the pulled coating. As a consequence, the measured coating strength has to be multiplied with this factor to get a tensile strength value, which is comparable to conventional tensile tests (MFT test). This was also proved experimentally by correlating strength values determined by the MFT and the TCT tests. Moreover, thanks to the current wide diffusion of the test, raw data are also used with the label *TCT strength* enabling a faster process control and optimization. Coating roughness and waviness can complicate the determination of the coating cross-sectional area and can influence the obtained coating strength value. If coating roughness or coating waviness is more than one fifth of the coating thickness, it is recommended that the coating surface be machined.

4.4.1.3 Bending

Measurements of bending strength and modulus of elasticity in bending should be made in principle for materials whose principal stressing mode is bending.

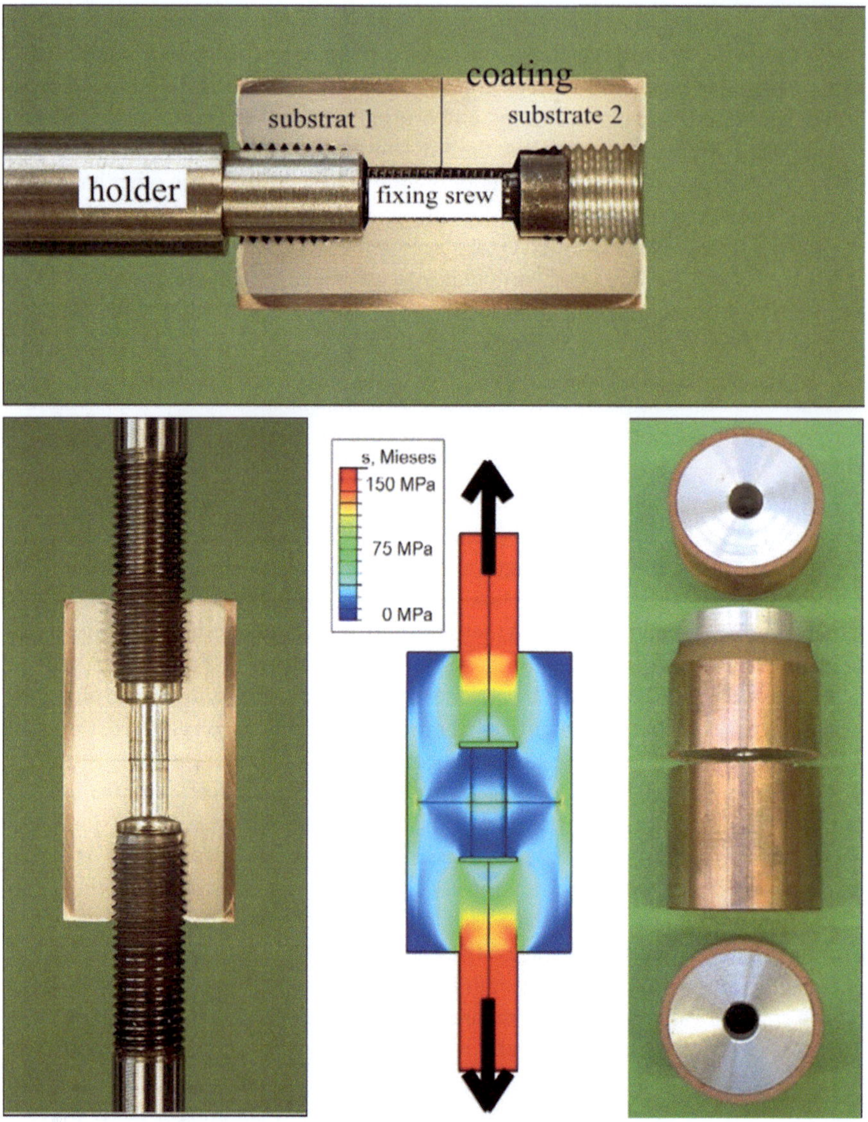

Fig. 4.39 Sample preparation and testing procedure for the TCT test according to Schmidt (2006a)

According to ASTM E855-08, three procedures are considered: (1) cantilever beam, (2) 3-point bending and (3) 4-point bending, where only (2) and (3) are employed in coated systems. The beam is positioned on a two-roller support and is subject to a normal load focussed in the central position (3-point test, procedure A—Fig. 4.40a or focussed in two positions at a fixed position close to the centre (4-point test, procedure B—Fig. 4.40a. The beam can be a freestanding coating or more generally a substrate coated on one side; in the last case, the position of the coated face

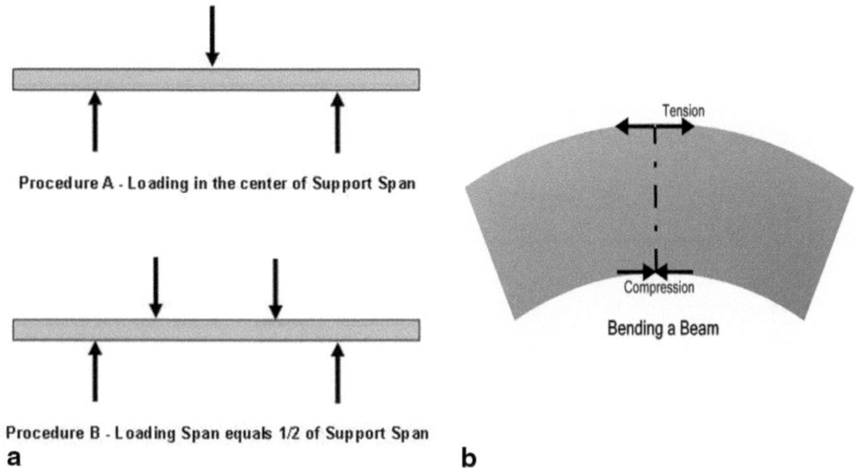

Fig. 4.40 Schematic representation of bending or flexural test **a** 3- and 4-point-bending configuration and **b** tension and compression stress state on bending a beam

will determine the sense of the load as represented in Fig. 4.40b: Towards the roller support, the coating is subject to tensile load, while its opposite face is under compression.

The preparation of coated specimens to perform a bending test is quick as well as the testing procedure by using a universal tensile equipment. A stress–strain curve of the whole specimen (coated specimen) can be obtained and an estimation of flexural strength and flexural strain according to the formula reported, for example, in Davis (2004).

4.4.1.4 Other Procedures

Shear strength is used sometimes to investigate the adhesion and cohesion of thick coatings by following commonly the guidelines of DIN EN 15340, and a specific application in the case of CS coating is reported, for example, in Binder (2011). The use of ring test is also reported in Coddet et al. (2014) even if high coating thickness must be produced.

4.5 Influence of CS Parameters on Coating Strength

The influence of CS deposition parameters on the strength of the deposited coatings have been extensively studied by Schmidt et al. (2006a) for pure copper coatings. A significant relation between the DE and coating strength has been first emphasized by mapping the variation of these properties as a function of process gas temperature

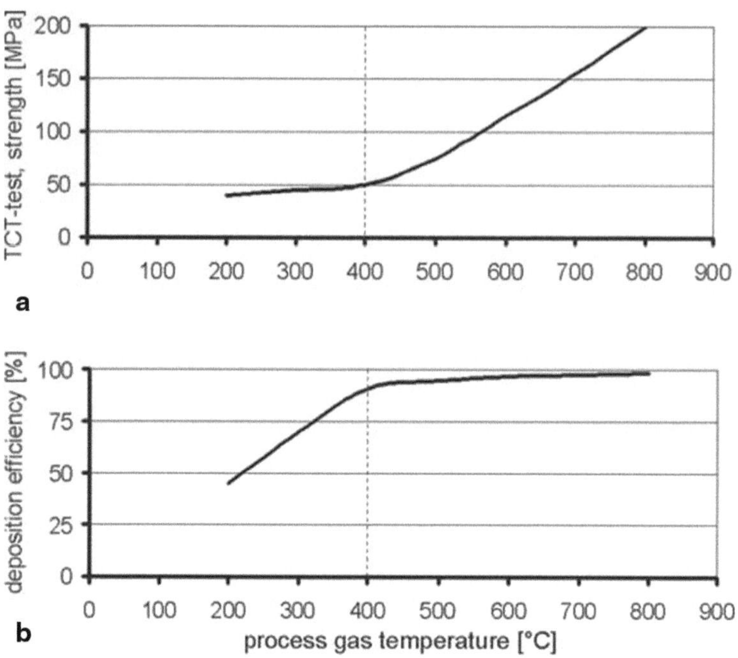

Fig. 4.41 **a** Coating strength, determined by tubular coating tensile *(TCT)* tests. **b** DE as a function of the process gas temperature. (Schmidt et al. 2006a)

as shown in Fig. 4.41. A −38+11-μm Cu powder feedstock was sprayed using nitrogen at 3.0 MPa as process gas. The DE is reported to grow linearly up to achieve a saturation limit, then the slope of the curve is strongly reduced, and only a slight increase is observed for further enhancement of the process gas temperature. On the contrary, the TCT strength trend is reported to grow slowly up to the DE saturation limit and then to increase suddenly its slope as the process gas temperature is further increased. This behaviour is essentially motivated by considering the particle–particle bonding mechanisms: While in the region of low process gas temperature the total impact energy (and momentum) of the incoming particles is mainly devoted to increase the amount of successfully stuck particles when the efficiency is saturated, further providing impact energy and momentum enable us to improve the quality of the bonding, enhancing the particle plastic deformation and promoting the well-known interfacial shear mechanisms responsible of coating adhesion and particle–particle bonding (Assadi et al. 2003).

The statements regarding the mechanical strength behaviour are further confirmed by comparing the stress–strain curves of copper coatings obtained with *standard* and *optimized* conditions (optimized conditions mean higher process gas temperature and powder injection in elongated preheated chambers as in currently manufactured stationary deposition equipment) or rather with spraying conditions before and after the critical point highlighted in Fig. 4.41. Stress–strain curves

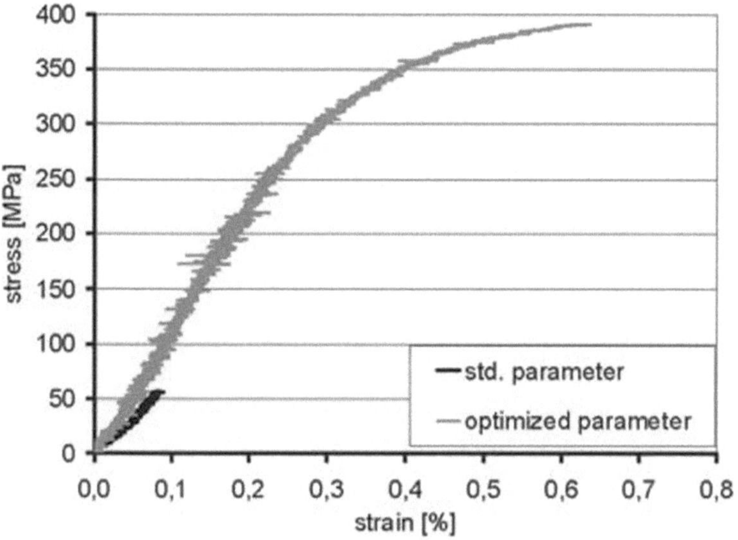

Fig. 4.42 Stress–strain curves of MFT tests for a coating sprayed using standard or optimized conditions. (Schmidt et al. 2006a)

are obtained by performing MFT test and are shown in Fig. 4.42. The specimen prepared with standard spraying conditions results an UTS of 57 MPa after reaching an elongation of only 0.08 %. The Young's modulus was determined to be 71 GPa, being much smaller than the reference data for copper from the literature (125 GPa). On the other hand, the specimen prepared with optimized conditions results an UTS of 391 MPa at corresponding elongation of 0.63 %, demonstrating properties close to highly deformed bulk material. In agreement, the measured Young's modulus of 117 GPa is similar to the literature value for copper. Again, the fractographic investigation revealed that optimized coatings show dimples and strong particle–particle bond strength, while standard coatings are essentially cleaved along the particle–particle boundaries as reported in Schmidt et al. (2006a).

It is then possible to summarize that cohesive strength progressively increases as the particle impact conditions exceed the critical conditions (i.e. critical velocity) to enable particle sticking and efficient coating growth. This is true as long as the conditions are within the deposition window as defined in (Schmidt et al. 2006b; Assadi 2011). On the contrary, if the particle velocity goes beyond the limit of the deposition window, then the mechanical properties fall down mainly due to the strong coating erosion and the development of noticeable residual stress combined with a loss of bond strength.

4.6 Influence of Powder Characteristics

CS is extremely sensitive to the quality and characteristics of powder feedstock deriving from the key role of particle impact temperature and velocity on the coating growth mechanism. As discussed above, the coating quality and the strength and mechanical properties in particular are also strongly related to particle impact parameters and plastic deformation phenomena. So that, as the powder feedstock characteristics influence the particle impact velocity and temperature, they also play a role on the cohesive strength of the deposited coating.

First of all, particle size and density are effective in changing the drag efficiency, the in-flight particle velocity and the interaction with the bow shock, hence the deceleration and deflection of the impacting particle close to the substrate zone. The effectiveness of coarser particle in the enhancement of the TCT strength is reported in Assadi et al. (2011) in the case of pure copper and pure titanium coatings. The strength is plotted against the particle velocity and the v_p/v_{cr} parameter, or rather the ratio between particle impact velocity and critical velocity, meaning the amount of particle velocity exceeding the critical velocity for the specific deposited material. Indeed, while the average particle size, studied in four different size distributions in the case of copper from −5+25 μm to −105+45 μm, and two different size distributions in the case of titanium 33 and 45 μm, seems to be not influent in changing the DE; it has a not-negligible effect in the coating strength behaviour (Figs. 4.43 and 4.44) or rather, a lower particle velocity is required to ensure the growth of high-strength coatings. However, this effect is still included in the function of the particle velocity when plotted against the v_p/v_{cr} resulting in a linear trend as detailed in Assadi et al. (2011).

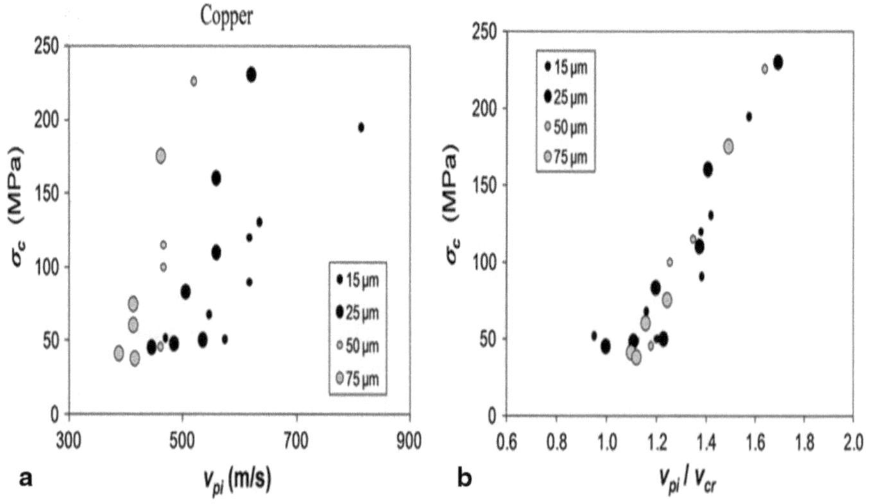

Fig. 4.43 Measured values of the cohesive strength of cold-sprayed copper coatings, as plotted against **a** particle impact velocity and **b** the ratio of particle impact velocity to critical velocity. (Assadi et al. 2011)

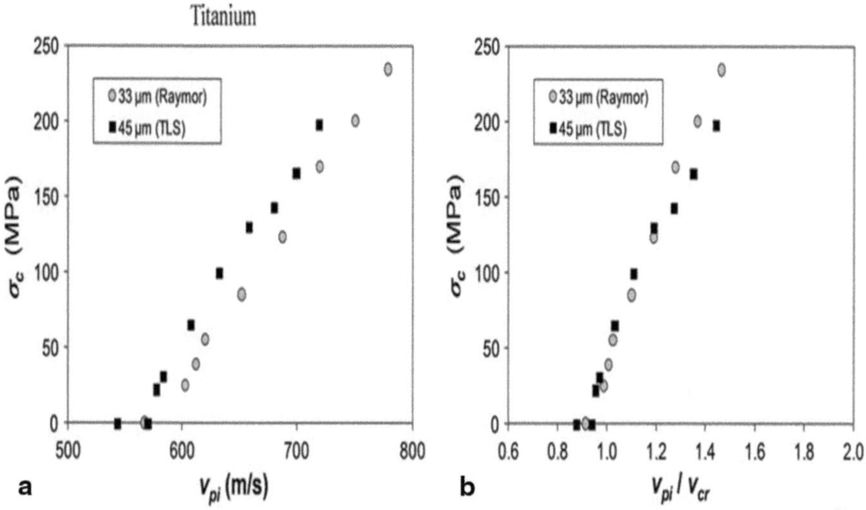

Fig. 4.44 Measured values of the cohesive strength of cold-sprayed titanium coatings, as plotted against **a** particle impact velocity and **b** the ratio of particle impact velocity to critical velocity. (Assadi et al. 2011)

The beneficial effect of using coarse particles is also reported in the case of pure Al coatings deposited with a size distribution of the powder feedstock as −105+63 μm (Van Steenkiste et al. 2002). Stress–strain curve is reported in Fig. 4.45 resulting in a mechanical behaviour similar to the corresponding bulk material and, in particular, 56 and 90 MPa, respectively, YS and UTS. YS ranges between the values reported in the case of both bulk aluminium, 35 MPa, and cold-worked aluminium, 106 MPa, certainly more close and representative of the CS deposition process and coating microstructure.

Fig. 4.45 Tensile testing of a kinetically (low-pressure cold spray) sprayed Al coating produced at a temperature of 288 °C. (Van Steenkiste et al. 2002)

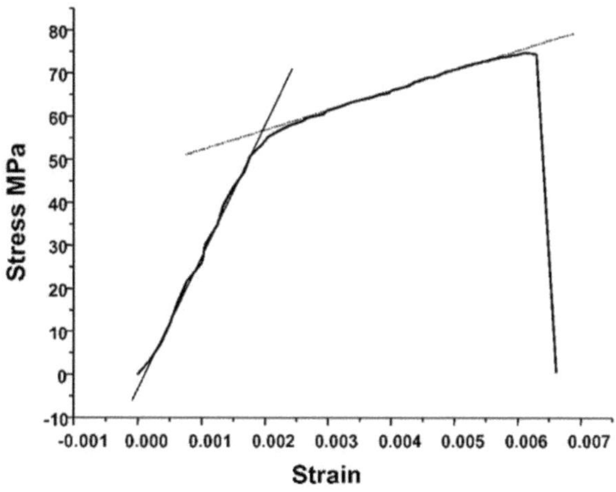

The particle shape as well can influence in particular the drag mechanism of the particle into the nozzle and as a consequence their exit velocity. This effect is put into evidence in Wong et al. (2013) comparing the performances of spraying commercially pure titanium (CP-Ti) powders with spherical, irregular and sponge feedstock. There are no results in terms of coating strength in this study; however, DE trends as well flattening ratio and microhardness are reported, considering that the previous discussion can represent a plausible basis to predict also the evolution of cohesive strength.

Particle surface oxidation must also be considered when talking about coating and the influence on cohesive strength (Jeandin et al. 2014). It is reported that the oxide layer that naturally covers the particle can be broken upon high-velocity impact (Li et al. 2010; Yin et al. 2012); however, part of the cracked oxide is entrapped in the coating microstructure mainly in the central zone of the plastically deformed particle because the outward metal jet only forms at the peripheral region of the interface (Yin et al. 2012). This hard oxide creates a barrier to the particle–particle bonding leading to a detrimental effect on coating strength and ductility. So that a careful selection of the initial feedstock, manufacturing and storage conditions as well as spraying parameters are fundamental to preserve as much as possible a low oxygen content that is mandatory to avoid the growth of surface oxide layers, in order to obtain high-strength coatings and ensure a good reliability on the obtained results.

4.7 Influence of Deposition Strategy

The influence of deposition strategy in terms of spray angle, gun transverse velocity and standoff distance on CS coating microstructural and mechanical properties have been extensively reported in the literature even if it is not yet completely clarified. There are several studies focused, for example, on standoff distance (Li et al. 2006), spray angle (Li et al. 2007a, b) or coating build-up (Rech et al. 2014); however, there are many parameters to take into account such as powder and substrate material and characteristics, spray parameters, nozzle type and shape, powder injection geometry, substrate size and thermal properties, etc. So, it is very hard to identify some generalized guidelines, and in the practice, each CS performer develops his own technical know-how based on specific deposition process and final application.

Within this scenario, some examples can be reported to introduce the discussion about the influence of the deposition strategy; for example, the effect of spray angle on the coating strength of titanium coatings is reported in Binder et al. (2011), where a reduction of TCT strength is observed as a function of the particle incidence angle. TCT strength is reported to be reduced from about 290 MPa (perpendicular incidence) down to about 90 MPa (45° incidence), and this detrimental effect is explained by the reduction of perpendicular component of impact velocity. Indeed, several coating characteristics, such as porosity, shear strength and TCT strength, plotted against the ratio v_{p90}/v_{cr}, where v_{p90} represents the perpendicular component

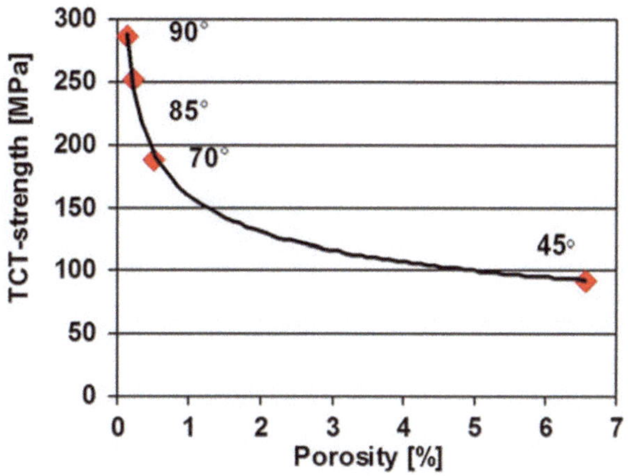

Fig. 4.46 Correlation between tubular coating tensile *(TCT)* strength of Ti coatings and porosity. The coatings were cold sprayed with nitrogen using a gas temperature of 1000 °C and a gas pressure of 4 MPa. (Binder et al. 2011)

of the particle velocity showing the expected linear trend. It is interesting to emphasize the behaviour of TCT strength as a function of porosity (Fig. 4.46), where the increasing porosity combined with a larger average size of the pores account for the decrease of the coating strength due to the promotion of particle–particle debonding and crack nucleation.

Coating thickness (0.5–2.0 mm) and gun transverse velocity (the 2.0-mm thick coatings were deposited with a slow single pass to four faster pass) effect on tensile properties of cold-sprayed A6061 coatings have been recently reported in Rech et al. (2014). Four-point bending test by following E855/90 guidelines have been used to obtain stress–strain curves. Despite the cohesive strength reported to be essentially unaffected by the coating thickness and the deposition strategy (i.e. number of pass to deposit the coating), a difference is emphasized in coating microstructure and fracture analysis. Once the applied load is sufficient to promote crack nucleation and first stage of propagation, the thick coating deposited with a single pass exhibits no opposition to the crack propagation until the interface with the substrate is reached and sudden fracture of the coating is observed. On the other hand, the coatings deposited by a multi-pass strategy exhibit barrier properties to the crack propagation, thanks to the presence of more interfaces between subsequent passes as observed in multilayer coatings deposited with other techniques (Tjong and Chen 2004).

4.8 Effect of Post-annealing on Strength

A tailored process optimization and an appropriate selection of feedstock material and deposition strategy generally represent a valid solution to achieve a coating cohesive strength sufficient for several industrial applications in particular regarding the deposition of ductile metals such as pure copper or aluminum. On the contrary this is certainly necessary but not sufficient when approaching high-strength coat-

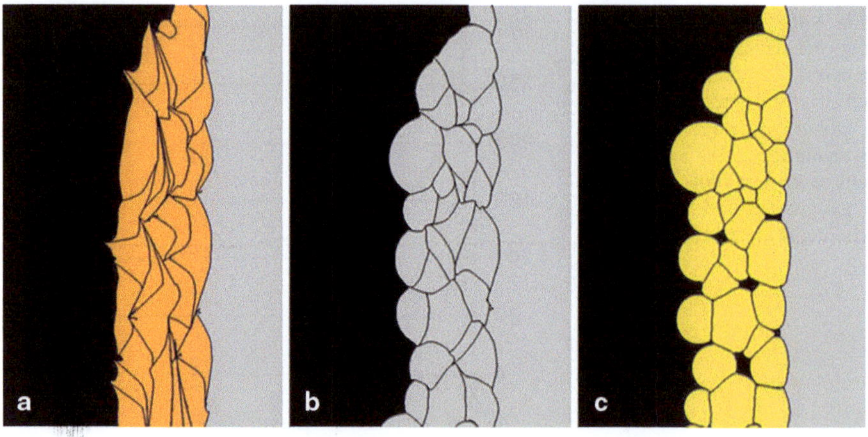

Fig. 4.47 2D simulation of a multi-impact scenario calculated for different material combinations under identical impact conditions. The initial impact temperature was set to 20 °C and particle impact velocities ranged between 400 and 650 m/s depending on particle diameters. Particle sizes were varied in range between 8 and 50 lm. **a** Cu on steel 316L. **b** Steel 316L on steel 316L. **c** Ti–6Al–4V on steel 316L. (Schmidt et al. 2009)

ing materials as Ni or Ti alloys. In this case, the particle plastic deformation upon impact is lower due to the higher YS, in particular in the range of impact temperature, leading to the development of more porosity and lower bonding at the particle–particle interface. These concepts are essentially represented in the simulated coating cross section obtained by a multi-impact calculation for different material combination under identical impact condition as shown in Fig. 4.47 (Schmidt et al. 2009).

While copper on steel is characterized by significant plastic deformation and compactness of the microstructure, this is not the case with AISI316L and especially Ti–6Al–4V coatings where significant porosity is observed and the negligible particle–particle deformation will be responsible for a low cohesive strength of the coating.

This preamble gives the basis to a twofold development in order to enable CS for the deposition of high-strength materials: on the one hand, the run to enhance the performance of the deposition equipment (i.e increase of particle velocity by allowing higher process gas pressure and temperature), on the other hand, the investigation about the opportunity to perform post-deposition annealing to consolidate the microstructure and promoting a sintering process. Regarding the second topic, several attempts are reported in the literature confirming the beneficial effect of post-deposition annealing on both cohesive strength and elongation properties of deposited coatings.

As for any powder metallurgy product, the sintering process of a CS coating can be performed but it must be done carefully: Vacuum or at least oxygen-free atmosphere (i.e. argon, nitrogen) heat treatment is generally operated in order to prevent the formation of an oxide layer outward the particle surface avoiding the formation of sintering necks. Annealing temperature as low as possible is advisable in order to

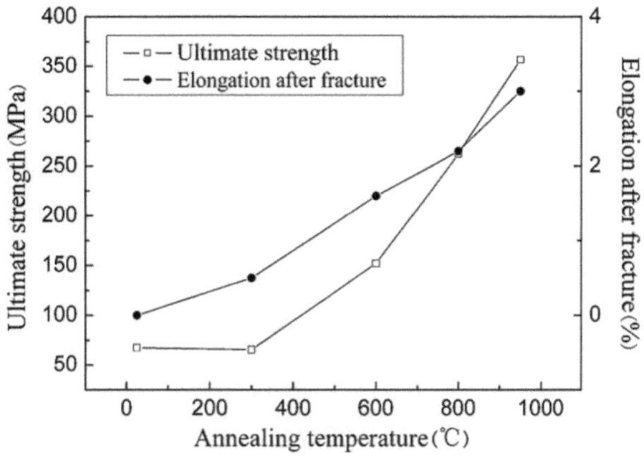

Fig. 4.48 Ultimate strength and the elongation of cold-sprayed coating and annealed coating at different temperatures. (Meng et al. 2011a, b)

avoid distortion and residual stress development and, last but not least, to preserve part of the beneficial effects of using a low-temperature deposition technique such as, for example, an average compressive residual stress, a very fine microstructure and superior coating hardness. Moreover, even if beneficial, the realization of a post-deposition thermal annealing can be difficult or often unaccepted by the industrial point of view due to the component size or base material or even specific production process.

Bearing this in mind, the beneficial effect of thermal annealing on strength and elongation properties is reported by many authors: Meng et al. (2011a) reported the enhancement of UTS and elongation after fracture of AISI304 CS coatings evaluated by MFT after 1 h annealing performed in vacuum at 10^{-3} Pa. Annealing treatment is reported to induce the atom diffusion through the interface between the particles so changing the particle–particle interface from pure mechanical interlocking bonding to metallurgically bonding through the progressive achievement of a sintering process as confirmed by a fractographic study. Moreover, the diffusion also reduced the potential crack nucleation sites which were present in the as-sprayed coatings enhancing the ultimate strength of the as-sprayed coating up to five times in the case of 900 °C annealing as shown in Fig. 4.48. However, the annealed coatings always contained some defects such as medium-size pores coming from the coalescence of coating microporosity and the agglomerated oxide particles, and these defects would induce the fracture taking place in advance. Therefore, the ultimate strength and the elongation of the annealed coating were lower than that of the bulk 304 stainless steel (SS). This behaviour is more or less observed also in the case of other coating materials such as Cu–0.5Cr–0.05Zr (Coddet et al. 2014) and Cu–4Cr–2Nb (Yu et al. 2011).

The stress–strain curves of pure copper coatings deposited before and after 1 h thermal annealing in vacuum are shown in Fig. 4.49 (Gärtner et al. 2006). In addition to the previous considerations, the initial coating microstructure is reported to be fundamental to lead the annealing treatment more effective in increasing cohesive strength. In this sense, cold-sprayed coatings processed with helium show a

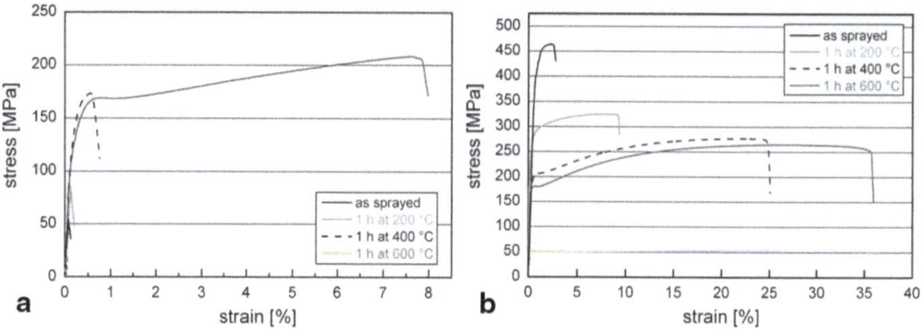

Fig. 4.49 Stress–strain curves of cold-sprayed coatings produced with before and after 1-h annealing in vacuum at different temperature. Coatings have been deposited using process gas **a** nitrogen and **b** helium. (Gartner et al. 2006)

similar performance as highly deformed bulk material and also after subsequent annealing, strength and elongation to develop failure in a similar manner as for cold-rolled sheets leading to elongation up to 35 %. Nevertheless, cold-sprayed coatings processed with nitrogen show brittle failure under relatively low tensile stress and also after thermal annealing; only the closure of particle–particle interfaces which are just under compressive contact is observed and therefore the higher elongation to failure in particular is (only) around 8 %.

Moving to higher-strength materials the pressureless sintering of Inconel718 CS coatings is discussed in Levasseur et al. (2012) and Wong et al. (2012). Again, the influence of initial coating microstructure is reported to be essential in order to promote metallurgical bonding at the particle–particle interface and the coatings deposited with higher impact velocity can benefit more effectively of the annealing treatment reaching ultimate tensile stress and elongation up to 763.6 MPa (62 % of the corresponding bulk material) and 24.7 %, respectively (Wong et al. 2012), as shown in the stress–strain curve in Fig. 4.50.

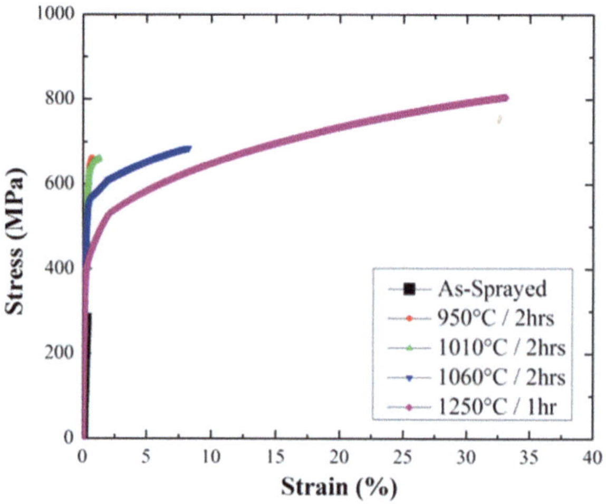

Fig. 4.50 Stress–strain curves of In718 cold spray coatings after different thermal annealing. (Wong et al. 2012)

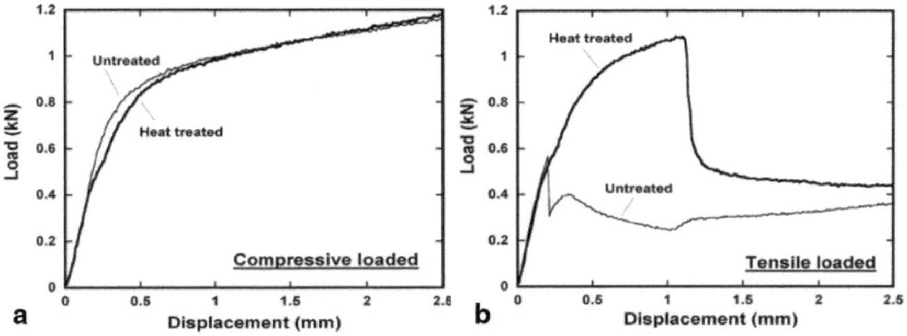

Fig. 4.51 Results of 4-point bending tests of heat-treated specimens. **a** Compressive loaded. **b** Tensile loaded. (Ogawa et al. 2008)

The influence of heat treatment on CP-Al coatings deposited by low-pressure CS as a function of the mechanical loading under 4-point bending test is reported in Ogawa et al. (2008). It is interesting to notice (Fig. 4.51) that under compressive loading, the effect of thermal annealing on stress–strain curve is negligible, while under tensile loading the effectiveness is evident. The explanation of this behaviour is based on the mechanism of cohesive failure that is due to the formation and propagation of vertical cracks starting from porosity or microstructural defects and propagating through the coating thickness; in this sense, under compressive load, this specific mechanism is not involved leading to a good mechanical behaviour also for as-sprayed materials.

Summarizing, a tailored annealing treatment is beneficial for the cohesive strength of a CS coating, thanks to the promotion of atomic diffusion at the particle–particle interface and consequent activation of a sintering process. The initial coating quality, in terms of microstructure compactness, low porosity and initial strength, is fundamental to enable the diffusion and preventing the oxidation at the particle–particle interface. So, all efforts to obtain a full-density as-deposited coating are twofold essentially to provide high-strength-coated materials both before and after thermal annealing. Finally, the coating performances in terms of strength are generally lower or much lower than the corresponding bulk materials due to the embedding of oxide and residual porosity that play an active role in the formation and propagation of cracks.

4.9 Residual Stresses

This section deals with a general description of residual stress phenomena during cold spraying. More specific information on residual stresses is also found in Chap. 5. Residual stress can be found in the surface of practically every material. Stress is the result of surface and bulk treatments by mechanical, thermal or chemical means, either alone or in combination. Residual stresses develop during most manufacturing processes involving material deformation, heat treatment, machining or processing operations that transform the shape or change the properties of a material. They arise from a number of sources and can be present in the unprocessed raw material, introduced during manufacturing or can arise from in-service loading. The residual stresses may be sufficiently large to cause local yielding and plastic deformation, cracking and local delamination in surface coatings, on a both microscopic and macroscopic level, and can severely affect component performance. For this reason, it is vital that some knowledge of the internal stress state can be deduced from either measurements or modelling predictions. Tensile residual stresses in the surface of a component or in a coating are generally undesirable since they can contribute to, and are often the major cause of, fatigue failure, quench cracking and stress-corrosion cracking. Compressive residual stresses in the surface layers are usually beneficial since they increase both fatigue strength and resistance to stress-corrosion cracking, and increase the bending strength of brittle ceramics and glass. However, excessive compressive stress can cause cohesive failure (spallation) in the case of a bulk material, and adhesive or cohesive failure in the case of a coating. In general, residual stresses are beneficial when they operate in the opposite direction of the applied load (e.g. a compressive residual stress in a component subjected to an applied tensile load).

According to Rickerby (1986), Rickerby and Burnett (1988) and Withers and Bhadeshia (2001), there are three types of residual stress: the macrostress (type I stress), which is distributed homogeneously over macroscopic areas that are higher than grain size in the case of polycrystalline coatings materials; microstress (type II stress), which is homogenous over microscopic areas such as one grain or subgrain; inhomogeneous microstress (type III stress), which is inhomogeneous even on a microscopic level. The dimensional scales characteristic of the three different type of stress are schematically shown in Fig. 4.52.

In general, the type I stress is the most prevailing contribution and is of particular interest from an engineering point of view, especially in material science and tribology. However, when comparing results from different techniques, some information must be given to the sampling volume and resolution of each measurement method in relation to the type of residual stress being measured, particularly when the type II and III micro-residual stresses are of interest. For example, it is important to

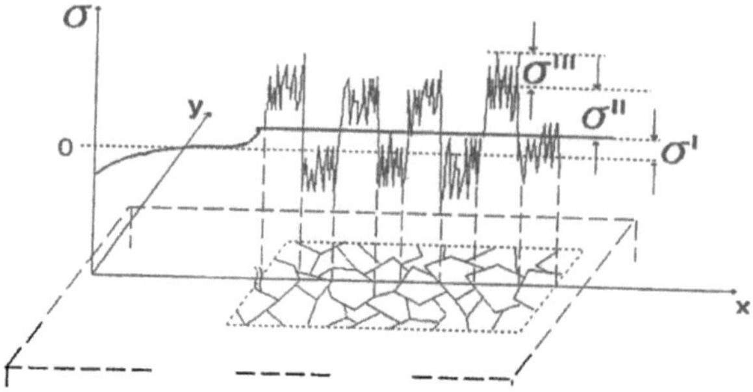

Fig. 4.52 Classification of stress according to length scales

consider the concept of the characteristic volume, which can be used to describe the volume over which a given type of residual stress averages to zero, and in this sense it is important to consider the relevance of the local (type II and III micro-residual stresses) variations arising from the presence of a composite material, or either a multiphase material, a specific texturing or a locally strained microstructure, etc. Most "material removal techniques" (e.g. hole drilling, layer removal) chip off large volumes of material over which type II and III stresses average to zero. Besides the formulas, relating the relaxed (due to material removal) and measured strains are based on the elastic theory that considers materials as a continuum medium, so that only the macro-residual stresses can be measured. On the other hand, diffraction methods, like X-ray diffraction, give a qualitative measurement of the micro strain by the broadening of the diffraction peak.

Among the origin of residual stresses, and especially moving into the field of surface coating, it is possible to focus the attention on two main contributions (Luzin et al. 2011):

3. The **deposition stresses,** σ_d, characteristics of the deposition process considered and related to the growth mechanism. For example, in thermal spray, this contribution is typically a tensile "quench" stress originating from shrinkage of a solidifying splat on the surface. For CS coatings, this stress is rather compressive, characteristic of a peening process.
4. The **thermal stresses,** σ_{th}, developed during cooling down of a composite substrate coating system from an elevated temperature of deposition. It may be qualitatively explained as follows: Upon imposition of a change in temperature, a difference in the expansion or contraction of the dissimilar layered materials results in a variation of the residual stress along the thickness direction of each layer. The stress between the coating and the substrate translates by shear at the interface, causing the coated systems to contract, elongate or bend. Practically, a biaxial thermal strain, ε_{th}, appears in films bonded to substrates having different thermal expansion coefficients, at a temperature higher or lower than the substrate or deposition temperature. Without plastic deformation in composite structure during temperature change, thermal stress is directly related to the elastic strain through Hooke's law:

$$\sigma_{th} = \left(\frac{E_c}{1-\upsilon_c}\right)\varepsilon_{th} = \left(\frac{E_c}{1-\upsilon_c}\right)(\alpha_s - \alpha_c)(T - T_0), \tag{4.4}$$

where E_c and υ_c are Young's modulus and Poisson's ratio of the coating, respectively; α_c and α_s are the film and the substrate thermal expansion coefficient, respectively; T_0 is the temperature of the free stress state, and T is the actual temperature.

Typically, high temperature deposition techniques such as thermal spray lead to a coating residual stress state dominated by the thermal stress contribution, while on the other hand, low temperature, plastic deformation techniques lead to negligible thermal stress contribution and the coating residual stress state is thus dominated by the deposition stress contribution which can be either tensile or compressive depending on the specific technique. In the case of CS, the high-speed impact and related peening effect represented the key factor influencing residual stress state resulting typically in compressive residual stress state.

4.9.1 Determination of Residual Stress

The determination of residual stresses on surfaces and coatings can be carried out by several approaches and techniques (Schajer 2013) as schematically summarized in Table 4.2, also discussed in Chap. 5. Each technique provides some advantages and disadvantages, and the selection must be performed taking into account the material and coating/specimen characteristics as well the target properties of interest.

The mechanical methods such as hole drilling and layer removal are essentially based on extensimetric determination; hole drilling consists of essentially two stages: (1) removal of the investigated material by drilling a hole (typically 2 mm diameter) and (2) measurement of the relaxation strains occurring around the hole by means of an extensimetric rosette. The theory is well known and the execution relatively easy to implement (ASTM E837-08) as recently summarized in Huang et al. (2013). However, both these tests are destructive, the spatial and depth resolution are relatively low and the methods are not sensitive to phase or structure of the material. They can be suitable to determine macrostress in quite homogeneous materials and coatings in order to have an average and quite accurate residual stress estimation. Both layer removal and hole drilling can be performed in *incremental* procedures in order to perform depth profiles as reported in the case of thermal spray coatings in Valente et al. (2005) and cold-sprayed A6061 coatings in Rech et al. (2011).

Table 4.2 Summary of the more used techniques to measure residual stress in surface-engineered materials and coating technology

Technique	Destructive/ nondestructive	Phase distinction	Accuracy	Spatial resolution	Depth resolution	Availability/ quickness
Hole drilling	Destructive	No	••	•	•	••••
MLRM	Destructive	No	••	•	••	••
Bending	Destructive	No	•••	•	/	••••
XRD	Nondestructive	Yes	•••	•••	•••	••
Neutron diffraction	Nondestructive	Yes	•••	••	••	•

MLRM multiple layer recursive matching, *XRD* X-ray diffraction

Bending method is based on the fact that the deposition of a layer leads to the development of residual stress which induces the substrate to curve. Then, from the variations of curvature, it is possible to calculate the related variations in stress as a function of film thickness and elastic properties. Curvature can be measured using contact methods (profilometry, strain gauges) or without direct contact (video, laser scanning) allowing curvatures down to about 10 mm^{-1} to be routinely detected. The Stoney (1909) equation is often used to relate the curvature radius to the biaxial stress in the plane of coating:

$$\sigma_c = \frac{E_s' \, t_s^2}{6 \, t_c} K, \tag{4.5}$$

where σ_c is the residual stress of the coating, E_s' is $E_s' = {}^{E_s}/_{1-\upsilon_s}$ and E_s and v_s are the elastic modulus and Poisson ratio of the substrate, t_s and t_c are the thickness of the substrate and of the coating, respectively, and K is the curvature. Bending method, similarly to other mechanical methods cited above, provides an estimation of the type (I) overall residual stress and may be used for many coating materials including multilayered and multi-structured materials At present, many theoretical models have been developed in order to predict the curvature of composite beam caused by residual stress (Brenner and Senderoff 1949; Masters and Salamon 1993). The applicability of Stoney equation is an important subject, and some checks are necessary to validate the use of this approach for the calculation of residual stress in thin and thick films. There are four major requirements in order to allow the use of Stoney equation:

5. Biaxial stress approximation
6. Small deflection (low K)
7. Narrow strip sample
8. $t_c \ll t_s$

The condition of biaxial stress approximation is not valid for monocrystals and materials characterized by strong texturing, while planar isotropic materials characterized by no preferential orientation are ideal for the application of the Stoney formula; however, there is no specification on the range of applicability of the equation for samples presenting texture. Small deflections (low curvatures, K) and narrow strip samples recommend a proper selection of specimen geometry and the condition that $t_f \ll t_s$ has been introduced in order to neglect the bending contribution and consider the planar strain mismatch as the only contribution which causes residual stress. This condition can be easily satisfied in thin-film technology (Vijigen and Dautzenberg 1995), but represented the main restriction to employ the bending method to determine stress with good accuracy in thermal spray coating. Approximately, to have an accuracy better than 5 % requires a t_c/t_s ratio lower than 0.02, meaning that with an Almen's plate 4–5 mm thick, the coating thickness must be lower than 0.08–0.10 mm. More recently, new models to enhance the method accuracy in the case of thicker coatings have been developed, further extending the opportunity to use bending method in thermal and CS coatings (Kõo and Valgur 2010; Wang et al. 2010a, b; Benabdi and Roche 1997).

In this field, the *progressive growth methods* must be mentioned: While Stoney equation considers the curvature of a static bilayer system after deposition and can be considered an ex-situ technique, these methods predict the residual stresses in progressively deposited coatings as in in situ technique; the deposition stress is introduced as the coating is formed layer by layer with a specified layer thickness, such as the misfit strain, coming from either the deposition stress or the different thermal contraction, is accommodated after each layer addition. Among these models, one of the more used is the Tsui and Clyne (1997) that is especially designed for layer by layer coating deposition techniques as in particular thermal spray and CS process. One of the major advantages of using this model is the capability to split the residual stress term and have a prediction of both the thermal stress contribution and the deposition stress contribution.

On the contrary, diffraction techniques can ensure high spatial resolution and specific sensitivity to the phase and structure of the material investigated. They are suitable for composite or finely structured materials and coatings and especially when the role of microstress/microstrain needs to be emphasized. The diffraction methods are described by Bragg's law:

$$2d_{\mathrm{hkl}} \sin\theta_{\mathrm{hkl}} = \lambda, \tag{4.6}$$

where d_{hkl} is the distance between the selected lattice planes hkl, λ is the wavelength and the angle θ_{hkl} is the scattering angle. When a material is under a compressive or tensile stress state, there is a lattice distortion or rather a variation in the *d*-spacing of its lattice associated with a *shift* in the position of the diffraction peak in the diffractogram. To find a connection between mechanical methods and diffraction methods, it is possible to consider that the crystal lattice is adopted as a natural and ever-present *atomic plane strain gauge* embedded in each crystallite or grain (Hutchings et al. 2005; Schajer 2013). In truth, the experimental techniques are not able to determine *d*-spacing variations (i.e. residual stresses) with atomic spatial resolution; however, the spatial resolution is typically some order of magnitude lower with respect to mechanical methods. By the experimental point of view, X-ray diffraction and neutron diffraction are the two techniques employed to determine residual stresses. Among these, X-ray is certainly the more diffused and employed due to the relatively simpler and cheaper equipment. The main difference between X-ray and neutron is related to the penetration depth of the incident beam and the corresponding investigated volume; X-ray is very surface sensitive with a few microns of penetration depth, while neutron can penetrate deep into the matter up to some millimetres enabling a better average measurement and for these reasons is preferentially employed to characterize thick coatings or directly part of components. In this sense, XRD is generally employed in combination with a layer-removal method to have a better average or also to perform residual stress profiles, while with neutron diffraction the condition of the incident beam can be properly tuned in order to define in a wide range the volume of investigation. The analytical model generally adopted to determine the values of the stresses is the $\sin^2\psi$ method that considers linear relation between stress and strain and a plane stress condition. The model assumes a linear relation between the lattice distance d and $\sin^2\psi$, where ψ is the angle subtended by the bisector of the incident and diffracted X-ray beam, according to the relation:

$$\sigma_\varphi = \left(\frac{E}{1+\nu}\right)_{\text{hkl}} \frac{1}{d_{0(\text{hkl})}} \left(\frac{\partial d_{\varphi\psi(\text{hkl})}}{\partial \sin^2\psi}\right), \tag{4.7}$$

where σ is the stress component in an assigned direction, E is the elastic modulus of the material, hkl are the lattice planes, d_0 is the lattice spacing of the planes hkl in the undeformed material, ψ is the angle subtended by the bisector of the incident and diffracted XR beam, and the $\left(\frac{\partial d_{\varphi\psi(\text{hkl})}}{\partial \sin^2\psi}\right)$ is the slope of the line d-$\sin^2\psi$

4.9.2 Residual Stresses in CS Coatings

The understanding of residual stress generation and evolution in CS deposits can be a useful tool to explain the coating growth mechanism. Deposition stress are typically compressive in CS and are originated by the peening effect and plastic deformation upon continuous high-velocity impact performed by the incoming particles flow. Matejicek and Sampath (2001) studied the impact of a single particle (*single splat*) and reported the residual stresses in cold-sprayed copper particles as a function of particle velocity and the resulting values are few tens of MPa in the case of particle velocity ranging from 500 to 700 m/s. Moving from single impact to a multiple particle deposition scenario, the final stress state is a very fine balance between the kinetic impact/shot peening effect upon particle impact and the thermal effect leading to annealing and stress relieving performed by the hot gas jet. In this context, Luzin et al. (2011) stated that the residual stress on CS coatings is almost entirely a deposition stress determined by the plastic deformation process of the spray material due to the high-velocity impact of the particles, while thermal effects do not play a notable role in changing the distribution of the induced stresses. They studied the systems copper/aluminium depositing copper and aluminium coatings on copper and aluminium substrates, and the residual stress profiles obtained by neutron diffraction are shown in Fig. 4.53. They also treat empirically the experimental results with Tsui and Clyne's progressive model and verify quantitatively the negligible contribution of thermal stress in the total residual stress.

They further estimate the impact parameters with the theoretical approach of linear momentum transfer on impact and defined the following relationship to predict the maximum residual stresses at the surface of a CS coating:

$$\sigma_{\max} = -(0.333 + 0.286\alpha\beta)(1-\alpha\beta)[(1-2\alpha\beta)\sigma_s + k\cdot\alpha\beta\cdot p_{\max}], \tag{4.8}$$

$$p_{\max} = \frac{2}{\pi}\left(\frac{5}{4}\pi E_*^4 \rho V^2\right)^{0.2}, \tag{4.9}$$

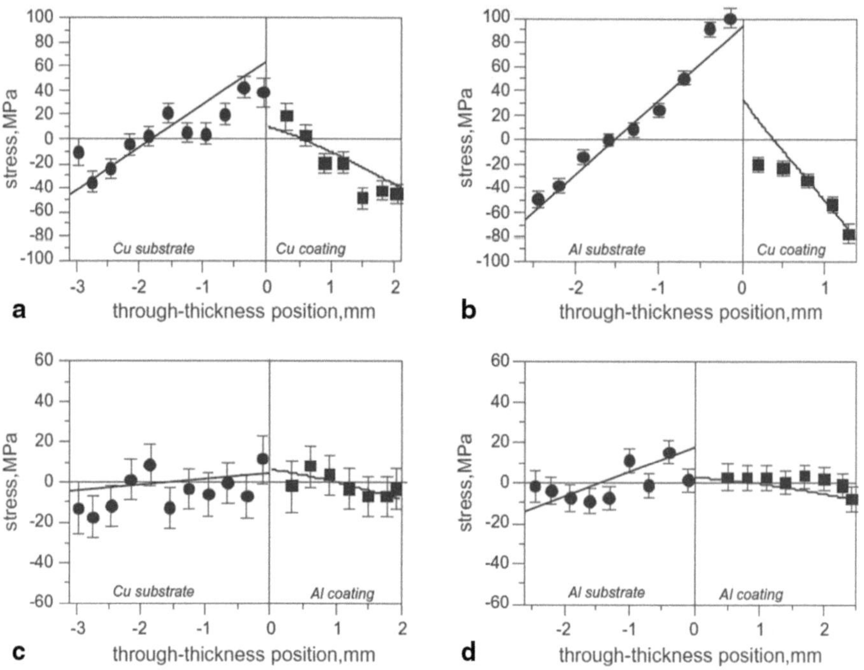

Fig. 4.53 Measurement *(symbols)* and model fit *(solid lines)* of the through-thickness in-plane stress distributions for **a** Cu/Cu sample, **b** Cu/Al sample, **c** Al/Cu sample and **d** Al/Al sample. (Luzin et al. 2011)

where p_{max} is the maximum pressure calculated in the assumption of Hertzian contact, σ_s is the yield stress of the material, ρ its density, V is the impact velocity, E^* is the equivalent modulus defined as $E_* = E/(1-\upsilon^2)$ with v Poisson coefficient and k is a constant close to 1. The two parameters α and β are coupled into a product that describe in simple terms the elastoplastic state of the deformed material: α is the ratio of the strain-hardening rate (tangent modulus) to the Young's modulus, and β is the ratio of the true plastic strain to the true elastic strain. The accuracy of residual stress prediction according to Eqs. (4.7) and (4.8) are verified in the case of Al/Cu system, Al/Mg.

Looking at Eqs. (4.7) and (4.8), one of the most important prediction is that plastic material properties, and in particular the YS at impact temperature (*effective YS*), are strongly related to residual stress development and in particular in the case of αβ small Eq. (4.7) reduces to σ_{max} =0.33* YS, meaning that residual stress is 1/3 of YS (at impact temperature; Spencer et al. 2012a, b). Indeed, Eqs. (4.7) and (4.8) reintroduce the important role of impact temperature through the effective YS term and linking the effective YS and the residual stress development with the flattening ratio and the critical velocity. In this sense, the flattening ratio can be assumed as an index of particle deformation upon impact and a way to evaluate the impact strain according to Luzin et al. (2011); this estimation of impact strain combined with

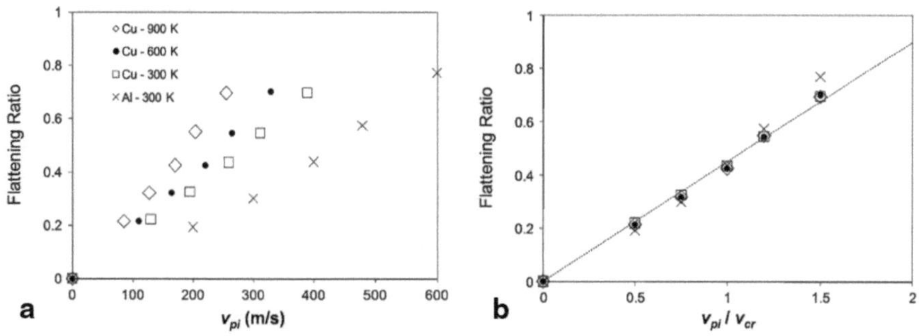

Fig. 4.54 Calculated flattening ratios of copper and aluminum as a function of **a** particle impact velocity and **b** the ratio of particle impact velocity to critical velocity. The *dashed line* in **b** shows the relation: $y = 0.46x$. (Assadi et al. 2011)

the impact velocity value can be used to calculate the impact duration by assuming the impacting particle decelerates linearly according to Taylor impact test (Meyers 1994) and giving also an estimation of average strain rate. Finally, the average impact pressure can be calculated based on the momentum transfer over the calculated impact time (Van Steenkiste et al. 2002), and this is in direct relation with residual stress according to Eqs. (4.7) and (4.8). On the other hand, as shown in Fig. 4.54a, the flattening ratio always increases with increasing particle impact velocity, though the rate of this increase depends strongly on material properties, as well as on particle temperature. Interestingly, the flattening ratio exhibits little dependence on material properties or temperature, when it is plotted against the ratio of the particle impact velocity (v_{pi}) to the critical particle impact velocity (v_{cr}; Fig. 4.54b) whose formula according to Schmidt et al. (2006a, b) also includes the dependency on effective yield stress. Consequently, all variations are embedded in the v_{pi}/v_{cr} function, and as a consequence the flattening ratio appears to be a unique function of v_{pi}/v_{cr}, regardless of the values of materials and process parameters (Assadi et al. 2011), confirming once more the strategic significance of the critical velocity parameter in the description of CS deposition.

The crucial role of the effective yield stress has a major influence in the case of material exhibiting yield stress versus temperature behaviour with large variation in the low-temperature zone as in the case of ductile and low-melting temperature metals such as in particular Al and Al alloys. Figure 4.55 reports the trends of YS of some common Al alloys, compared with Ni cold drawn and a Ni superalloy as a function of temperature. It is evident that impact temperature of few hundreds of degrees that are typically achievable and used with CS deposition, can reduce the YS to a value even lower than 100 MPa, while they are substantially ineffective in the case of HP Ni cold drawn and totally ineffective for high-strength superalloys.

Again, Eqs. (4.7) and (4.8) in combination with an experimental determination of residual stress can also be used as a tool to back calculate the effective YS as reported in Spencer et al. (2012a, b) in the case of pure Al and A6061, A7075 Al alloys coatings deposited by CS with different conditions and equipment. As

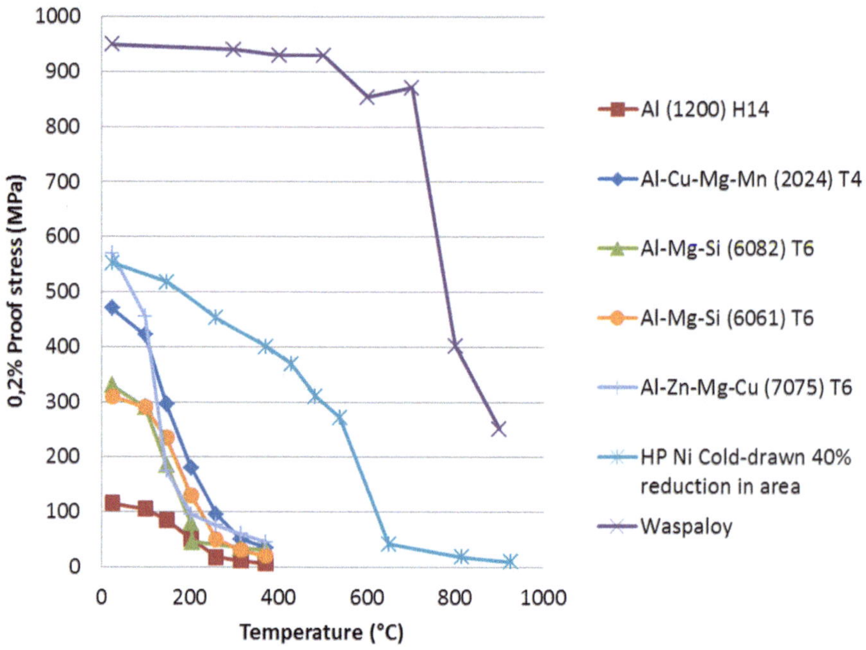

Fig. 4.55 Yield strength as a function of temperature of several Al alloys, Ni and waspaloy. (Data from Journal of NBS; Jenkins)

would be expected, the effective yield stress decreases by a factor ~2 for the sample sprayed at lower temperature (Al, 100–400 °C), but it can be as high as ~7 for samples sprayed at high temperature (550 °C).

If the enhancement of particle impact temperature has a beneficial effect in increasing the projected material ductility at the impact and hence its sprayability, it must be taken into account that it is also responsible of quenching stresses that can promote coating delamination. In particular, it is related to the discontinuity of residual stress profile at the interface between substrate and coating: This can depend also on substrate material and temperature (material size, deposition strategy, etc.) and can affect noticeably the coating adhesion leading to crack formation. Regarding the particle impact temperature and effective YS of the sprayed material, it must be taken into account that there is not a homogeneous temperature distribution in the metal particle during the impact, and these variations can produce different plastic behaviour of the material as a function of their position and local differences in residual stress distribution. Formation of shear instabilities is historically reported in CS deposition, and temperature rises high enough to induce local melting at the edge of the splat are also reported by many authors (Assadi et al. 2003). Recently Saleh et al. (2014) developed a smooth particle hydrodynamic(SPH) model to describe the particle–particle impact interaction during CS deposition of A6061 aluminium alloy and the nature of inter- and intra-layer adhesion. An interesting result is related to the significant variation in the extent of plastic deformation between the core of an impacted particle and its periphery with the core exhibiting a lower degree of plastic deformation, while at the periphery plastic deformation as severe to induce local microwelding events is observed. Accordingly, the behaviour of the residual stress profiles has been studied confirming the fine balance between thermal and kinetic effect on the final stress state and including the relevance of local variations within the coating microstructure due to the nonhomogeneous plastic deformation. A further proof of the agreement between experimental results and simulated profiles analysed with Tsui and Clyne method is reported confirming once more the dominance of kinetic over thermal effects of the CS deposition process. The evolution of residual stress trends for Ti, Cu and Al CS coatings deposited on carbon steel (S355), SS (AISI316) and aluminium alloy (A6061) is studied in Suhonen et al. (2013) further evaluating the effect of various pretreatments such as grit blasting and CS blasting. The crucial role of the first layer deposition on coating adhesion has been emphasized, while by the point of view of residual stress mainly compressive stresses are reported due to the nature of CS. However, the possibility to generate either tensile or compressive test depending on the combination of coating and substrate material is reported, or rather depending on the variation of the thermal stresses contribution, directly proportional to the difference of thermal expansion coefficient between the substrate and the coating.

A further parameter influencing the residual stress is the coating thickness, and in particular, the deposited coating showed a lower stress value at the interface with the substrate to grow up along the depth from the interface to the surface. In this sense, the peening effect of bombarding particles "accumulates" with repeated impacts as confirmed also by the difference on residual stress observed on single splat with respect to multiple impact (Matejicek and Sampath 2001). The pure peening effect on residual stress profile on Al-based coatings have been empirically studied by comparing the residual stress profiles of pure Al coating with Al/Al_2O_3 coatings obtained by spraying different Al and Al_2O_3 powder mixtures confirming that the additional peening performed by the impinging ceramic particles induce an increase in the whole amount of (compressive) stress (Rech et al. 2009).

The effect of thermal input on residual stress evolution on A6061 alloy coatings has been experimentally assessed in Rech et al. (2011), where the use of substrate preheating in the range 24–375 °C as well a different deposition strategy (*single pass* and *multi pass*) have been considered. The residual stresses, measured by using XRD, bending method and modified layer removal methods, are reported to be compressive in all cases with a slight trend to reduce the amount of compressive stress with the increase of preheating temperature.

The similarities of CS process with shot peening regarding the kinetic aspects and the influence on residual stress generation have been studied considering both the shot peening of impacting particle during the coating growth (Ghelichi et al. 2014a, b) and the effect of impacting particles on the residual stress profile induced on the substrate (Shayegan et al. 2014).

Considering the effect on coating growth, the main statement is that kinetic impact plays the more significant role in the determination of residual stresses in CS coatings; however, a new model has been developed to further include a term to describe the thermal annealing effect on residual stresses. Summarizing, this model involved a two-step approach in which the first step accounted for the peening effect of the impinging particle, while the second step accounted for the annealing effect responsible of stress relaxation mechanism. This second step is the novelty of this model and the Zenner–Wer–Avrami function is employed by the authors to calculate the stress relaxation contribution as a function of annealing time at a fixed temperature. The model accuracy has been assessed by comparing simulated results with experimental determination of residual stress by means of XRD in the case of A5053 aluminum alloy coatings.

The effect of impacting particle on substrate residual stress profile is reported in Shayegan et al. (2014) in the case of deposition of Al1100 alloy on extruded AZ31B magnesium alloy. A new model is developed which used Cowper–Symonds model to describe the higher strain rate of CS coating and Johnson–Cook material model to describe the particle impact. A parametric study performed on the single particle model has been developed to evaluate the effect of velocity, particle shape and diameter, impact angle and friction between the particle and the substrate, on the residual stress induced on the substrate. The main results are shown in Fig. 4.56. The typical shape of the profiles are in agreement with residual stress profiles exhibited by shot-peened surfaces or rather a slight compressive state in the surface layer (substrate/coating interface in the case of CS deposition) followed by an increasing residual stress up to a maximum compressive stress observed at a critical depth

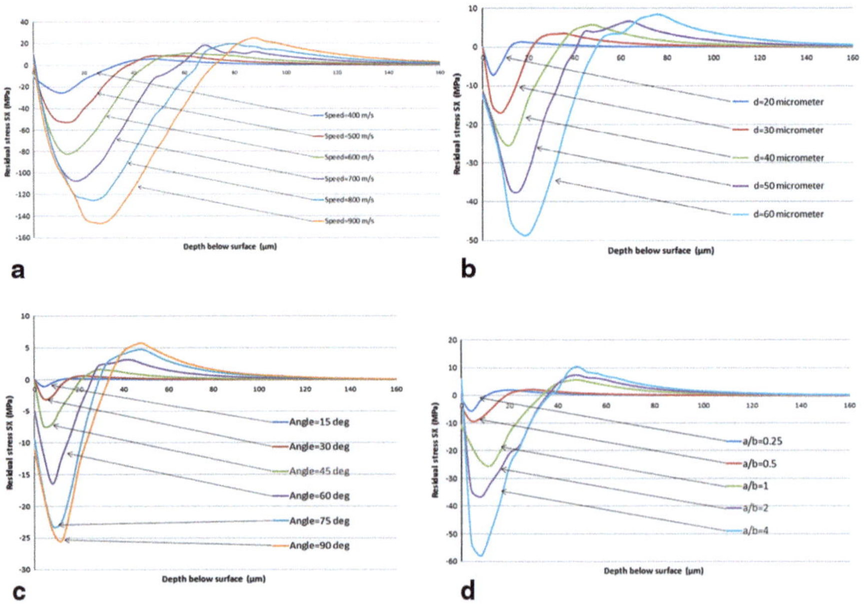

Fig. 4.56 Residual stress profiles generated in an AZ31B substrate when impacted by an Al1100 particle as a function of **a** particle speed, **b** particle diameter, **c** impact angle and **d** particle aspect ratio. (Shayegan et al. 2014)

inside the substrate. This critical depth resulted in a function of spray conditions and particle characteristics as shown by the calculated profiles shown in Fig. 4.56. Residual stress progressively reduces into the substrate up to zero and to a further tensile peak, balancing the total stress into the material and whose intensity is related to the intensity of compressive peak.

4.10 Fatigue

A specific discussion on the effect of residual stresses on fatigue is found in Chap. 5 and Sect. 4.10.2. Since fatigue accounts for about 90 % of all mechanical failures, fatigue behaviour of materials and structural components has been of great importance to be fully understood for a reliable mechanical design. Fatigue failure occurs because generally the ongoing repetition of identical or similar loads strongly reduces the loads the material can bear. Furthermore, the failure is not preceded by large plastic deformation even in ductile materials (that is to say in the elastic-linear field) rendering it more difficult to detect component damage than under static loads—the danger of catastrophic failure is thus rather large (Rösler 2007).

The introduction of some basic concepts to the description of fatigue strength is necessary to understand the potential effect of a specific surface coating. Fatigue is encountered under a time-dependent cyclic loading as schematically represented in Fig. 4.57; the time dependence is described by the *period T* defined as the time for a cycle or alternation of the load. Sinusoidal or triangular cycling are the most frequently considered and replicated in laboratory testing. The load is described by the stress amplitude, and the mean stress, defined as:

$$\sigma_a = \frac{\sigma_{max} - \sigma_{min}}{2}; \sigma_m = \frac{\sigma_{max} + \sigma_{min}}{2}. \tag{4.10}$$

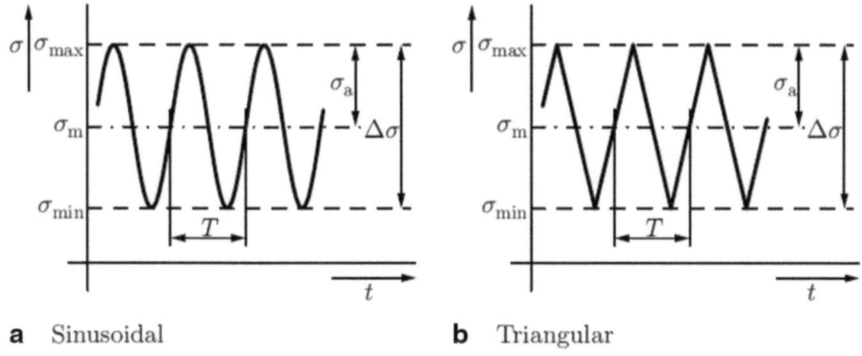

Fig. 4.57 Time-dependent cyclic loading. **a** Sinusoidal. **b** Triangular

pulsating in compression	$\sigma > 0$ / $\sigma < 0$	$\sigma_{\mathrm{max}} < 0$	$R > 1$
zero-to-compression	$\sigma > 0$ / $\sigma < 0$	$\sigma_{\mathrm{max}} = 0$	$R = -\infty$
reversed	$\sigma > 0$ / $\sigma < 0$	$\sigma_{\mathrm{m}} < 0$	$-\infty < R < -1$
fully reversed	$\sigma > 0$ / $\sigma < 0$	$\sigma_{\mathrm{m}} = 0$	$R = -1$
reversed	$\sigma > 0$ / $\sigma < 0$	$\sigma_{\mathrm{m}} > 0$	$-1 < R < 0$
zero-to-tension	$\sigma > 0$ / $\sigma < 0$	$\sigma_{\mathrm{min}} = 0$	$R = 0$
pulsating in tension	$\sigma > 0$ / $\sigma < 0$	$\sigma_{\mathrm{min}} > 0$	$0 < R < 1$
static	$\sigma > 0$ / $\sigma < 0$	$\sigma_{\mathrm{min}} = \sigma_{\mathrm{max}}$	$R = 1$

Fig. 4.58 Typical load curves and *R* ratios

Another quantity commonly used to define the fatigue cycle is the so-called fatigue stress ratio (or simply stress ratio), defined as:

$$R = \sigma_{\mathrm{min}} / \sigma_{\mathrm{max}}, \tag{4.11}$$

where σ_{max} and σ_{min} are, respectively, the maximum stress and minimum stress in the cycle. The stress ratio, *R* defined in Eq. (4.10) is a second parameter commonly used to further describe the load. Finally, alternating or reversed stress is considered when a change of sign during the cycle is observed, while when the load is completely tensile (positive) or compressive (negative) through the cycle we talk of fluctuating or pulsating stress (Fig. 4.58).

By the experimental point of view, the fatigue strength of a material is typically described first by a stress-cycle diagram (also stress–life or S–N or Wohler diagram) as schematically shown in Fig. 4.59. The number of cycles, *N*, are plotted in the *x*-axis always in logarithmic scale, while the stress, *σ*, can be plotted in the *y*-axis in linear or logarithmic scale. Two separated regimes can be considered or rather the high cycle fatigue (HCF) and the low cycle fatigue (LCF) depending on the total number of cycles completed to have the final fracture. There is no well-defined number of cycles to distinguish the two regimes even if generally 10^4 (sometimes 5E4) is used. A stress amplitude that causes failure in the LCF regime or HCF regime is called the LCF strength and HCF strength, respectively. The damage mechanisms leading to fatigue failure are different in the LCF with respect to HCF, since in the first case plastic strain are involved in the fatigue cycles, while the HCF failure happens below the yield stress. It must be noticed that the slope of the S–N curve is usually much smaller in the LCF regime than in the HCF so that a small

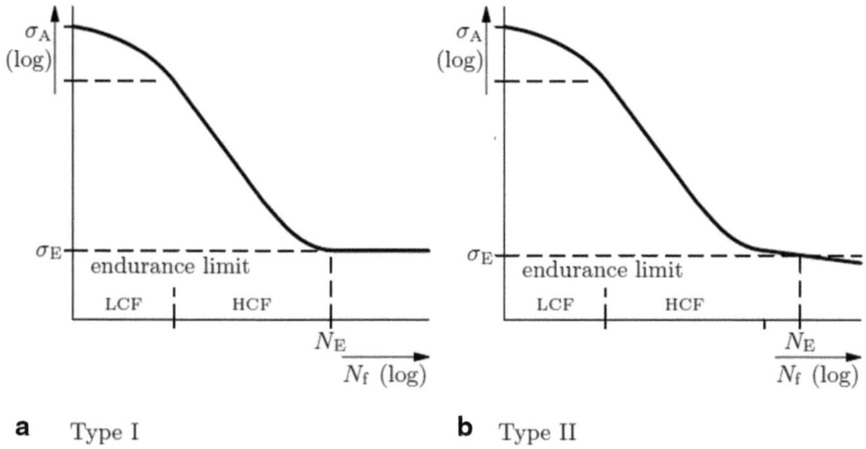

Fig. 4.59 The characteristics type of type-I and type-II S–N curves. (Rösler 2007)

change in the stress amplitude has a large effect on the number of cycles; for this reason, in the LCF regime (where elasto-plastic deformation take place), the strain ε is the parameter used to relate the number of cycle to failure to the severity of the fatigue cycle. These ε N curves are called Coffin–Manson curves and are the basic tools for LCF design. Generally, the scatter of the cycles at the same stress/strain to failure is rather large, meaning that fatigue strength is very sensitive to possible defects of the material. In particular, fatigue damage initiation involves a small volume of material and usually starts from the free surface of the material; that is to say that the surface state is critical and that fatigue strength is strongly influenced by the surface roughness, the residual stresses and possible surface work hardening. All these factors justify the large scatter of the fatigue test results making necessary the use of statistical methods to describe fatigue strength and draw limiting curves that represents a certain probability of failure. Some materials exhibit a true fatigue limit (sometimes also called the endurance limit). In this case, there exist a limiting number of cycles NE, with the S–N curve being almost horizontal at a larger number of cycles. In this case, the S–N diagram is of type I (Fig. 4.59a). A specimen that has survived NE cycles is never supposed to fail, and the stress level that corresponds to NE in the S–N curve is called the fatigue strength, endurance limit or fatigue limit σ_E. In many materials, there is no horizontal part of the S–N curve (type II, Fig. 4.59b). Although the slope of the S–N curve becomes smaller beyond a certain number of cycles, failure can still occur even with smaller fatigue amplitudes. These materials thus have no true fatigue limit. To ensure safety of the component, a limiting number of cycles of 10^8 is often used, ten times larger than the usual value for materials with a true fatigue limit. To state explicitly that a fatigue strength corresponds only to a certain number of cycles, not to a true fatigue limit, the number of cycles can be added to the subscript, as in $\sigma_{E(10E8)}$ (Rösler 2007).

Looking at the S–N curve plotted in double-logarithmic scale, it can be noticed that a straight line can fit the trend in a wide range of number of cycles and this linear trend is described by the Basquin equation:

$$\sigma_{A} = \sigma_{f}^{'}\left(2N_{f}\right)^{-a}, \tag{4.12}$$

where the fatigue strength coefficient $\sigma_f^{'}$ is related to the tensile strength. Some typical approximate values for $\sigma_f^{'}$ are 1.5 UTS for steels and 1.67 UTS for aluminum and titanium alloys. The fatigue strength exponent depends on the material and the specimen geometry; on smooth specimens, it ranges typically between 0.05 and 0.12 (Rösler 2007).

4.10.1 Fatigue Damage Mechanisms

The mechanisms of fatigue development and failure can be different, depending on material type, surface state, cycling conditions, environment, etc., and can be complex to analyse. However, in the case of metals, the fatigue failure is generally originated by initiation and growth of surface cracks. Only if the material has been previously hardened by means of some thermo or thermo-chemical treatment (carburizing, nitriding, induction hardening, etc.), the fatigue crack starts from an internal defect, generally a nonmetallic inclusion.

Apart from these cases, the failure mechanism is a three-step process: The first step is the *crack initiation,* the second step the *crack growth and propagation* under cyclic loads and the last step the *final catastrophic failure.* The understanding of these steps is fundamental to understand how a surface treatment and, in particular, in this case a CS coating can influence the fatigue life of a component. The crack initiation stage is very sensitive to the mechanical properties of the base material and its surface state. Surface defects, notches, cracks and microcracks are generally present on the surface of a metal; they came from production processes, machining or manufacturing steps or also by simple handling of the materials and components. Indeed, even if any micro defect is appreciable, the continuous sliding of adjacent grains subjected to the maximum shear stress originates the so-called persisting slip bands, a series of surface peaks and valleys that increase their dimension with the number of cycle till they form a crack that will grow and propagate upon continuous cyclic loading, with a crack path that is a function of the applied loads, finally leading to the final catastrophic failure. This damage mechanism is very sensitive to the surface state, in terms of both roughness and surface residual stresses, not significant under static loads but crucial in fatigue strength. The mechanical properties and the physical state of the surface are hence the key factors to control the crack initiation stage and enhance the fatigue life of the component.

This very schematic picture can highlight the potentiality of a surface treatment on determining the fatigue strength. The first target is the reduction of the surface defects and roughness; this can be traditionally achieved by metal working and further machining of the surface, for example, rolling and forging are well-known post-processing procedures to close micro-cavities and pores on the surface of the metal hence increasing the fatigue strength. On the other hand, the second target to enhance the fatigue strength is the surface hardening responsible to reduce the plastic deformation at low-yield stress, in this field surface treatment as shoot peening or nitriding in steels are generally considered for the scope. Finally, compressive residual stress on the surface is also known from the literature to be beneficial on fatigue strength providing a mechanical opposition to both crack initiation and propagation.

4.10.2 Fatigue Strength on CS Coatings

Despite the importance of fatigue strength on the real performances of a mechanic component in industrial application and the relevance of surface engineering treatments on influencing fatigue strength, the literature about the influence of CS coating on fatigue strength is far from being abundant. Furthermore, the testing procedures and conditions are numerous and the interpretation of results generally complex; so that, today the available experimental results are mainly diverse and sometimes also contradictory.

The encountered procedures to evaluate fatigue strength on CS coated metal parts are mainly the ASTM B593 "Standard Test Method for Bending Fatigue Testing for Copper–Alloy Spring Materials" and the ISO 1143 "metallic materials—rotating bar bending fatigue test" involving, respectively, pure bending and rotating bending stresses. The studied materials as both substrates and coatings are light alloys, mainly Al and Ti alloys according to the increasingly consolidating applications of CS in aeronautics and defence where these materials are strongly employed (Jones et al. 2011).

The main factors influencing the fatigue strength of a metallic material coated by CS can be summarized in:

- Bond strength—substrate/coating interface
- Coating material and quality (microstructure, porosity, mechanical properties)
- Residual stress state
- Surface roughness

While the position and experimental results from the different studies are unanimous (i.e. concerning bond strength), there are often some contradictions (i.e. residual stress state, coating material) due to the relatively new technological issue for CS coatings and the continuous rise of new experimental results as well to the complexity of the problem itself; the scope of this section is to report the main results available in the literature, also trying to identify some common and summarizing trends.

4.10.2.1 Effect of Bond Strength on Fatigue Properties

The whole literature available is unanimous in confirming the strong influence of bond strength on fatigue performances of a cold-sprayed specimen (Ghelichi et al. 2012; Sansoucy et al. 2007; Price et al. 2006) The crucial role of bond strength is necessary to avoid crack initiation directly at the substrate surface totally excluding the influence of the coating; or worse, preserving only the detrimental effect of the increase of substrate roughness and notch effect achieved by the peening of the first layer deposited. This phenomenon is shown, for example, in Ghelichi et al. (2012) in the case of CP-Al and A7075 coatings on A5052 substrates: the poor adhesion of CP-Al (Fig. 4.60a) is the first reason of the negligible influence of the coating presence on fatigue strength, while on the other hand, the noticeable bond strength of A7075 coating (Fig. 4.60b) is one of the main reasons of the significant enhancement of fatigue limit with respect to the bare substrate. Furthermore, the lack of adhesion at the substrate/coating interface seems to have a dominant role, having the capability to mask or nullify the other effects either beneficial or detrimental. For example, in Ghelichi (2014a, b), the effect of A7075 coating obtained by using a spherical gas-atomized microstructured powder as feedstock is compared with a A7075 coating obtained by using a nanostructured cryomilled powder with the following result: While the mechanical characteristics of nanostructured feedstock can lead to superior coating properties, the lack of bond strength as shown in cross-sectional fractographies (Fig. 4.61) is the main factor, combined with the porous microstructure,

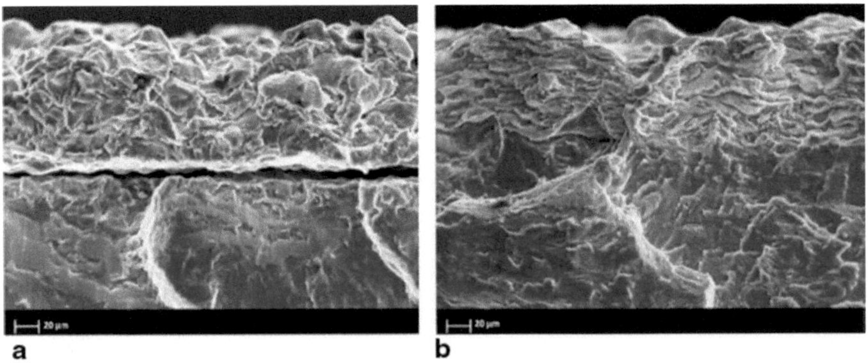

Fig. 4.60 Cross-sectional fractography of **a** CP-Al and **b** A7075 cold spray coating on A5052 substrate. (Ghelichi et al. 2012)

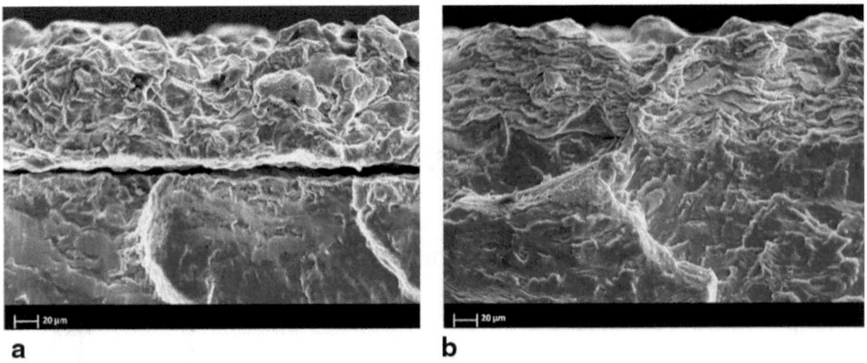

Fig. 4.61 Cross-sectional fractography of A7075 cold spray coating on A5052 substrate. The coating has been obtained by using **a** gas-atomized microstructured powders and **b** cryomilled nanostructured powders. (Ghelichi et al. 2014a, b)

responsible for the essentially negligible influence of the coating presence on final fatigue strength. So, “traditional” A7075 coatings resulted in an increase of fatigue limit up to 30 %, while harder nanostructured coatings have no influence on fatigue.

Excellent bond strengths (adhesion strength of 61 ± 4 MPa) are reported to be fundamental in determining one order of magnitude rise in number of cycles to failure of A2024 specimens coated with Al–Co–Ce with respect to bare substrate (Sansoucy et al. 2007).

Increasing bond strength by surface pretreatment can have a significant role on final fatigue performances. For example, combining grit-blasting pretreatment with subsequent CS coating is reported to have multiple beneficial effects on fatigue life enhancement (Ziemann et al. 2014). First, grit blasting is itself beneficial on fatigue life inducing a compressive residual stress on the surface; then the increased surface roughness is able to improve coating adhesion by making more effective the mechanical anchorage at the substrate/coating interface. On the other hand, combining shot peening with CS leads to a completely different final result. While the effect of shot peening as in the case of grit blasting is beneficial to fatigue strength, it has a detrimental effect on subsequent coating adhesion. So, the majority of shot-peened/coated specimens as obtained in Ziemann et al. (2014) show a failure initiation at the substrate/coating interface (Fig. 4.62a, b); on the other hand, the excellent

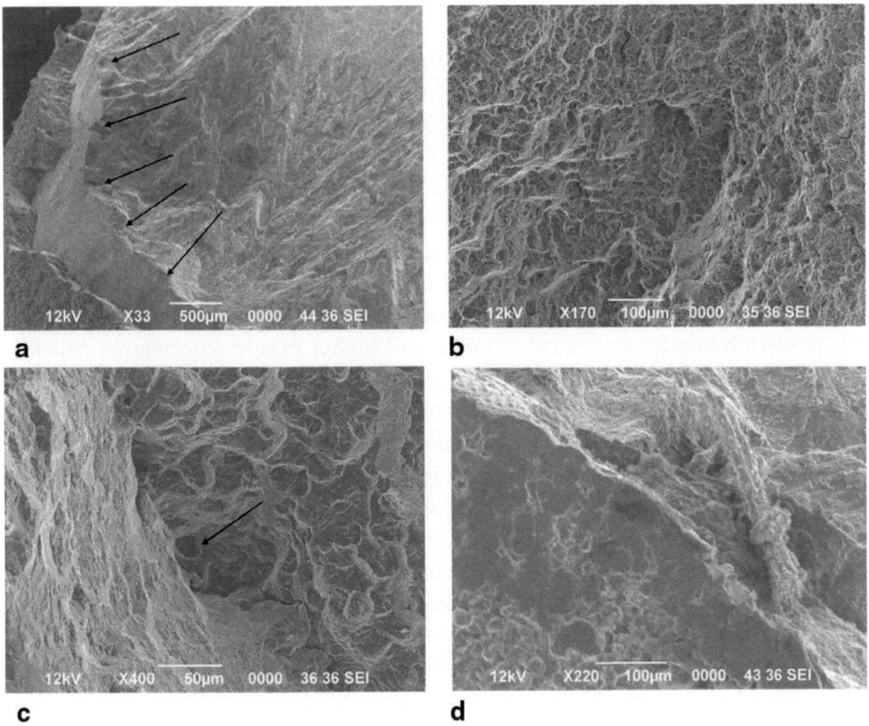

Fig. 4.62 Fractographies of representative cracks of **a**, **b** shot-peened/coated specimen and **c**, **d** grit-blasted/coated specimens. (Ziemann et al. 2014)

adhesion of grit-blasted/coated specimens seems to be the reason for the increased performance, and in that case the failure occurred within the substrate (Fig. 4.62c, d) and is reported to be due only to the excessive number of cycles experienced.

Bond strength can also play a negative role on fatigue strength as in the case of CP-Ti coatings on Ti6Al4V substrates as reported in Clzek et al. (2013); here, the high adhesion is combined with a low average quality of the coating surface in particular regarding high surface roughness and significant porosity; this situation in the first step promoted the crack initiation phenomena by notch effect and in the second step, thanks to the good adhesion of the coating to the substrate, it allowed the transfer of the formed vertical cracks from the coating to the substrate leading to a fast deterioration of the specimen (Clzek et al. 2013).

Summarizing, ensuring a sufficient bond strength at the substrate/coating interface is a mandatory requirement to allow the deposited coating to play a role on fatigue strength. The influence of all other characteristics, such as stress state or surface roughness, are secondary with respect to bond strength because not coating adhesion means that the crack initiation can start directly at the substrate surface excluding any (beneficial or detrimental) influence of the coating.

4.10.2.2 Effect of Coating Quality on Fatigue Strength

Regarding the role of coating material and quality, it is certainly proved that a good coating quality in terms of compact microstructure and high cohesive strength increase the average mechanical behaviour of the deposited material (Schmidt et al. 2006a, b; Assadi et al. 2011). In particular, in the case of fatigue resistance, the presence of surface roughness, porosity in the microstructure or lack of cohesive strength at the particle–particle boundaries is extremely critical improving the crack initiation by notch effect (Wong et al. 2012). Furthermore, the importance of deposition process optimization and the role of impact parameters on determining the coating quality is also well described in the literature and the better CS coating qualities are obtained with the more ductile materials such as pure metals (i.e. Cu, Al, Ni) exhibiting lower critical velocity and higher plastic deformation capability (Schmidt et al. 2006a, b; Assadi et al. 2011). However, these materials have also poor or almost low intrinsic mechanical properties so that it is not expected to obtain a significant increase in fatigue strength of the coated part. Therefore, *is it better to have a high-strength coating material deposited with poor quality or a low-strength coating material with excellent quality?* Unfortunately, there is neither a univocal answer nor a clear correlation between fatigue strength and coating properties (i.e. residual stress, surface roughness, porosity, YS, bond strength) to be able to trace some basic guidelines. In this sense, three examples of CS-coated Al alloys can help to depict how they spread the current scenario: the significant influence of coating material and its intrinsic mechanical properties in determining the fatigue behaviour of CS-coated A5052 substrates is reported in Ghelichi et al. (2012); a noticeable increase of fatigue life by using a good quality but humble CP-Al coating is reported in Ziemann et al. (2014) and finally the influence of performing CS

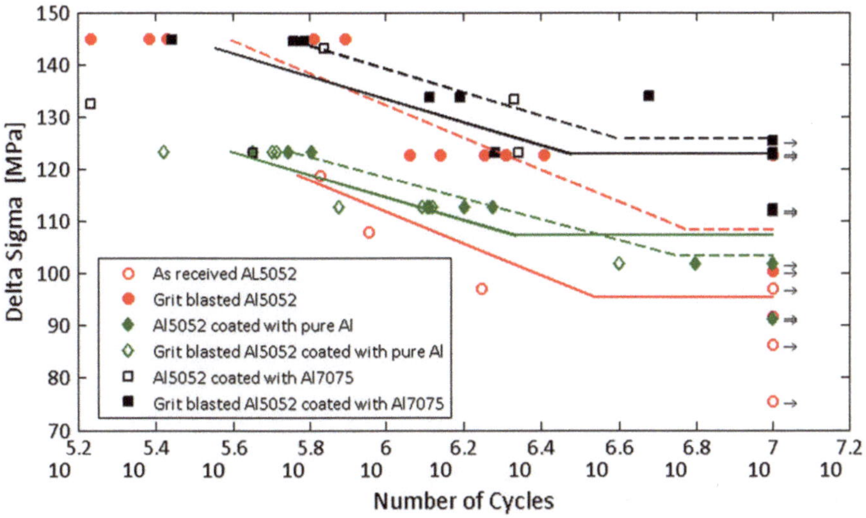

Fig. 4.63 S–N curves of A5052 as-received and grit-blasted substrates coated with CP-Al and A7075 alloy by low-pressure cold spray. (Ghelichi et al. 2012)

with the same material as the substrate, A6082, has been studied in Moridi et al. (2014a, b). Figure 4.63 showed the results reported in Ghelichi et al. (2012) supporting the primary effect of coating material on increasing both fatigue limit and the slope in the low-cycle regime.

Fatigue behaviour is reported to follow the fatigue strength of the stronger material (if a certain bond strength and coating quality can be ensured) among coating and substrate, strongly recommending the use of high-performance coating material. In this sense, up to 30 % improvement in fatigue limit is reported, for example, with A7075 coating on A5052 (Ghelichi et al. 2012) as shown in Fig. 4.63. Similar considerations are attributed to be the basis of the improvement of the HCF limit of AZ91D magnesium alloy by the deposition of a composite Al/Al_2O_3 coating (Xiong and Zhang 2014).

Moridi et al. (2014a, b) discuss the influence of a A6082 CS coating on fatigue strength of A6082 material; this study aims to exclude the influence of the coating material on fatigue behaviour focussing the attention on the characteristics induced by a CS-deposited layer. The coating exhibits good bond strength and compact microstructure so that its anchorage with base material is ensured. An ~15 % increase of the fatigue limit is reported supporting the thesis that CS technology can play a role on fatigue strength enhancement. On the other hand, the number of cycles to failure are almost unchanged with respect to uncoated specimens as well the slope of the curve in the low-cycle regime as emphasized in the S–N diagram in Fig. 4.64. In this sense, the observed propagation mechanism of macrocracks is quite the same in as-received and coated specimen. So, the coating is able to increase the threshold for crack propagation to be started, but once the crack starts to propagate, it results in the final fracture in more or less the same number of cycles of as-received specimens (Moridi et al. 2014a, b).

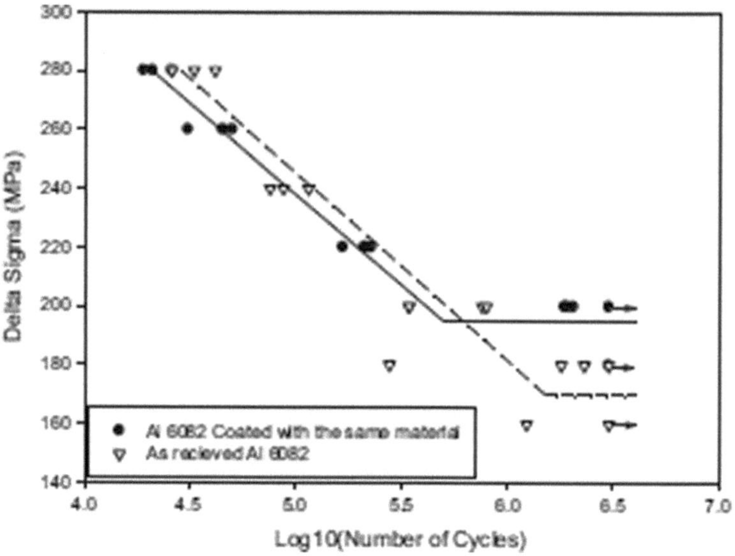

Fig. 4.64 S–N curves of as-received and A6082 cold-sprayed coated, A6082 substrates. (Moridi et al. 2014a, b)

Finally, a noticeable increase of the number of cycles to failure is reported in Ziemann et al. (2014) by combining grit-blasting and shot-peening pretreatment with CP-Al CS coating. While the coating material (CP-Al) exhibit lower performances with respect to the base material (A2024), a significant enhancement on fatigue life is reported.

Arouse one's interest in Ti alloy base material, the poor intersplat bonding and low modulus of a CP-Ti coating deposited by CS have been reported to be detrimental on fatigue strength of Ti–6Al–4V coatings (Price et al. 2006) Indeed, upon cycling, the intersplat decohesion leads to the formation of vertical cracks in which propagation and subsequent transfer to the substrate is the main cause responsible for premature failure (Clzek et al. 2013)

A first investigation on the fatigue life of stand-alone CS deposit is reported in AL-Mangour et al. (2013) in the case of AISI316L in comparison with bulk AISI316L. The CS coating has been post-annealed to promote diffusion and close particle–particle interfaces. The results, in agreement with the evolution of YS and ultimate strength, show that the cold-sprayed material exhibits a lower fatigue strength even after post-deposition annealing mainly due to the stiffness and the residual presence of pores, defects and morphological crack initiation sites at the particle–particle boundaries with respect to the bulk material.

Summarizing, both coating material and average coating quality play an active role in determining fatigue strength and fatigue life even if the numerous number of parameters involved and the wide field of investigation give back, at present, a scenario not yet clarified; hopefully, the support of more experimental results that are progressively enriching the literature will bring in clarity to this topic.

4.10.2.3 Effect of Residual Stresses on Fatigue Strength

The assignment of a compressive residual stress on the surface of a metal component is traditionally considered responsible for a beneficial effect on fatigue strength and life of the component (Schijve 2001). This represented the basis to develop shot-peening and grit-blasting surface treatments as currently already employed in several industrial procedures. By now, it is well known that CS deposition is able to ensure a compressive residual stress, thanks to the peening effect performed by the impacting particles as also reported in the previous section of this chapter. Therefore, the logical consequence is that residual stresses induced by CS are beneficial on fatigue strength; this is in general the truth, however, some factors must be considered: First of all, the compressive stress in the coating is generally balanced by an induction of tensile stresses on the substrate surface as reported, for example, in Rech et al. (2011). Such induced tensile stresses could lead to an earlier crack initiation and reduce the whole fatigue life as reported in the case of coated Ti–6Al–4V (Cizek et al. 2013; Price et al. 2006). Second, excessive compressive residual stress can affect bond strength leading to premature interfacial debonding and spallation. Then, the compressive residual stress state of a CS coating is mainly beneficial on fatigue strength of the metal component according to the results reported in the case of Al–Co–Ce on A2024 (Sansoucy et al. 2007) and A6082 (Moridi et al. 2014a, b) or CP-Al on A2024 (Ziemann et al. 2014). However, in order to have the capability to benefitfrom this effect, some requirements in terms of coating quality must be ensured. It is not coincidence that bond strength is reported to be excellent in all the cases previously mentioned (i.e. adhesion strength of 61 ± 4 MPa is reported in Sansoucy et al. (2007)) as well coating compactness and morphology (i.e. coating porosity in the range of 0.2–0.5 % (Ziemann et al. 2014)).

4.10.2.4 Effect of Surface Roughness on Fatigue Strength

The surface topography and roughness influences the fatigue behaviour in its first step, by providing a superior density of sites for crack initiation. The presence of corner, sharp edges, cavities or other topographic features induce an intensification of the local stress that promote crack nucleation and propagation, hence representing a drawback for the fatigue strength of a metal component. In this field, as-deposited thermal spray and CS coating exhibit a typically rough surface (average surface roughness, R_a, higher than 0.01 mm are reported) depending on employed feedstock powders and material characteristics (Papyrin et al. 2007). Indeed, surface roughness represented the other side of the coin when considering the effects on fatigue strength and must be minimized as much as possible to reduce its negative influence. CS of ductile materials can lead to greater particle deformation upon impact, hence promoting a low surface roughness and presence of defects; according to this trend, CP-Al coating is reported to improve the surface quality of grit-blasted A2024, improving its fatigue strength (Ziemann et al. 2014). At the same time, high surface roughness combined with relatively high porosity and surface/coating quality is reported to nullify the potential beneficial effects of A7075 nanostructured coatings (Ghelichi et al. 2014a, b). Post-deposition polishing or machining on as-deposited CS coatings can be useful to reduce the surface roughness and the surface defects as well as a post-deposition treatment such as shot peening as reported in Bageri et al. (2010); however, no literature up to now reports a deep investigation on this subjcct.

A summarizing table is reported collecting the main experimental procedures and results related to fatigue strength investigation on CS coatings (Table 4.3).

Table 4.3 Characteristics of cold spray deposits having a significant role in influencing the fatigue strength

Substrate material	Coating material	Test procedure	Performance	Reference
A2024-T3	Al–Co–Ce	ASTM B593	One order of magnitude enhancement CtF with respect to bare and Alclad (200 MPa stress)	Sansoucy et al. (2007)
A2024-T351	CP-Al	ISO 1143	+850% CtF (in combination with grit blasting—210 MPa stress)	Ziemann et al. (2014)
	CP-Al	ISO 1143	No significant influence (in combination with shot peening, 180 and 210 MPa)	Ziemann et al. (2014)
A5052	CP-Al	ASTM B593	No significant influence	Ghelichi et al. (2012)
	A7075	ASTM B593	+30% FL (in combination with grit blasting)	Ghelichi et al. (2012)
	A7075 (cryomilled nanostructured powders)	ASTMB593	No significant influence	Ghelichi et al. (2014a, b)
A6082	A6082	ISO 1143	+15% FL; no significant influence in FS	Moridi et al. (2014a, b)
AZ91D	Al/Al_2O_3	3-point bending	+20 MPa fatigue limit	Xiong and Zhang (2014)
Ti6Al4V	CP-Ti		−9% fatigue life	Cizek et al. (2013)
Ti6Al4V	CP-Ti	Rotating bending	Reduction 30–100 MPa in fatigue limit	Price et al. (2006)

CtF cycles to failure, *FL* fatigue limit, *CP* commercially pure

4.10.3 *Adhesive/Cohesive Strength*

Adhesion is a fundamental property in any type of coating/substrate system because it is closely related to the durability and longevity of the whole system. The performance and reliability of a coating depend on the mechanical integrity of coating/substrate systems, that is, the adhesion of coatings to their substrates. Therefore, there is a need to achieve and, at the same time, to evaluate the coating adhesion with simple and reliable methods (Chen et al. 2014). The experimental measurement of adhesion can be achieved by defining the force of adhesion (often defined as the maximum force per unit area) exerted when two materials are separated; or the work of adhesion, namely the work in detaching two materials from one another (Rickerby and Stern 1996). Many theories and mechanisms for thermally sprayed coatings have been proposed; none, however, covers all situations, and no adhesion test satisfies all requirements. Therefore, the best test method is often the one that simulates practical stress conditions (Lin and Berndt 1994; Mittal 1978). The main methods used to test thermally sprayed coatings are as follows (Pawloski 2008):

- A family of tests, based on the force of adhesion, including tensile adhesion test (TAT), also known as the "pull-off" method, pin test, shear test
- A family of tests based on fracture mechanics, including bending tests, the double cantilever beam (DCB) method and indentation test
- Other methods, such as scratch test and "laser-shock" test

4.10.3.1 TAT or "Pull-off" Method

The TAT has been used widely as a routine quality control tool for thermal spray and CS coatings. The TAT arrangement is illustrated in Fig. 4.65; at the center, a coated specimen is attached to a support fixture by epoxy so that a tensile force can be applied. Tensile strength results from division of the maximum load applied at rupture by the cross-sectional area. The adhesion strength of the coating is given if the failure occurs only at the coating–substrate interface. The cohesive strength of

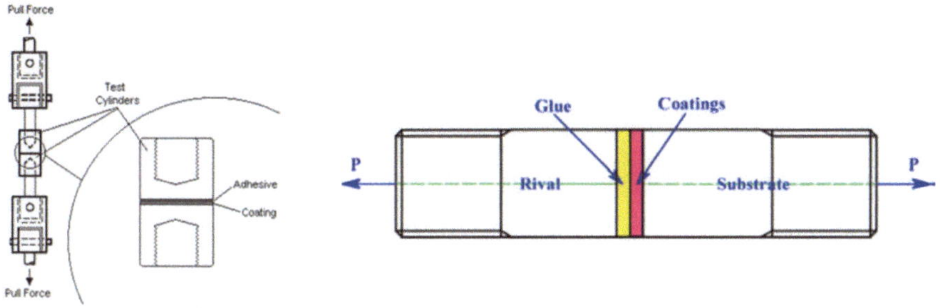

Fig. 4.65 Schematic representation of TAT arrangement

the coating is given if the rupture is entirely within the coating. Mixed-mode failure may also occur, making the interpretation of results difficult. Users generally apply the ASTM C633 Standard, but there are also other procedures and other national standards (for example EN 582 and JIS H8664). However, the TAT procedure has several shortcomings, including the penetration of epoxy and alignment of test fixtures. Another limitation of the ASTM test results from its limited strength, which does not exceed $p = 80–100$ MPa, which makes it impossible to test well-adhering coatings (Lin and Berndt 1994; Mittal 1978; Pawloski 2008).

Unlike other thermal spray techniques, quite thick coatings can be obtained by the process of CS. The adhesive strength of thick coatings can be measured using a novel testing method described in Huang and Fukanuma (2012) and shown in Fig. 4.66. First, thick coatings of more than 5 mm are deposited on a conventional tensile specimen with a diameter of 25 mm, as shown in Fig. 4.66a. Then, the test piece is machined to obtain a specimen as shown in Fig. 4.66b and mounted as represented in Fig. 4.66c. The part near the coating/substrate interface is cut thin to ensure that the rupture happens in that area during the tensile test. In spite of the arc transition used at the inner corner near the interface of coating/substrate, the stress concentration may cause the failure of some specimens near the inner corner; even though the geometry of a specimen has not been optimized, the lower limit of adhesive strength can be obtained.

Other pull-off tests that allow the measurement of coating adhesion on flat-plate specimens are currently employed for CS coatings (Marrocco et al. 2006;

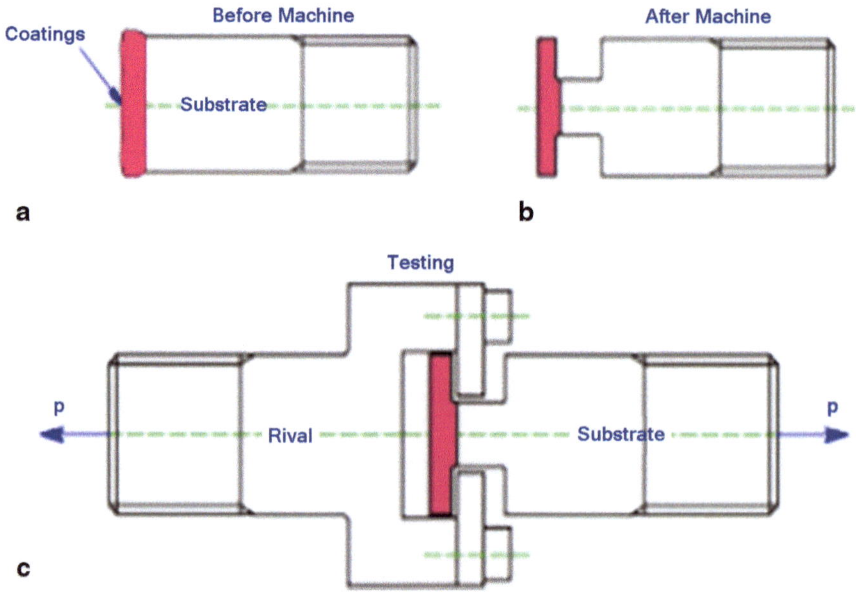

Fig. 4.66 The novel method to test the adhesive strength of coatings according to Huang and Fukanuma (2012)

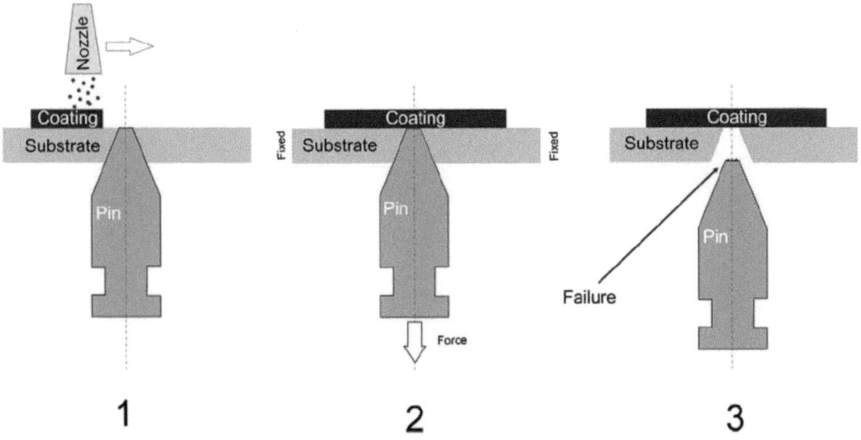

Fig. 4.67 Pin adhesion test method. *1* deposition of coating on substrate with flush-mounted pin, *2* strong fixation of substrate and application of force to a bottom end of the pin, *3* increase of the force until the coating ruptures. (Sova et al. 2013b)

Price et al. 2006; Van Steenkiste et al. 2002; Tao et al. 2009). The general pull-off test (described, e.g. in ASTM Standard D4541, Standard Test Method for Pull-Off Strength of Coatings Using Portable Adhesion Testers) is performed by securing a loading fixture (dolly, stud) normal (perpendicular) to the surface of the coating with an adhesive. After the adhesive is cured, a testing apparatus is attached to the loading fixture and aligned to apply tension normal to the test surface. The force applied to the loading fixture is then gradually increased and monitored until either a plug of material is detached or a specified value is reached.

Coating adhesion can also be measured using pin method as described, for example, in Smurov et al. (2010) and Sova et al. (2013b). Schematically, the method is illustrated in Fig. 4.67. A coating is deposited on the pin flush-mounted to the substrate. The pin has a shape of truncated cone with the diameter of the top base equal to 2 mm. Coating is deposited on the top base of the pin as well as on the substrate surface. After spraying, the substrate is fixed and the mechanical force is applied axially to the pin. The value of force necessary to detach the pin from the coating is suggested as adhesive or cohesive strength depending on the zone where the rupture takes place.

Shear test methods were developed for rapid evaluation of the adhesion/cohesion strength of a coating on its substrate without the need of gluing and curing, and better describes the behaviour of the coating when subjected to shearing loads (Spencer et al. 2012a, b; Wang et al. 2010a, b; Yandouzi et al. 2009). In this technique, a coating sample is submitted to shear loading in a direction parallel to the substrate/coating interface using a commercial hard metal plate as a punch, thus pressing against the coating while a sample holder maintains the substrate fixed during testing. The most common shear test standards are EN 15340 and ASTM F1044. The fracture mechanics approach to the evaluation of crack propagation is based on defining adhesion in terms of a stress intensity factor, *K*, or strain energy release

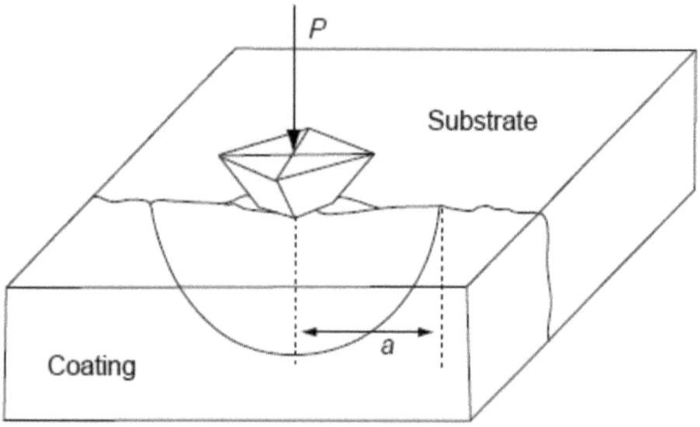

Fig. 4.68 Schema of the interfacial indention test. (Chicot et al. 1996)

rate, *G*. Methods of measurement include the DCB test, double torsion test, bending test (3- or 4-point), single-edge notched test, and compact tension test, but there are little experimental data on CS coatings in the literature (Ziemann et al. 2014).

Finally, an indentation test, shown schematically in Fig. 4.68, enables the determination of apparent interfacial toughness (Demarecaux et al. 1996; Marot et al. 2006; Chicot et al. 1996). The indentation test is realized on a polished cross section of a sprayed coating. A Vickers indentation is made at the interface of the coating with the substrate, using an appropriate alignment of the indenter. The crack generated by the penetration is localized along the interface and has a semicircular shape (Fig. 4.68). An appropriate mathematical treatment enables determination of the apparent interfacial toughness, K_c. Knowledge of the Young's moduli and hardness of the coating and substrate is necessary in the calculations.

The scratch test, originally studied by Benjamin and Weaver (1960), is often used to characterize thin, hard coatings. The scratch test consists of using an indenter which is moved linearly on the film surface under an increasing load. The load which corresponds to the detachment of the coatings is defined as a critical one. An example of using scratch test on cold-sprayed coating is reported in Seo et al. (2012a, b).

4.10.3.2 Adhesion of CS Coatings

Adhesion of CS coatings primarily depends on the bonding of the particles to the substrate surface. The bonding of particles in cold gas spraying is presumed to be the result of extensive plastic deformation and related phenomena at the interface (Assadi et al. 2003). Thus, most of the factors influencing the adhesion strength are the same influencing the cohesive strength of the coatings, already described in the previous sections. Higher bond strength and particle adhesion are obtained enhancing the particle deformation upon impact, for example, by increasing the particle impact velocity as reported in Fukanuma and Ohno (2004) and Huang and Fukanuma (2012), and, in this sense, the use of helium is more effective to obtain higher adhesion as reported in Vezzu et al. (2014 in the case of Waspaloy coatings.

For the same reasons, thanks to the more significant plastic deformation, coatings of ductile metals with lower melting points typically exhibit higher bond strength than coatings of high-strength alloys once the spraying parameters are defined as reported in Li et al. (2007a, b) and Koivuluoto et al. (2008a, b).

Second, bond strength certainly depends on substrate characteristics in terms of both mechanical properties and surface characteristics (roughness, chemical). As for coating material, ductile substrates able to provide significant plastic deformation upon impact often result in higher adhesion values (Fukanuma and Ohno 2004; Gärtner et al. 2006, Stoltenhoff et al. 2006, Fukanuma and Huang 2009); for the same reasons, a partial preheating of the substrate can be employed as a useful procedure to lower the YS of the material enhancing its plastic deformation and, as a consequence, the adhesion of the coating (Suo et al. 2012; Rech et al. 2011) even if excessive preheating can develop detrimental quenching stress in particular when substrate and coating materials have different thermal expansion properties. Looking at the surface state, the effect of the substrate preparation on the adhesion strength is a complex issue, and in this field different results are obtained depending on the combination of considered materials and spray parameters. For example, grit blasting often improves coating adhesion (Irissou et al. 2007; Danlos et al. 2010; Vezzu et al. 2014), thanks to the cleaning effect and crack of the surface oxide layer and also to the enhanced mechanical interlock provided by the surface roughening. Nevertheless, surface roughness is also reported to have no significant effect on the bond strength (Wu et al. 2006), or even to induce a slight decrease of the bond strength, for example, for a Ti6Al4V substrate coated with pure titanium (Price et al. 2006). This negative effect is due to the work hardening associated to the grit-blasting treatment of the substrate surface responsible for the reduced plastic deformation capability of the sprayed coating to bond to the substrate. Other preparation methods can be employed, and their effect has been reported such as polishing, shoot peening, chemical etching (Irissou et al. 2007) or laser ablation and heating (Danlos et al. 2010). Laser ablation in particular is reported to be effective on surface cleaning as well as on promoting intimate bonding between coating and substrate even if the local thermal input can lead to the development of thermally affected zones as well as of significant quenching stress also in case of temperature-sensitive materials as aluminum alloys, as reported in Danlos et al. (2010). However, it must be taken into account that this process modifies also the surface structure and morphology leading to the loss of one essential characteristic of CS deposition.

A further way to influence and improve the bond strength is the possibility to blend the initial powder feedstock with some weight percent of hard, often ceramic, filler such as aluminum oxide or silicon carbide. Indeed, the presence of hard particles in the powder feedstock typically has a beneficial effect on coating adhesion, as reported by many authors (Lee et al. 2005; Irissou et al. 2007; Tao et al. 2009; Wang et al. 2010a, b; Spencer et al. 2012a, b) mainly because of the increased plastic deformation of the ductile particles due to the additional peening effect performed by the hard and essentially un-deformable impinging particles. However, by further increasing the volume fractions of hard particles in the powder feedstock, and as a consequence in the coating microstructure, this gain in bond strength must be balanced with an increased ratio of weak hard–hard and hard–ductile interfaces in respect to strong ductile–ductile interfaces resulting in a progressively reduced coating cohesion and adhesion (Wang et al. 2010a, b; Sevillano et al. 2013). The impinging of the substrate surface with hard particles during the coating growth is reported to be effective to clean the surface removing low adhered stick particles, native oxide and surface defects hence promoting adhesion (Tao et al. 2009). Finally, the thermal evolution of the substrate/coating interface and the local substrate temperature during the deposition of the first layer is fundamental to determine interface reactivity with (eventual) oxide formation, residual stress state and the particle plastic deformation so that some further parameters such as substrate shape and clamping as well as gun velocity and standoff distance can influence the thermal input on the growing coating having a role in the determination of final bond strength.

A review of the bond strength reported in the literature for a wide range of coating and substrate materials is reported in Tables 4.4 and 4.5 concerning coatings obtained using nitrogen or helium as process gas, respectively.

Table 4.4 Summary of bond strength for several substrate/coating pairs. All coatings are deposited with process gas nitrogen

Coating material	Substrate material	Substrate preparation	Bond strength		Reference
Al	AZ91 Mg	Grit blasting	18	Stud-pull	Tao et al. (2009)
Al	AA6061		24	?	Lee et al. (2005)
Al	Al7075	Grit blasting	40	ASTM C-633	Irissou et al. (2007)
Al				Stud-pull	Van Steenkiste et al. (2002)
Al	AA2024-T351	Grit blasting,	40	ASTM C-633	Ziemann et al. (2014)
Al	Al7075-T651	Polished, grit blasted, shot peening, chemical etching	30–40	ASTM C-633	Irissou et al. (2007)
Al	ZE41A-T5 Mg		>43	ASTM C-633	DeForce et al. (2011)
AA2319	Steel	Grit blasting	34	ASTM C-633	Li et al. (2007a, b)
AA6061	AA2017	Degreased surfaces	28	ASTM C-633	Danlos et al. (2010)
AA6061	AA2017	Grit blasting	36	ASTM C-633	Danlos et al. (2010)
AA6061	AA2017	Ablated by laser	51	ASTM C-633	Danlos et al. (2010)
AA6061	AA2017	Heated and ablated by laser	65	ASTM C-633	Danlos et al. (2010)
AA6082	AA6082	Grit blasting	24	ASTM C-633	Moridi et al. (2014a, b)
Al–5Mg	ZE41A-T5 Mg	Grit blasting	51.7	ASTM C-633	DeForce et al. (2011)
Al–12Si	Steel	Grit blasting	>50	ASTM C-633	Li et al. (2007a, b)

Table 4.4 (continued)

Coating material	Substrate material	Substrate preparation	Bond strength		Reference
Al–12Si	Steel	Grit blasted	20–70	Stud-pull	Wu et al. (2006)
Al–12Si	Steel	Grit blasting	>50	ASTM C-633	Li et al. (2007a, b)
$Al+Al_2O_3$	AZ91 Mg	Grit blasting	32	Stud-pull	Tao et al. (2009)
$Al+Al_2O_3$	AA6061		45	–	Lee et al. (2005)
$Al+Al_2O_3$	Al7075	Grit blasting	>60	ASTM C-633	Irissou et al. (2007)
Cu	Copper	Grit blasting	17	JIS H 8664	Fukanuma and Ohno (2004)
Cu	Aluminum	Grit blasting	24	JIS H 8664	Fukanuma and Ohno (2004)
Cu	Aluminum	Grit blasting	>40	ASTM C-633	Gärtner et al. (2006)
Cu	Steel	Grit blasting	10–20	ASTM C-633	Gärtner et al. (2006)
Cu	Aluminum, Copper	Grit blasting	40	EN 582	Stoltenhoff et al. (2006)
Cu	Steel	Grit blasting	10	EN 582	Stoltenhoff et al. (2006)
Cu	Copper, AA5052, AA6063		>100	Modified tensile test	Huang and Fukanuma (2012)
$Cu+Al_2O_3$	Copper, steel	Grit blasting	20–23	EN582	Koivuluoto et al. (2008a, b)
Fe+Al	Steel	Grit blasting	38	ASTM C-633	Yang et al. (2011)
Mg	Aluminum	Grit blasting	10	ASTM C-633	Suo et al. (2012)
Ni	Steel	Grit blasting	25	ASTM C-633	Li et al. (2007a, b)
Ni	Aluminum		>50		Fukanuma and Huang (2009)
Ni	Copper		40		Fukanuma and Huang (2009)
Ni	Stainless steel		35		Fukanuma and Huang (2009)
$Ni+Al_2O_3$	Copper, steel	Grit blasting	8–9	EN582	Koivuluoto et al. (2008a, b)
$Ni–Cr_3C_2$	Steel	Grit blasting	27.5–39.5	ASTM C-633	Wolfe et al. (2006)
NiCoCrAlTaY	Steel	Grit blasting		ASTM C-633	Li et al. (2007a, b)
Stellite 6	Steel	Ground	53 (cohesive)	ASTM C-633	Cinca et al. (2013a)
Ti	Steel	Grit blasting	15	ASTM C-633	Li et al. (2007a, b)
Ti	Stainless steel	Grit blasting	19	JIS H 8664	Fukanuma and Ohno (2004)
Ti	AA 7075-T6	Degreased and scratched	34	ASTM C-633	Cinca et al. (2010)

Table 4.4 (continued)

Coating material	Substrate material	Substrate preparation	Bond strength		Reference
Ti6Al4V	Steel	Grit blasting	10	ASTM C-633	Li et al. (2007a, b)
WC–Co	AA7075 T6	SiC paper	76	ASTM F1147	Dosta et al. (2013)
WC–Co	AA7075 T6	SiC paper	19–26	ASTM C-633	Couto et al. (2013)
$Zn+Al_2O_3$	Copper, steel	Grit blasting	33–38	EN582	Koivuluoto et al. (2008a, b)

Table 4.5 Summary of bond strength for several substrate/coating pairs. All coatings are deposited with process gas helium

Coating material	Substrate material	Substrate preparation	Bond strength		Reference
Al	AZ91 Mg	Grinding	20	Shear test	Wang et al. (2010a, b)
Al, AA6061	AZ91 Mg	SiC paper	30–36	ASTM C-633	Spencer et al. (2009)
AA4047	ZE41A-T5 Mg		>37	ASTM C-633	DeForce et al. (2011)
AA5356	ZE41A-T5 Mg		>35	ASTM C-633	DeForce et al. (2011)
AA2224	AA2224		65	EN 582	Stoltenhoff and Zimmermann (2009)
AA6061	AlSi1		28	EN 582	Stoltenhoff and Zimmermann (2009)
AA7075	AA7075		30	EN 582	Stoltenhoff and Zimmermann (2009)
Al–5Mg	ZE41A-T5 Mg	Grit blasting	60	ASTM C-633	DeForce et al. (2011)
Al–12Si	AA6061-T6	Grit blasted	49	ASTM C-633	Sansoucy et al. (2008)
Al–12Si	AA6061-T6	Grit blasting	21	Shear test (EN15340)	Yandouzi et al. (2009)
Al–12Si+SiC	AA6061-T6	Grit blasting	43	ASTM C-633	Sansoucy et al. (2008)
Al–12Si+SiC	AA6061-T6	Grit blasting	16–20	Shear test (EN15340)	Yandouzi et al. (2009)
$Al+Al_2O_3$	AZ91 Mg	Grinding	40	Shear test	Wang et al. (2010a, b)
CP $Al+Al_2O_3$ $AA6061+Al_2O_3$	AZ91 Mg	SiC paper	40	ASTM C-633	Spencer et al. (2009)

Table 4.5 (continued)

Coating material	Substrate material	Substrate preparation	Bond strength		Reference
Cu	Aluminum	Grit blasting	30–35	ASTM C-633	Taylor et al. (2006)
Cu	Copper, AA5052, AA6063		>150	Modified tensile test	Huang and Fukanuma (2012)
SS + Al_2O_3	AZ91 Mg	SiC paper	25–60	Shear test	Spencer et al. (2012a, b)
Ti	stainless steel	Grit blasting	50	JIS H 8664	Fukanuma and Ohno (2004)
Ti	Ti6Al4V	Polished, ground	22	PAT	Marrocco et al. (2006)
Ti	Ti6Al4V	As-received, grit blasted	32–37	PAT	Price et al. (2006)

CP commercially pure, *SS* stainless steel, *PAT* Process Analytical Technology, *SiC* silicon carbide

4.10.3.3 Laser Shock Adhesion Test: LASAT®

The LAser Shock Adhesion Test, namely LASAT®, has been especially developed in France to thermally sprayed coatings for the past two decades (Berthe et al. 2011). This development followed a former US work by Gupta (1995) for planar interfaces and thin coatings only. The best stimulus for LASAT® was to accommodate for weaknesses of conventional adhesion tests, primarily "pull-off" testing (Sect. 4.3.5.1). Well-known limitations of the latter relate to the application to high bond strength or porous coatings due to the use of glue. In addition, pull-off testing is rather tedious, time consuming and discrepant, in a limited range of experimental conditions.

- *The Coating substrate bond strength* can be determined through the exploitation of laser shock effects in LASAT® (Arrigoni et al. 2012; Fig. 4.69). First, a (primary) compression wave due to a laser pulse of a few GW.cm^{-2} for a few ns typically. This wave is followed by a release wave. Both waves propagate through the materials system till the primary wave reflects from the free surface of the coating surface in another rarefaction wave. When the two rarefaction waves intersect, a tensile stress forms, to which the coating–substrate interface is submitted. The stress level from which the interface fails corresponds to the bond strength. Stresses can be calculated through the simulation of shock wave propagation using numerical codes such as SHYLAC or RADIOSS. Debonding can be detected from real-time monitoring by interferometry or post-mostem observation of the materials. Two laser shock loading regimes are conventionally used, that is, direct irradiation (Fig. 4.69a) when irradiating the bare material, and the so-called confined regime (Fig. 4.69b) in which the plasma due to laser beam–surface interaction is confined by a transparent medium such as water or glass. Confining the plasma increases laser shock pressure therefore stress loading.

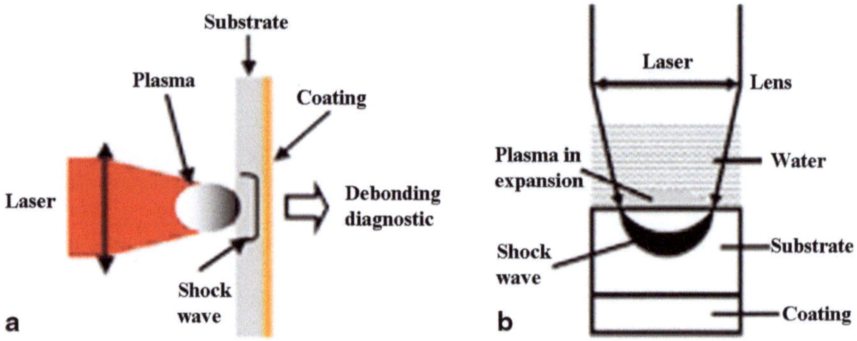

Fig. 4.69 Schematic illustrations of LASAT **a** in the direct irradiation regime and **b** in the confined regime. (After Berthe et al. 2011)

- In addition to the determination of the coating–substrate bond strength in conventional conditions, LASAT® can be performed at a temperature different from room temperature, in a specific atmosphere and/or in a liquid (Guipont et al. 2010). Multilayered materials can also be characterized. LASAT® is very suitable for CS coatings (Blochet et al. 2014; Koivuluoto et al. 2013; Giraud et al. 2012, for the most recent publications), which could highlight the role of oxidation on bond strength in particular (Christoulis et al. 2010; Barradas et al. 2005). Relevant successful advances relate to the feasibility of applying LASAT® to the surface of the coating rather than to the back of the substrate (Begue et al. 2013) and to small-sized materials systems such as splats (Jeandin 2011; Fig. 4.70). The latter was very attractive for investigation into CS in so far as a local approach to phenomena and properties is powerful (see Sects. 4.1 and 4.2). The role of particle oxidation state on adhesion could be exhibited. The bond strength of a splat (for given spraying conditions and material) can decrease dramatically with oxide layer thickness, for example, from 415 to 280 MPa typically for CS Cu onto Al (Fig. 4.70).

- *Coating cohesion*, that is, splat–splat adhesion strength can also be determined using LASAT® using a similar testing process (Barradas et al. 2005).

Two LASAT® variants, still based on laser shock testing, were also developed to go into phenomena which can be considered as very relevant to CS and related understanding of the process.

- *LASERFLEX* consists of light amplification by stimulated emission of radiation (LASER) flier experiments in which the flier can be a particle or a foil. Since in-flight conditions can be similar to those encountered in CS, LASERFLEX is claimed to be a tool for experimental simulation of CS (Barradas et al. 2007; Jeandin 2011). Advantages are that this simulation is easier to be controlled and implemented compared to CS.

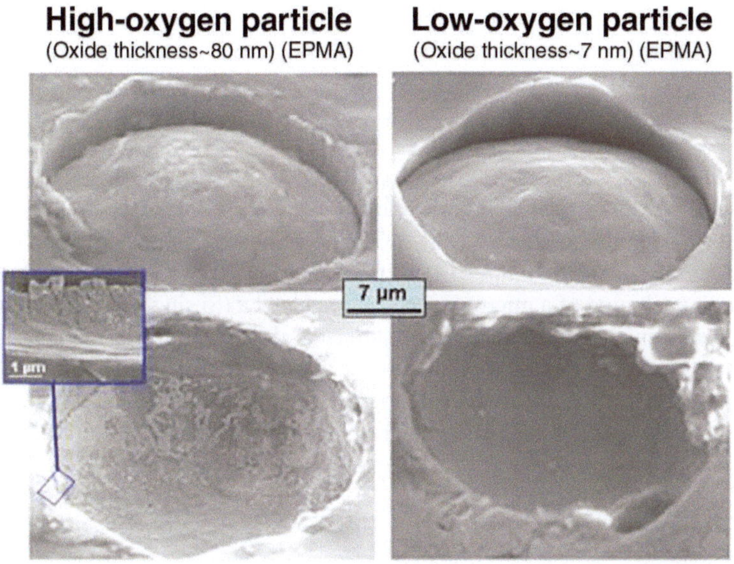

Fig. 4.70 SEM images of the Cu splat—Al substrate system, *(top)* before and *(bottom)* after LASAT® for two-particle oxidation levels (after Jeandin 2011). The *upper pictures* show embedded splats, the *lower pictures* show the corresponding craters after splat removal due to LASAT®. *EPMA* electron probe microanalysis

- *CLASS, that is, Cold LAser Shock Spray,* is based on the use of a laser shock to spall off a coating to respray a coating, which could have been obtained by CS (Fig. 4.71; Jeandin et al. 2014). CLASS could be developed. First, this can be expected to be an advanced test to characterize the property gradient within a cold-sprayed coating, though the CLASS process efficiency parameter, for example. Second, CLASS could be a new spraying process using particulate material which would exhibit unique starting properties since made of already-sprayed powder.

From the development of LASAT® and its variants, one may expect that of a novel integrated laser-shock control chain to test powders, for example, for cost-effective,

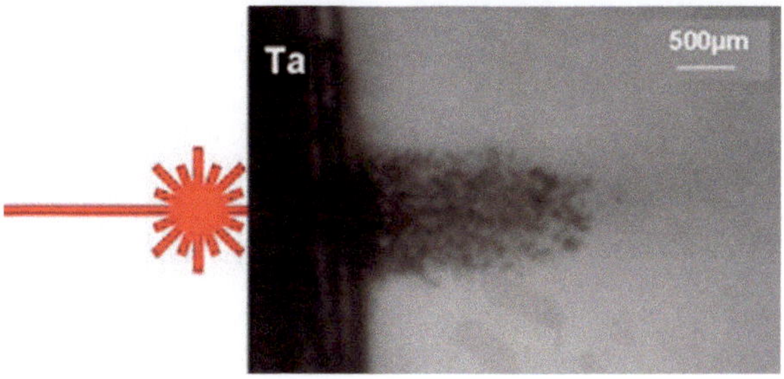

Fig. 4.71 Shadowgraph image of CLASS'ed Ta cold-sprayed coating. The laser pictogram schematizes the laser shock irradiation. (After Jeandin et al. 2014)

rapid and powerful control at reception of batches. This would consist in LASA-Testing of splats obtained from LASERFLEX'ed particles onto substrate samples.

4.10.3.4 Microtensile Test

Microtensile testing was recently developed using focussed ion beam (FIB) facilities and successfully applied to cold-sprayed coatings for characterization of interface strength (Ichikawa et al. 2014). FIB is used for various operations, that is, in situ micromachining of a microtensile specimen and a (Si single crystal) microcantilever beam, microsample picking up, tungsten deposition to fix the specimen to the cantilever and to the microprobe, and scanning ion microscopy (SIM) for real-time monitoring of the test in the chamber (Fig. 4.72).

The applied load, *F,* is calculated from the displacement, *d,* of the cantilever beam, that is,

$$F = \frac{Ewt^3}{4l^3}d,$$

where E is the elastic modulus, l thelength, w the width and t the thickness of the cantilever.

The rupture stress is equal to the load divided by the fracture area. This area is determined through SEM observation of the fracture surface of the specimen after testing.

Microtensile testing is quite suitable for comparing the strength of the core of splats to that of splat–splat interfaces within the coating and to that of the coating itself, the latter being determined using conventional pull-off testing. Great differences could be shown for cold-sprayed Cu, for example. These were about 670, 180 and 350 MPa when considering interfaces which exhibited microvoids (Ichikawa et al. 2014). The latter were therefore acertained to be major influencing factors on the macroscopic strength of the coating. Micotensile testing shows a high potential to correlate smal-sized microstructure features (Sect. 4.1.) to mechanical properties.

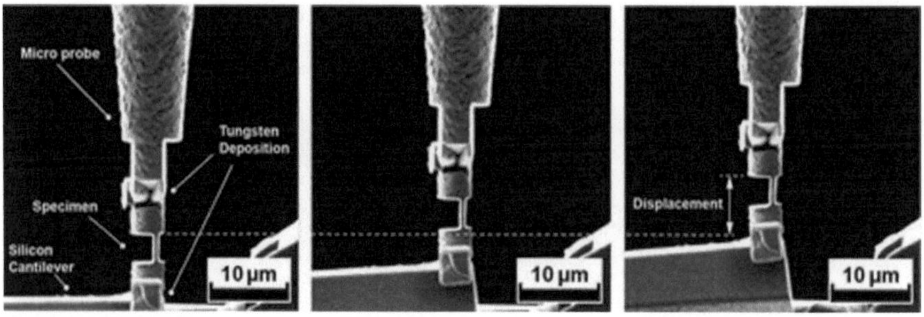

Fig. 4.72 SIM images of the microtensile test set-up (from *left* to *right*), before testing, during testing and just before the rupture of the specimen. (After Ichikawa et al. 2014)

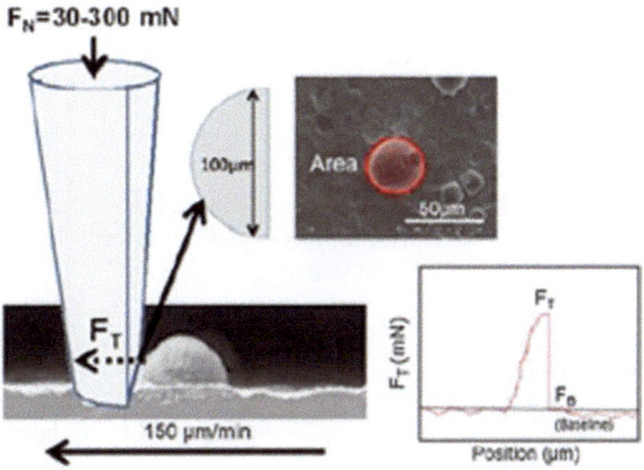

Fig. 4.73 Schematic illustration of splat adhesion test including cross-sectional and top views of the splat in the tested area plus a corresponding force–displacement diagram when shearing the splat. (After Goldbaum et al. 2011)

4.10.3.5 Modified Ball-bond Shear Test

Ball-bond shear testing can be modified to be a splat adhesion technique using micro-scratch tester. As described by Goldbaum et al. (2012), the technique consists in applying a normal force, F_N, onto a stylus that is placed on the substrate at a certain distance of the splat, the adhesion of which has to be measured. The substrate is then moved horizontally below the stylus (Fig. 4.73). The splat–substrate bond strength is calculated as a function of the tangential force, F_T, exerted on the stylus, the baseline force, F_B, and the splat area, A, according to the following equation:

$$\frac{F_N - F_B}{A}.$$

The splat area is determined using optical observation prior to splat shearing. This test can be employed for rather rapid characterization of the influence of various spray processing parameters including the particle diameter. This can complement LASAT® (Sect. 4.3.6.3) due to a different loading mode. Both tests are required as an approach to splat adhesion since assessing the splat-flattening ratio is not enough actually.

4.11 Wear Resistance

Wear properties of the cold-sprayed coatings are reported rather briefly in the literature. These studies are mostly concentrated on friction of coefficient and sliding wear studies. Some examples of the wear properties of cold-sprayed coatings are shown in Table 4.6.

In one research, cold-sprayed nanocrystalline Cu coatings were successfully sprayed, and they showed lower wear rate compared to conventional cold-sprayed

Table 4.6 Selection of wear properties of cold-sprayed coatings

Author	Coating/substrate	Wear properties
Guo et al. 2015	HPCS (Cu–8wt%Sn) +9.5/36.8/57.6 vol.% AlCuFeB on mild steel	Improved CoF with mixturing quasicrystal particles to CuSn powder. Wear resistance improved with CuSn+9.5%QC compared with CuSn coating
Li et al. 2011	Heat-treated (950 °C, 5 h) HPCS Fe60Al40 on stainless steel	Higher abrasive wear resistance of heat-treated FeAl coating at room temperature and at elevated temperature compared with stainless steel
Pitchuka et al. 2014	Al amorphous/anocrystalline (Al–4.4Y–4.3Ni–0.9Co–0.35Sc (at.%)) alloy coatings on Al6061	As-sprayed coatings had higher CoF value (0.55) compared to 0.38 for heat-treated coatings
Melendez et al. 2013	LPCS WC–12Co+Ni on low carbon steel	Comparable abrasive wear resistance of LPCS WC-based coating compared with HVOF and HPCS WC-based coatings
Shockley et al. 2013	C<s Al+10wt% Al_2O_3	Stable friction coefficient behaviour and lower dry sliding wear of Al+Al_2O_3 coating compared to Al coating
Guo et al. 2009	Tin–bronze/Tin and tin–bronze/quasicrystal (AlCuFeB) composite coatings	Composite coatings present lower friction efficient than the pure bronze coating. This can be attributed to the higher hardness of the composite coatings
Shockley et al. 2014	Cold-sprayed Al, Al+Al_2O_3 coating on mild steel substrate	Dry sliding friction of sapphire against cold-sprayed Al–22 wt% Al_2O_3 coating was lower and more stable than the cold-sprayed pure Al

COF coefficient of friction, *LPCS* low-pressure cold spray *HPCS* high-pressure cold spray, *HVOF* high-velocity oxygen fuel, *WC* tungsten carbide

coatings (Fig. 4.74). Nanocrystalline Cu has the potential to be used as a coating material in the bearing parts (Liua et al. 2012).

In the other study, Li et al. (2011) have shown lower abrasive wear rate of heat-treated cold-sprayed FeAl coating compared with 2520 heat-resistant SS at room temperature and at high temperature (800 °C). In this case, dominant abrasive wear mechanism was suggested to be a microcutting (Li et al. 2011). On the other hand, dry sliding wear behaviour of as-sprayed and heat-treated cold-sprayed Al amorphous/nanocrystalline alloy coatings was studied. Heat-treated coatings exhibited higher wear resistance and lower coefficient of friction (CoF) than as-sprayed coatings (Pitchuka et al. 2014), whereas cold-sprayed Al-based coatings showed better dynamic friction properties than Al bulk material (Attia et al. 2011). Furthermore, improved wear properties of cold-sprayed Al–Al_2O_3 composite coatings have been reported by Shockley et al. (2013). CoF were found to be lower with increased

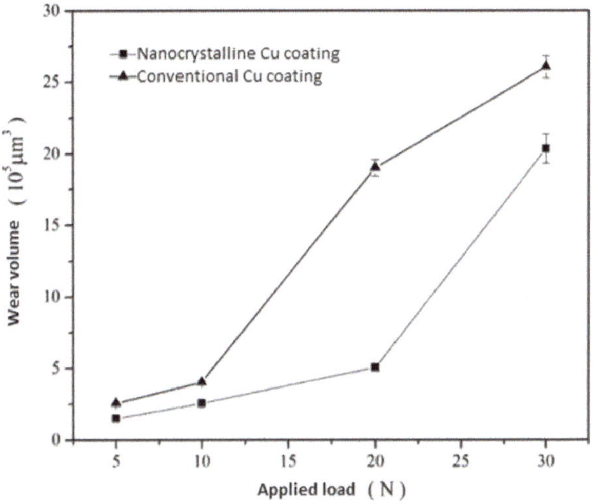

Fig. 4.74 Wear volume versus the applied load for the cold-sprayed nanocrystalline Cu and the cold-sprayed conventional Cu coatings. (Liua et al. 2012)

amount of Al_2O_3 in the coating. In addition, wear rates were decreased with increased Al_2O_3 content as presented in Fig. 4.75 (Shockley et al. 2013).

Spencer et al. (2009) have studied wear properties of cold-sprayed (kinetic metallization) Al and Al+Al_2O_3 coatings. Sliding wear rate was significantly decreased with increasing amount of Al_2O_3 particles of as-sprayed and heat-treated Al and 6061 Al alloy coatings, Fig. 4.76. Wear type was changed from adhesive to abrasive while using Al+75 Al_2O_3 instead of Al coatings. Figure 4.77 presents SEM images of worn surfaces of cold-sprayed Al and Al+75% Al_2O_3 coating surfaces after sliding wear test (Spencer et al. 2009).

Wear properties of cold-sprayed WC-based metal matrix composite (MMC) coatings were reported in Melendez et al. (2013). LPCS WC–12 wt% Co blended with Ni on low carbon steel showed decreasing abrasive wear resistance with increasing

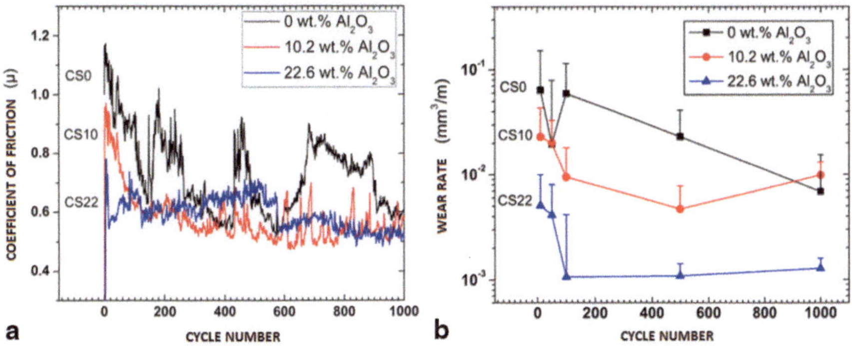

Fig. 4.75 Results of cold-sprayed Al, Al–10.2 wt% Al_2O_3 and Al–22.6 wt% Al_2O_3 coatings in in situ tribometry tests. **a** Average friction of coefficient versus cycles. **b** Dry sliding wear rate versus cycles. (Shockley et al. 2013)

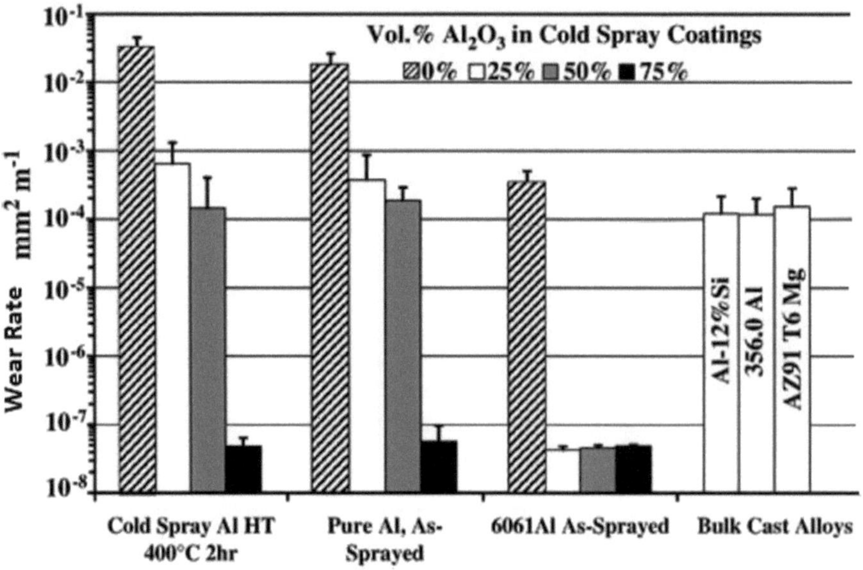

Fig. 4.76 Wear rate of cold-sprayed Al and Al+Al_2O_3 coatings after ball-on-disk wear studies compared to bulk alloys. (Spencer et al. 2009)

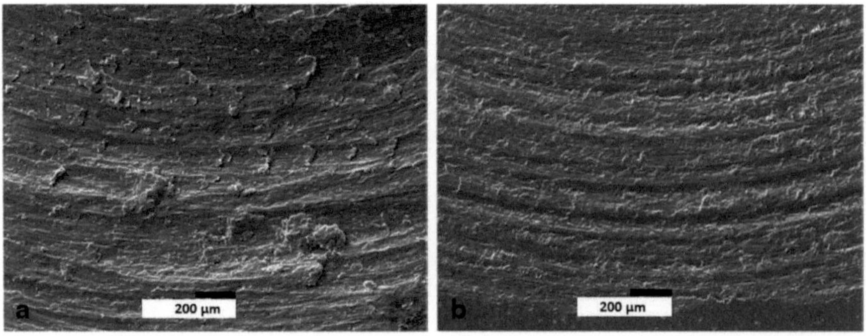

Fig. 4.77 Wear tracks after ball-on-disk wear tests of **a** cold-sprayed Al and **b** cold-sprayed Al+75 % Al_2O_3 coatings. SEM images. (Spencer et al. 2009)

amount of WC particles in the powder mixture (Melendez et al. 2013). On the contrary, Luo et al. (2014a, b) have found that cold-sprayed 20 vol.% cBN–NiCrAl nanocomposite coating had comparable two-body abrasion wear rate with HVOF-sprayed WC–12Co coating. Furthermore, heat-treated (750 °C, 5 h) cBN–NiCrAl nanocomposite coatings increased wear resistance by 33 % compared with as-sprayed coating.

Some researchers have been focused on self-lubricating cold-sprayed coatings. For example, Stark et al. (2012) have found that embedded hBN particles in Ni and Ni–phosphorous coating reduced 40 % of CoF and the wear volume of 25 % during a reciprocating wear test compared with pure Ni cold-sprayed coatings. In addition,

Guo et al. (2015) have studied mechanically blended mixtures of AlCuFeB quasicrystal and bronze powders sprayed by CS process in order to tailor quasicrystal-reinforced MMC coatings. It is stated that, quasicrystal materials have unique properties, such as low surface energy, high hardness, low CoF and good wear and corrosion resistance and thus interesting materials sprayed with CS processes. Porosity level has been decreased with increasing amount of quasicrystal particles in the MMC coating. In addition, CoF decreased slightly with increasing amount of quasicrystal particles in the CuSn powder, whereas wear resistance improved only with low amount of quasicrystal particles due to their reinforcing effect (Guo et al. 2015).

4.12 Corrosion Resistance

During the past years, corrosion studies and research of cold-sprayed coatings and the publications have been rapidly increased. In this part, corrosion is shortly described as phenomena. Cold-sprayed coatings have shown their potential to be used as corrosion barrier coatings due to their dense and impermeable coating structures. Therefore, selection of corrosion properties of cold-sprayed coatings is presented.

4.12.1 Corrosion

Corrosion is related to the chemical or electrochemical reaction between material (e.g. metal, coating) and its environment (Jones 1996). Corrosion resistance is necessary in several industries, in, for example, chemical and process equipment, and energy production systems. Generally, corrosion protection of metals is based on anodic protection by passivity or cathodic protection by sacrificial anode behaviour. Basically, corrosion starts if protection fails or breaks down, making a metal vulnerable to attacks of corrosion (Talbot and Talbot 1998).

Corrosion can occur in different forms, for example, uniform, pitting, crevice and galvanic corrosion are typical forms of corrosion for coatings. The most common forms of localized corrosion are pitting and crevice corrosion in which corrosive conditions could penetrate relatively rapidly (Jones 1996). *Pitting corrosion* occurs if the passive layer of protecting material is locally damaged, pits form on the surface and underlying metal is open for the attack (Frankel 2003). Therefore, pitting corrosion causes highly localized damages (Schweitzer 1996). For coatings, where structures are not totally adhering and uniform, pitting corrosion takes place quite easily. Porosity in the anodically protective coatings accelerates pitting type corrosion by opening the way for aqueous solution to penetrate inside the coating structure (Chatterjee et al. 2001). *Crevice corrosion* is said to be one of the most damaging forms of corrosion, and it causes localized corrosion (Kelly 2003). *Galvanic corrosion* occurs when two dissimilar metals form an electrical couple in the same electrolyte. In the electrical couple, corrosion starts in the less corrosion-resistant material (more active) which becomes the anode, whereas the more resistant material (nobler) will be cathode (Baboian 2003). Metallographic structure and microstructural properties affect corrosion behaviour (Jones 1996). Cathodic materials protect less noble materials anodically, whereas anodic materials give cathodic protection to the nobler material. The cathodic protection is based on sacrificial behaviour of anodic material, for example, Zn-based coating on steel substrate. In that case, porosity is not critical, whereas in the anodically protective coatings, impermeability of the coating and passive layer on its surface is critical for corrosion resistance.

4.12.1.1 Corrosion Tests

Open-cell potential measurements and salt spray (ASTM B117 standard) tests as wet corrosion tests are relevant methods to evaluate the denseness (density, impermeability) of coatings on corrodible substrate (e.g. carbon steel in saltwater conditions). Additionally, salt spray (fog) testing is a commonly used test method to evaluate the quality of various coatings. This particular test enables the use of different corrosive solutions and different test temperatures in a controlled test condition (B117-90 1992). Corrosion protectiveness and corrosion rates can be estimated with polarization behaviour of the coatings (Schweitzer 1996). Polarization measurements are widely used in the corrosion studies of cold-sprayed coatings. Furthermore, corrosion properties can be studied with more application-related tests, for example, hot corrosion tests and electrochemical corrosion tests in certain specific exposure conditions.

Open-Cell Potential Behaviour

Existing interconnected porosity (through porosity or open porosity) can be evaluated by using open-cell potential measurements. While microscopic characterization reveals structural details of the coatings, denseness can be analysed by corrosion studies. Figure 4.78 illustrates the potential behaviour of coating, substrate and porous coating. If the coating contains interconnected porosity, potential behaviour is composed of potentials of both coating and substrate which is seen as mixed potential.

If the open-cell potential value of the coating approaches the value of the corresponding bulk material, it indicates impermeable and dense coating structure. However, if the value of the coating approaches the value of the substrate material, it reflects the through porosity in the coating structure. In such situations, testing liquid has an open access to penetrate from the surface of the coating to the interface between coating and substrate, it will corrode the substrate, and corrosion products will come up to the surface. Figure 4.79 presents the open-cell potential behaviour of dense cold-sprayed coatings (Ta, Cu, Ni) and cold-sprayed coating with open porosity (NiCr).

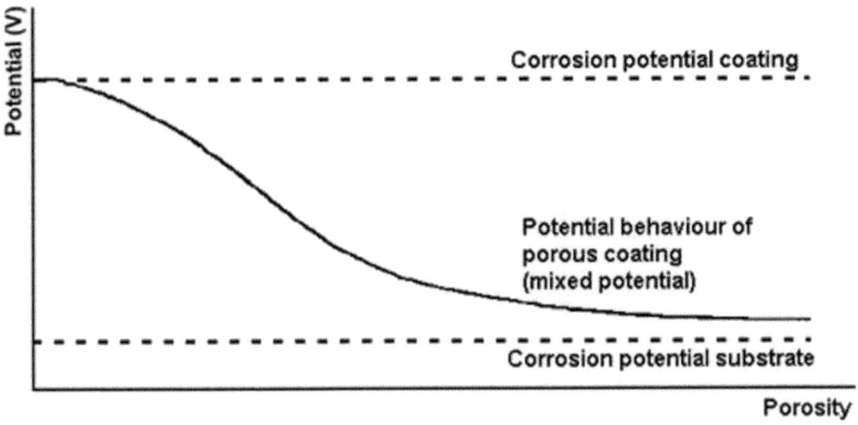

Fig. 4.78 Potential behaviour of the coating as a function of porosity. (Vreijling 1998)

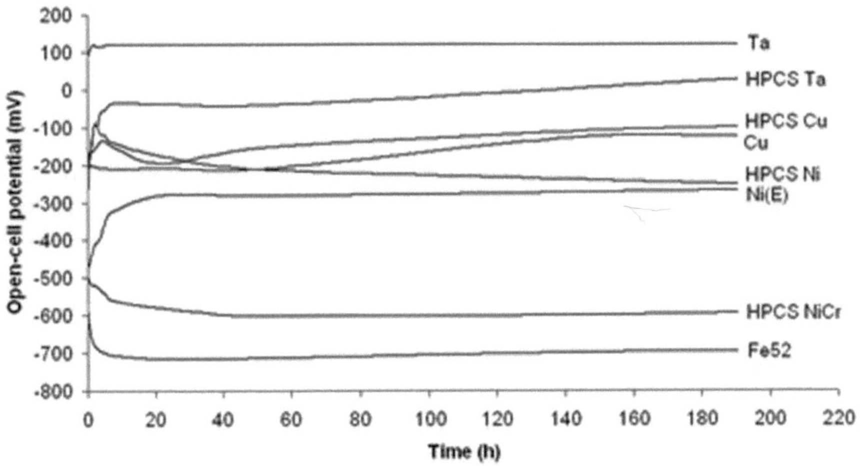

Fig. 4.79 Open-cell potential behaviour of Ta and Cu bulk materials, electrolytically prepared Ni (Ni(E)), cold-sprayed Ta, Cu, Ni and NiCr coatings and Fe52 substrate material in 3.5 % NaCl solution (Ag/AgCl reference electrode). *HPCS* high-pressure cold spray. (Modified after Koivuluoto et al. 2008b)

4.12.2 Corrosion Properties of Cold-Sprayed Coatings

Corrosion properties and behaviour of the cold-sprayed coatings are increasingly reported in the literature during the past years. Recently, review papers concerning corrosion properties of cold-sprayed coatings have been published by Bala et al. (2014) and Koivuluoto and Vuoristo (2014). Most of the corrosion studies of cold-sprayed coatings are concentrated on the corrosion protection of materials which

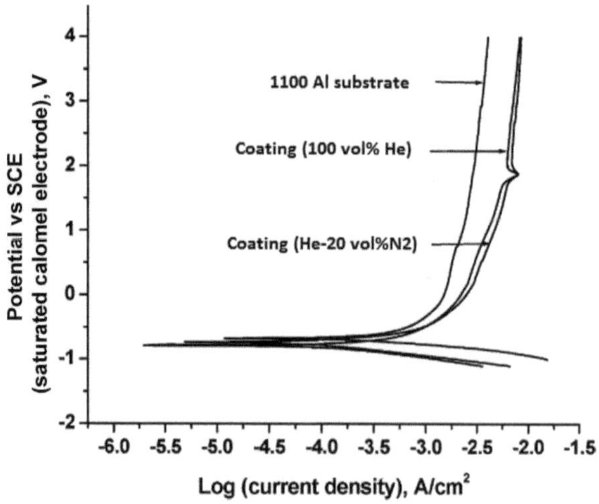

Fig. 4.80 Polarization behaviour of cold-sprayed 1100 Al coatings and 1100 Al bulk material in 0.5 M H_2SO_4. (Balani et al. 2005b)

give cathodic protection based on sacrificial behaviour, for example, Zn, Al and Al-based composites (Maev and Leshchynsky 2006; Champagne 2007; Djordjevic and Maev 2006; Karthikeyan et al. 2004; Blose et al. 2005; Xiong et al. 2009; Villafuerte et al. 2009; Kroemmer and Heinrich 2006; Chavan et al. 2013; Bu et al. 2012b; DeForce et al. 2011; Spencer et al. 2009; Dzhurinskiy et al. 2012). For instance, Blose et al. (2005) have reported the successful corrosion protection of steel substrates with cold-sprayed Zn, Al, and Zn–Al coatings against wet corrosion. Karthikeyan et al. (2004) have shown that corrosion resistance of the cold-sprayed Al coatings was higher than that of Al bulk material analysed using polarization measurements. Figure 4.80 shows polarization behaviour of cold-sprayed Al coatings and Al bulk material. Passivation of the coatings is first linear and then curving slightly followed by linear behaviour again. This indicates repassivation of the coatings (Karthikeyan et al. 2004).

In the other study, cold-sprayed Al coating on Mg–alloy substrate had dense structure and due to that, sufficient corrosion protection in NaCl solution (Tao et al. 2010). In addition, the cold-sprayed Al+Al_2O_3 coatings performed improved anti-corrosion ability compared with Al bulk material in the polarization measurements (Xiong et al. 2009). LPCS Al coatings act as sacrificial anode and protect the AA2024 substrate, reflecting possibilities for corrosion protection in atmospheric and seawater conditions (Villafuerte et al. 2009). In addition, dense cold-sprayed Al coating has shown to protect sintered NdFeB magnets for the corrosion (Ma et al. 2014). Dense structures and good corrosion resistances of low-pressure cold-sprayed Al, Al+Al_2O_3, Al+Zn+Al_2O_3 coatings have been also achieved by Dzhurinskiy et al. (2012). Figure 4.81 shows corrosion potential behaviour of these coatings. Cold-sprayed Al coating (CP1) had similar corrosion potential behaviour than Al bulk material. Additionally, corrosion potential moves noble with increasing exposure time in the case of Al coatings with Al_2O_3 (CP2 and CP3). This can be caused by the growth of protective oxide layer. On the other hand, corrosion potential decreases Al+Al_2O_3+Zn

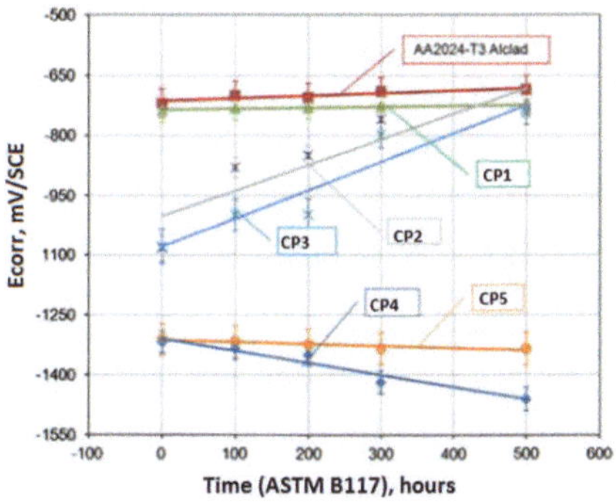

Specimen designation	Composition, vol.%		
	Al	α-Al_2O_3	Zn
CP1	100	0	0
CP2	75	25	0
CP3	50	50	0
CP4	30	50	20
CP5	35	40	25

Fig. 4.81 Corrosion potential (*E*corr) as a function of salt fog corrosion test. Sample codes are presented in the table. *ASTM* American Society for Testing and Materials. (Dzhurinskiy et al. 2012)

coatings with increasing exposure time which can be explained by sacrificial behaviour of Zn and dissolution of the coating layer (Dzhurinskiy et al. 2012).

There is high potential to use CS processes in the production of corrosion barrier coatings (low porosity; Papyrin et al. 2007). The cold-sprayed Al and Zn coatings (Kroemmer and Heinrich 2006; Bu et al. 2012b) and LPCS Zn/Al/ Al_2O_3 coatings (Djordjevic and Maev 2006) are able to be used for corrosion protection. Zinc coatings are widely used as corrosion protection of steel components in aqueous and marine environments. Zinc gives sacrificial protection over steel due to its more negative corrosion potential. Chavan et al. (2013) have studied polarization behaviour of as-sprayed and heat-treated cold-sprayed Zn coatings in 3.5 %NaCl solution. Coatings' lifetime was increased as sacrificial coating due to the as-sprayed and heat-treated cold-sprayed Zn coatings formed protective passive layer which improved corrosion resistance of Zn coatings. Furthermore, heat-treated Zn coating had reduced corrosion current density, indicating its longer corrosion protection (Chavan et al. 2013).

One potential for corrosion protection by cold-sprayed coatings is Al-based coatings on Mg-based substrate material. Mg-based alloys are widely used in structural components due to their low weight and high strength. However, corrosion and wear properties are not very good. In order to improve the corrosion resistance, Mg–alloy components are coated with corrosion resistance materials, for example, Al and Al-based composite coatings. Improvement of the corrosion properties of Al coatings on Mg substrate was achieved by adding ceramic particles to the metallic powder. Cold-sprayed $Al+Al_2O_3$ coatings had similar corrosion resistance than bulk Al alloys (Spencer et al. 2009). Addition of hard particles to the powder mixture has three main functions: (i) to keep the nozzle clean (eliminate nozzle clogging), (ii) to activate the sprayed surface and (iii) to hammer the coating structure (densification; Koivuluoto et al. 2008a). In the other study, denseness of cold-sprayed Al coatings was improved by using $Al+Mg_{17}Al_{12}$ blended powders. Addition of hard intermetallic particles decreased porosity and improved corrosion resistance. Behaviour of the coatings is closer to the Al bulk, reflecting corrosion protection behaviour over substrate material (Bu et al. 2012b).

Also, corrosion properties of cold-sprayed Al–5Mg coatings have been studied (DeForce et al. 2011). Coatings have shown their potential for corrosion protection due to the fact that they behave very well in corrosion study, and after 1000 h exposure in salt spray test, there was not any sign of Mg corrosion observed on the coating surfaces. In addition, coating had minimal galvanic reaction coupled with Mg, indicating galvanic compatibility (DeForce et al. 2011).

Denseness, density and impermeable structures are the first criterion for the corrosion resistance of coatings which are nobler than substrate material. For example, Cu, Ta, Ni and Ni-based alloys are nobler than steel substrate. Denseness means impermeability of the coatings, indicating coating structures without existing through porosity. Denseness or, on the other hand, existing through porosity is identified by using corrosion tests. Cold-sprayed Ni, Ta, Ti, SS, and brass coatings with low porosity in their structures are reported to have potential for corrosion resistance applications (Koivuluoto et al. 2010b; Koivuluoto and Vuoristo 2010a; Marx et al. 2005; Hoell and Richter 2008; Bala et al. 2010b; Wang et al. 2006; Wang et al. 2008; AL-Mangour et al. 2013).

The open-cell potential measurements and salt spray tests showed that the LPCS Cu and Cu+Al_2O_3 coatings contained through porosity in their structures. Open-cell potentials of the coatings were close to the values of Fe52 substrate material (Koivuluoto et al. 2008a; Koivuluoto and Vuoristo 2010b). Regardless, in the CS process, powder type and composition had strong influence on the denseness of the coatings. Denseness of the coatings was improved with Al_2O_3 particle addition. Optimal composition of metallic and ceramic particles in the powder mixture depends on sprayed material combination and powder type of metallic particles (Koivuluoto and Vuoristo 2010b). Figure 4.82 shows overall dense coating structures of the HPCS Cu and Ta coatings, having a similar open-cell potential behaviour with corresponding bulk materials. The coatings remained stable in the long-time exposure, indicating their structural durability (Koivuluoto et al. 2010b). These materials are under the high-interest category due to the fact that tantalum has extraordinary corrosion resistance (ASM Handbook 13B, ASM Metals Handbook 2005) due to the formation of a highly stable passivating layer (Schweitzer 1996). Tantalum resists corrosion effectively in acids (not HF), salts and organic chemicals even at elevated

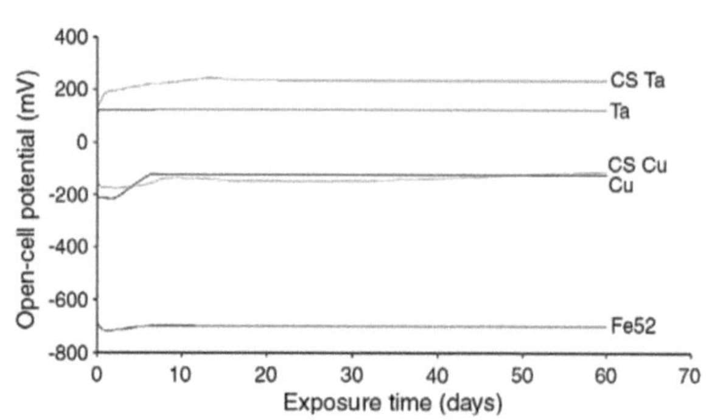

Fig. 4.82 Open-cell potentials of high-pressure cold spray *(HPCS)* Ta and Cu coatings, Ta and Cu bulk materials and Fe52 substrate material as a function of exposure time in 3.5 %NaCl solution. Ag/AgCl, reference electrode. (Koivuluoto et al. 2010b)

Table 4.7 Corrosion potential E_{corr}, corrosion current density i_{corr}, passivation potential E_{pp} and passivation current density i_{pp} of tantalum bulk material and CS coatings in 3.5 wt% NaCl and 40 wt% H_2SO_4 solutions analysed by Tafel extrapolation. (Koivuluoto et al. 2009)

Sample	Solution	T (°C)	E_{corr} (V)	i_{corr} (μA/cm^2)	E_{pp} (V)	i_{pp} (μA/cm^2)
Ta bulk	NaCl	22	−0.66	1.1	0	16
HPCS Ta	NaCl	22	−0.67	1.1	0.05	11
Ta bulk	NaCl	80	−0.68	0.5	−0.25	20
HPCS Ta	NaCl	80	−0.66	0.6	0.05	13
Ta bulk	H_2SO_4	22	−0.32	0.4	0.08	12
HPCS Ta	H_2SO_4	22	−0.33	0.3	0.10	12
Ta bulk	H_2SO_4	80	−0.34	0.8	0.04	15
HPCS Ta	H_2SO_4	80	−0.30	2.0	0.05	15

HPCS high-pressure cold spray

temperatures (ASM Handbook 13B, ASM Metals Handbook 2005). Furthermore, tantalum as a dense coating acts like corrosion barrier coating on a steel substrate, providing high corrosion resistance in many environments (Jones 1996).

Ta bulk material and dense HPCS Ta coating get passivated rapidly, and, above their passivation potential, corrosion rate falls to a very low value in the passive area due to the stable passive layer (Jones 1996). Table 4.7 shows the corrosion characteristics of the tantalum bulk material and the HPCS Ta coating in NaCl and H_2SO_4 solutions at 22 and 80 °C. Coatings and bulk material had similar corrosion characteristics (Koivuluoto et al. 2009). Dense HPCS Ta coating behaved like the bulk material, indicating real corrosion resistance of the coating.

In addition, the electrochemical tests in 1 M KOH solution revealed stable passive behaviour of the HPCS Ta coatings due to the fully dense coating structures. Figure 4.83 shows similar anodic corrosion behaviour of Ta bulk and cold-sprayed Ta coatings on different substrates (CSTa1 on Al, CSTa2 on Cu and CSTa3 on steel; Koivuluoto et al. 2010a).

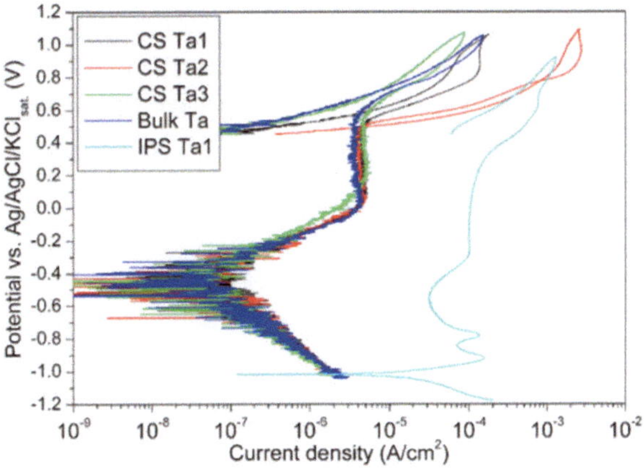

Fig. 4.83 Polarization behaviour of cold-sprayed Ta coatings (CSTa1 on Al, CSTa2 on Cu and CSTa3 on steel), Ta bulk material and inert plasma-sprayed *(IPS)* Ta coating in 1 M KOH. (Koivuluoto et al. 2010a)

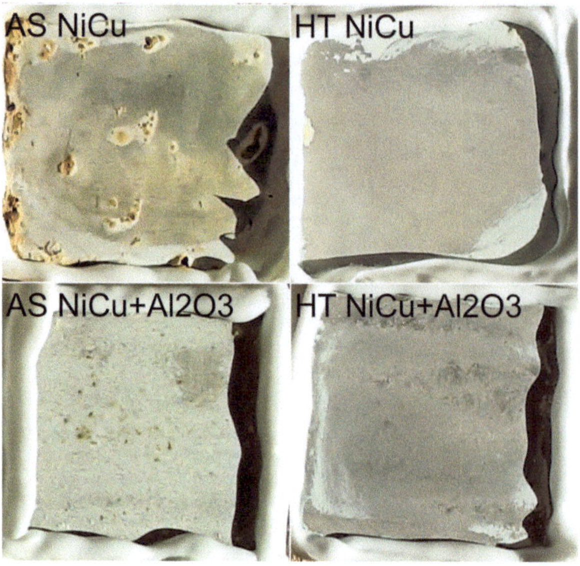

Fig. 4.84 As-sprayed and heat-treated HPCS NiCu and NiCu+Al_2O_3 coating surfaces after an 80-h exposure in Corrodkote test. (After Koivuluoto et al. 2015)

Open-cell potentials of as-sprayed and heat-treated HPCS Ni and Ni–20Cu coatings have been reported to be closer to the bulk materials (Ni and Ni-30Cu) than to the substrate material (Fe52), indicating high coating quality (Koivuluoto and Vuoristo 2010a). The denseness of Ni and Ni–Cu coatings was improved with heat treatments. Heat treatment densified the coating structures due to the recovery, recrystallization and void reduction by the softening and rearrangement of grains (Koivuluoto et al. 2007, 2015; Koivuluoto and Vuoristo 2010a). The denseness of HPCS Ni coatings was also improved by using optimized spraying parameters (Koivuluoto and Vuoristo 2010a). On the other hand, one way to improve the denseness of the coatings was to add hard particles into the metallic powder (Koivuluoto and Vuoristo 2009, 2010a; Koivuluoto et al. 2015). For example, denseness of cold-sprayed NiCu was improved with the addition of Al_2O_3 particles. Figure 4.84 presents coating surfaces after an 80-h exposure in Corrodkote test. Denser structures were achieved with heat treatment and Al_2O_3 addition. Corrodkote test is 100 times more aggressive than salt spray fog test (Koivuluoto et al. 2015).

Generally speaking, Ni–Cr alloys have high oxidation and corrosion resistance in high temperatures, and, therefore, they are used in boilers and electrical furnaces. The suitability of cold-sprayed Ni–50Cr coatings for these conditions has been studied with accelerated hot corrosion tests in molten salt Na_2SO_4–60 %V_2O_5 paste (900 °C, 1 h) (Bala et al. 2010b). These coatings had dense and oxygen-free structures, and they showed better corrosion resistance than uncoated boiler steels (Fig. 4.85; Bala et al. 2010b). Furthermore, cold-sprayed Ni–20Cr coating had better corrosion-erosion resistance than uncoated T22 boiler steel substrate (Bala et al. 2012). However, in the comparison between cold-sprayed and HVOF-sprayed Ni–20Cr coatings, HVOF coatings behave better due to the Cr_2O_3 layer formed on the surface (Bala et al. 2012). Hot corrosion resistance of cold-sprayed Ni–20Cr

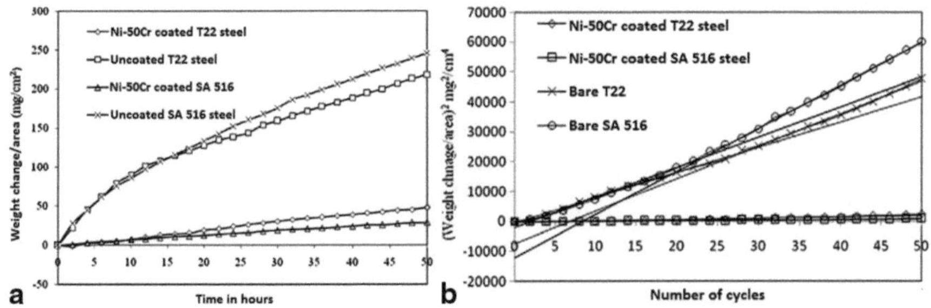

Fig. 4.85 Hot corrosion studies of cold-sprayed Ni–50Cr coatings and uncoated substrates. **a** Weight change/area versus time. **b** Weight change/area versus number of cycles in Na_2SO_4–60% V_2O_5 environment at 900 °C. (Bala et al. 2010b)

coating was better than uncoated steel substrate which is explained by the formation of protective oxides (Bala et al. 2010a). In the other study, Ni-based amorphous coatings with high corrosion resistance were prepared by using kinetic metallization (cold spraying). For example, cold-sprayed NiNbTiZrCoCu coating had an extremely low passive current density and wide passive region, indicating extreme corrosion resistance (Wang et al. 2006).

Ti and Ti alloys have good corrosion properties, and they are widely used in marine environments (Wang et al. 2008). Corrosion properties of cold-sprayed Ti coatings and denseness improvements have been reported (Wang et al. 2008). Cold-sprayed Ti coating sprayed with high pressure and high temperature had the lowest porosity and thus better corrosion properties. As-polished Ti coating sprayed with optimized spray parameters showed similar behaviour than Ti bulk material. Only, polarization current density was higher, caused by differences between densities of coating (Wang et al. 2008).

Titanium is also used in biomedical applications due to its good corrosion resistance and biocompatibility. Corrosion protection is based on the formation of passive titanium dioxide layer. Furthermore, addition of hydroxyapatite (HAP) improves bioactivity. Reportedly, corrosion resistance of cold-sprayed Ti + 50HAP was better than cold-sprayed Ti + 20HAP (Zhou and Mohanty 2012). In addition, heat treatment improved corrosion behaviour of the Ti + 20HAP coating. All coatings had passivation range; Ti + 50HAP was the most stable (Zhou and Mohanty 2012). Additionally, cold-sprayed Ti coatings were densified by vacuum heat treatment as posttreatment (Hussain et al. 2011). On the other hand, Marrocco et al. (2011) have improved corrosion resistance of cold-sprayed Ti coatings by post-treating the coatings with laser. They densified the top of the coatings and that way eliminated interconnected porosity. As the results show, posttreated cold-sprayed Ti coatings had a similar corrosion resistance that corresponding bulk Ti had, which is shown in the polarization tests (Fig. 4.86; Marrocco et al. 2011).

Additionally, Dosta et al. (2013) have studied corrosion properties of cold-sprayed WC–25Co coatings. Dense coatings were produced, and any sign of corrosion was not detected in the coatings after electrochemical studies in NaCl solutions

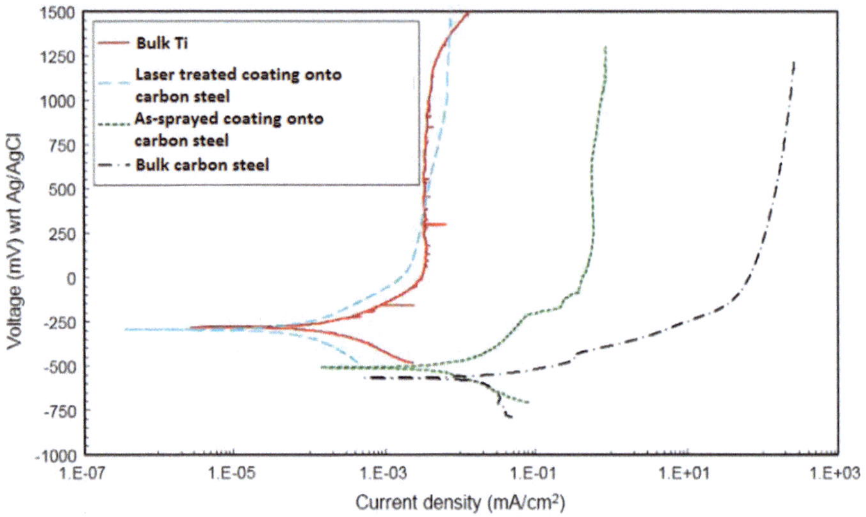

Fig. 4.86 Potentiodynamic polarization scans in 3.5 % NaCl of bulk Ti, carbon steel and as-sprayed Ti coating (on carbon steel and laser-treated Ti coating). (Marrocco et al. 2011)

(Dosta et al. 2013). In the other study, corrosion properties of cold-sprayed SS coatings were studied (AL-Mangour et al. 2013). SSs and Co–Cr alloys have been used in medical applications due to their high corrosion resistance. AL-Mangour et al. (2013) have cold-sprayed SS mixed with Co–Cr particles. They reduced porosity by optimizing composition of the composite (33 %Co–Cr) and heat treating the coatings. Improvement of the corrosion resistance was detected with the polarization behaviour. Composite coatings had lower corrosion rate compared with pure SS (Fig. 4.87; AL-Mangour et al. 2013).

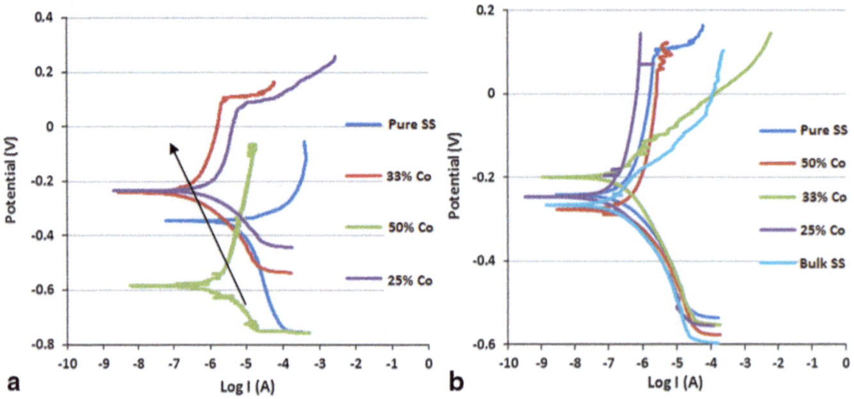

Fig. 4.87 Polarization behaviour of cold-sprayed stainless steel *(SS)* + CoCr coatings **a** as-sprayed coatings and **b** annealed (1100 °C) coatings. (AL-Mangour et al. 2013)

Cold-sprayed SS coatings had similar polarization behaviour with corresponding bulk material with optimized powder properties. Spencer and Zhang (2011) have found the highest corrosion resistance of cold-sprayed SS coating with mixed powders (− 10 and − 22 μm). Two 316L powders with different particle size distributions were mixed, and this way improved the corrosion resistance of coating close to that of bulk material (Spencer et al. 2011). Corrosion properties were also detected to coating thickness; coating with higher thickness had higher corrosion resistance analysed by polarization behaviour (Spencer et al. 2011) and surface finishing (polished surfaces had higher corrosion resistance than as-sprayed coatings with certain roughness (Wang et al. 2008)).

Acknowledgement Chapter authors would like to thank Mrs Odile Adam from MINES ParisTech for efficient help in the achievement of the list of references, all the Ph.D students and research scientists who worked at MINES ParisTech/C2P in the field of cold spay for relevant research results, Prof Yuji Ichikawa for helpful discussion and providing with pictures, Enrico Vedelago and Silvano Rech from Veneto Nanotech for their kind support and Prof Mario Guagliano from Politecnico di Milano for helpful discussion.

References

Ajaja, J., D. Goldbaum, and R. R. Chromik. 2011. Characterization of Ti cold spray coatings by indentation methods. *Acta Astronautica* 69 (11–12): 923–928.

Ajdelsztajn, L., J. M. Schoenung, B. Jodoin, et al. 2005. Cold spray deposition of nanocrystalline aluminum alloys. *Metallurgical and Materials Transactions A* 36:657–666.

Ajdelsztajn, L., B. Jodoin, and J. M. Schoenung. 2006. Synthesis and mechanical properties of nanocrystalline Ni coatings produced by cold gas dynamic spraying. *Surface & Coatings Technology* 201:1166–1172.

AL-Mangour, B., Mongrain, R., Irissou E., and S. Yue. 2013. Improving the strength and corrosion resistance of 316 L stainless steel for biomedical applications using cold spray. *Surface and Coatings Technology* 216:297–307.

Amsellem, O., F. Borit, D. Jeulin, et al. 2012. Three-dimensional simulation of porosity in plasma sprayed alumina using microtomography and electrochemical impedance spectrometry for finite element modeling of properties. *Journal of Thermal Spray Technology* 21:193–201.

Andreola, F., C. Leonelli, M. Romagnoli, et al. 2000. Techniques used to determine porosity. *American Ceramic Society Bulletin* 79:49–52.

Arrigoni, M., M. Boustie, C. Bolis, et al. 2012. Shock mechanics and interfaces. In *Mechanics of solid interfaces,* eds. M. Braccini and M. Dupeux, 211–248. London: Wiley.

Ashgriz, N. 2011. *Handbook of atomization and sprays: Theory and applications*. New York: Springer.

ASM Metals Handbook. 2005. *Corrosion: Materials, corrosion of nonferrous alloys and speciality products, corrosion of tantalum and tantalum alloys,* vol. 13B. Materials Park: ASM International.

Assadi, H., F. Gartner, T. Stoltenhoff, et al. 2003. Bonding mechanism in cold gas spraying. *Acta Materialia* 51:4379–4394.

Assadi, H., T. Schmidt, H. Richter, J. O. Kliemann, K. Binder, F. Gaertner, T. Klassen, and H. Kreye. 2011. On parameter selection in cold spraying. *Journal of Thermal Spray Technology* 20 (6): 1161–1176.

Assadi, H., T. Klassen, F. Gartner, et al. 2014. Modelling of impact and bonding of inhomogeneous particles in cold spraying. In International thermal spray conference, Barcelona, 21–23 May 2014, 203–207.

Attia, H., M. Meshreki, A. Korashy, V. Thomson, and V. Chung. 2011. Fretting wear characteristics of cold gas-dynamic sprayed aluminum alloys. *Tribology International* 44:1407–1416.

Au, P., et al. 1980. *Flow property measurements from instrumented hardness Tests. Non-destructive evaluation in the nuclear industry,* vol. 10, 597–610. New York: ASM.

B117-90. 1992. *Standard test method of salt spray (Fog) testing. Annual book of ASTM standards,* 19–25. Philadelpia: ASTM.

Baboian, R. 2003. *Galvanic corrosion, corrosion fundamentals, testing, and protection, ASM handbook,* vol. 13A, 210–213. Materials Park: ASM International.

Bae, G., K. Kanga, H. Naa, J. J. Kimb, and C. Lee. 2010. Effect of particle size on the microstructure and properties of kinetic sprayed nickel coatings. *Surface and Coatings Technology* 204 (20): 3326–3335.

Bae, G., J. I. Jang, and C. Lee. 2012. Correlation of particle impact conditions with bonding, nanocrystal formation and mechanical properties in kinetic sprayed nickel. *Acta Materialia* 60 (8): 3524–3535.

Bageri, S., P. I. Fernàndez, R. Ghelichi, M. Guagliano, and S. Vezzù. 2010. Effect of shot peening on residual stresses and surface work-hardening in cold sprayed coatings. *Key Engineering Materials* 417–418:397–400.

Bala, N., H. Singh, and S. Prakash. 2010a. High temperature corrosion behavior of cold spray Ni-20Cr coating on boiler steel in molten salt environment at 900 °C. *Journal of Thermal Spray Technology* 19 (1–2): 110–118.

Bala, N., H. Singh, and S. Prakash. 2010b. Accelerated hot corrosion studies of cold spray Ni–50Cr coating on boiler steels. *Materials & Design* 31:244–253.

Bala, N., H. Singh, S. Prakash, and J. Karthikeyan. 2012. Investigations on the behavior of HVOF and cold sprayed Ni–20Cr coating on T22 boiler steel in actual boiler environment. *Journal of Thermal Spray Technology* 21 (1): 144–158.

Bala, N., H. Singh, J. Karthikeyan, and S. Prakash. 2014. Cold spray coating process for corrosion protection: A review. *Surface Engineering* 30 (6): 414–421.

Balani, K., A. Agarwal, S. Seal, and J. Karthikeyan. 2005a. Transmission electron microscopy of cold sprayed 1100 aluminum coating. *Scripta Materialia* 53:845–850.

Balani, K., T. Laha, A. Agarwal, J. Karthikeyan, and N. Munroe. 2005b. Effect of carrier gases on microstructural and electrochemical behavior of cold-sprayed 1100 aluminum coating. *Surface and Coatings Technology* 195:272–279.

Barradas, S., R. Molins, M. Jeandin, et al. 2005. Application of laser shock adhesion testing to the study of the interlamellar strength and coating substrate adhesion in cold sprayed. *Surface and Coatings Technology* 197:18–27.

Barradas, S., V. Guipont, R. Molins, et al. 2007. Laser shock flier impact simulation of particle substrate interactions in cold spray. *Journal of Thermal Spray Technology* 16:548–556.

Beghini, M., L. Bertini, and V. Fontanari. 2002. On the possibility to obtain the stress-strain curve for a strain-hardening material by spherical indentation. International *Journal of Computer Applications in Technology* 15 (4): 168–175.

Beghini, M., L. Bertini, and V. Fontanari. 2006. Evaluation of the stress–strain curve of metallic materials by spherical indentation. *International Journal of Solids and Structures* 43 (7–8): 2441–2459.

Begue, G., G. Fabre, V. Guipont, et al. 2013. Laser shock adhesion test (LASAT) of EB-PVD TBCs: towards an industrial application. *Surface and Coatings Technology* 237:305–132.

Benabdi, M., and A. Roche. 1997. Mechanical properties of thin and thick coatings applied to various substrates, Part I, an elastic analysis of residual stresses within coating materials. *Journal of Adhesion Science and Technology* 11 (2): 281–299.

Benjamin, P., and C. Weaver. 1960. Measurement of adhesion of thin films. Proceedings of the Royal Society of London. Series A. *Mathematical and Physical Sciences* 254 (1277): 163–176.

Berthe, L., M. Arrigoni, J. P. Boustie, et al. 2011. State of the art laser adhesion test (LASAT). *Nondestructive Testing and Evaluation* 26:303–317.

Binder, K., J. Gottschalk, M. Kollenda, F. Gartner, and T. Klassen. 2011. Influence of impact angle and gas temperature on mechanical properties of titanium cold spray deposits. *Journal of Thermal Spray Technology* 20 (1–2): 234–240.

Blochet, Q., F. Delloro, F. Borit, et al. 2014. Influence of spray angle on the cold spray of Al for the repair of aircraft components. In International thermal spray conference, Barcelona, 21–23 May 2014, 69–74.

Blose, R., D. Vasquez, and W. Kratochvil. 2005. *Metal passivation to resist corrosion using the cold spray process. Thermal Spray 2005: Explore its surfacing potential!*. Basel: ASM International.

Bolelli, G., et al. 2010. Depth-sensing indentation for assessing the mechanical properties of cold-sprayed Ta. *Surface and Coatings Technology* 205 (7): 2209–2217.

Borchers, C., F. Gartner, T. Stoltenhoff, and H. Kreye. 2005. Formation of persistent dislocation loops by ultra-high strain rate deformation during cold spraying. *Acta Materialia* 53:2991–3000.

Brenner, A., and S. Senderoff. 1949. National Bureau of Standards. Research Paper RP1954, vol. 42, 105–123.

Bu, H., M. Yanodouzi, C. Lu, and B. Jodoin. 2012a. Post-heat treatment effects on cold sprayed aluminium coatings on AZ91D magnesium substrates. *Journal of Thermal Spray Technology* 21 (3–4): 731–739.

Bu, H., M. Yandouzi, C. Lu, D. MacDonald, and B. Jodoin. 2012b. Cold spray blended $Al+Mg_{17}Al_{12}$ coating for corrosion protection of AZ91D magnesium alloy. *Surface and Coatings Technology* 207:155–162.

Champagne, V., ed. 2007. *The cold spray materials deposition process: fundamentals and applications,* 362 Cambridge: Woodhead Publishing Ltd.

Champagne, V. K., and D. J. Helfritch. 2014. Mainstreaming cold spray—Push for applications. *Surface Engineering* 30 (6): 396–403.

Champagne, V. K., D. Helfritch, P. Leyman, et al. 2005. Interface material mixing formed by the deposition of copper on aluminum. *Journal of Thermal Spray Technology* 14:330–334.

Chatterjee, U., S. Bose, and S. Roy. 2001. *Environmental degradation of metals*. New York: Marcel Dekker Inc.

Chavan, N. M., B. Kiran, A. Jyothirmayi, P. P. Sudharshan, and G. Sundararajan. 2013. The corrosion behavior of cold sprayed zinc coatings on mild steel substrate. *Journal Thermal Spray Technology* 22 (4): 463–470.

Chen, Z., et al. 2014. A review on the mechanical methods for evaluating coating adhesion. *Acta Mechanica* 225 (2): 431–452.

Chicot, D., P. Démarécaux, and J. Lesage. 1996. Interfacial indentation test for the determination of adhesive properties of thermal sprayed coatings. *Thin Solid Films* 283:151–157.

Choi, W. B., L. Li, V. Luzin, R. Neiser, T. Gnäupel-Herold, H. J. Prask, S. Sampath, and A. Gouldstone. 2007. Integrated characterization of cold sprayed aluminum coatings. *Acta Materialia* 55:857–866.

Christoulis, D. K., S. Guetta, E. Irissou, et al. 2010. Cold spraying coupled to nano-pulsed Nd-YAG laser surface pre-treatment. *Journal of Thermal Spray Technology* 19:1062–1073.

Christoulis, D. K., S. Guetta, V. Guipont, et al. 2011. The influence of the substrate on the deposition of cold sprayed titanium: an experimental and numerical study. *Journal of Thermal Spray Technology* 20:523–533.

Cinca, N., and J. M. Guilemany. 2013. Structural and properties characterization of stellite coatings obtained by cold gas spraying. *Surface and Coatings Technology* 220:90–97.

Cinca, N., M. Barbosa, S. Dosta, et al. 2010. Study of Ti deposition onto Al alloy by cold gas spraying. *Surface and Coatings Technology* 205:1096–1102.

Cinca, N., E. López, S. Dosta, and J. M. Guilemany. 2013a. Study of stellite-6 deposition by cold gas spraying. *Surface and Coatings Technology* 232:891–898.

Cinca, N., J. M. Rebled, S. Estradé, et al. 2013b. Influence of the particle morphology on the cold gas spray deposition behavior of titanium on aluminum light alloys. *Journal of Alloys and Compounds* 554:89–96.

Cizek, J., O. Kovarik, J. Siegl, K. A. Khor, and I. Dlouhy. 2013. Influence of plasma and cold spray deposited Ti layers on high-cycle fatigue properties of Ti6Al4V substrates. *Surface and Coatings Technology* 217:23–33.

Cochelin, E., F. Borit, G. Frot, et al. 1999. Oxidation and particle deposition modeling in plasma spraying of Ti–6Al–4V/SiC fiber composites. *Journal of Thermal Spray Technology* 8:117–124.

Coddet, P., C. Verdy, C. Coddet, F. Lecoutrier, and F. Debray. 2014. Comparison of cold sprayed Cu–0.5Cr–0.05Zr alloys after various heat treatments versus forged and vacuum plasma sprayed alloys. *Journal of Thermal Spray Technology* 23 (3): 486–491.

Couto, M., S. Dosta, M. Torrell, J. Fernández, and J. M. Guilemany. 2013. Cold spray deposition of WC-17 and 12Co cermets onto aluminum. *Surface and Coatings Technology* 235:54–61.

Ctibor, P., R. Lechnerova, and V. Benes. 2006. Quantitative analysis of pores of two types in a plasma sprayed coating. *Materials Characterization* 56:297–304.

Danlos, Y., S. Costil, X. Guo, H. Liao, and C. Coddet. 2010. Ablation laser and heating laser combined to cold spraying. *Surface and Coatings Technology* 205 (4): 1055–1059.

Davis, J. R. 2004. *Tensile testing*. Materials Park: ASM International.

DeForce, B. S., T. J. Eden, and J. K. Potter. 2011. Cold spray Al–5 % Mg coatings for the corrosion protection of magnesium alloys. *Journal of Thermal Spray Technology* 20 (6): 1352–1358.

Delloro, F., M. Faessel, H. Proudhon, et al. 2014a. A morphological approach to the modeling of the cold spray process. In International thermal spray conference, Barcelona, 21–23 May 2014, 221–226.

Delloro, F., M. Faessel, D. Jeulin, et al. 2014b. X ray micro-tomography and modeling of cold sprayed coated powders. In International thermal spray conference, Barcelona, 21–23 May 2014, 886–891.

Demarecaux, P., D. Chicot, and J. Lesage. 1996. Interface indentation test for the determination of adhesive properties of thermal sprayed coatings. *Journal of Materials Science Letters* 15 (16): 1377–1380.

Descurninges, L. L., L. T. Mingault, V. Guipont, et al. 2011. Influence of powder particles oxidation on properties of cold sprayed tantalum. In *Thermal spray 2011: Proceedings of the international thermal spray conference,* ed. J. Jerzembeck, 60–65. Düsseldorf: DVS.

DIN 50359-1. 1997. Universal hardness testing of metallic materials, test method.

Djordjevic, B., and R. Maev. 2006. SIMATTM application for aerospace corrosion protection and structural repair. In *Thermal spray 2006: building on 100 years of success,* eds. B. Marple, M. Hyland, Y.-C. Lau, R. Lima, and J. Voyer. Seattle: ASM International.

Dosta, S., M. Couto, and J. M. Guilemany. 2013. Cold spray deposition of a WC-25Co cermet onto Al7075-T6 and carbon steel substrates. *Acta Materialia* 61:643–652.

Dupuis, P., Y. Cormier, A. Farjam, B. Jodoin, and A. Corbeil. 2014. Performance evaluation of near-net pyramidal shaped fin arrays. *International Journal of Heat and Mass Transfer* 69:34–43.

Dykhuizen, R. C., M. F. Smith, D. L. Gilmore, et al. 1999. Impact of high velocity cold spray particles. *Journal of Thermal Spray Technology* 8:559–564.

Dzhurinskiy, D., E. Maeva, E. Leshchinsky, and R. Maev. 2012. Corrosion protection of light alloys using low pressure cold spray. *Journal of Thermal Spray Technology* 21 (2):304–313.

Field, J. S., and M. V. Swain. 1995. Determining the mechanical properties of small volumes of material from submicrometer spherical indentations. *Journal of Materials Research* 10 (1):101–112.

Fischer-Cripps, C. A. 1997. Elastic–plastic behaviour in materials loaded with a spherical indenter. *Journal of Materials Science* 32 (3): 727–736.

Fischer-Cripps, A. C. 2000. *Introduction to contact mechanics*. New York: Springer.

Fischer-Cripps, A. C. 2005. *The IBIS handbook of nanoindentation*. Forestville: Fischer-Cripps Laboratories Pty Ltd.

Fischer-Cripps, A. C. 2011. *Nanoindentation,* 3rd ed. New York: Springer.

Frankel, G. 2003. *Pitting corrosion, corrosion fundamentals, testing, and protection, ASM Handbook,* vol. 13A, 236–241. Materials Park: ASM International.

Fukanuma, H., and R. Huang. 2009. Development of high temperature gas heater in the cold spray coating system. In Thermal spray 2009: Proceedings of the international thermal spray conference, 267–272.

Fukanuma, H., and N. Ohno. 2004. A study of adhesive strength of cold spray coatings. In Proceedings of the international thermal spray conference. ASM International.

Gan, J. A., and C. C. Berndt. 2014. Surface roughness of plasma sprayed coatings: a statistical approach. In International thermal spray conference, Barcelona, 21–23 May 2014, 599–604.

Ganesan, A., J. Affi, M. Yamada, et al. 2012. Bonding behavior studies of cold sprayed copper coating on the PVC polymer substrate. *Surface and Coatings Technology* 207:262–269.

Gärtner, F., T. Stoltenhoff, T. Schmidt, and H. Kreye. 2006. The cold spray process and its potential for industrial applications. *Journal of Thermal Spray Technology* 15 (2): 223–232.

Ghelichi, R., D. MacDonald, S. Bagherifard, H. Jahed, M. Guagliano, and B. Jodoin. 2012. Microstructure and mechanical behavior of cold spray coated Al5052. *Acta Materialia* 60:6555–6561.

Ghelichi, R., S. Bagherifard, D. Mac Donald, M. Brochu, H. Jahed, B. Jodoin, and M. Guagliano. 2014a. Fatigue strength of Al alloy cold sprayed with nanocrystalline powders. *International Journal of Fatigue* 65:51–57.

Ghelichi, R., S. Bagherifard, D. Mac Donald, M. Brochu, I. Fernandez-Pariented, B. Jodoin, and M. Guagliano. 2014b. Experimental and numerical study of residual stress evolutionin cold spray coating. *Applied Surface Science* 288:26–33.

Giraud D., F. Borit, V. Guipont, et al. 2012. Polymer metallization using cold spray: Application to aluminum coating of polyamide 66. In *International thermal spray conference ITSC 2012,* ed. R. S. Lima, 265–270. Houston: ASM.

Giraud, D., M. H. Berger, M. Jeandin, et al. 2015. TEM study of cold sprayed titanium particle bonding onto Ti–6Al–4V. *Surface and Coatings Technology* submitted for publication.

Goldbaum, D., J. Ajajaa, R. R. Chromik, W. Wonga, S. Yue, E. Irissou, and J. G. Legoux. 2011. Mechanical behavior of Ti cold spray coatings determined by a multi-scale indentation method. *Materials Science and Engineering A* 530:253–265.

Goldbaum, D., J. M. Shockley, R. R. Chromik, et al. 2012. The effect of deposition conditions on adhesion strength of Ti and Ti6Al4V cold spray splats. *Journal of Thermal Spray Technology* 21:288–303.

Grujicic, M., C. L. Zhao, W. S. DeRosset, et al. 2004. Adiabatic shear instability based mechanism for particles/substrate bonding in the cold gas dynamic spray process. *Materials and Design* 25:681–688.

Gu, S. 2013. Computational modelling to assist the development of advanced thermal spray technologies. Cold Spray Club meeting, Paris, 11 Oct 2013. http://www.mat.ensmp.fr/clubcold-spray. Accessed 4 Sept 2014.

Guetta, S., M. H. Berger, F. Borit, et al. 2009. Influence of particle velocity on adhesion of cold sprayed splats. *Journal of Thermal Spray Technology* 18:331–342.

Guipont, V., M. Jeandin, S. Bansard, et al. 2010. Bond strength determination of hydroxyapatite coatings on Ti–6Al–4V substrates using the laser shock adhesion test (LASAT). *Journal of Biomedical Materials Research Part A* 95:1096–1104.

Guo, X, G. Zhang, W. Y. Li, L. Dembinskia, Y. Gao, H. Liao, and C. Coddet. 2007 Microstructure, microhardness and dry friction behavior of cold-sprayed tin bronze coatings. *Applied Surface Science* 254 (5): 1482–1488.

Guo, X., G. Zhang, W. Li, Y. Gao, H. Liao, and C. Coddet. 2009. Investigation of the microstructure and tribological behavior of cold-sprayed tin-bronze-based composite coatings. *Applied Surface Science* 255:3822–3828.

Guo, X., J. Chen, H. Yu, H. Liao, and C. Coddet. 2014. A study on the microstructure and tribological behavior of cold-sprayed metal matrix composites reinforced by particulate quasicrystal. *Surface and Coatings Technology* 268:94–98.

Gupta, V. 1995. System and method for measuring the interface tensile strength of planar interfaces, US Patent 5,438, 402, 1 Aug.

Halterman, T. 2013. General electric developing cold spray 3D painting technology. http://www.3dprinterworld.com/article/general-electric-developing-cold-spray-3d-painting-technology. Accessed 4 Sept 2014.

Heimann, R. B., J. I. Kleiman, E. Litovski, S. Marx, R. Ng, S. Petrov, M. Shagalov, R. N. S. Sodhi, and A. Tang. 2014. High-pressure cold gas dynamic (CGD)-sprayed alumina-reinforced aluminum coatings for potential application as space construction material. *Surface and Coatings Technology* 252:113–119.

Hoell, H., and P. Richter. 2008. KINETIKS® 4000—New perspective with cold spraying. In *Thermal spray 2008: thermal spray crossing borders,* ed. E. Lugscheider, 479–480. Maastricht: DVS.

Hryha, E., et al. 2009. An application of universal hardness test to metal powder particles. *Journal of Materials Processing Technology* 209 (5): 2377–2385.

Huang, R., and H. Fukanuma. 2012. Study of the influence of particle velocity on adhesive strength of cold spray deposits. *Journal of Thermal Spray Technology* 21 (3–4): 541–549.

Huang, X., Z. Liu, and H. Xie. 2013. Recent progress in residual stress measurement techniques. *Acta Mechanica Solida Sinica* 26 (6): 570–583.

Huber, N., D. Munz, and C. Tsakmakis. 1997. Determination of Young's modulus by spherical indentation. *Journal of Materials Research* 12 (9): 2459–2469.

Hussain, T. 2013. Cold spraying of titanium: A review of bonding mechanisms, microstructure and properties. *Key Engineering Materials* 533:53–90.

Hussain, T., D. G. McCartney, P. H. Shipway, and T. Marrocco. 2011. Corrosion behavior of cold sprayed titanium coatings and free standing deposits. *Journal of Thermal Spray Technology* 20 (1–2): 260–274.

Hutchings, M. T., P. J. Withers, T. M. Holden, and T. Lorentzen. 2005. *Introduction to the characterization of residual stress by neutron diffraction*. New York: CRC Press.

Ichikawa, Y., Y. Watanabe, I. Nonaka, et al. 2014. Micro-scale interface strength evaluation of cold sprayed deposit. In International thermal spray conference, Barcelona, 21–23 May 2014, 707–710.

Irissou, E., and B. Arsenault. 2007. Corrosion study of cold sprayed aluminum coatings onto Al 7075 alloy. In *Thermal Spray 2007: Global coating solutions,* eds. Basil R. Marple, Margaret M. Hyland, Yuk-Chiu Lau, Chang-Jiu Li, Rogerio S. Lima, and Ghislain Montavon, vol. 6, 549–554. Materials Park: ASM International.

Irissou, E., J. G. Legoux, B. Arsenault, and C. Moreau. 2007. Investigation of Al–Al_2O_3 cold spray coating formation and properties. *Journal of Thermal Spray Technology* 16 (5–6): 661–668.

Itoh, Y., and S. Suyama. 2007. Microstructure, thermal and electrical properties of aluminium coatings produced by cold spray. *Journal of the Society of Materials Science* 56 (11): 1022–1027 (in Japanese).

Jeandin, M. 2011. A Socratic approach to surface modification: the example of thermal spray. In *SMT 24, 24th international conference on surface modification technologies,* ed. E. Beyer, 3–20. Dresden: Valardocs.

Jeandin, M., G. Rolland, L. L. Descurninges, et al. 2014. Which powders for cold spray? *Surface Engineering* 30:291–298.

Jenkins, W. D., T. G. Digges, C. R. Johnson. 1957. Tensile properties of copper, nickel, and 70-percent-copper-30-percent-nickel and 30-percent-copper-70-percent-nickel alloys at high temperatures. Journal of Research of the National Bureau of Standards 58 (4). Research Paper 2753.

Jodoin, B., L. Ajdelsztajn, E. Sansoucy, A. Zúñiga, P. Richer, and E. J. Lavernia. 2006. Effect of particle size, morphology, and hardness on cold gas dynamic sprayed aluminum alloy coatings. *Surface and Coatings Technology* 201:3422–3429.

Jones, D. 1996. *Principles and prevention of corrosion*. Upper Saddle River: Prentice-Hall.

Jones, R., N. Matthews, C. A. Rodopoulos, et al. 2011. On the use of supersonic particle deposition to restore the structural integrity of damaged aircraft strcutures. *International Journal of Fatigue* 33:1257–1267.

Jones, R., L. Molent, S. A. Barter, N. Matthews, and D. Z. Tamboli. 2014. Supersonic particle deposition as a means for enhancing the structural integrity of aircraft structures. *International Journal of Fatigue [P]* 68:260–268. Elsevier, London.

Karthikeyan, J., T. Laha, A. Agarwal, and N. Munroe. 2004. Microstructural and electrochemical characterization of cold-sprayed 1100 aluminum coating. In *Thermal spray 2004: Advances in technology and application*. Osaka: ASM International.

Kelly, R. 2003. *Crevice corrosion, corrosion fundamentals, testing, and protection, ASM handbook,* vol. 13A, 242–247. Materials Park: ASM International.

Kikuchi, S., S. Yoshino, M. Yamada, M. Fukumoto, and K. Okamoto. 2013. Microstructure and thermal properties of cold sprayed Cu–Cr composite coatings. *Journal of Thermal Spray Technology* 22 (6): 926–931.

Kim, H. J., C. H. Lee, and S. Y. Hwang. 2005. Fabrication of WC–Co coatings by cold spray deposition. *Surface and Coatings Technology* 191:335–340.

Kim, K. H., M. Watanabe, J. Karakita, and S. Kuroda. 2008. Grain refinement in a single titanium powder particle impacted at high velocity. *Scripta Materialia* 59:768–771.

Kim, B. M., C. J. Lee, and J. M. Lee. 2010. Estimation of work hardening exponents of engineered metals using residual indentation profiles of nano-indentation. *Journal of Mechanical Science and Technology* 24:73–76.

Kim, D. Y., J. J. Park, J. G. Lee, et al. 2013. Cold spray deposition of copper electrodes on silicon and glass substrates. *Journal of Thermal Spray Technology* 22:1092–1102.

Ko, K. H., J. O. Choi, and H. Lee. 2014. Pretreatment effect of Cu feedstock on cold-sprayed coatings. *Journal of Materials Processing Technology* 214:1530–1535.

Koivuluoto, H., and P. Vuoristo. 2009. Effect of ceramic particles on properties of cold-sprayed Ni–20Cr+Al_2O_3 coatings. *Journal of Thermal Spray Technology* 18 (4):555–562.

Koivuluoto, H., and P. Vuoristo. 2010a. Structural analysis of cold-sprayed nickel-based metallic and metallic-ceramic coatings. *Journal of Thermal Spray Technology* 19 (5): 975–989.

Koivuluoto, H., and P. Vuoristo. 2010b. Effect of powder type and composition on structure and mechanical properties of Cu+Al_2O_3 coatings prepared by using low-pressure cold spray process. *Journal of Thermal Spray Technology* 19 (5): 1081–1092.

Koivuluoto, H., and P. Vuoristo. 2014. Structure and corrosion properties of cold sprayed coatings: a review. *Surface Engineering* 30:404–414.

Koivuluoto, H., J. Lagerbom, and P. Vuoristo. 2007. Microstructural studies of cold sprayed copper, nickel, and nickel–30 % copper coatings. *Journal of Thermal Spray Technology* 16 (4): 488–497.

Koivuluoto, H., J. Lagerbom, M. Kylmälahti, and P. Vuoristo. 2008a. Microstructure and mechanical properties of low-pressure cold-sprayed coatings. *Journal of Thermal Spray Technology* 17 (5–6): 721–727.

Koivuluoto, H., M. Kulmala, and P. Vuoristo. 2008b. Structural properties of high-pressure cold-sprayed and low-pressure cold-sprayed coatings. In *Surface modification technologies 22,* eds. T. Sudarshan and P. Nylen, 65–72. Sweden: Trollhättan.

Koivuluoto, H., J. Näkki, and P. Vuoristo. 2009. Corrosion properties of cold-sprayed tantalum coatings. *Journal of Thermal Spray Technology* 18 (1): 75–82.

Koivuluoto, H., G. Bolelli, L. Lusvarghi, F. Casadei, and P. Vuoristo. 2010a. Corrosion resistance of cold-sprayed Ta coatings in very aggressive conditions. *Surface Coatings and Technology* 205 (4): 1103–1107.

Koivuluoto, H., M. Honkanen, and P. Vuoristo. 2010b. Cold-sprayed copper and tantalum coatings—Detailed FESEM and TEM analysis. *Surface and Coatings Technology* 204 (15): 2353–2361.

Koivuluoto, H., A. Coleman, K. Murray, M. Kearns, and P. Vuoristo. 2012. High pressure cold sprayed (HPCS) coatings prepared from OFHC Cu feedstock: overview from powder characteristics to coating properties. *Journal of Thermal Spray Technology* 21 (5): 1065–1075.

Koivuluoto, H., et al. 2013. Coating performance and durability of Zn-based composite materials prepared by using low pressure cold spraying. In Thermal Spray 2013: Proceedings of the international thermal spray conference, May 2013, 252–257. ASM International.

Koivuluoto, H., G. Bolelli, A. Milanti, L. Lusvarghi, and P. Vuoristo. 2015. Microstructural analysis of high-pressure cold-sprayed Ni, NiCu and NiCu+Al_2O_3 coatings. *Surface and Coatings Technology*, 268:224–229.

Kõo, J., and J. Valgur. 2010. Seventh International DAAAM Baltic conference "INDUSTRIAL ENGINEERING" 22–24 April 2010, Tallinn, Estonia.

Kroemmer, W., and P. Heinrich. 2006. Cold spraying—Potential and new application ideas. In *Thermal spray 2006: building on 100 years of success,* eds. B. Marple, M. Hyland, Y.-C. Lau, R. Lima, and J. Voyer. Seattle: ASM International.

Kuhn, H., and D. Medlin. 2004. *ASM Metals handbook, mechanical testing and evaluation,* vol. 8. Materials Park: ASM International.

Lee, H. Y., S. H. Jung, S. Y. Lee, Y. H. You, and K. H. Ko. 2005. Correlation between Al_2O_3 particles and interface of Al–Al_2O_3 coatings by cold spray. *Applied Surface Science* 252 (5): 1891–1898.

Lee, H. Y., H. Shin, S. Y. Lee, and K. H. Ko. 2008. Effect of gas pressure on Al coatings by cold gas dynamic spray. *Materials Letters* 62:1579.

Levasseur, D., S. Yue, and M. Brochu. 2012. Pressureless sintering of coldsprayed Inconel718 deposit. *Materials Science and Engineering A* 556:343–350.

Li, Chang-Jiu, and Wen-Ya Li. 2003. Deposition characteristics of titanium coating in cold spraying. *Surface and coating Technology* 167 (2–3): 278–283.

Li, C. J., and G. J. Yang. 2013. Relationships between feedstock structure, particle parameter, coating deposition, microstructure and properties for thermally sprayed conventional and nanostructured WC-Co. *International Journal of Refractory Metals and Hard Materials* 39:2–17.

Li, W.Y., C. Zhang, X. P. Guo, G. Zhang, H. L. Liao, C. J. Li, C. Coddet. 2006. Effect of standoff distance on coating deposition characteristics in cold spraying. *Materials and Design* 29:297–304.

Li, W. Y., C. Zhang, X. P. Guo, G. Zhang, H. L. Liao, C. J. Li, and C. Coddet. 2007a. Deposition characteristics of Al–12Si alloy coating fabricated by cold spraying with relatively large powder particles. *Applied Surface Science* 253 (17): 7124–7130.

Li, W. Y., C. Zhang, X. P. Guo, G. Zhang, H. L. Liao, C. J. Li, and C. Coddet. 2007b. Study on impact fusion at particle interfaces and its effect on coating microstructure in cold spraying. *Applied Surface Science* 254 (2): 517–526.

Li, W. Y., C. J. Li, and G. J. Yang. 2010. Effect of impact induced melting on interface microstructure and bonding of cold sprayed zinc coating. *Applied Surface Science* 257:1516–1523.

Li, C.-J., H.-T. Wang, G.-J. Yang, and C.-G. Bao. 2011. Characterization of high-temperature abrasive wear of cold-sprayed FeAl intermetallic compound coating. *Journal of Thermal Spray Technology* 20 (1–2): 227–233.

Li, W., C. Huang, M. Yu, and H. Liao. 2013. Investigation on mechanical property of annealed copper particles and cold sprayed copper coating by a micro-indentation testing. *Materials and Design* 46:219–226.

Liang, Y., et al. 2011. Microstructure and nano-mechanical property of cold spray Co-base refractory alloy coating. *Acta Metallurgica Sinica (English Letters)* 24 (3): 190–194.

Lin, C. K., and C. C. Berndt. 1994. Measurement and analysis of adhesion strength for thermally sprayed coatings. *Journal of Thermal Spray Technology* 3 (1): 75–104.

Litovski, E., J. I. Kleiman, M. Shagalov, and R. B. Heimann. 2014. Measurement of the thermal conductivity of cold gas dynamically sprayed alumina-reinforced aluminium coatings between −150 °C and +200 °C. New test method and experimental results. *Surface and Coatings Technology* 242:141–145.

Liu, J., Zhou, X., Zheng, X., Cui, H., and J. Zhang. 2012. Tribological behavior of cold-sprayed nanocrystalline and conventional copper coatings. *Applied Surface Science* 258:7490–7496.

Luo, X.-T., C.-X. Lia, F.-L. Shangb, G.-J. Yanga, Y.-Y. Wanga, and C.-J. Lia. 2014a. High velocity impact induced microstructure evolution during deposition of cold spray coatings: A review. *Surface and Coatings Technology* 254:11–20.

Luo, X.-T., E.-J. Yang, F.-L. Shang, G.-J. Yang, C.-X. Li, and C.-J. Li. 2014b. Microstructure, mechanical properties, and two-body abrasive wear behavior of cold-sprayed 20 vol.% cubic BN-NiCrAl nanocomposite coating. *Journal of Thermal Spray Technology*, 23(7): 1181–1190.

Lupoi, R., and W. O'Neill. 2011. Powder stream characteristics in cold spray nozzles. *Surface and Coatings Technology* 206:1069–1076.

Luzin, V., K. Spencer, and M. X. Zhang. 2011. Residual stress and thermo-mechanical properties of cold spray metal coatings. *Acta Materialia* 59:1259–1270.

Lynden-Bell, R. M. 1995. A simulation study of induced disorder, failure and fracture of perfect metal crystals under uniaxial tension. *Journal of Physics Condensed Matter* 7:4603–4624.

Ma, C., X. Liu, and C. Zhou. 2014. Cold-sprayed Al coating for corrosion protection of sintered NdFeB. *Journal of Thermal Spray Technology* 23 (3): 456–462.

Maev, R., and V. Leshchynsky. 2006. Air gas dynamic spraying of powder mixtures: theory and application. *Journal of Thermal Spray Technology* 15 (2): 198–205.

Marot, G., et al. 2006. Interfacial indentation and shear tests to determine the adhesion of thermal spray coatings. *Surface and Coatings Technology* 201 (5): 2080–2085.

Marrocco, T., et al. 2006. Production of titanium deposits by cold-gas dynamic spray: Numerical modeling and experimental characterization. *Journal of Thermal Spray Technology* 15 (2): 263–272.

Marrocco, T., T. Hussain, D. G. McCartney, and P. H. Shipway. 2011. Corrosion performance of laser posttreated cold sprayed titanium coatings. *Journal of Thermal Spray Technology* 20 (4): 909–917.

Marx, S., A. Paul, A. Köhler, and G. Hüttl. 2005. Cold spraying—Innovative layers for new applications. In *Thermal spray 2005: Explore its surfacing potential!*, ed. E. Lugscheider 209–215. Basel: ASM International.

Masters, C. B., and N. J. Salamon. 1993. Geometrically nonlinear stress-deflection relations for thin film/substrate systems. *International Journal of Engineering Science* 31:915–925.

Matejicek, J., and S. Sampath. 2001. Intrinsic residual stress in single splats produced by thermal spray processes. *Acta Materialia* 49:1993–1999.

Mc Cune, R. C., W. T. Donlon, O. O. Popoola, et al. 2000. Characterization of copper layers produced by cold gas dynamic spraying. *Journal of Thermal Spray Technology* 9:73–82.

Melendez, N., Narulkar, V., Fisher, G., and A. McDonald. 2013. Effect of reinforcing particles on the wear rate of low-pressure cold-sprayed WC-based MMC coatings. *Wear* 306:185–195.

Meng, X., J. Zhang, J. Zhao, Y. Liang, and Y. Zhang. 2011a. Influence of gas temperature on microstructure and properties of cold spray 304SS coating. *Journal of Material Science and Technology* 27 (9): 809–815.

Meng, X., J. Zhang, W. Han, J. Zhao, and Y. Liang. 2011b. Influence of annealing treatment on the microstructure and mechanical properties of cold sprayed 304 stainless steel coating. *Applied Surface Science* 258:700–704.

Meyers, M. A. 1994. *Dynamic behavior of materials*. New York: Wiley.

Mittal, K. L. 1978. Adhesion measurement: Recent progress, unsolved problems, and prospects. In *Adhesion measurement of thin films, thick films, and bulk coatings, STP 640,* ed. K. L. Mittal, 5–17. Philadelphia: ASTM.

Moridi, A., S. M. Hassani-Gangaraj, M. Guagliano, et al. 2014a. Cold spray coating: review of material systems and future perspectives. *Surface Engineering* 30:369–395.

Moridi, A., S. M. H. Gangaraj, S. Vezzu, and M. Guagliano. 2014b. Number of passes and thickness effect on mechanical characteristics of cold spray coating. *Procedia Engineering* 74:449–459.

Morris, M. A., E. Sauvain, and D. G. Morris. 1987. Post compaction heat treatment response of dynamically compacted Inconel 718 powder. *Journal of Material Science* 22:1509–1516.

Nayebi, A., et al. 2001. New method to determine the mechanical properties of heat treated steels. *International Journal of Mechanical Sciences* 43 (11): 2679–2697.

Nayebi, A., et al. 2002. New procedure to determine steel mechanical parameters from the spherical indentation technique. *Mechanics of Materials* 34:243–254.

Ogawa, K., K. Ito, Y. Ichimura, Y. Ichikawa, S. Ohno, and N. Onda. 2008. Characterization of low pressure cold sprayed aluminium coatings. *Journal of Thermal Spray Technology* 17 (5–6): 728–735.

Oliver, W. C., and G. M. Pharr. 1992. An improved technique for determining hardness and elastic modulus using load and displacement sensing indentation experiments. *Journal of Materials Research* 7 (6): 1564–1583.

Oliver, W. C., and G. M. Pharr. 2004. Measurement of hardness and elastic modulus by instrumented indentation: advances in understanding and refinements to methodology. *Journal of Materials Research* 19:3–20.

Ortner, H. M., P. Ettmayer, and H. Kolaska. 2014. The history of the technological progress of hardmetals. *International Journal of Refractory Metals and Hard Materials* 44:148–159.

Papyrin, A., Kosarev, V., Klinkov, S., Alkimov, A., and V. Fomin. 2007. *Cold spray technology,* 1st ed., 328. Amsterdam: Elsevier (printed in the Netherlands).

Pattison, J., S. Celotto, R. Morgan, et al. 2007. Cold gas dynamic manufacturln a non thermal approach to freeform fabrication. *International Journal of Machine Tools and Manufacture* 47:627–634.

Paul, H., J. H. Driver, and C. Maurice, et al. 2007. The role of shear banding on deformation texture in low stacking fault energy metals as characterized on model Ag crystals. *Acta Materialia* 55:575–588.

Pawloski, L. 2008. *The science and engineering of thermal spray coatings,* 2nd ed. Hoboken: Wiley.

Perton, M., S. Costil, W. Wong, D. Poirer, E. Irissou, J. G. Legoux, A. Blouin, and S. Yue. 2012. Effect of pulsed laser ablation on the adhesion and cohesion of cold sprayed Ti–6Al–4V coatings. *Journal of Thermal Spray Technology* 21 (6): 1322–1332.

Pitchuka, S., B. Boesl, C. Zhang, D. Lahiri, A. Nieto, G. Sundararajan, and A. Agarwal. 2014. Dry sliding wear behavior of cold sprayed aluminum amorphous/nanocrystalline alloy coatings. *Surface and Coatings Technology* 238:118–125.

Poza, P., C. J. Múnez, M. A. Garrido-Maneiro, S. Vezzù, S. Rech, and A. Trentin. 2014. Mechanical properties of Inconel 625 cold-sprayed coatings after laser remelting. Depth sensing indentation analysis. *Surface and Coatings Technology* 243:51–57.

Price, T. S., P. H. Shipway, and D. G. McCartney. 2006. Effect of cold spray deposition of a titanium coating on fatigue behaviour of a titanium alloy. *Journal of Thermal Spray Technology* 15 (4): 507–512.

Rech, S., A. Trentin, S. Vezzu, J. G. Legoux, E. Irissou, C. Moreau, and M. Guagliano. 2009. Characterization of residual stresses in Al and Al/Al_2O_3 cold sprayed coatings. *Proceedings of ITSC* 2009:1012–1017.

Rech, S., A. Trentin, S. Vezzu, J. G. Legoux, E. Irissou, and M. Guagliano. 2011. Influence of pre-heated Al 6061 substrate temperature on the residual stresses of multipass Al coatings deposited by cold spray. *Journal of Thermal Spray Technology* 20 (1–2): 243–251.

Rech, S., A. Trentin, S. Vezzu, E. Vedelago, J. G. Legoux, and E. Irissou. 2014. Different cold spray deposition strategies: single- and multi-layers to repair aluminium alloy components. *Journal of Thermal Spray Technology* 23(8): 1237–1250.

Revankar, G. 2000. *ASM Handbook: Mechanical testing and evaluation, hardness testing,* vol. 8, 416–614. Materials Park: ASM International.

Rickerby, D. S. 1986. Internal stress and adherence of titanium nitride coatings. *Journal of Vacuum Science and Technology A* 4: 2809–2814.

Rickerby, D. 1996. Measurement of coating adhesion. In *Metallurgical and ceramic protective coatings,* ed. K. Stern, 306–333. Amsterdam: Springer.

Rickerby, D. S., and P. J. Burnett. 1988. Correlation of process and system parameters with structure and properties of physically vapour-deposited hard coatings. *Thin Solid Films* 157: 195–223.

Rolland, G., F. Borit, V. Guipont, et al. 2008. Three dimensional analysis of cold sprayed coatings using microtomography. In *International thermal spray conference,* ed. E. Lugscheider, 607–610. Düsseldorf: DVS.

Rolland, G., Y. Zeralli, V. Guipont, et al. 2012. Lifetime of cold sprayed electrical contacts. In 26th international conference on electrical contacts (ICEC-ICREPEC 2012), 338–345. 14–17 May 2012, Beijing.

Rösler, J., H. Harders, and M. Bäker. 2007. *Mechanical behaviour of engineered materials*. New York: Springer.

Saleh, M., V. Luzin, and K. Spencer. 2014. Analysis of the residual stress and bonding mechanism in the cold spray technique using experimental and numerical methods. *Surface and Coatings Technology* 252:15–28.

Sansoucy, E., G. E. Kim, A. L. Moran, and B. Jodoin. 2007. Mechanical characteristics of Al–Co–Ce coatings produced by the cold spray process. *Journal of Thermal Spray Technology* 16 (5–6): 651–660.

Sansoucy, E., P. Marcoux, L. Ajdelsztajn, and B. Jodoin. 2008. Properties of SiC-reinforced aluminum alloy coatings produced by the cold gas dynamic spraying process. *Surface and Coatings Technology* 202 (16): 3988–3996.

Schajer, G. S. 2013. *Practical residual stress measurement methods*. Chichester: Wiley.

Schijve, J. 2001. *Fatigue of structures and materials*. Boston: Kluwer.

Schmidt, T., T. Gartner, H. Assadi, and H. Kreye. 2006a. Development of a generalized parameter window for cold spray deposition. *Acta Materialia* 54:729–742.

Schmidt, T., T. Gartner, and H. Kreye. 2006b. New developments in cold spray based on higher gas and particle temperatures. *Journal of Thermal Spray Technology* 15 (4): 488–494.

Schmidt, T., H. Assadi, F. Gartner, et al. 2009. From particle acceleration to impact and bonding in cold spraying. *Journal of Thermal Spray Technology* 18:794–808.

Schweitzer, P. A., ed. 1996. *Corrosion engineering handbook*. New York: Marcel Dekker.

Seo, D., K. Ogawa, K. Sakaguchi, N. Miyamoto, and Y. Tsuzuki. 2012a. Parameter study influencing thermal conductivity of annealed pure copper coatings deposited by selective cold spray process. *Surface and Coating Techonology* 206:2316–2324.

Seo, D., K. Ogawa, K. Sakaguchi, N. Miyamoto, and Y. Tsuzuki. 2012b. Influence of crystallite size and lattice spacing on thermal conduction of polycrystalline copper deposited by solid particle impingment: contribution of electron and phonon conduction. *Surface and Coatings Technology* 207:233–239.

Seo, D., M. Sayar, and K. Ogawa. 2012c. SiO_2 and $MoSi_2$ formation on Inconel 625 surface via SiC coating deposited by cold spray. *Surface and Coatings Technology* 206 (11–12): 2851–2858.

Sevillano, F., P. Poza, C. J. Múnez, S. Vezzù, S. Rech, and A. Trentin. 2013. Cold-sprayed Ni–Al_2O_3 coatings for applications in power generation industry. *Journal of Thermal Spray Technology* 22:772–782.

Shabel, B., and R. Young. 1987. A new procedure for the rapid determination of yield and tensile strength from hardness tests. In *Nondestructive characterization of materials II,* eds. J. Bussière, et al., 335–343. Boston: Springer.

Shayegan, G., H. Mahmoudi, R. Ghelichi, J. Villafuerte, J. Wang, M. Guagliano, and H. Jahed. 2014. Residual stress induced by cold spray coating of magnesium AZ31B extrusion. *Materials and Design* 60:72–84.

Shockley, J., H. Strauss, R. Chromik, N. Brodusch, R. Gauvin, E. Irissou, and J.-G. Legoux. 2013. In situ tribometry of cold-sprayed Al–Al_2O_3 composite coatings. *Surface and Coatings Technology* 215:350–356.

Shockley, J., S. Descartes, E. Irissou, J.-G. Legoux, and R. Chromik. 2014. Third body behavior during dry sliding of cold-sprayed Al–Al_2O_3 composites: In situ tribometry and microanalysis. *Tribology Letters* 54:191–206.

A. B. Shorey, S. D. Jacobs, W. I. Kordonski, and R. F. Gans , 2001. Experiments and observations regarding the mechanisms of glass removal in magnetorheological finishing. Applied Optics 40 (1)20–33.

Smurov, I., V. Ulianitsky, S. Zlobin, and A. Sova. 2010. Comparison of cold spray and detonation coatings properties. Proceedings of Thermal Spray: Global Solutions for Future Application, Singapore, May 2010.

Sova, A., S. Grigoriev, A. Okunkova, et al. 2013a. Potential of cold gas dynamic spray as additive manufacturing technology. *International Journal of Advanced Manufacturing Technology* 69:2269–2278.

Sova, A., et al. 2013b. Cold spray deposition of 316 L stainless steel coatings on aluminium surface with following laser post-treatment. *Surface and Coatings Technology* 235:283–289.

Spencer, K., and Zhang, M.-X. 2011. Optimisation of stainless steel cold spray coatings using mixed particle size distributions. *Surface and Coatings Technology* 205:5153–5140.

Spencer, K., D. M. Fabijanic, and M.-X. Zhang. 2009. The use of Al–Al_2O_3 cold spray coatings to improve the surface properties of magnesium alloys. *Surface and Coatings Technology* 204:336–344.

Spencer, K., D. M. Fabijanic, and M. X. Zhang. 2012a. The influence of Al_2O_3 reinforcement on the properties of stainless steel cold spray coatings. *Surface and Coatings Technology* 206 (14): 3275–3282.

Spencer, K., V. Luzin, N. Matthews, and M. X. Zhang. 2012b. Residual stresses in cold spray Al coatings: the effect of alloying and of process parameters. *Surface and Coatings Technology* 206:4249–4255.

Stark, L., I. Smid, A. Segall, T. Eden, and J. Potter. 2012. Self-lubricating cold-sprayed coatings utilizing microscale nickel-encapsulated hexagonal boron nitride. *Tribology Transactions* 55:624–630.

Stoltenhoff, T., and F. Zimmermann. 2009. Cold spray coatings for aluminum aerospace components exposed to high dynamic stresses. Proceedings of 8th HVOF Kolloquim Erding 5–6 November 2009, 135–143.

Stoltenhoff, T., H. Kreye, and H. J. Richter. 2001. An analysis of the cold spray process and its coatings. *Journal of Thermal Spray Technology* 11:542–550.

Stoltenhoff, T., C. Borchers, F. Gärtner, and H. Kreye. 2006. Microstructures and key properties of cold-sprayed and thermally sprayed copper coatings. *Surface and Coatings Technology* 200 (16–17): 4947–4960.

Stoney, G. 1909. The Tension of Metallic Films Deposited by Electrolysis. Proceedings of the Royal Society of London. Series A, Containing Papers of a Mathematical and Physical Character (1905-1934). 1909-05-06. 82 (553):172–175.

Sudharshan, P.P., V. Vishnukanthan, and G. Sundararajan. 2007. Effect of heat treatment on properties of cold sprayed nanocrystalline copper alumina coatings. *Acta Materialia* 55:4741–4751.

Suhonen, T., A. Varis, S. Dosta, M. Torrell, and J. M. Guilemany. 2013. Residual stress development in cold sprayed Al, Cu and Ti coatings. *Acta Materialia* 61:6329–6337.

Sun, J., Y. Han, and K. Cui. 2008. Innovative fabrication of porous titanium coating on titanium by cold spraying and vacuum sintering. *Materials Letters* 62:3623–3625.

Sundararajan, G., P.P. Sudharshan, A. Jyothirmayi, A. Ravi, and C. Gundakaram. 2009. The influence of heat treatment on the microstructural, mechanical and corrosion behaviour of cold sprayed SS 316 L coatings. *Journal of Materials Science* 44 (9): 2320–2326.

Suo, X. K., M. Yu, W. Y. Li, M. P. Planche, and H. L. Liao. 2012. Effect of substrate preheating on bonding strength of cold-sprayed Mg coatings. *Journal of Thermal Spray Technology* 21 (5): 1091–1098.

Tabor, D. 1951. *The hardness of metals*. Oxford: Clarendon Press.

Talbot, D. and J. Talbot. 1998. *Corrosion science and technology*. Boca Raton: CRC Press LLC.

Taljat, B., T. Zacharia, and F. Kosel. 1998. New analytical procedure to determine stress-strain curve from spherical indentation data. *International Journal of Solids and Structures* 35 (33): 4411–4426.

Tao, Y., et al. 2009. Effect of α-Al_2O_3 on the properties of cold sprayed Al/α-Al_2O_3 composite coatings on AZ91D magnesium alloy. *Applied Surface Science* 256 (1): 261–266.

Tao, Y., T. Xiong, C. Sun, L. Kong, X. Cui, T. Li, and G.-L. Song. 2010. Microstructure and corrosion performance of a cold sprayed aluminium coating on AZ91D magnesium alloy. *Corrosion Science* 52:3191–3197.

Taylor, K., B. Jodoin, and J. Karov. 2006. Particle loading effect in cold spray. *Journal of Thermal Spray Technology* 15 (2): 273–279.

Tjong, S. C., and H. Chen. 2004. Nanocrystalline materials and coatings. *Materials Science and Engineering R* 45:1–88.
Tran-Cong, S., M. Gay, and E. E. Michaelides. 2004. Drag coefficients of irregularly shaped particles. *Powder Technology* 139:21–32.
Tsui, Y. C., and T. W. Clyne. 1997. An analytical model for predicting residual stresses in progressively deposited coatings Part1: Planar geometry. *Thin Solid Films* 306:23–33.
Valente, T., C. Bartuli, M. Sebastiani, and A. Loreto. 2005. Implementation and development of the incremental hole drilling method for the measurement of residual stress in thermal spray coatings. *Journal of Thermal Spray Technology* 14 (4): 462–470.
Van Steenkiste, T. H., J. R. Smith, and R. E. Teets. 2002. Aluminum coatings via kinetic spray with relatively large powder particles. *Surface and Coatings Technology* 154:237–252.
Venkatesh, L., N. M. Chavan, and G. Sundararajan. 2011. The influence of powder particle velocity and microstructure on the properties of cold sprayed copper coatings. *Journal of Thermal Spray Technology* 20 (5): 1009–1021.
Vezzu, S., S. Rech, E. Vedelago, G. P. Zanon, G. Alfeo, A. Scialpi, and R. Huang. 2014. On deposition of Waspaloy coatings by cold spray. *Surface Engineering* 30:342–351.
Vezzu, S., C. Cavallini, S. Rech, E. Vedelago, and A. Giorgetti. 2015. Development of high strength, high thermal conductivity cold sprayed coatings to improve thermal management in hybrid motorcycles. *SAE International Journal of Materials Manufacturing* 8 (1): 180–186. doi:10.4271/2014-32-0044
Vijgen, R. O. E., and J. H. Dautzenberg. 1995. Mechanical measurement of the residual stress in thin PVD films. *Thin Solid Films* 270:264–269.
Villa, M., S. Dosta, and J. M. Guilemany. 2013. Optimization of 316 L stainless steel coatings on light alloys using cold gas spray. *Surface and Coatings Technology* 235:220–225.
Villafuerte, J., D. Dzhurinskiy, R. Ramirez, E. Maeva, V. Leshchynsky, and R. Maev. 2009. Corrosion behavior and microstructure of the Al–Al_2O_3 coatings produced by low pressure cold spraying. In *Thermal spray 2009: expanding thermal spray performance to new markets and applications,* eds. B. Marple, M. Hyland, Y.-C. Lau, C.-J. Li, R. Lima, and G. Montavon, 908–913. Las Vegas: ASM International.
Vreijling, M. 1998. Electrochemical characterization of metallic thermally sprayed coatings, Doctoral Thesis, printed in the Netherlands, 143.
Wang, A. P., T. Zhang, and J. Q. Wang. 2006. Ni-based fully amorphous metallic coating with high corrosion resistance. *Philosophical Magazine Letters* 86 (1): 5–11.
Wang, H.-R., B.-R. Hou, J. Wang, W.-Y. Li. 2008. Effect of process conditions on microstructure and corrosion resistance of cold-sprayed Ti coatings. *Journal of Thermal Spray Technology* 17 (5–6): 736–741.
Wang, Q., et al. 2010a. The influence of ceramic particles on bond strength of cold spray composite coatings on AZ91 alloy substrate. *Surface and Coatings Technology* 205 (1): 50–56.
Wang, T. G., S. S. Zhao, W. G. Hua, J. B. Li, J. Gong, and C. Sun. 2010b. Estimation of residual stress and its effects on the mechanical properties of detonation gun sprayed WC-Co coatings. *Material Science and Engineering A* 527:454–461.
Wang, Q., et al. 2013. Microstructure characterization and nanomechanics of cold-sprayed pure Al and Al–Al_2O_3 composite coatings. *Surface and Coatings Technology* 232:216–223.
Withers, P. J., and H. K. D. H. Bhadeshia. 2001. Residual stress. Part 1–measurement techniques. *Materials Science and Technology* 17:355–365.
Wolfe, D. E., T. J. Eden, J. K. Potter, and A. P. Jaroh. 2006. Investigation and characterization of Cr_3C_2-based wear-resistant coatings applied by the cold spray process. *Journal of Thermal Spray Technology* 15 (3): 400–412.
Wong, W, A. Rezaeian, E. Irissou, J. G. Legoux, and S. Yue. 2010. Cold spray characteristics of commercially pure Ti and Ti–6Al–4V. *Advanced Materials Research* 89–91:639–644.
Wong, W., E. Irissou, P. Vo, M. Sone, F. Bernier, J.-G. Legoux, H. Fukanuma, and S. Yue. 2012. Cold spray forming of inconel 718. *Journal of Thermal Spray Technology* 22 (2-3): 413–421.

Wong, W., P. Vo, E. Irissou, A. N. Ryabinin, J. G. Legoux, and S. Yue. 2013. Effect of particle morphology and size distribution on cold-sprayed pure titanium coatings. *Journal of Thermal Spray Technology* 22 (7): 1140–1153.

Wu, J., J. Yang, H. Fang, S. Yoon, and C. Lee. 2006. The bond strength of Al–Si coating on mild steel by kinetic spraying deposition. *Applied Surface Science* 252 (22): 7809–7814.

Xie, J., D. Nelias, H. Walter-Le Berre, et al. 2013. Numerical modeling for cold sprayed particle deposition. In 40th Leeds-Lyon symposium on tribology & tribochemistry forum, 4–6 Sept 2013, Lyon.

Xiong, Y., and M. X. Zhang. 2014. The effect of cold sprayed coatings on the mechanical properties of AZ91D magnesium alloy. *Surface and Coatings Technology* 253:89–95.

Xiong, T., Y. Tao, C. Sun, H. Jin, H. Du, and T. Li. 2009. Study on corrosion behavior of cold sprayed Al/α–Al_2O_3 deposit on AZ91D alloy. In *Thermal spray 2009: expanding thermal spray performance to new markets and applications,* eds. B. Marple, M. Hyland, Y.-C. Lau, C.-J. Li, R. Lima, and G. Montavon, 669–672. Las Vegas: ASM International.

Xiong, Y., X. Xiong, S. Yoon, et al. 2011. Dependence of bonding mechanisms of cold sprayed coatings on strain rate induced non equilibrium phase transformation. *Journal of Thermal Spray Technology* 20:860–865.

Yamada, M., Y. Kandori, S. Kazumori, et al. 2009. Fabrication of titanium dioxide photocatalyst coatings by cold spray. *Journal of Solid Mechanics and Materials Engineering* 3:210–216.

Yan, W., C. Lun Pun, and G. P. Simon. 2012. Conditions of applying Oliver–Pharr method to the nanoindentation of particles in composites. *Composites Science and Technology* 72:1147–1152.

Yandouzi, M., P. Richer, and B. Jodoin. 2009. SiC particulate reinforced Al–12Si alloy composite coatings produced by the pulsed gas dynamic spray process: microstructure and properties. *Surface and Coatings Technology* 203 (20–21): 3260–3270.

Yang, G. J., H. T. Wang, C. J. Li, and C. X. Li. 2011. Effect of annealing on the microstructure and erosion performance of cold-sprayed FeAl intermetallic coatings. *Surface and Coatings Technology* 205 (23–24): 5502–5509.

Yin, Shuo, Xiao Fang Wang, Wenya Li, Hanlin Liao, and Hongen Jie. 2012. Deformation behavior of the oxide film on the surface of cold sprayed powder particle. *Applied Surface Science* 259:294–300.

Yin, S., P. He, H. Lia, et al. 2014. Deposition features of Ti coating using irregular powders in cold spray. *Journal of Thermal Spray Technology* 23:984–990.

Yu, Min, Wen-Ya Li, Chao Zhang, and Hanlin Liao. 2011. Effect of vacuum heat treatment on tensile strength and fracture performance of cold-sprayed Cu-4Cr-2Nb coatings. *Applied Surface Science* 257 (14): 5972–5976.

Zeralli, Y., G. Rolland, M. Jeandin, et al. 2014. Novel in-situ gradient heat treatment during cold spray. In International thermal spray conference, Barcelona, 21–23 May 2014, 923–928.

Zhou, X., and P. Mohanty. 2012. Electrochemical behavior of cold sprayed hydroxyapatite/titanium composite in Hanks solution. *Electrochimica Acta* 65:134–140.

Ziemann, C. W., M. M. Sharma, B. D. Bouffard, T. Nissley, and T. J. Eden. 2014. Effect of substrate roughnening and cold spray coating on the fatigue life of AA2024 specimens. *Materials and Design* 54:212–221.

Zou, Y., W. Qin, E. Irissou, et al. 2009. Dynamic recrystallization in the particle/particle interfacial region of cold sprayed nickel coating: electron backscatter diffraction characterization. *Scripta Materialia* 61:899–902.

Zou, Y., et al. 2010. Microstructure and nanohardness of cold-sprayed coatings: electron backscattered diffraction and nanoindentation studies. *Scripta Materialia* 62 (6): 395–398.

Chapter 5
Residual Stresses and Fatigue Life Enhancement of Cold Spray

H. Jahed and R. Ghelichi

5.1 Introduction

Residual stresses are referred to as stresses remaining in a structure in the absence of external mechanical or thermal loads. Trapped stresses such as stress in a preloaded bolt, which in most cases are due to elastic deformation (e.g., will be completely freed upon removal of preload), are also referred to as residual stresses. However, a more precise reference to residual stress is when elastic and plastic zones coexist within a structure free of external forces. Such stresses are induced where a stress gradient exists in a body and the stresses are larger than the elastic limit. Upon removal of external forces and to maintain the permanent deformation caused by the stress raiser, a self-balanced system of internal forces (residual stresses) remains in the structure. Therefore, residual stresses are usually local (next to the location of the stress concentration) and bounded by a large elastic region. Residual stresses are caused by most manufacturing processes such as welding, machining, and forming or are intentionally produced by processes like peening and autofrettage.

Tensile residual stresses are normally unwanted stresses that are induced because of manufacturing and/or joining processes, service overloads, and material defects. Tensile residual stress puts initial material flaws into an opening mode of fracture and hence promotes crack initiation and propagation, leading to a significant reduction in the life of the components. This type of residual stress is detrimental to the structure, and designers are normally trying to minimize or completely remove tensile residual stresses. Standard processes such as thermal stress relief are often used after manufacturing processes to remove unwanted residual stresses.

H. Jahed (✉)
Mechanical and Mechatronics Engineering Department, University of Waterloo, Waterloo, ON N2L 5H5, Canada
e-mail: hjahedmotlagh@uwaterloo.ca

R. Ghelichi
Department of Mechanical Engineering, Massachusetts Institute of Technology, Cambridge, MA 02139, USA

J. Villafuerte (ed.), *Modern Cold Spray*, DOI 10.1007/978-3-319-16772-5_5

Compressive residual stresses, on the other hand, are beneficial. These types of stresses work against mode I crack initiation and delay formation and propagation of cracks. In most cases, the presence of the compressive residual stresses leads to the possibility of application of larger external loads and longer lives. Compressive residual stresses are normally intentionally made to enhance fatigue life. Established industrial processes like shot peening of aerospace and automotive parts, and autofrettage of pressure vessels are common examples of methods for creating beneficial compressive residual stresses.

Residual stresses can be produced by nearly every mechanical, chemical, and thermal process including coating technologies. Cold spray coating uses high particle velocity and impact energy to enable material coating on a substrate at a relatively low temperature. A by-product of cold spray coating is the formation of residual stress due to the peening effect of the particles' collision with the substrate. The high particle impact velocity causes high local stresses which lead to plastic deformation in the substrate in the proximity of the particle–substrate interface. Figure 5.1 depicts such plastic zone formation schematically. Due to the localized effect of the collision, the area that went under plastic deformation was surrounded by a large elastic domain resulting in the formation of local residual stresses. The dent resulting from the particle impact causes tension at the surface which upon unloading (adhesion of particle to surface) creates compressive residual stresses at and near the surface (Fig. 5.1). Such stresses are beneficial for fatigue life of the coated part. Due to the self-equilibrated nature of residual stress (i.e., zero net internal forces) associated with the compressive residual stress at and near the surface, there is tensile residual stress through the depth to balance off the residual force. A schematic pattern of the distribution of residual stress due to collision of a single particle with the substrate is shown in Fig. 5.1.

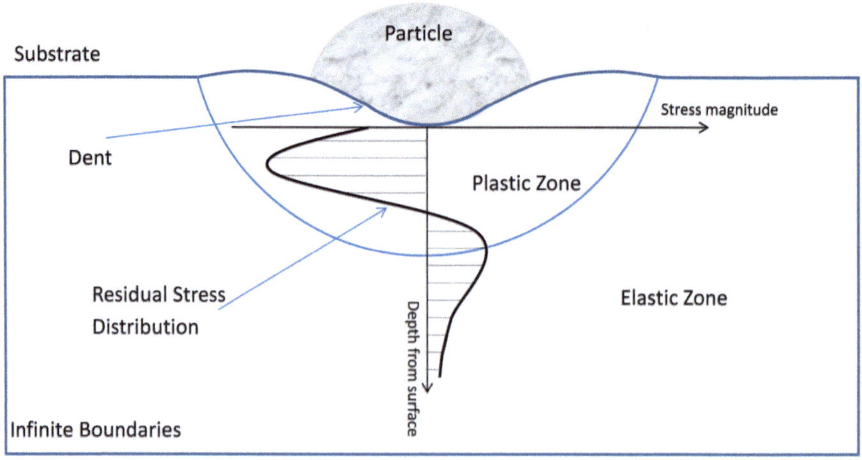

Fig. 5.1 Schematic of particle collision with the substrate in cold spray leading to formation of beneficial residual stress. Typical residual stress distribution at and near the surface is also shown. (Shayegan et al. 2014)

There have been a number of studies on the residual stress induced by cold spray coating. The recent literature is briefly reviewed here. McCune et al. (2000) have measured residual stresses for copper coatings, and Bagherifard et al. (2010a) performed X-ray diffraction (XRD) measurements on stresses induced in aluminum coatings. Ghelichi et al. (2012) measured the residual stress of different Al-alloy-coated samples and studied their effect on fatigue behavior. An interesting common observation in all studies is the relaxation of the compressive residual stresses at the interface of the coating and the substrate; conversely, the major difference is the reported stress in the substrate that is shown close to zero in McCune et al. (2000), whereas it is of a considerable amount, with respect to that of the deposited material, in other studies (Bagherifard et al. 2010a; Ghelichi et al. 2012). Price et al. (2006) studied the effect of the deposition of titanium on titanium. A Ti6Al4V alloy was coated with pure titanium by cold spray. Coatings were performed on as-received and grit-blasted (GB) samples. Fully reversed rotating bending fatigue of samples before and after coating revealed a 15 % reduction in fatigue endurance limit of the as-received samples, but no significant reduction was observed on the GB substrate. The reduction in fatigue endurance limit was attributed to the residual stress induced by the coating. However, in a study by Cizek et al. (2013), the cold spray of titanium on Ti6Al4V substrates was studied, and the average fatigue lives of cold-sprayed samples were reported to be shorter by 9 % when compared to the as-received uncoated specimens. Luzin et al. (2011) studied the residual stress in Cu- and Al-coated samples by neutron powder diffraction stress measurement. They used the Tsui and Clyne's progressive model (Tsui and Clyne 1997) that was originally developed to model the residual stress accumulation in thermal spray coatings to interpret their empirical results. Luzin et al. (2011) concluded that the residual stress is determined almost entirely by the plastic deformation process of the spray material due to the high-velocity impact of the particles and that the thermal effects do not play a notable role in changing the distribution of the induced stresses. Spencer et al. (2012) also followed the same approach using Tsui and Clyne's progressive model (Tsui and Clyne 1997) to understand the residual stress induced by cold-spraying Al and an Al alloy on Mg substrates. The residual stress distribution in their work was obtained by measurements using neutron diffraction with high spatial resolution. They concluded that the residual stress profile was more dependent on the alloy content, that is, intrinsic resistance to plastic deformation, than the spray-processing conditions. Sansoucy et al. (2007) studied the cold spray of Al–13Co–26Ce alloy particles on AA 2024-T3 substrate. They reported the fatigue and the bond strength of the Al–Co–Ce coatings. The results show that the Al–Co–Ce coatings improved the fatigue behavior of AA 2024-T3 specimens when compared to uncoated specimens. They attributed the increase in the fatigue properties to the residual compressive stresses induced in the coatings. Jeong and Ha (2008) investigated the effect of cold-spraying aluminum alloy A356 powder on the substrate made of the same material. They reported a significant 200 % improvement in the fatigue strength of the coated samples. The effect of cold spray coating on microstructure, residual stresses distribution, and fatigue life of Al5052 substrate coated with pure aluminum and Al7075 powders was studied by Ghelichi et al. (2012).

They reported a 30 % increase in the fatigue life of coated samples as compared to as-received samples when tested under a fully reversed cantilever-bending test. The increase in life was attributed to the presence of residual stress after the cold spray. The residual stress measurement in their work showed the peening effect of the coating on the substrate and verified the importance of the compressive residual stress on the fatigue life of the components. Moridi et al. (2014) studied the effect of cold spray of Al6082 by the same material on its fatigue life. They showed that the fatigue strength of coated samples was increased by 15 %. Ghelichi et al. (2014b) have studied, by experimental measurement and numerical simulation using the finite element method, the physical phenomenon of the residual stress inducement in the aluminum samples. They have shown that the process temperature has an annealing effect; that is, the gas temperature relieves the residual stress induced by particle impact. The numerical simulation by considering the annealing effect of the gas temperature matches with the experimental measurement. Shayegan et al. have studied the effect of aluminum powders on AZ31 by experiments and simulation. In the numerical simulation, both Ghelichi et al. (2014b) and Shayegan et al. (2014) have considered the randomness of the impacts. Shayegan et al. reported a 10 % increase in fatigue endurance and a maximum of 40 % life enhancement in the low-cycle fatigue of AZ31B extrusion when coated with AL1100.

A number of factors influence the residual stress level in cold spray coatings, including the impact velocity, particle size and hardness, particle temperature, quenching of the sprayed material due to high cooling rate, and thermal mismatch between the coating and the substrate. The recent studies on the residual stresses induced by cold spray coating can be categorized into: (1) experimental measurements of residual stress to quantify the level and deepness of these stresses; (2) numerical modeling of the impact and residual stresses; and (3) the influence of cold spray coating on the fatigue life. A brief review of approaches and details of the experimental measurement and numerical models of these three topics are provided in the following sections.

5.2 Experimental Measurement of Residual Stress in Cold Spray Coating

Constant bombardment of particles on the substrate and coating area is known as a major feature that results in increasing the residual stresses. The cold spray process is principally different from other types of thermal spray coating processes in terms of its lower process temperature and higher particle impact velocity, both of which directly affect the resulting residual stress distribution. The studies on quantifying residual stress in cold spray coating differ from one another because of either the method of measurements (layer removal, X-ray diffraction, or neutron diffraction) or the substrate/particle material (Al, Ti, Cu, or Mg).

McCune et al. (2000) have studied the residual stress of the cold spray coating of copper powders. The residual stress in the samples was measured using the layer removal technique developed by Greving et al. (1994). The residual stress based on the depth from the outer surface was measured for different coated samples. To calculate the residual stresses from the released residual strain due to the removal of a layer, the modulus of elasticity of the material is needed. The measurement of Young's modulus was performed using the composite cantilever method of Rybicki et al. (1995). As expected, they found the residual stress to be compressive at and near the surface and tensile away from the surface. They have further noticed for the first time the effect of the temperature of the coating process as the annealing effect on the residual stress induced by the powders' impacts; that is, the residual stress has been reduced by increasing the coating temperature. Figure 5.2 shows the residual stress measurement of the coated samples presented by McCune.

Ghelichi et al. (2012, 2014a, b) have performed a series of studies on the coating residual stress measurement. To study the residual stress in-depth profile before and after the cold spray process, the XRD method was used. The analysis of the surface layer was performed using a diffractometer with radiation CrK_{α}, a circular irradiated area of 1 mm diameter, $Sin^2(\psi)$ method, and diffraction angle 2θ of 139° scanned between −45 and 45°. The effect of penetration depth of the radiation was approximately 5 μm. The in-depth measurements have been carried out step by step by removing a very thin layer of material (0.01/0.02 mm) using an electro-polishing device in order to obtain the in-depth profile of residual stresses. Different measurements have been performed in three rotation angles (0, 45, and 90°). They measured the residual stress induced by different aluminum particles on different aluminum alloy substrates. The coating samples included Al5052 substrate coated with Al7075 particles, and pure aluminum (size ranging from 10 to 40 μm). Figure 5.3 shows three different measurements of the coated samples for different aluminum alloy coating.

Ghelichi et al. (2014b) have performed a specific test to verify the annealing phenomenon in the coated samples. GB samples have gone under the coating process with the exact same temperature and pressure condition but with no powders. The coating temperature and time of the coating process for samples coated as with

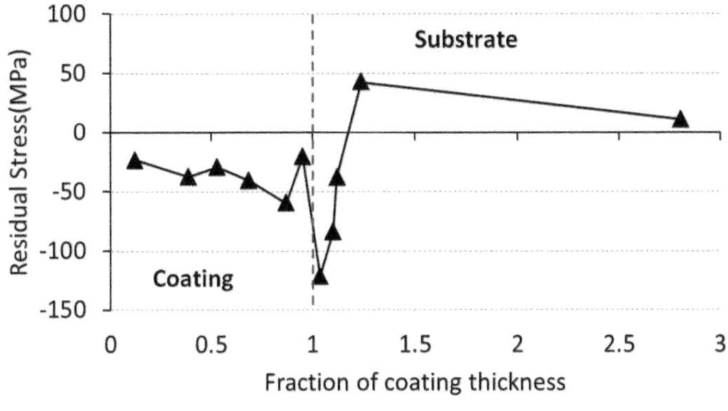

Fig. 5.2 Experimental measurement of the residual stress on copper-coated samples. (McCune et al. 2000; Ghelichi 2012)

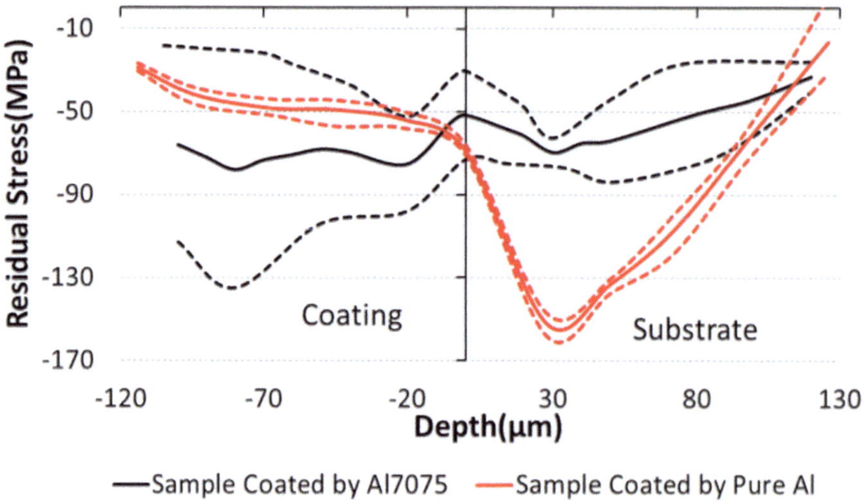

Fig. 5.3 Residual stress measurement for Al5052 samples coated with Al7075 and pure Al; *dashed lines* show the lower and upper bounds of measurements, and *solid lines* show an average value. (Ghelichi et al. 2014a)

Al7075 were 500 °C and 420 s, respectively, while for samples coated with pure Al were 350 °C and 90 s. Grit blasting will cause residual stress in the samples. The results show more time or temperature relieves more of the residual stress in the samples. The final results are shown in Fig. 5.4. By using this figure, Ghelichi et al. (2014a) estimated the annealing constants.

Shayegan et al. (2012), using the same XRD technique, have measured the residual stress for magnesium alloy substrate coated by aluminum powders. An air-quenched AZ31B extrusion piece was used as the substrate. The samples were first stress relieved using the American Society for Metals (ASM)-recommended method of thermal treatment of 260 °C for 15 min (1989) to remove residual stresses of the extrusion process. The substrate was then coated by commercially available

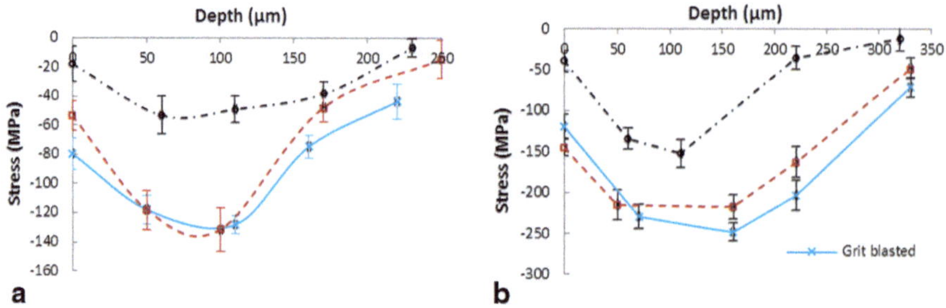

Fig. 5.4 The effect of coating temperature on relieving the residual stress in the samples: **a** Al5052, **b** Al6061. (Ghelichi et al. 2014b)

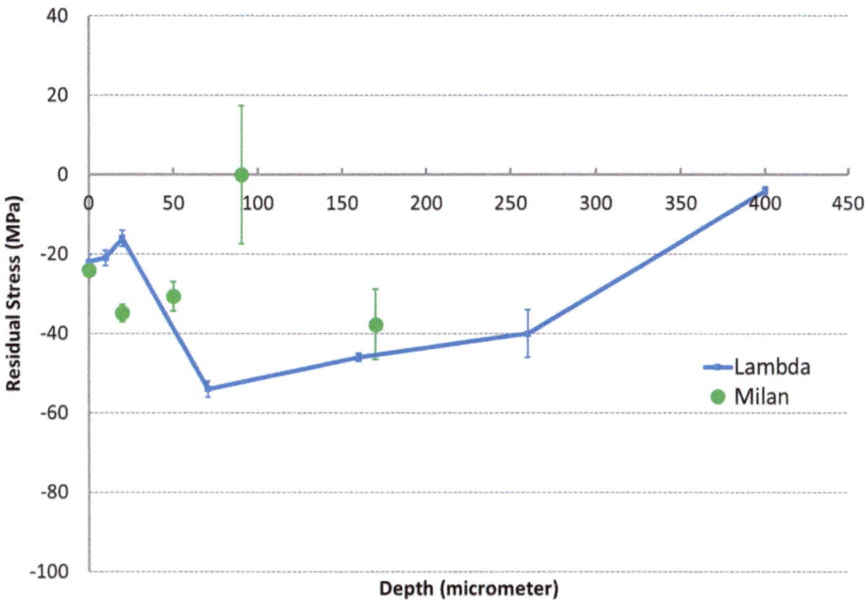

Fig. 5.5 Residual stress measurement for Al-coated AZ31 extrusion (Shayegan et al. 2014); the two sets of measurements were performed at Lambda Technologies and Polytechnic University of Milan

pure aluminum (Al > 99.7 %) powders with an average particle size of 25 μm. The sample surface was grit blasted before the cold spray with aluminum oxide. The following processing settings were used: temperature of 350 °C, pressure of 250 psi, and standoff distance of 12 mm. The coat displayed bond strength of 22 MPa, hardness of 34–37 HB, and density greater than 99.5 %. XRD method was performed to measure the residual stresses at nominal depths of 10, 20, 70, 160, 260, and 400 μm in the extrusion direction. Figure 5.5 shows the experimental measurement of the residual stress.

The common aspects of all these measurements can be summarized in a few important facts. All different measurements show that compressive residual stresses are induced at the substrate surface. The level of this stress depends on material and coating parameters. The residual stress remains compressive beneath the surface and turns into tensile beyond that. The maximum compressive stress is the order of magnitude higher than the maximum in-depth tensile residual stress. The residual stress fades away beyond a few hundred microns of the substrate. The results also show that the residual stress will be relieved on the substrate; that is, the interface of the deposited materials and the substrate will experience less compressive residual stress as compared to both substrate and coating. This may be due to the high local plastic deformation and local grain size refinement of the interface (Ghelichi 2012). Finally, increasing the coating temperature will decrease the induced residual stress due to an annealing effect.

In this regard, the mathematical modeling using numerical simulation has been used in order to have a better perspective of this complex physical phenomenon. In the following section, two different models are reviewed in detail.

5.3 Numerical Modeling of a Cold Spray Process

Numeral simulations are used to provide a better understanding of different physical phenomena and to provide a tool for sensitivity and parameter analysis. They help qualitatively and, with an appropriate fitting, quantitatively to monitor different aspects of the process, especially when the size is small and deformation happens in a fraction of seconds where the experiment cannot capture every single detail. Simulation also helps in reducing the production cost by minimizing the number of experimental trials by providing some in-depth information on the effect of parameter changes on the final results.

In cold spray coating, the high-velocity impact of microparticles induces high local plastic deformation, which in turn induces compressive residual stress at and near the surface of the coated samples. Due to the high rate of impact velocity and flux of the particles on the substrate, the substrate experiences a very complex rate-dependent plastic deformation, which includes softening and localization in the material behavior (Champagne 2007).

There have been a number of efforts by researchers in simulating cold spray coating. Most of these researches are dedicated to the study of a single-particle impact in the coating process. These studies are focused on the effect of plastic behavior of the metallic substrate under a high-rate large local deformation due to the impact. Assadi et al. (2003) and Grujicic et al. (2014a), by choosing the Johnson–Cook plastic model, studied the single-particle impact. They have shown that the localization in stress–strain curvature can be the main reason for bonding the particles on the substrate. This theory has been backed up by experimental observation later on Ghelichi et al. (2011). Their study has been continued and modified by Kumar et al. (2009), Li et al. (2006), and Ghelichi et al. (2011). These studies were aimed at approximating the value of "critical velocity" which is the minimum velocity required to guarantee particle-to-surface bond. The latter references give a very good insight of the behavior of a particle in a single-particle impact shot. Few studies are focused on the response of the substrate due to multiple particle impacts. These studies have shown that the size and velocity of the particles are major factors altering the results in the process. In practice and in the real coating process, there is usually a broad distribution of the particle sizes with different velocities involved, which may be captured in simulation.

Here, in more detail, two studies are being reviewed. These two studies have considered the randomness of the particle size impact and different physical parameters that affect the plastic deformation and consequently the residual stress of the process. The studies by Ghelichi et al. (2014b) and Shayegan et al. (2014) are discussed here.

5.3.1 Residual Stress Due to Cold Spray of Aluminum Substrate

Ghelichi et al. (2014b) studied the residual stress of the two different aluminum alloys (Al5052 and Al6061) coated by Al7075. Due to the intrinsic similarity of the cold spray coating to shot peening, the model introduced by Bagherifard et al. (2010b) for shot peening simulation has been used.

One of the main constituents of cold spray simulation is the particle size distribution. To introduce the randomness of particle size, the Rosin–Rammler model (RosIn 1933; Ramakrishnan 2000; Eq. 5.1), which has been often used to estimate the particle distribution (Allen 2003; Li et al. 2006), was adopted. In Eq. 5.1, the cumulative density function (CDF), D_p is the particle diameter, and D_{rr} and Q are the Rosin–Rammler coefficients. The actual particle morphology of the powder used in the experiments was obtained through scanning electron microscope (SEM) images and then used to find the fitting parameters of the Rosin–Rammler model. Figure 5.6 shows a representative SEM image that was employed to obtain the particle histogram (Fig. 5.6b), which in turn yielded the Rosin–Rammler model parameters through fitting.

$$R(\mathrm{CDF}) = \left\{ 1 - \exp\left[0 \left(\frac{\ln(D_p)}{\ln(D_{rr})} \right)^{Q} \right] \right\} \times 100\%. \tag{5.1}$$

A uniform distribution was assumed for the position of the particles in the impact zone as shown by a red circle in Fig. 5.7. The number of particles has been calculated based on the thickness of the coated samples and average particle size.

Another major parameter in cold spray simulation is the particle velocity. In cold spray coating, which is a process with specified pressure and temperature, the particle velocity is a function of its diameter. To obtain an estimation of the impact velocity of particles with different sizes, based on the Dykuizen and Smith model (Grujicic et al. 2003), the relationship (Eq. 5.2) presented by Grujicic et al. (2004b) and Ghelichi et al. (2011) was adopted.

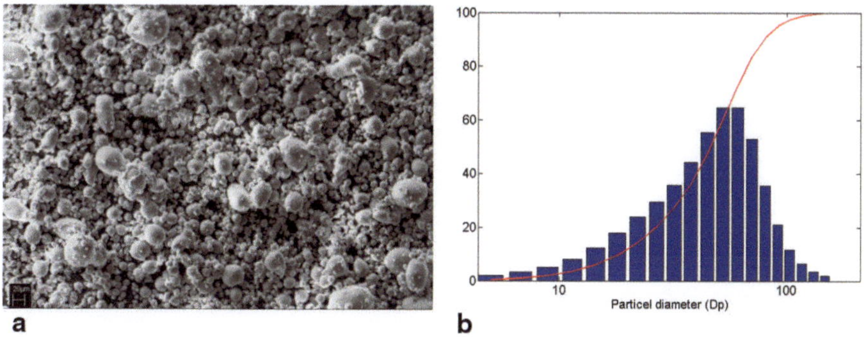

Fig. 5.6 **a** SEM image of Al7075 particle morphology and size distribution, **b** histogram and the CDF of the particle size distribution based on Rosin–Rammler (D_{rr}=45.95; Q=21.46). (Ghelichi et al. 2014b)

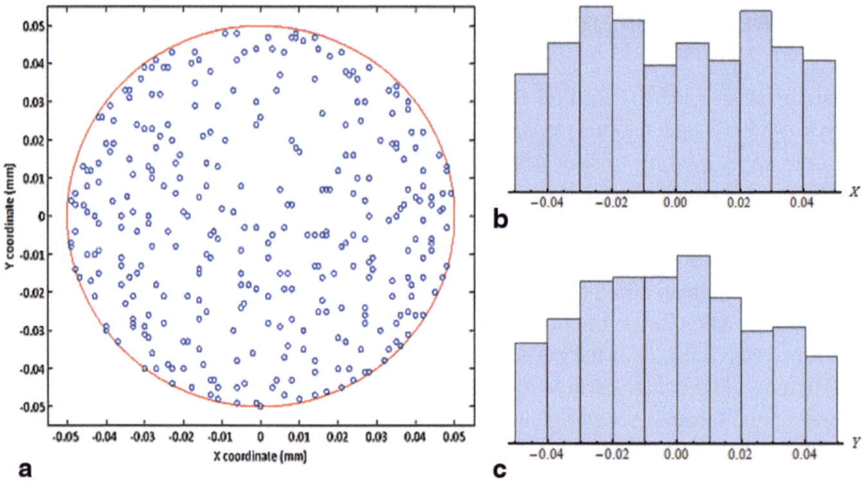

Fig. 5.7 **a** The center of the particles *(blue dots)* in one of the models with respect to the borders of the impact zone *(red circle)*; **b** and **c** show the histogram of distribution of the particles in *x* and *y* direction. (Ghelichi et al. 2014b)

$$\begin{aligned} V_{\text{impact}} &= V_{\text{p}} e^{-\rho_{\text{st}} * L_{\text{st}} / 4 \rho_{\text{p}} D_{\text{p}}} \\ \rho_{\text{st}} &= Re(-1.04 + 2.27M - 0.21M^2) \\ L_{\text{st}} &= Re(0.97 - 0.02M) \end{aligned} \tag{5.2}$$

In Eq. 5.2, V_{impact} is the velocity of the particles at the surface of the substrate at the time of impact and Re is the Reynolds number. The Mach number (M) and particle velocity (V_{p}) at the end of the cold spray nozzle can be calculated by Eqs. 5.3 and 5.4:

$$\begin{aligned} M &= \left[\kappa_1 \frac{A}{A^*} + (1 - \kappa_1 \right)]^{k_2} \\ \kappa_1 &= 218.0629 - 243.5764\gamma + 71.7925\gamma^2 \\ \kappa_2 &= -0.122450 + 0.28130\gamma \end{aligned} \tag{5.3}$$

and

$$\begin{aligned} \frac{V_{\text{p}}}{V_{\text{e}}} &= 0.5\left(\frac{V_{\text{p}}}{V_{\text{e}}}\right)_{18} + 0.5\left(\frac{V_{\text{p}}}{V_{\text{e}}}\right)_{20} \\ \left(\frac{V_{\text{p}}}{V_{\text{e}}}\right)_{20} &= -e^{\frac{-9\mu x}{\rho_{\text{p}} D_{\text{p}}{}^2 V}} + 1 \\ \left(\frac{V_{\text{p}}}{V_{\text{e}}}\right)_{18} &= -e^{-\sqrt{\frac{3\rho_0 C_{\text{D}} x}{\rho_{\text{p}} D_{\text{p}}}}} + 1. \end{aligned} \tag{5.4}$$

The gas velocity at the nozzle exit (V_e) and density of the gas (ρ_0) can be calculated by Eqs. 5.5 and 5.6. Gas viscosity (μ) may be evaluated using Sutherland's formula (Robert 1984; Dykhuizen and Smith 1998).

$$\begin{aligned}\rho_e &= \frac{\rho_0}{(1+\frac{\gamma-1}{2}M^2)^{1/\gamma-1}}\\ V_e &= M\sqrt{\gamma R T_e}\\ T_e &= \frac{T_0}{1+\frac{1-\gamma}{2}M^2}.\end{aligned} \tag{5.5}$$

$$\begin{aligned}\rho_0 &= \frac{P_0}{RT_0}\\ \mu &= \mu_0\left(\frac{a}{b}\right)\left(\frac{T_e}{T_{S0}}\right)^{3/2}\\ a &= 0.555T_0 + C; b = 0.555T + C.\end{aligned} \tag{5.6}$$

Another major constituent of cold spray particle impact simulation is the particle temperature. Papyrin et al. (2006) presented an analytical formulation for the particle temperature at the contact surface:

$$\begin{aligned}T_p &= T_0 + C\exp\left(\frac{N_u\frac{6k}{d_p^2}x}{\rho_p \vartheta_p C_P}\right)\Bigg|_{x=\text{impact dist}}\\ N_u &= 2a + 0.459bRe^{0.55}\,\mathrm{Pr}^{0.33}\\ a &= \exp(-M)\left(1+\frac{17M}{Re}\right)^{-1}, b = 0.666 + 0.333\exp\left(-\frac{17M}{Re}\right),\end{aligned} \tag{5.7}$$

where k stands for thermal conductivity; Re is the Reynolds number for particles; C_P represents the heat capacity of the particles; $\mathrm{Pr} = \frac{c_p\mu}{k}$ is the Prandtl number, where c_p is the specific heat, and μ is the dynamic viscosity obtained from Eq. 5.6.

Another significant constituent of the simulation is the material model for the substrate and particles. Considering that there is appreciable plasticity with high strain rate (impact) involved, proper plasticity models need to be adopted to mimic the characteristics of the deformation. In this study, a combined isotropic and kinematic hardening model (Lemaitre and Chaboche 1990) was chosen for the substrate and the Johnson–Cook model (Johnson and Cook 1983) presented in Eq. 5.8 was adopted for the powders. The material constants for the Al particles are presented in Table 5.1 (Brar et al. 2009).

Table 5.1 Mechanical properties of the material. (Johnson and Cook 1983)

Material	Hardness (Rockwell)	Density (kg/m³)	Specific heat (J/kg °K)	Melting point (°K)	B_1 (MPa)	B_2 (MPa)	n	C	M
Al 7075-T6	87	2800	960	910	546	674	0.72	0.059	1.56

$$\sigma_{eq} = [B_1 + B_2(\epsilon_p)^n][1 + Cln(\dot{\epsilon}_p / \dot{\epsilon}_{p0})] \times \left[1 - \left(\frac{T - T_{init}}{T_{melt} - T_{init}}\right)^m\right] \tag{5.8}$$

Finally, a model for the annealing process of the particle and substrate after collision is required for the simulation. Due to the effect of the high temperature (close to annealing temperature) in the coating process, a phenomenon very similar to annealing may happen which releases the stress in the substrate (Totten 2002). The effect of time, as a main factor of annealing process, on residual stress relaxation can be modeled by adopting the Zener–Wert–Avrami function (Fine 1964):

$$\sigma^{rs} / \sigma_0^{rs} = \exp[-(St_a)^m] \tag{5.9}$$

where m is a numerical parameter depending on the relaxation mechanism. Based on Eq. 5.9, in samples with the same geometry, volume, and material, when annealing parameters including time and temperature are the same, the fraction of $\sigma_0^{rs} / \sigma^{rs}$ remains constant. This constant has been applied as the annealing effect on numerical simulations' results to evaluate the final distribution of residual stresses.

5.3.1.1 Numerical Model Description

The commercial finite element software Abaqus/Explicit 6.12–1 Abaqus 6.12-1 (2012) was utilized to investigate multiple particle impact effects. A finite element (FE) model of the substrate and particles was created. Target mesh was set up by C3D8R 8-node linear brick elements with reduced integration and hourglass control. Due to the incompatibility of the very small size of the particles and large size/volume of the substrate, and the localized effect of the impact, the bottom-side face of the substrate (target) was meshed by the half-infinite elements that provide quiet boundaries by minimizing the reflection of dilatational and stress waves back into the region of interest.

Figure 5.8a, b shows different views of the FE model. In this figure, the arrangement and the particle size distribution have been presented with respect to the red circle that represents the impact zone. The experimentally measured in-depth residual stress on the GB samples before the coating process has been considered as predefined stress field in the FE model before the particles impact. The numerical model was tuned for Al7075 powders and both Al5052 and Al6061 as the substrates.

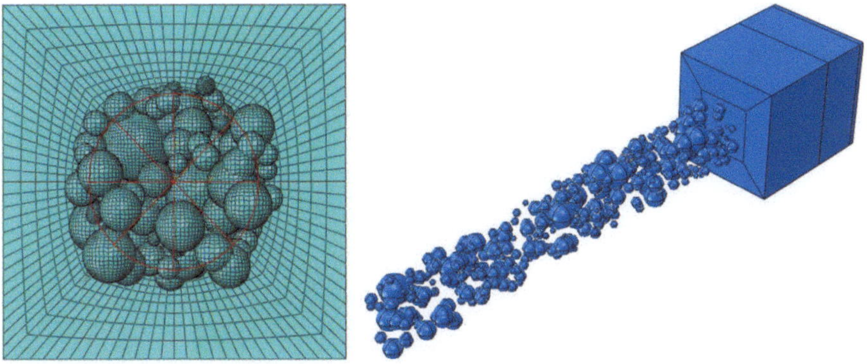

a b

Fig. 5.8 **a** Top and **b** isometric view of the model developed for residual stress calculations (the *red* area represents the "impact-zone"). (Ghelichi et al. 2014b)

The results of the numerical simulation before and after considering the annealing effect of the temperature are presented in Fig. 5.9. It is clearly shown that results are improved when the effect of the process temperature on relieving the stress is considered. The numerical simulation and the experimental measurements confirm that in the cold spray coating process constant bombarding of particles induces residual stresses in the substrate. Although the process gas temperature is less than the melting point of the materials involved, it was shown to have a negative annealing effect on the favorable compressive residual stress induced by previous grit blasting and the particles' impact. This effect was clearly noted by means of experiments on samples formerly grit blasted and then submitted to the cold spray process without using powders. The results of the experimental measurements show the considerable effect of the process temperature on promoting residual stress relaxation.

A two-step approach aimed at realistically simulating the cold spray process was developed to obtain the final residual stress state in the substrate. It considers the residual stresses induced by grit blasting, particle size distribution, random impact position, and the corresponding velocity and temperature for the particles as a function of their size. The first step is based on an FE/explicit simulation and considers different material models compatible with the phenomenon. Even if some approximation is considered with respect to the real process (e.g., the bonding of the particles on the substrate is not considered in the modeling), the numerical model allows for studying the peening effect of the particles. The results obtained from the FE simulation do not consider the effect of the annealing. In this regard, in the second analytical step, the Zener–Wert–Avrami method (Fine 1964) was utilized to introduce the annealing effect on the numerical results using the experimental measurements performed on uncoated samples.

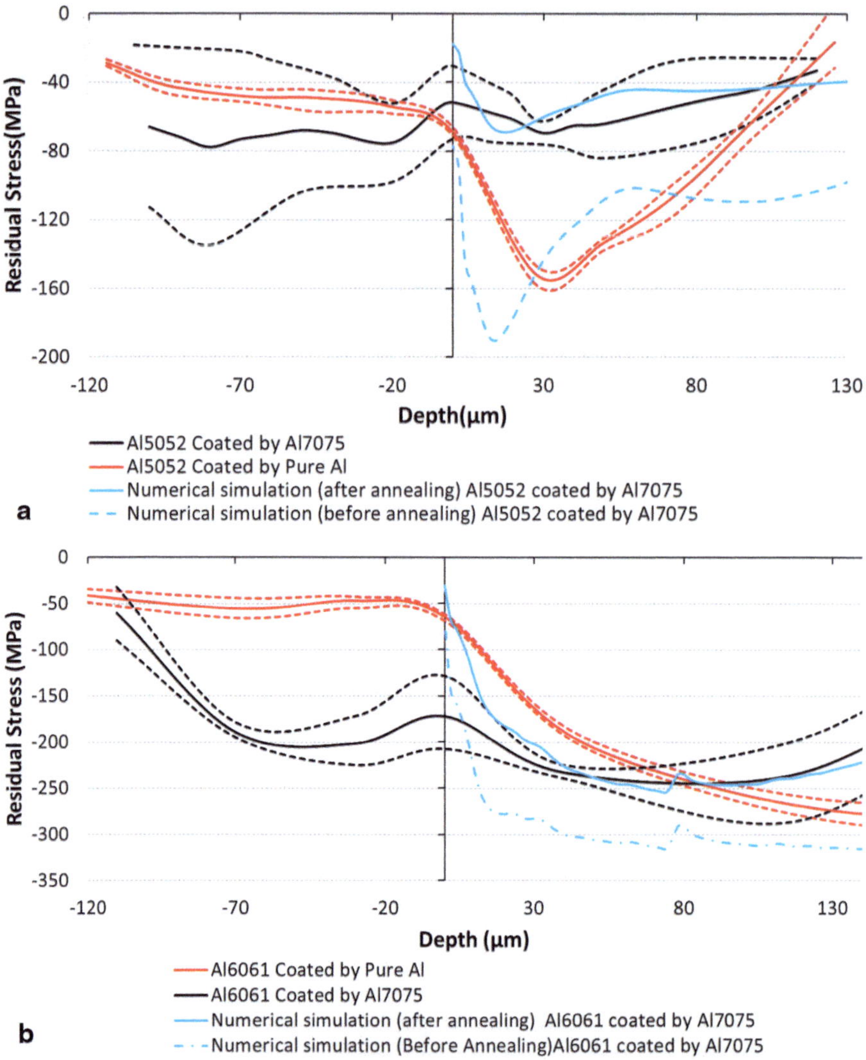

Fig. 5.9 Residual stress measurement for the coated samples with Al7075 and pure Al compared to the numerical simulation. **a** Al5052, **b** Al6061. (Ghelichi et al. 2014b)

5.3.2 Residual Stress Due to Cold Spray of Magnesium Substrate

Shayegan et al. (2014) presented the most recent numerical model. They simulated the coating of aluminum particles on magnesium (AZ31) extrusion substrate using LS-DYNA (LS-DYNA3D 1999) software. The focus of their study was the peening effect of the spraying and the resulting residual stress.

Magnesium extrusion has two main mechanical characteristics: (1) yield asymmetry and (2) directional anisotropy. Because of limited slip system in hexagonal close-packed (HCP) magnesium, deformation twinning is a major deformation mechanism when extension along the *c*-axis (basal axis) is induced. This results in a different deformation mechanism in tension and compression known as yield asymmetry. LS-DYNA (LS-DYNA3D 1999) material library provides MAT-124 that captures yield asymmetry by accepting two independent tensile and compressive stress–strain curves. Noting that yield asymmetry is the dominant deformation factor, and by ignoring the directional anisotropy of substrate, Shayegan et al. (2014) adopted MAT-124 in conjunction with the Cowper–Symonds (Eq. 5.11) strain rate model.

$$\frac{\sigma}{\sigma_0} = 1 + \left(\frac{\dot{\varepsilon}}{D}\right)^{\frac{1}{p}}, \tag{5.10}$$

where $\dot{\varepsilon} = \sqrt{\dot{\varepsilon}_{ij}\dot{\varepsilon}_{ij}}$ is the strain rate, σ is the von Mises stress, σ_0 is the semi-static yield stress and $D = 24{,}124$ and $p = 3.09$ (Najafi and Rais-Rohani 2011) are the strain rate parameters.

The tension and compression curves of AZ31B shown in Fig. 5.10 and other alloy parameters including density $(1770\,\mathrm{kg}/\mathrm{m}^3)$, Young's modulus (45 GPa), and Poisson's ratio (0.35) are used in the simulation.

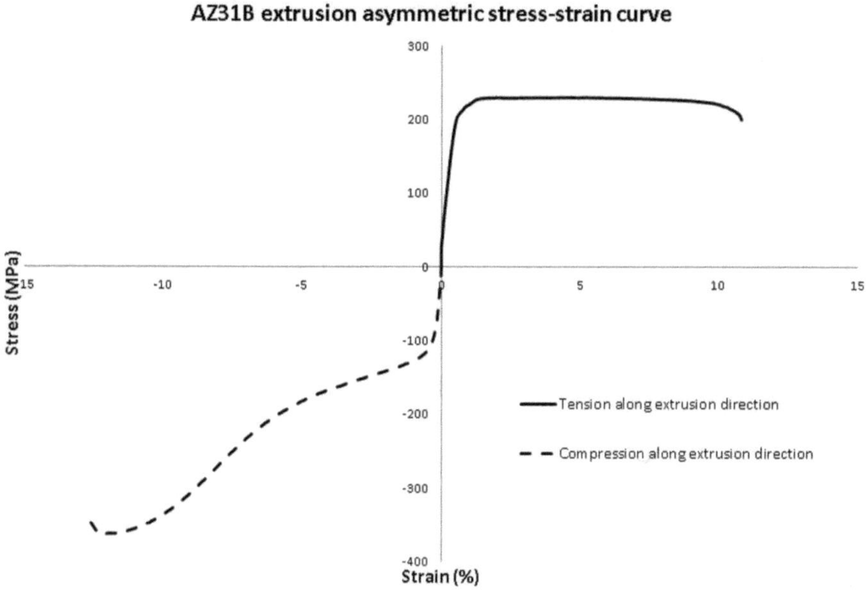

Fig. 5.10 Tension–compression curve of AZ31B. (Albinmousa et al. 2011)

Table 5.2 Johnson–Cook parameters for AL 1100-O. (Benck 1976; Pierazzo et al. 2008)

Value	Parameter
49	A (MPa)
157	B (MPa)
0.167	n
0.016	C
1.7	m
293	$T_{\text{room}}(K)$
933	$T_{\text{melt}}(K)$

The same Johnson–Cook material model (LS-DYNA MAT-15) as in Ghelichi et al. (2011) was used for aluminum particles which is suitable for problems with a high rate of strain with proper material softening due to adiabatic temperature increase (Table 5.2).

For the dissemination of stress waves due to high-velocity collision of particles to the work piece, the following Gruneisen (Albinmousa et al. 2011) equation is used:

$$p = \frac{\rho_0 C_0^2 \mu \left[1 + \left(1 - \frac{\gamma_0}{2}\right)\mu - \frac{a}{2}\mu^2\right]}{\left[1 - (S_1 - 1)\mu - S_2 \frac{\mu^2}{\mu + 1} - S_3 \frac{\mu^3}{(\mu + 1)^2}\right]^2} + (\gamma_0 + d\mu)E; \; \mu = \frac{\rho}{\rho_0} - 1, \quad (5.11)$$

where C_0 is the bulk speed of sound, ρ_0 is the initial density, ρ is the current density, S_1, S_2, and S_3 are the coefficients of the slope of the $U_s - U_p$ curve, U_s is the shock wave velocity, U_p is the particle velocity, γ_0 is Gruneisen's gamma at the reference state, d is the first-order correction coefficient per volume for γ_0, and E is the internal energy per unit reference volume. Proper parameters of Gruneisen equation for Al 1100-O have been published by Group GMX-6, Los Alamos (1969; Table 5.3).

Other material constants of AL 1100-O particles that are used are density (2710 kg/m^3), Young's modulus (70 GPa), Poisson's ratio (0.33), shear modulus (26 GPa), and specific heat capacity (890 J/kg K) (Fig. 5.11).

In this study, the randomness in the size and the shape of the particles have been considered. A distribution has been chosen based on the average size of the particles obtained from the powder morphology in the coating process and different shapes are given to the particles. The final results of the numerical model showing the distribution of residual stress at and near the surface of the substrate is presented in Fig. 5.12. Also shown in the same figure are the results of residual stress measurements of two sets of coated samples using XRD. Given the variability in the measurements, the simulation results are comparable to the measurements, especially at the surface of the substrate.

Table 5.3 Gruneisen equation of state parameters for AL 1100-O (1969)

Value	Parameter
5328	C (m/s)
1.338	S_1
0	S_2
0	S_3
2	γ_0
0	A

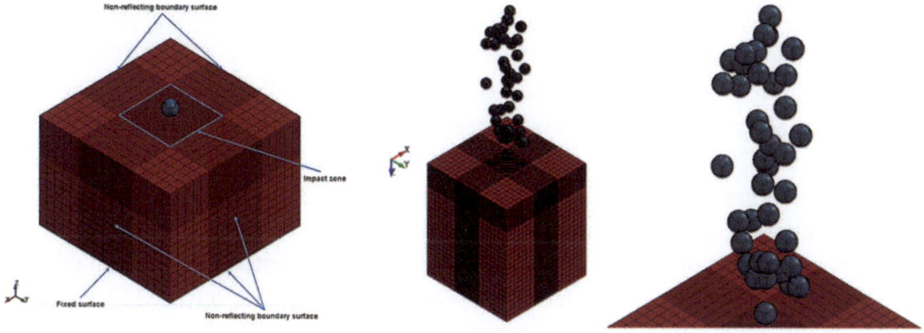

Fig. 5.11 Mesh configuration and developed finite element model. (Shayegan et al. 2014)

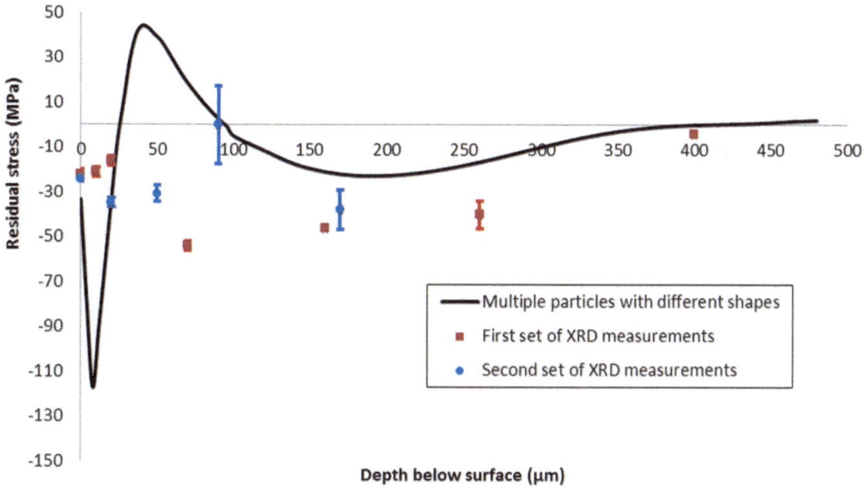

Fig. 5.12 Numerical results with Shayegan et al. (2014) model compared to experiments. *XRD* X-ray diffraction

5.4 Life Enhancement Resulting from Cold Spray Coating

The compressive residual stress on the surface will delay crack initiation and propagation and increase the fatigue endurance of the metallic components. Due to the presence of initial compressive stress, under service cyclic load, the maximum tensile stress stays at a very low value, and in some cases even at zero. Therefore, initial surface cracks are not subjected to the opening mode of crack propagation. Hence, the presence of the beneficial compressive residual stress prolongs the life or can allow for higher external loads to be applied. The formation of compressive residual stresses due to the peening effect of cold spray was discussed in details in the previous sections; we now focus on the life enhancement resulting from cold spray coating.

Ghelichi et al. (2012) studied the effect of cold spray coating on Al alloys using different powders and treatments. The specimen suggested by ASTM-B93 (2009) has been used for performing fatigue tests on samples shown in Fig. 5.13. The load control pure bending fatigue tests were performed using the specimen suggested by ASTM B-593 (2009). This standard suggests a fixed-cantilever, constant-deflection-type machine. The specimens are designed for the bending test based on the approach suggested by standard as presented in Fig. 5.13a along with a picture of the prepared specimen (Fig. 5.13b). The specimens were tested using a load control pure bending fatigue machine. Bending fatigue tests (stress ratio $R = -1$) have been carried out at room temperature at a nominal frequency of 90 Hz on as-received, GB, and other coated series of specimens. The specimens that passed ten million cycles were considered as run-out. The brief staircase method presented by Dixon and Massey (1969) has been used for performing the tests with a stress step of 10 MPa, and the Hodge–Rosenblatt (Dixon and Massey 1969) approach has been considered for calculating the fatigue endurance corresponding to 10 MPa. The fatigue test data have been elaborated based on ASTM standard E739–91 (2010) to obtain the S–N diagram for the different applied treatments.

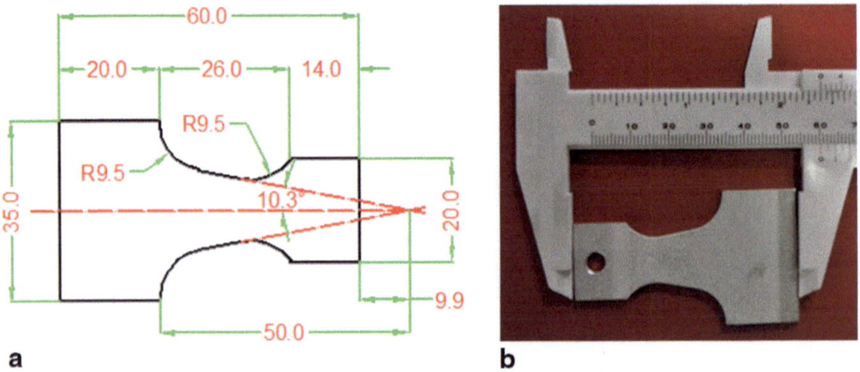

Fig. 5.13 Test specimen suggested by ASTM Standard B593-96 (2009), which has a uniform stress distribution under bending. **a** Detailed design of the specimen (all units in millimeter). **b** A view of a prepared specimen. (Ghelichi et al. 2012)

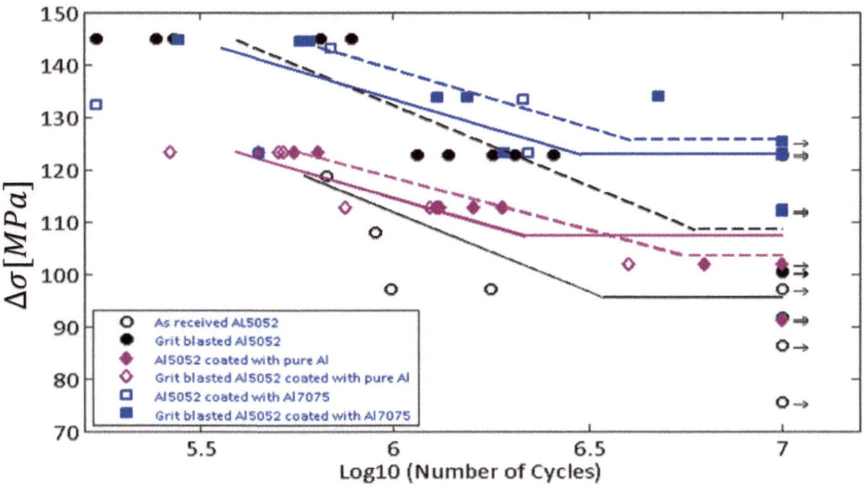

Fig. 5.14 The S–N diagram obtained from fatigue tests of coated and uncoated samples. (Ghelichi et al. 2012)

Two different types of aluminum powders have been coated by low-pressure cold spray coating (Cadney et al. 2008; Maev and Leshchynsky 2008) on both GB and as-received Al5052 in order to study the effect of the treatments on the fatigue endurance of the substrate. The specimens have been tested through a pure bending fatigue test. Residual stress measurement by XRD, microhardness tests, and SEM observations of fracture surfaces has been performed on the specimens. The results, presented in Fig. 5.14, show that the improvement in the fatigue life depends on the powders. The results prove that the residual stress due to grit blasting improves the fatigue endurance regardless of the powder and substrate, and an extra advantage can be obtained by choosing the appropriate powders.

Two main reasons have been mentioned for the life improvement, hardness of the particles, and the bonding of the deposited materials on the substrate. The microhardness test shows a great improvement in the hardness of the Al7075 with respect to the substrate. Incidentally, the grit blasting of the sample helped to increase the fatigue life; this improvement is mainly due to inducing extra residual stress in the substrate. SEM images (Fig. 5.15) show that the samples with higher fatigue endurance have a stronger bonding between the substrate and deposited materials.

Mahmoudi et al. (2012) and Kalatehmollaei et al. (2014) worked on fatigue improvement achieved by cold spray coating of magnesium substrate with aluminum alloys. For this purpose, cylindrical hourglass specimens of AZ31B extrusion have been used. Three groups of specimens including: as-received, stress-relieved, and stress-relieved/coated specimens have been tested by a rotating bending machine (RBM) to prepare S–N curves for each group. Comparing S–N curves provides an

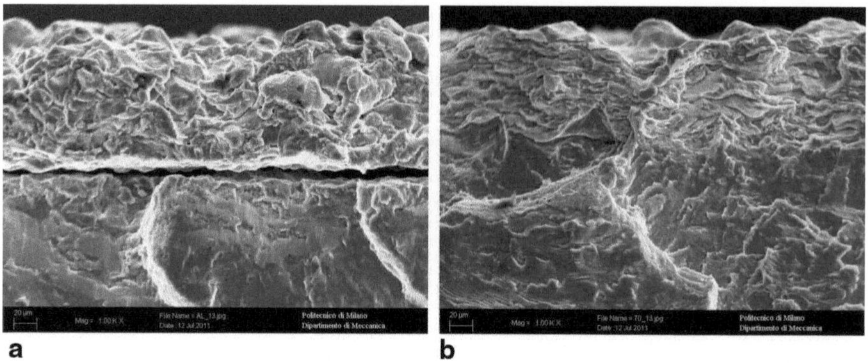

Fig. 5.15 SEM observation of the interface of the deposited material and the substrate in broken samples. **a** Sample coated by pure Al. **b** Sample coated by Al7075. (Ghelichi et al. 2012)

estimation of the effect of cold spray coating on the fatigue strength of AZ31B. All tests were performed at standard laboratory conditions. Tests were performed at frequencies between 50 Hz for low-cycle to 100 Hz for high-cycle tests. Specimens used for the test are shown in Fig. 5.16. The overall length of the specimens is 70 mm, and the diameter of the neck is 7 mm.

There was an initial compressive residual stress in as-received specimens due to the extrusion process of AZ31B. Using XRD measurement on a cylindrical specimen in the hourglass zone shows the amount of this compressive stress to be 43 MPa. As-received specimens have been used for fatigue tests to correlate the initial residual stress to fatigue life and to be able to compare results with other groups. A stress relief process was applied on the second group of AZ31B specimens. The ASM-recommended stress relief process, 500 °F (260 °C) for 15 min, was used for these specimens. The purpose is to evaluate the fatigue strength in absence of any residual stress. XRD measurement on stress-relieved cylindrical specimens on an hourglass area showed that the surface stress was removed. The purpose of selecting this group for fatigue tests is to evaluate the residual stress that is induced due to cold spray coating and its effect on the fatigue life of AZ31B. Stress-relieved

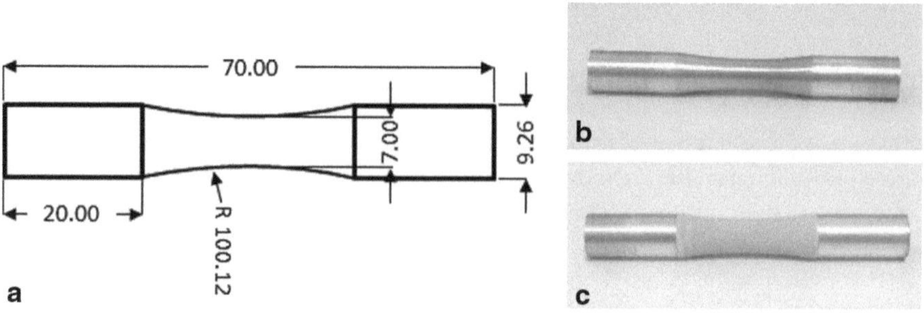

Fig. 5.16 Round specimens used for extracting S–N curve of AZ31B; **a** drawing, **b** uncoated, and **c** coated samples. (Kalatehmollaei et al. 2014)

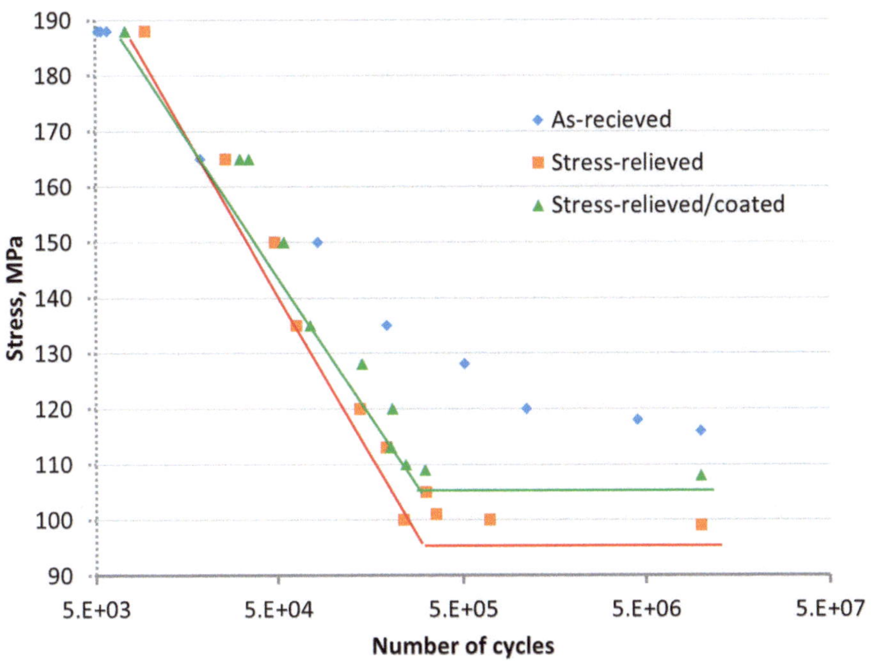

Fig. 5.17 S–N curves of as-received, stress-relieved, and stress-relieved/coated specimens of AZ31B. (Kalatehmollaei et al. 2014)

specimens have been coated with the following specifications and parameters: particle velocity =400 m/s, particle average diameter =40 μm, particle material of aluminum alloy, series 1100, grit blasting before cold spray coating, and coating layer thickness =0.1 mm. The coating has been applied on the hourglass area only. Figure 5.17 shows the S–N curve for three groups of specimens.

Figure 5.18 shows the fracture surface of a stress-relieved/coated specimen that has been tested under nominal stress of 188 MPa. As Fig. 5.18a presents, the crack initiation region is extended from the middle right to middle bottom of the image. Except for the crack initiation point (shown in the center of Fig. 5.18a), no coating layer delamination is observed around the fracture surface of the specimen; this observation proves strong bonding between coating particles and the specimen surface. Figure 5.18c shows the coating layer and the fracture surface.

As expected, there is a correlation between surface residual stress and fatigue strengths. Table 5.4 shows the values of surface residual stress, measured by XRD, and endurance limit for the three groups of specimens.

In low-cycle fatigue, there is a considerable drop in the fatigue life after stress relief. This is due to the considerable loss of compressive residual stress. Cold spray coating process improves the fatigue strength slightly. This means that the residual stress induced by cold spray coating is much lower than the residual stress induced by the initial extrusion process forming of the as-received specimens. The as-received specimens show better fatigue strength among the three groups. The S–N

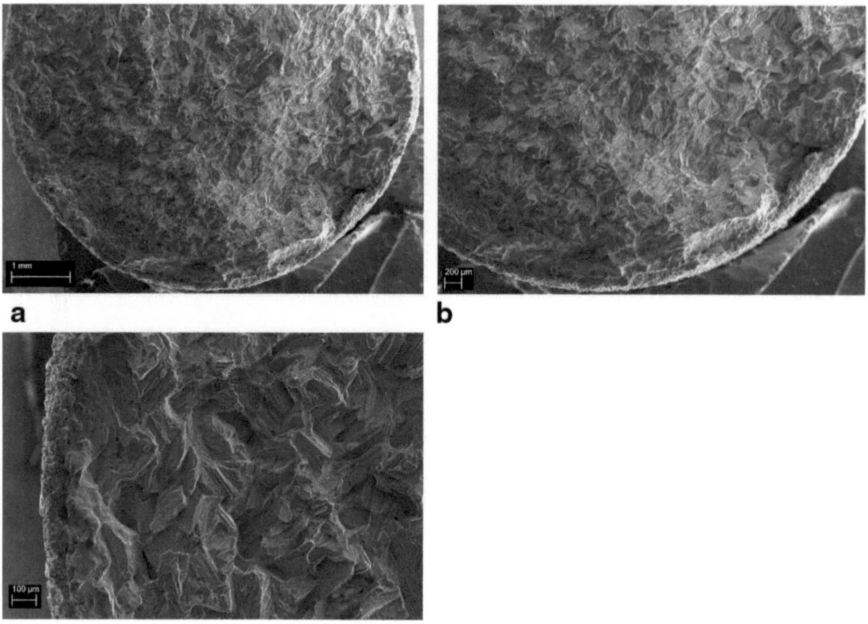

Fig. 5.18 Fracture surface of stress-relieved/coated specimen with the nominal stress of 188 MPa; **a** and **b** crack initiation region, **c** coating layer. (Mahmoudi 2012)

Table 5.4 XRD results on cylindrical specimens. (Cadney et al. 2008)

Group	Residual stress on surface (MPa)	Endurance limit (MPa)	Remarks
As-received	−43	116	Measured on hourglass area
Stress relieved	~0	99	Measured on hourglass area
Stress relieved/coated	−22	108	Measured on coated flat sample

curve for this group has a smooth transition to the endurance limit while the S–N curves of the stress-relieved and stress-relieved/coated specimens have sharp transitions to the endurance limit. The stress relief process lowers the endurance limit of the as-received specimens from 116 to 99 MPa, equal to 14.6 %. The coating process improves the fatigue strength of stress-relieved specimens and increases their endurance limit from 99 to 108 MPa, equal to 9 %.

In a different study, Cizek et al. (2013) performed fatigue tests for four different sets of samples. The same conclusion has been drawn regarding the grit blasting which can improve the fatigue limit of the samples. The 4-mm-thick Ti6Al4V sheets were cut into the shape shown in Fig. 5.19. Chemical degreasing was carried out in order to remove any oil or contaminant from the substrate surfaces. The samples have been grit blasted with SiC and Al_2O_3 particle.

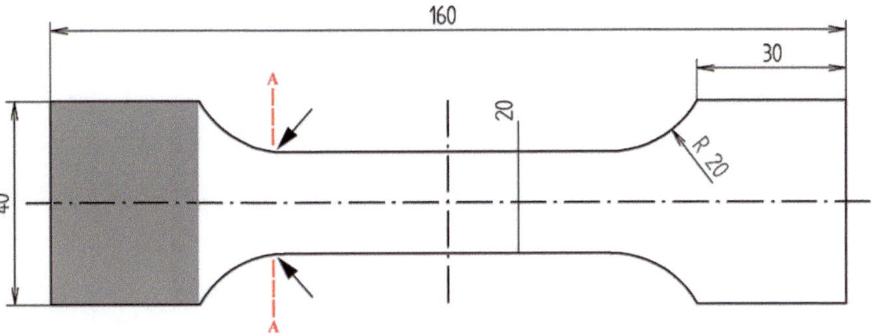

Fig. 5.19 Ti6Al4V samples used for cold spray coating. (Cizek et al. 2013)

Using the measured resonance frequencies from the fatigue experiments, the coating elastic moduli were obtained by a method described in Kovářík et al. (2008). Figure 5.20 shows the morphology of the surface of the different sets of the samples. The grit-blasting procedure produced a roughened surface (Ra~3 μm as compared to Ra = 0.82 μm of the as-received samples) and introduced abrasive angular Al_2O_3 and SiC particles into the Ti6Al4V material.

The final results show that cold spray deposition of Ti layers reduces fatigue life and causes the lowest values among the tested sets (91 % of the as-received set); it

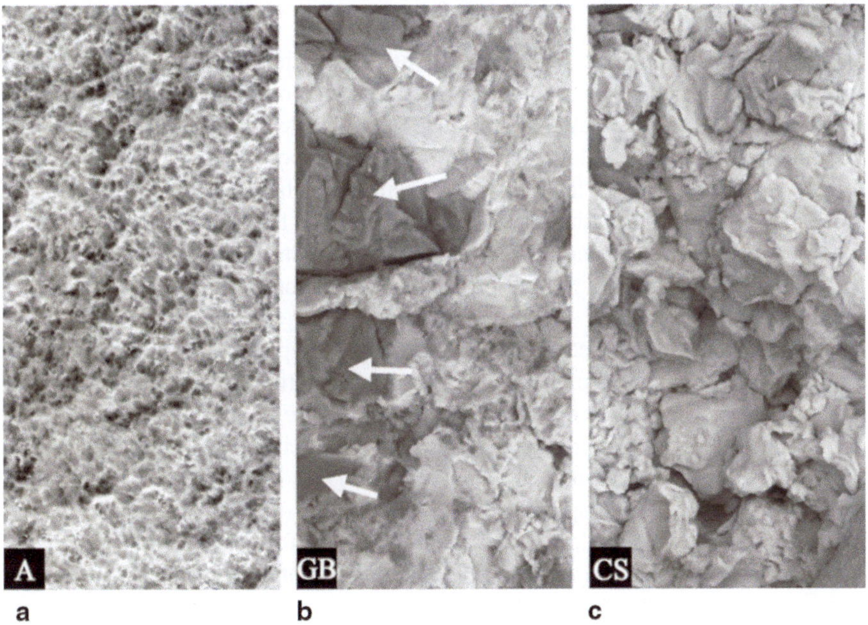

Fig. 5.20 Surface morphology of the **a** as-received *(A)*, **b** grit-blasted *(GB)*, and **c** cold-sprayed *(CS)* samples. (Cizek et al. 2013)

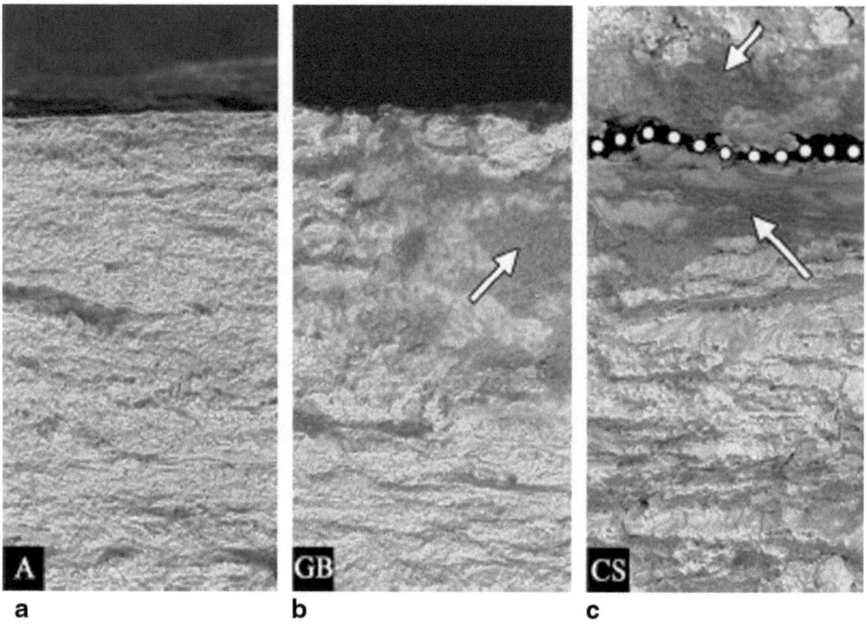

Fig. 5.21 Fatigue crack morphologies close to the substrate surfaces at approximately 3 mm from the sample edges for **a** as-received *(A)*, **b** grit-blasted *(GB)*, and **c** cold-sprayed *(CS)* samples. The fracture morphology of *GB* and *CS* substrates was obliterated by crushed abrasive particles near the substrate surface (contact wear-like, indicated by *arrows*). (Cizek et al. 2013)

is believed that the decrease in life may be due to a vertical cracking of the coatings followed by the crack transfer into the substrates and compensation of the residual peening stresses in the coatings by induction of tensile stresses in the substrates. Fractography analysis, as shown in Fig. 5.21, of the deposits enabled differentiating the areas of the fracture surfaces that underwent the cyclic opening/closing from the areas of the final rupture; the areas were characterized by signs of contact wear of the fracture surfaces, causing obliteration of the initial structure; due to the obliteration, the crack initiation sites and propagation directions could not be accurately identified in the coatings.

Price et al. (2006) studied the effect of cold spray deposition of titanium coating on Ti6Al4V on fatigue life of the coated samples. The fatigue lives of the as-received and GB materials, both before and after coating, were experimentally studied. A 15 % reduction in fatigue endurance was observed after application of the coating on the substrate; however, no significant reduction was observed on its application to the GB substrate. The compressive residual stresses reported through cold spray titanium coatings were too low to prevent fatigue crack formation and advancement. It should be considered that Price et al. (2006) have reported negligible compressive residual stresses on the deposited materials and the coated samples. The SEM observations also showed delamination in deposited material after fatigue test, which represented no or a little contribution of the deposited materials to the fatigue test.

On the other hand, Sansoucy et al. (2007), in particular, worked on bending fatigue and the bonding strength of the Al–Co–Ce coatings. The results show that Al–Co–Ce coatings improved the fatigue behavior of Al-2024-T3 specimens compared to uncoated specimens.

5.5 Concluding Remarks

The mechanism of formation of residual stress in a cold spray process was discussed. Methods for quantifying the magnitude of these stresses were reviewed, and experimental measurements for the case of aluminum and magnesium were presented. The theoretical foundation for modeling the impact of particles with substrate was studied in detail, and two approaches were laid out thoroughly. Finally, the benefit of residual stresses induced by cold spray coating on fatigue life enhancement of the substrate was explored.

Following general conclusions may be drawn from above presentation:

- Compressive residual stresses are formed because of an elastically bounded local plasticity induced by high-impact energy of coating particles with the substrate.
- Parameters such as impact velocity, particle size, material properties of particles and substrate, process temperature, standoff distance, initial as-received residual stresses, and particle shape are the controlling factors of the resulting residual stress due to impact.
- The magnitude of residual stress induced by cold spray depends heavily on the choice of the substrate and coating powder. While harder aluminum particle coating on aluminum substrate created a large residual stress leading to a significant fatigue life enhancement, the case of titanium showed the opposite.
- Coating temperature can result in an annealing process after the particle adhered to the surface and cooled down to normal temperature. This annealing process results in relaxation of residual stress. An after-simulation annealing correction calculation was proposed which showed better results when compared to measurement.

References

Abaqus 6.12–1. 2012. Analysis user's manual, simulia. http://asm.matweb.com/search/SpecificMaterial.asp?bassnum=MA7075T6. Accessed 1 June 2015.

ASM. 1989. *ASM Metal Handbooks*. Properties and Selection: Nonferrous Alloys and Special-Purpose Materials, vol. 2.

ASTM standard B593–96. 2009. Standard test method for bending fatigue testing for copper-alloy spring materials.

ASTM standard E739–10. 2010. Standard practice for statistical analysis of linear or linearized stress life (S–N) and strain life (e–N) fatigue data.

Albinmousa, J., H. Jahed, and S. Lambert. 2011. Cyclic axial and cyclic torsional behaviour of extruded AZ31B magnesium alloy. *International Journal of Fatigue* 33 (11): 1403–1416.

Allen, T. 2003. *Powder sampling and particle size determination*. 1st ed. Elsevier Science. ISBN: 044451564X.

Assadi, H., F. Gärtner, T. Stoltenhoff, and H. Kreye. 2003. Bonding mechanism in cold gas spraying. *Acta Materialia* 51 (15): 4379–4394.

Bagherifard, S., I. Fernàndez Parienete, R. Ghelichi, M. Guagliano, and S. Vezzù. 2010a. Effect of shot peening on residual stresses and surface work-hardening in cold sprayed coatings. *Key Engineering Materials* 417:397–400.

Bagherifard, S., R. Ghelichi, and M. Guagliano. 2010b. A numerical model of severe shot peening (SSP) to predict the generation of a nanostructured surface layer of material. *Surface and Coatings Technology* 204 (24): 4081–4090.

Benck, R. F. 1976. Quasi-static tensile stress strain curves–II, rolled homogeneous armor, DTIC Document.

Brar, N. S., V. S. Joshi, B. W. Harris, M. Elert, M. D. Furnish, W. W. Anderson, W. G. Proud, and W. T. Butler. 2009. *Constitutive model constants for Al7075-T651 and Al7075-T6*. Aip conference proceedings.

Cadney, S., M. Brochu, P. Richer, and B. Jodoin. 2008. Cold gas dynamic spraying as a method for freeforming and joining materials. *Surface and Coatings Technology* 202 (12): 2801–2806.

Champagne, V. K. 2007. *The cold spray materials deposition process: Fundamentals and applications*. 1st ed. Cambridge: Woodhead Publishing Limited. ISBN: 9781845691813.

Cizek, J., O. Kovarik, J. Siegl, K. A. Khor, and I. Dlouhy. 2013. Influence of plasma and cold spray deposited Ti Layers on high-cycle fatigue properties of Ti6Al4V substrates. *Surface and coatings technology* 217:23–33.

Dixon, W. J., and F. J. Massey. 1969. *Introduction to statistical analysis*. New York: McGraw-Hill.

Dykhuizen, R. C., and M. F. Smith. 1998. Gas dynamic principles of cold spray. *Journal of Thermal Spray Technology* 7 (2): 205–212.

Fine, M. E. 1964. *Introduction to phase transformations in condensed systems. Macmillan Materials Science Series*. New York: Macmillan Comp.

Ghelichi, R. 2012. Cold spray coating aimed nanocrystallization: Process characterization and fatigue strength assessment, PhD Thesis, Polytechnic University of Milan.

Ghelichi, R., S. Bagherifard, M. Guagliano, and M. Verani. 2011. Numerical simulation of cold spray coating. *Surface and Coatings Technology* 205 (23): 5294–5301.

Ghelichi, R., D. MacDonald, S. Bagherifard, H. Jahed, M. Guagliano, and B. Jodoin. 2012. Microstructure and fatigue behavior of cold spray coated Al5052. *Acta Materialia* 60 (19): 6555–6561.

Ghelichi, R., S. Bagherifard, D. Mac Donald, M. Brochu, H. Jahed, B. Jodoin, and M. Guagliano. 2014a. Fatigue strength of Al alloy cold sprayed with nanocrystalline powders. *International Journal of Fatigue* 65:51–57.

Ghelichi, R., S. Bagherifard, D. MacDonald, I. Fernandez-Pariente, B. Jodoin, and M. Guagliano. 2014b. Experimental and numerical study of residual stress evolution in cold spray coating. *Applied Surface Science* 288:26–33.

Greving, D. J., E. F. Rybicki, and J. R. Shadley. 1994. Through-thickness residual stress evaluations for several industrial thermal spray coatings using a modified layer-removal method. *Journal of thermal spray technology* 3 (4): 379–388.

Grujicic, M., C. Tong, W. S. DeRosset, and D. Helfritch. 2003. Flow analysis and nozzle-shape optimization for the cold-gas dynamic-spray process. *Proceedings of the Institution of Mechanical Engineers, Part B: Journal of Engineering Manufacture* 217 (11): 1603–1613.

Grujicic, M., C. L. Zhao, W. S. DeRosset, and D. Helfritch. 2004a. Adiabatic shear instability based mechanism for particles/substrate bonding in the cold-gas dynamic-spray process. *Materials & design* 25 (8): 681–688.

Grujicic, M., C. L. Zhao, C. Tong, W. S. DeRosset, and D. Helfritch. 2004b. Analysis of the impact velocity of powder particles in the cold-gas dynamic-spray process. *Materials Science and Engineering: A* 368 (1): 222–230.

Jeong, C. Y., and S. Ha. 2008. Fatigue properties of Al–Si casting alloy with cold sprayed Al/SiC coating. *International Journal of Cast Metals Research* 21 (1–4): 235–238.

Johnson, G. R., and W. H. Cook. 1983. *A constitutive model and data for metals subjected to large strains, high strain rates and high temperatures*. Proceedings of the 7th international symposium on ballistics.

Kalatehmollaei, E., H. Mahmoudi-Asl, and H. Jahed. 2014. An asymmetric elastic–plastic analysis of the load-controlled rotating bending test and its application in the fatigue life estimation of wrought magnesium AZ31B. *International Journal of Fatigue* 64:33–41.

Kovářík, O., J. Siegl, and Z. Procházka. 2008. Fatigue behavior of bodies with thermally sprayed metallic and ceramic deposits. *Journal of thermal spray technology* 17 (4): 525–532.

Kumar, S., G. Bae, K. Kang, S. Yoon, and C. Lee. 2009. Effect of powder state on the deposition behaviour and coating development in kinetic spray process. *Journal of Physics D: Applied Physics* 42 (7): 075305.

Lemaitre, J., and J.-L. Chaboche. 1990. *Mechanics of solid materials*. New York: Cambridge University Press.

Li, C.-J., W.-Y. Li, and H. Liao. 2006. Examination of the critical velocity for deposition of particles in cold spraying. *Journal of Thermal Spray Technology* 15 (2): 212–222.

LS-DYNA3D. 1999. User's manual. Ver. 950. Livermore software technology corporation, Livermore, California.

Los Alamos. 1969. "Selected Hugoniots", Group GMX-6, Los Alamos Scientific Lab., LA-4167–MS.

Luzin, V., K. Spencer, and M. X. Zhang. 2011. Residual stress and thermo-mechanical properties of cold spray metal coatings. *Acta Materialia* 59 (3): 1259–1270.

Maev, R. G., and V. Leshchinsky. 2008. *Introduction to low pressure gas dynamic spray: Physics & technology*. Wiley-VCH Verlag GmbH.

Mahmoudi, H. 2012. MSc. thesis. Mechanical and Mechatronics Engineering Department, University of Waterloo.

Mahmoudi, H., H. Jahed, and J. Villafuerte. 2012. The effect of cold spray coating on fatigue life of AZ31B. 9th International conference on magnesium alloys and their applications, Vancouver, BC.

McCune, R. C., W. T. Donlon, O. O. Popoola, and E. L. Cartwright. 2000. Characterization of copper layers produced by cold gas-dynamic spraying. *Journal of Thermal Spray Technology* 9 (1): 73–82.

Moridi, A., S. M. Hassani-Gangaraj, M. Guagliano, and S. Vezzu. 2014. Effect of cold spray deposition of similar material on fatigue behavior of Al 6082 alloy. *Fracture and Fatigue* 7:51–57 (Springer).

Najafi, A., and M. Rais-Rohani. 2011. Mechanics of axial plastic collapse in multi-cell, multi-corner crush tubes. *Thin-Walled Structures* 49 (1): 1–12.

Papyrin, A., V. Kosarev, K. V. Klinkov, A. Alkhimov, and V. M. Fomin. 2006. *Cold spray technology*. (A. Papyrin. ed.). Elsever.

Pierazzo, E., N. Artemieva, E. Asphaug, E. C. Baldwin, J. Cazamias, R. Coker, G. S. Collins, D. A. Crawford, T. Davison, D. Elbeshausen, K. A. Holsapple, K. R. Housen, D. G. Korycansky, and K. WÜNnemann. 2008. Validation of numerical codes for impact and explosion cratering: Impacts on strengthless and metal targets. *Meteoritics & Planetary Science* 43 (12): 1917–1938.

Price, T. S., P. H. Shipway, and D. G. McCartney. 2006. Effect of cold spray deposition of a titanium coating on fatigue behavior of a titanium alloy. *Journal of Thermal Spray Technology* 15 (4): 507–512.

Ramakrishnan, K. N. 2000. Modified Rosin Rammler equation for describing particle size distribution of milled powders. *Journal of Materials Science Letters* 19 (21): 1903–1906.

Robert, C. 1984. *CRC handbook of chemistry and physics*. Boca Raton: CRC Press.

RosIn, P. 1933. The laws governing the fineness of powdered coal. *Journal of the Institute of Fuel* 7:29–36.

Rybicki, E. F., J. R. Shadley, Y. Xiong, and D. J. Greving. 1995. A cantilever beam method for evaluating Young's modulus and Poisson's ratio of thermal spray coatings. *Journal of Thermal Spray Technology* 4 (4): 377–383.

Sansoucy, E., G. E. Kim, A. L. Moran, and B. Jodoin. 2007. Mechanical characteristics of Al-Co-Ce coatings produced by the cold spray process. *Journal of Thermal Spray Technology* 16 (5–6): 651–660.

Shayegan, G., H. Mahmoudi, R. Ghelichi, J. Villafuerte, J. Wang, M. Guagliano, and H. Jahed. 2014. Residual stress induced by cold spray coating of magnesium AZ31B extrusion. *Materials & Design* 60:72–84.

Spencer, K., V. Luzin, N. Matthews, and M. X. Zhang. 2012. Residual stresses in cold spray Al coatings: The effect of alloying and of process parameters. *Surface and Coatings Technology* 206 (19–20): 4249–4255.

Totten, G. E. 2002. *Handbook of residual stress and deformation of steel*. Ohio: ASM International, Materials Park.

Tsui, Y. C., and T. W. Clyne. 1997. An analytical model for predicting residual stresses in progressively deposited coatings part 1: Planar geometry. *Thin Solid Films* 306 (1): 23–33.

Weast R. C., and D. R. Lide. 1984. *CRC handbook of chemistry and physics. University of Rhode Island. Coastal Resources Center*. Cleveland: CRC Press.

Chapter 6
Commercial Cold Spray Equipment and Automation

J. Villafuerte, W. Birtch and J. Wang

6.1 Commercial Cold Spray Equipment

6.1.1 Cold Spray Nozzles

The spray nozzle is an important design component in a cold spray gun. It plays the crucial role of converting high-enthalpy, high-pressure, and low-velocity gas into the low-enthalpy, low-pressure, and high-velocity gas jet necessary for particle acceleration. The ultimate objective of a cold spray nozzle is to create favorable gas flow conditions at the nozzle exit in order to maximize the ability of particles to effectively consolidate upon impact with the substrate. The physics of the process was well explained in Chap. 2. Typically, a convergent–divergent type (DeLaval) nozzle is used to generate such favorable supersonic gas flow at the nozzle exit. The characteristics of the supersonic flow, including its kinetic and thermal energy content, are a function of both nozzle geometry and gas parameters (type of gas, pressure, and temperature). The amount of energy (thermal and kinetic) actually transferred to the particles strongly depends on the physical characteristics of the powder (e.g., density, shape, size distribution, etc), where and how the particles are injected into the gas stream, and the powder to gas mass ratio. The key characteristics of powdered materials for cold spray and their role in the process were explained in Chap. 3.

Some important parameters of the nozzle geometry (Fig. 6.1) are the divergence ratio (D/d), divergent shape, and length of the diverging section. These geometrical

J. Villafuerte (✉)
Corporate—Supersonic Spray Technologies, CenterLine Windsor Ltd.,
Windsor, ON, Canada
e-mail: Julio.villafuerte@cntrline.com

W. Birtch . J. Wang
Supersonic Spray Technologies, CenterLine (Windsor) Limited, Windsor, ON, Canada
e-mail: wally.birtch@cntrline.com

J. Wang
e-mail: Jianfeng.wang@cntrline.com

J. Villafuerte (ed.), *Modern Cold Spray,* DOI 10.1007/978-3-319-16772-5_6

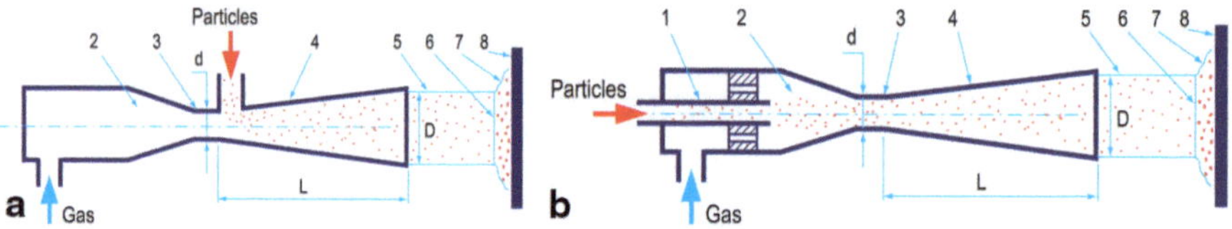

Fig. 6.1 Downstream and upstream injection nozzle parameters. *1* tube for injection of particles, *2* pre-chamber, *3* nozzle throat (*d*=diameter), *4* divergent supersonic section, *5* free jet, *6* bow shock wave, *7* compressed layer, *8* substrate. (Papyrin et al. 2007; Papyrin 2001)

features determine the characteristics of the gas flow inside and at the nozzle exit. When optimizing the nozzle geometry to maximize gas jet velocity, designers often use computational fluid dynamic (CFD) simulation software. As the gas jet leaves the nozzle, other conditions prevail including bow shock wave and compressed layers which may affect particle impact conditions on the substrate. In practice, round nozzle exit diameters range from 2 to 12 mm, while throat diameters range from 1.0 to 3.0 mm. Large deposition areas are often achieved by using a raster approach of subsequent passes with a step-over pattern, which is typically 25 % of the total width of the pass. Because of the sharp threshold for adhesion versus rebounce conditions, the width of any over-spray is relatively small. Subsequently, a spray pass generally displays well-defined sharp edges with a width close to the nozzle exit diameter.

Repeated use of nozzles may eventually cause nozzle clogging and/ or internal erosion wear. Nozzle clogging is more predominant with pure metals such as tin, aluminum, nickel, and indium. Nozzle clogging mechanisms are not fully understood, but it is apparent that nozzle surface finish, surface temperature, and surface chemistry play fundamental roles. In order to minimize or avoid nozzle clogging, manufacturers use two approaches: (a) fabricate nozzles out of high-temperature polymers which display nonclogging characteristics and/or (b) fabricate nozzles out of metals but with the addition of continuous water cooling, and with highly polished internal surfaces. In the former approach, fabricators note that polymers typically show a maximum service temperature of around 400 °C. Therefore, nozzle designs are made hybrid with the convergent section of the nozzle made of metal while limiting the polymeric part of the nozzle to the divergent section (cooler) as far as possible from the nozzle throat. External nozzle water cooling is also effective in avoiding clogging; however, this adds complexity to the design which limits accessibility while increasing the cost of acquisition and operation of the spray gun.

6.1.2 Downstream Injection Equipment

In downstream injection, air or nitrogen, at low to medium pressures (4–34 bar), is preheated up to 550 °C. This high-enthalpy gas is then forced through the converging–diverging nozzle to allow the conversion of enthalpy into kinetic energy by rapidly expanding the gas to the supersonic regime (typically around 300–900 m/s).

During this energy transformation process, the gas accelerates throughout the divergent nozzle length, while its temperature drops significantly. The unique characteristic of this family of equipment is that the powder feedstock is introduced downstream into the diverging section of the nozzle (Fig. 6.2).

One benefit of the downstream injection method is its ability to operate without the need of a high-pressure powder feeder, when operating at pressures below 9 bar. This is because, at these low pressures, there is sufficient negative pressure (below atmospheric) created at the divergent side of the nozzle to suck in spray powder into the jet stream.

Another benefit is the fact that only the divergent section of the nozzle is subject to erosion by the direct impact of the spray powder. Therefore, designers often split the nozzle into two distinct components: the *nozzle holder,* which contains the converging section plus the nozzle throat or orifice, and the *nozzle tube,* which is the divergent section of the nozzle. Because the nozzle holder only sees clean gas, the life of this component is practically unlimited. On the other hand, the nozzle tube is subject to erosion by the spray powder, requiring regular replacement during service. The life of this component is associated with the makeup of the tube as well as the nature of the spray powder.

When designed for low gas pressures and temperatures, downstream injection systems can be compact, portable, and very economical. However, at low gas pressures and temperatures, the maximum attainable particle velocities may limit the range of spray-able materials to include low melting point ductile metals such as pure aluminum, zinc, and tin. Fortunately, the spray-ability of these materials, at low gas pressures and temperatures, can significantly be enhanced by the addition of harder particles into the powder mix (typically ceramics). Ceramic particles

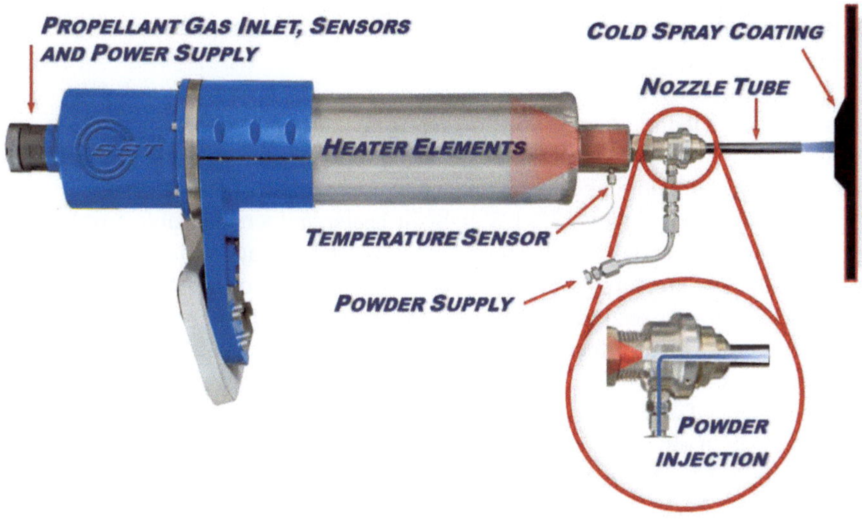

Fig. 6.2 A schematic drawing showing the operating principle of downstream injection equipment. (Villafuerte et al. 2010)

Fig. 6.3 Portable downstream injection equipment (3.5 kW/8 bar/600 °C). (Courtesy of DYMET Corporation)

in the mix produce a micro-hammering effect that helps not only to compact the underlying layers but also constantly clean (activate) the surfaces and increase surface roughness for better adhesion (Maev and Leshchynsky 2008). Also noting that a percentage of these hammering particles end up embedded as a dispersoid in a fully deformed metal matrix, resulting in a composite coating. Depending on the application, such composite microstructure may be beneficial or pose a limitation; for example, in some cases, improved mechanical properties derived from dispersion strengthening are favorable, while in other cases the presence of mechanically bonded ceramic–metal interfaces may represent a challenge when exposed to extremely corrosive environments.

One of the early downstream injection commercial systems was developed in Russia by the Obninsk Center for Powder Spraying Ltd. (OCPS) under the brand name DYMET (Dymet Corporation 2014; Fig. 6.3). This system uses compressed air (5–8 bar) at 400 l/min heated in the gun to a maximum of 600 °C.

The spray gun includes a 3.5-kW light air heater and a replaceable nozzle tube. The nozzle life is about 1 spray-hour for a powder feed rate of about 0.5 g/s. Because of the low pressure required, this downstream injection system uses a simple gravimetric powder feeder. This system has been used extensively in Russia and other countries, mainly for field maintenance including corrosion repair, and dimensional restoration using commercially pure aluminum, copper, zinc, nickel, tin, and lead blended with appropriate amounts of ceramic particles (Alumina) to maximize deposition efficiency. Typical deposition efficiency is 20–30 % with a deposition rate of 3–10 g/min for these materials.

As the technology progressed, the maximum operating pressures of downstream injection equipment have been increased, resulting in increased deposition efficiencies and extended the range of spray-able materials to other materials such as

Fig. 6.4 Downstream injection system (15 kW/34 bar/550 °C) integrated with cabinet, ventilation, and auxiliary systems for mechanized operation. (Photos courtesy of CenterLine (Windsor) Limited)

stainless steels, titanium, and nickel-based alloys. In 2006, CenterLine (Windsor) Limited acquired exclusive rights from OCPS to commercialize this technology for the North American market (CenterLine Limited 2015). CenterLine's SST™ industrial units are production-ready downstream injection machines, based on the same principle as Dymet machines (Kashirin et al. 2002), but operating at a wider range of gas pressures (4–34 bar) using air, nitrogen, or helium and providing up to 550 °C gas heating with a power consumption of 3.8, 4.2, and 15 kW, depending on the specific model. These systems (Figs. 6.4 and 6.5a) were engineered and manufactured to comply with all North American industrial standards. This equipment is commercially used for corrosion repair, dimensional restoration, metallization, and

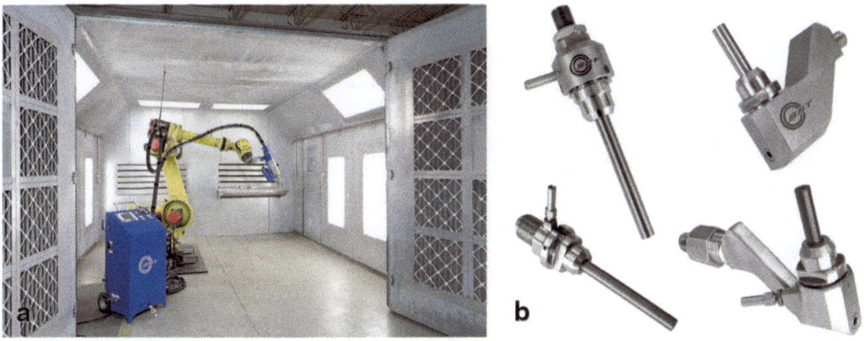

Fig. 6.5 **a** Downstream injection system (15 kW/34 bar/550 °C) integrated with spray booth, ventilation, and auxiliary systems for robotic operation. **b** Modular straight and 90° downstream injection spray nozzles for multiple applications. (Photo courtesy of CenterLine (Windsor) Limited)

other applications in the field or in controlled environments for either manual or fully automated operations. In order to optimize production uptime, the Supersonic Spray Technologies (SST) modular nozzle assemblies come with a bayonet twist lock for quick tube change which can accommodate all kinds of nozzle configurations (straight or 90°) and nozzle tubes including anti-clogging nozzles and wear-resistant nozzles (Fig. 6.5b).

6.1.3 Upstream Injection Equipment

In upstream injection cold spray, helium or nitrogen at high pressures (up to 70 bar) is preheated (up to 1100 °C). Similar to downstream injection, this high-enthalpy gas is then forced through a converging–diverging nozzle to allow the conversion of enthalpy into kinetic energy by rapidly expanding the gas to the supersonic regime (typically around 1000 m/s). In this type of equipment, the powder feedstock is injected axially into the gas stream upstream section of the nozzle throat, using a high-pressure powder feeder (Fig. 6.6).

One main benefit of this approach is that, because of the higher gas pressure and earlier injection point, spray particles can attain higher velocities than with downstream injection systems, simply due to the longer interaction between particles and the gas jet. Another benefit is the powder preheating effect, where the feedstock particles have a chance of being preheated in the high-pressure section side of the nozzle, leading to higher particle temperature than in downstream injection cold spray, consequently having a favorable effect on the materials' critical velocity, as explained in Chap. 2. These combinations make upstream injection equipment suitable for depositing higher melting point materials such as nickel-based and tantalum alloys, particularly if the carrier gas is helium. These materials are typically more difficult to spray with downstream injection equipment.

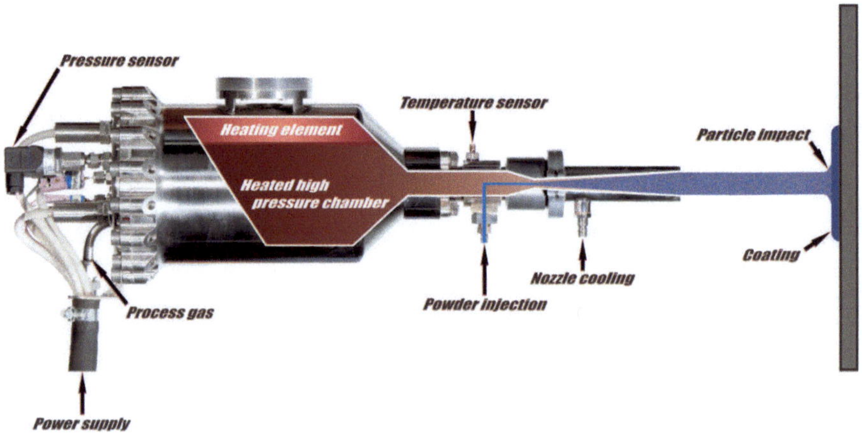

Fig. 6.6 Schematic drawing showing the operating principle of upstream injection equipment. (Courtesy of Impact Innovations)

On the down side, the high cost and restricted availability of helium adversely affects the economics and long-term planning for the use of upstream injection applications that require this gas. Although, these limitations have been mitigated to some extent by helium recovery systems, the reality is that nitrogen has become the gas of choice for most cold spray applications that require upstream injection equipment, evidently forcing applicators to compensate for the benefits of the low density of helium by operating with nitrogen at extreme high temperatures and pressures. Another challenge in upstream injection includes the tendency of particles to erode the nozzle throat as they are propelled through by the hot gas. Therefore, nozzle throats ought to be made out of high wear-resistant materials, which eventually wear off. Additionally, because of the higher nozzle operating temperatures, nozzle clogging must be tightly controlled by water cooling of the nozzle tube.

Heating up pressurized gases at relatively high gas flow volumes requires special attention to the engineering of the gas heater. Because of the required size of the heater, typically, gas heaters in upstream injection systems are external to the gun or work in conjunction with an auxiliary heater mounted inside the spray gun. One challenge with a remote external heater is the design of flexible conduits necessary to deliver the hot pressurized gas to the gun; a gas that may be in the 70 bar and 1000 °C range, not to mention heat and pressure losses which would occur during the transfer. Therefore, at those parameters, a split heater system with a reasonable sized heater right on the gun appears as the best option. A large heater in the gun renders the tool difficult to maneuver even robotically. This also limits the ability of the gun to get into hard-to-access spaces. Special nozzle configurations are designed to access locations such as tight inside diameters. Impact Innovations (2014), Plasma Giken (2014), Oerlikon Metco (2014), and VRC Metal Systems (VRC 2014) all manufacture commercial upstream injection systems with different features and capacities. These systems operate at pressures up to 70 bar providing gas temperatures of up to 1100 °C with power consumptions from about 34 to 70 kW.

Upstream injection equipment is typically less mobile than downstream injection hardware, simply because of the complexity of the equipment specifications, including the gun size required to operate at higher pressures and temperatures. Consequently, upstream injection has traditionally been used as a stationary tool in order to spray specialty materials requiring high impact velocities for bonding. More recently, some manufacturers developed a portable upstream cold spray system by downgrading design pressures (20 bar) and temperatures (400 °C), so that limiting the range of spray-able materials. others developed upstream injection systems with a lighter spray gun by completely removing the gas heater from the gun body. Examples of commercially available upstream injection systems depicting various design configurations are depicted in Figs. 6.7, 6.8, and 6.9.

Fig. 6.7 Upstream injection cold spray system (40 kW/50 bar/1100 °C) showing control unit, powder feeder, and water-cooled spray gun with a 40-kW gun-mounted gas heater. (Photo courtesy of Impact Innovations 2014)

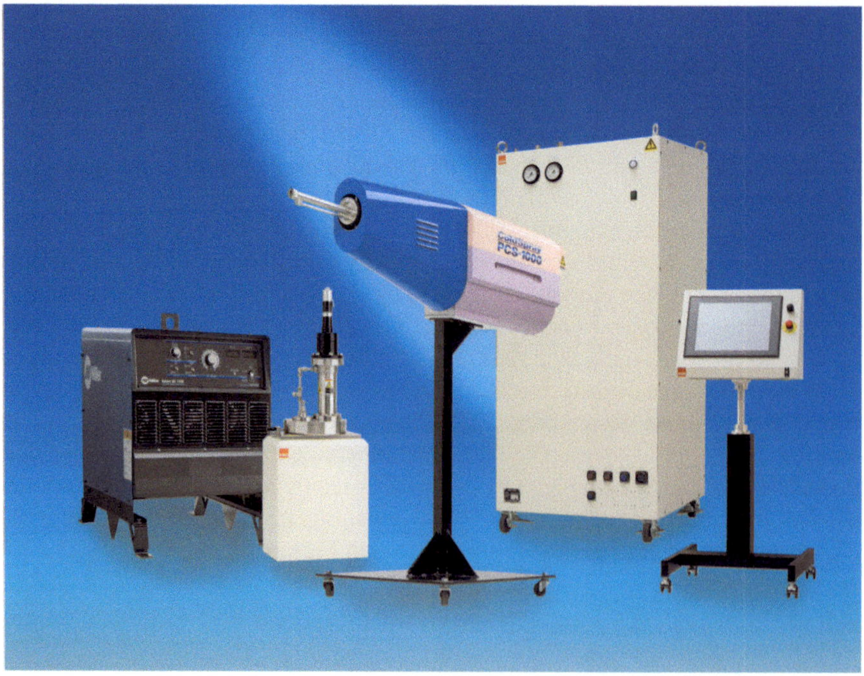

Fig. 6.8 Upstream injection system (70 kW/50 bar/1000 °C) showing water-cooled spray gun with a 70-kW gun-mounted gas heater. (Photo courtesy of Plasma Giken Co.)

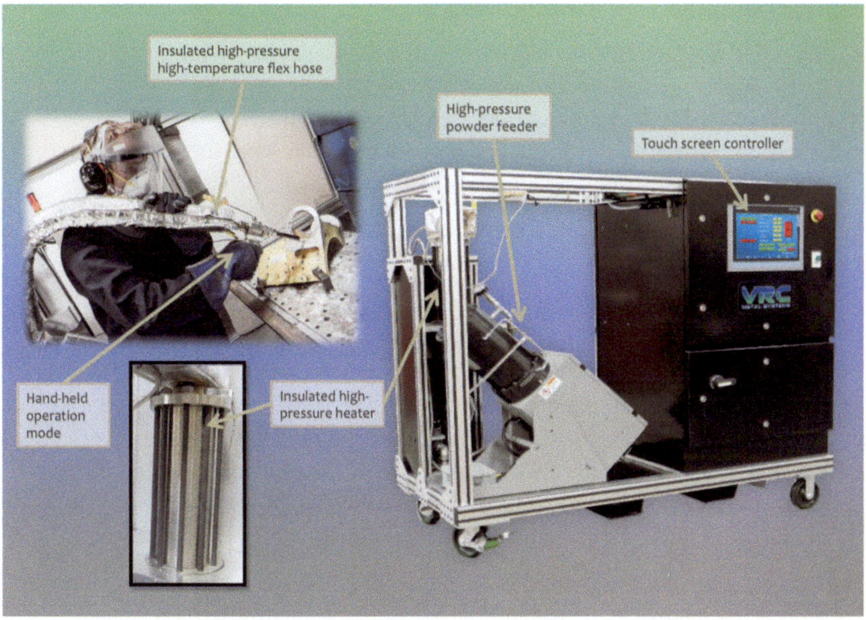

Fig. 6.9 Upstream injection system (15–45 KW/70 bar/900°C) showing spray gun connected to remote gas heater for both robot and handheld operations. (Photo courtesy of VRC Metal Systems)

6.2 Automation of the Cold Spray Process

6.2.1 Industrial Automation

Industrial automation refers to the use of mechanical devices without or with the control of a human operator to complete repetitive manufacturing tasks. For many years now, since the implementation of the first computer systems in manufacturing, industrial automation has steadily become a fundamental component of today's manufacturing processes. One strong driver of this trend is the fact that computers or programmable logic controllers (PLC) driving mechanical devices are capable of performing recurring tasks faster and more effectively than human operators, thus leading to increased productivity and consistency.

Automation can adopt many forms from a simple single-axis linear drive to complicated multi-axis computer numerical control (CNC) machining centers. In this section, we also make reference to dedicated automation and flexible automation. Dedicated automation is generally designed to perform single or multitasks aimed at performing a specific process following a specific pattern. On the other hand, flexible automation has the capability of being reprogrammed to do many other functions completely different from its original tasks.

The main advantages of automation are:

- Increased productivity by eliminating the dependency of production on how skilled and/or the emotional state of the operator.
- Improved quality by removing the element of human error.
- Increased process and/or product consistency by allowing the automation to do repetitive motions or process more consistently.
- Reduce direct labor expenses.

The main disadvantages of automation are:

- An automated system may have a limited level of intelligence, and is therefore more susceptible to committing errors outside of its immediate scope of knowledge.
- Unpredictable development costs.
- Automation cost for a new product or plant typically requires a very large initial investment in comparison with the unit cost of the product, although the cost of automation may be spread among many products and over time.
- Increased indirect labor to maintain more automation.

In manufacturing, the purpose of automation has shifted to issues broader than productivity, cost, and time. It has shifted to focus on quality with consistency and repeatability of the process. Therefore, users have exerted pressure on automation suppliers to build automation components that are more accurate and consistent. This trend has also been reflected in the case of cold spray, as users of this technology are increasingly demanded to produce more consistent and higher quality cold spray deposits. This can only be achieved by having more control over the cold spray process parameters as well as tighter control of the characteristics of the feedstock material, as further described in the following sections.

6.2.2 Automation of Cold Spray Process Controls

Control of an automated cold spray process can be attained by monitoring and controlling process parameters such as gas pressure, gas temperature, feedstock feed rate, and gun travel speed. Commercial cold spray systems use a fixed diameter of the DeLaval nozzle throat (or "orifice"), which is contained within the spray gun itself. This orifice operates in the choked condition for the various gases.

The actual mass flow rate of the carrier gas is determined by the gas density, gas pressure, and gas temperature. Therefore, to control gas mass flow for a given gas, pressure and temperature must be controlled within certain tolerances. This is generally achieved with the implementation of closed-loop control using pressure transducers and thermocouples in the gas circuit. Other alternatives include the implementation of closed-loop control using a mass flow controller and thermocouples.

Feedstock rate control can be attained by volumetric feeding or weight-loss delivery of the spray powder. Volumetric powder feeding is the most common technique as it is more economical. However, volumetric feeding is typically not consistent,

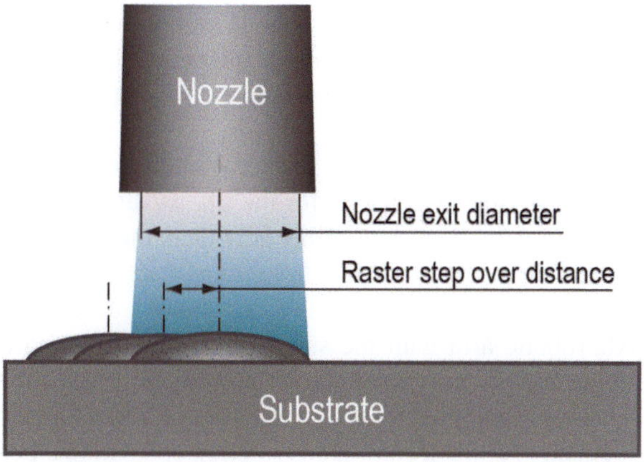

Fig. 6.10 Illustration of the raster step over method to attain smooth surface finish with cold spray

and monitored flow rates may vary up to 10 % depending on powder characteristics. A better method for controlling feed rate includes the weight-loss technique, which uses load cells to monitor feed rate and then controls the feedstock delivery rate to maintain a more constant feed rate. Other proposed methods include the control of particle count, being distributed within the system.

Monitoring and control of gun travel speed and raster step are desirable in order to produce smooth finish and consistent thickness of the deposit. The gun travel speed, together with the powder feed rate, determines the thickness of the deposit during one pass. It is generally desirable that the deposited thickness be within 0.13–0.5 mm per pass. Gun motion need to be generated from an automated drive, which may be capable of closed-loop feedback control.

The raster step over is important to define the thickness variation per gun raster. Depending on the surface smoothness required, the raster step over is typically set from 6 to 50 % of the nozzle exit diameter (Fig. 6.10).

6.2.3 Gun Manipulators

There are many possible combinations of automating the cold spray process. The following are some of the most common combinations:

- Single linear drive for the gun with linear movement of the substrate or part
- Single linear drive for the gun with rotational movement of the substrate or part
- Multiple axis robotic manipulation of the gun or the substrate with stationary substrate or part
- Multiple axis robotic manipulation of the gun with substrate manipulation by auxiliary axis

For a smooth and consistent coating on cylindrical substrates, the use of a variable speed linear drive for the gun and a variable speed rotational drive for the substrate is the most economical method for best results. This method is only useful when the coating thickness permits a large variability tolerance. However, when thickness variability has a tight window of tolerance, then a two-axis servo drive system for both the gun and substrate would be required along with a weight-loss powder-feeding system.

For substrates that have surface contour, multiple axis robotic manipulation of the gun would be desirable for the tilt, travel, and raster movements (Fig. 6.11). If the coating thickness variability is critical, then a weight-loss powder feeding would be required.

A more sophisticated level of automation consists of a gun mounted on a multi-axis robotic arm with the substrate or part mounted on a single or two-axis manipulator with coordinated axis control from the robot controller. Generally, the substrate manipulator has a rotational axis and a tilt axis. The rotational axis may be operated in either a continuous rotation or rotational position. The tilt axis generally has a 90° of freedom from vertical to horizontal positioning as a coordinated axis with the robot controller (Fig. 6.12). This scenario would also require the use of a volumetric or a weight-loss feedback powder-feeding system.

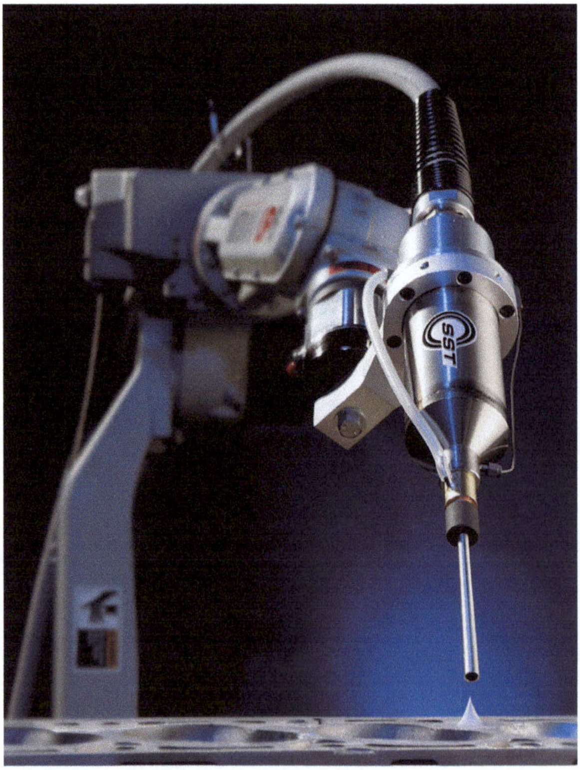

Fig. 6.11 Robotic cold spray gun mounted in a commercial multi-axis manipulator. (Courtesy of CenterLine (Windsor) Limited)

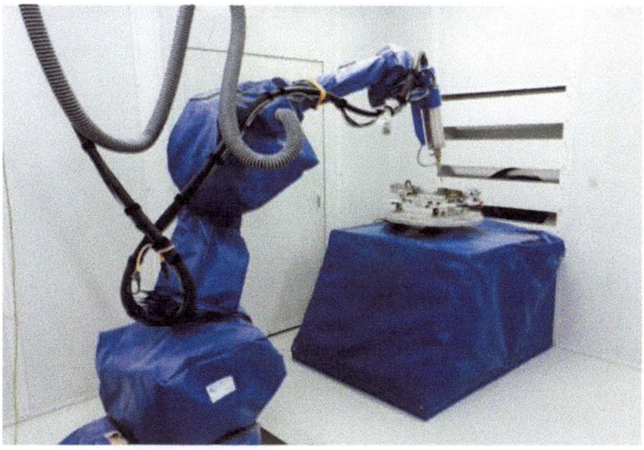

Fig. 6.12 Cold spray gun mounted on a multi-axis manipulator with the substrate or part mounted on a two-axis manipulator. (Courtesy of Able Engineering)

6.2.4 Automation Safety

6.2.4.1 Interlocks

Operator safety is one important consideration during the operation of any cold spray system for manual or fully automated operation. In particular, for any cold spray system to operate within a confined area there is a requirement to have interlocks installed for operator safety. In practice, both manual and automatic cold spray systems are typically interlocked with the dust collection system; in this configuration, the cold spray system will not operate unless the dust collection system is operating correctly. This could be achieved in a number of ways, but the most accepted way is to monitor the pressure differential in the upstream side of the filtering system. This can be achieved by monitoring the pressure differential between the dust collector and ambient pressure with the use of an appropriate differential pressure sensor. If the pressure differential is above or below certain threshold, the cold spray system would not operate.

For any kind of automation, there are always mandated safety regulations for the region where the cell operates, which restrict direct human intervention in the cell. Implementation of these regulations is attained by proper installation of either physical or electronic/light barriers, including physical doors, fences, and/or opto-electronic devices (light curtains) used to safeguard personnel in the vicinity of moving machinery that can cause harm. All forms of enclosure doors and openings must be interlocked with approved safety devices. These devices, normally safety switches, must be interlocked with the automation "E" stop circuitry (Fig. 6.13).

The powder-feeding system needs to be interlocked with the cold spray control system to ensure that the adequate start/stop sequence of operation can be controlled. The startup sequence of operation consists of initiating gas flow followed by energizing gas heating. When the gas reaches operational temperature, then powder feeding can be initiated. If this procedure is not adhered to, there is a possibility of

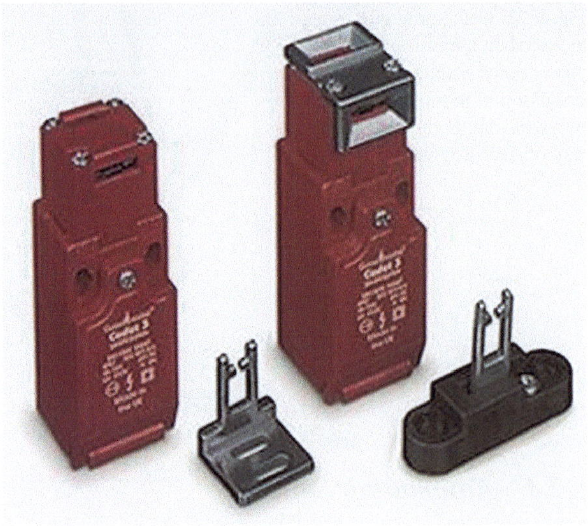

Fig. 6.13 Safety switch used for safeguarding entrance of personnel to active automated machinery

powder back feeding into the gas heater which can cause significant damage to the system. The shutdown procedure should be in reverse order of the startup.

Finally, the dust collection system needs to be interlocked with the cold spray control. The cold spray control shall not operate until the dust collection system is activated and operating up to specified performance level. This is achieved by using a differential pressure switch/gauge to monitor the pressure differential at upstream section of the dust collector. Most dust collectors run efficiently in the 0.010–0.017 bar (4–7″ water column). A cold spray system should not operate if the pressure differential falls outside this envelope.

6.2.4.2 Safety and Regulatory Compliance

6.2.4.2.1 Explosion Hazards Associated with Metal Dust

An explosion can result from a sudden chemical reaction burst between any combustible material and oxygen present in the normal atmosphere in a confined environment. Organic and/or inorganic combustible dusts suspended in air are prone to fire or explosions as the amount of surface area for oxidation is maximized. Except precious metals, most metals have a tendency to react with the oxygen present in air, at various levels. Therefore, metallic dust suspended in the atmosphere may represent a fire or explosion hazard, if the following three basic conditions are met.

1. The local concentration of the flammable material must be high enough to produce an ignitable mixture.
2. The amount of oxygen in the area must be in sufficient quantity around the flammable material to initiate a fire.

Fig. 6.14 Fundamental factors for the occurrence of fires or explosions

3. A source of ignition by exposed flame, spark, or high heat must be present.

Additionally, for an explosion to occur, the following additional elements are required:

4. Metallic dust in high enough density.
5. Dust cloud must be present in a confined area.

The presence of these five elements makes up the sides of what is known as the explosion pentagon. If any one of the five elements is missing, an explosion will not occur. All five elements must exist simultaneously for an explosion to occur (Fig. 6.14).

The process of cold spray delivers a metal powder in a carrier gas at temperatures below the melting point of the material sprayed. Carrier gases can be either helium, nitrogen, or air. Spray powder is normally sprayed in an air atmosphere. Except for precious metals, all metal powders are flammable, meaning that they could react with oxygen, holding a combustion process at some level. If the density of fugitive powder cloud around the spraying area is above a threshold (minimum explosible concentration, MEC), there is enough oxygen in the surrounding atmosphere, the location is confined, and there is a source of ignition, then the mixture could explode.

Cold spray work zones that have a potential for explosion are classified as hazardous areas by the various safety codes. In North America, these areas are designated as Class 2 Division 1 Group E, where gas, vapor, or mist will be present or expected to be present for long periods of time under normal operating conditions. Electrical equipment that must operate in such environments must be especially designed and tested to ensure that it does not initiate an explosion, due to a source of ignition or high surface temperature.

In Europe and the rest of the global community, hazardous locations for cold spray are designated by Atlantic trade wind experiments (ATEX) and International Electrotechnical Commission scheme for certification to standards relating to equipment for use in explosive atmospheres International Electrotechnical Commission Explosive (IECEx) as Zone 21 Group IIIC, depending on the circumstances, where the zone designation refers to an area in which an explosive mixture is likely to exist during normal operation.

To eliminate the risk of fires or explosions, cold spray users may follow the guidelines of the National Fire Protection Association (NFPA), an international, nonprofit group with 75,000 members from 100 nations. The following guidelines apply to cold spray operations: NFPA 654, Standard for the Prevention of Fire and Dust Explosions from the Manufacturing, Processing, and Handling of Combustible Particulate Solids; NFPA 484, Standard for Combustible Metals-2015; NFPA 33, Standard for Spray Application using Flammable or Combustible Material 2011; NFPA 69, Standard on Explosion Prevention Systems; and NFPA 70, the National Electric Code®. There are other groups that publish regulation and guidelines for fire prevention and safety including Occupational Safety and Health Administration (OSHA), Uniform Fire Code, and the American National Standards Institute. However, the local fire and safety regulatory bodies of the region in which a cold spray system is to be installed have the ultimate say on what specific precautions, norms, or design codes must be followed in order to allow the operation of a cold spray installation in the region. In the European Union, the ATEX 94/9/EC directive must be followed for any cold spray equipment installation to operate in the region.

The prime objective of a safe and successful cold spray installation is to remove, at least, one of the conditions that may lead to potential explosions. One method is to reduce the concentration level of metal dust in the hazardous zone down to well below its minimum explosion concentration (MEC) level. This can be achieved by diluting the concentration of metal dust in the hazardous zone by using enough ventilation and proper dust collection strategy; for example, integrators could size the dust collector's extraction volume to produce a safety factor, at least, ten times below the MEC level of the powder being sprayed. Another related factor to consider when designing a dust collection system is the velocity of the air movement at the work surface and in the ducting to the dust collector. The velocity in duct work needs to be sufficient enough to keep the dust particles airborne and not settle in the duct work.

An additional recommendation is to eliminate sources of ignition, especially in electrical equipment containing switches, by using intrinsic barriers that mitigate the possibility of creating sparks when contacts are made or broken.

The following recommendations are from NFPA 484–15:

A.9.4.10.2 Typically, the minimum conveying velocities range from 1078 m/min (3500 ft/min) to 1372 m/min (4500 ft/min) depending on the material being conveyed (ACGIH 2013).

A.9.4.10.3 U.S. Bureau of Mines, RI 6516, "Explosibility of Metal Powders," reports the results of tests conducted on 89 samples of metal powders of various grades and sizes. Minimum ignition energies (MIEs) for dust clouds ranged up to 15 mJ, whereas MIEs for dust layers ranged upward from 15 mJ. Ignition temperatures ranged upward from 320 °C (608 °F). MECs ranged upward from 40 g/m^3 (0.040 oz/ft^3). Maximum explosion pressures can exceed a gauge pressure of 620 kPa (90 psi).

The minimum velocity at the work surface has been determined to be 46 m/min (150 ft/min), but a safe design practice is to strive for 76.2 m/min (250 ft/min).

Fig. 6.15 Example of a differential pressure switch/gauge to monitor the differential pressure for safe operation of a cold spray system

The use of either sealed or pressurized control cabinets approved for hazardous areas will mitigate any possibility of metal powder making their way into their internal electrical components. Pressure levels in pressurized control cabinets and guns should be monitored with a differential pressure switch/gauge (Fig. 6.15). The minimum pressure differential threshold should be 2.54 cm (1″) water (0.002 bar). Below this level, a cold spray control system should shut down.

Last but not least, manufacturers and installers of cold spray systems should minimize buildup of static charge by grounding all components, where applicable, especially in critical zones. For example, all nonmetallic powder feed lines must be made out of static dissipating tubing that is properly grounded. The work piece and the work surface should also be grounded. All pneumatic conveying components from the work surface to the dust collector must also be grounded.

Dust Collection

In Chap. 9, details of powder reclamation and filtration methods are described, which are applicable to most thermal spray processes. Unlike most thermal spray processes, the cold spray process does not operate based on combustion or electric arcs. Therefore, smoke, vapors, or fumes are not present in the process. Consequently, in addition to regular air filtration-type dust collectors, cold spray can take advantage of wet dust collectors. One advantage of wet collection is that the use of water as a collector of fugitive particles can further mitigate the chance of an explosion. This generally means a wet dust collector can be placed at the work area with a very short connection point to the work surface, eliminating the need for controlled ducting to an outside paper cartridge or a bag type dust collector. Wet collection for cold spray may eliminate the need for isolation valves, additional controls, and fire suppression in the conveying system. Also, there is no need for air makeup systems as, in most cases, filtered air from the dust collector does not need to be exhausted outside. Additionally, depending on air quality testing at the exhaust, there would be no need for high efficiency particulate air (HEPA) filters at the exit port.

Noise Abetment

During the operation of cold spray systems, especially when using high pressure and temperatures to maximize jet flow velocities, the attenuation of noise levels may require special attention. Operating at low pressures (6–10 bar/ 90–150 psi) generates sound levels at about 100–105 dBA, whereas operating at high pressures (34–50 bar/500–750 psi) can generate sound levels in the range of 120–130 dBA. In both cases, it is recommended that the spray gun operates inside a soundproof enclosure. Otherwise, operators must wear hearing protection with a noise reduction rating (NRR) of, at least, 34 dB for high-pressure operation and at least 20 dB for low-pressure operation.

6.2.5 *Work Stations and Enclosures*

Cold spraying within a confined space must be carried out within a booth, enclosure, or cabinet. For manual and/or automatic spraying, there are multiple options available in the marketplace, including:

- Glove boxes of various sizes
- Three-sided downdraft or backdraft enclosures of various sizes
- Fully enclosed spray booths

The glove box approach is very similar to traditional grit-blasting cabinets. Their construction shall be made so that the enclosure and the work area both display anti-spark characteristics with proper grounding. Access gloves must be made out of antistatic materials. Any lighting inside the cabinet must be constructed to operate in hazardous locations as explained in the section "Explosion Hazards Associated with Metal Dust". Access doors must be interlocked with the cold spray system so that the system does not operate when the doors are open. The viewing glass must be rated accordingly for safety. The cabinet can be made of any size that is acceptable for the operator's working comfort (Fig. 6.16).

Three-sided downdraft or backdraft enclosures provide handheld applicators with more flexibility as far as the shape and size of the parts to be cold sprayed (Fig. 6.17). Again, with this configuration, both the work surface and enclosure are made of spark-resistant materials, and all conductive components of the enclosure

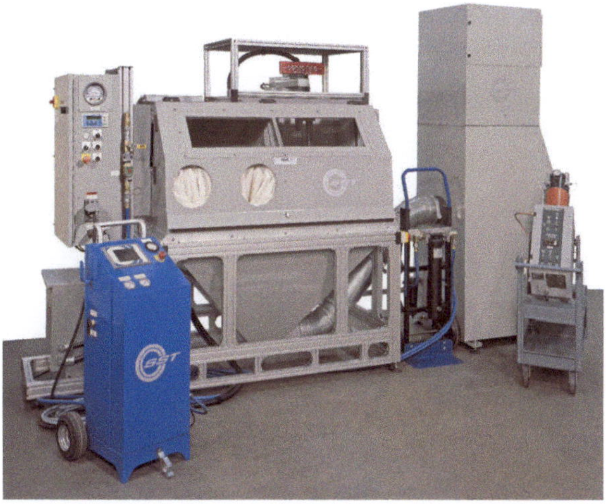

Fig. 6.16 Example of a dual glove cabinet designed for both handheld and mechanized spraying of parts. (Courtesy of CenterLine Ltd)

are grounded. In order to assist manual manipulation and reduce operator fatigue, the gun may be mounted on a tool balancer.

Fully enclosed acoustic enclosures or spray booths are recommended for dedicated robotic operation of the cold spray process, in particular, when spray parameters include high gas pressures and temperatures (Fig. 6.18). The design and configuration of the enclosure and its access doors are often custom built to the end user's requirements. Most acoustical enclosures designed for thermal spray applications include walls with a *noise reduction rating* (NRR) of 35 dB, which also safeguard operator from the automation. Access to the enclosure is interlocked with all key automation controls for operator's safety.

Automation in fully enclosed spray booths can be configured in many different ways; the most versatile method includes the use of a six-axis robot arm working in harmony with a two-axis indexing turntable. Control of tilt and rotation of the turntable can be obtained from the robot controller, which usually comes with auxiliary seventh and eighth axis for controlling additional devices. Anchoring of the robotic

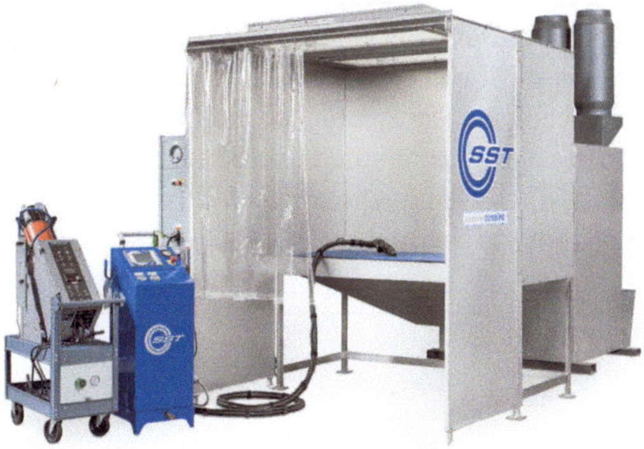

Fig. 6.17 Example of a three-sided downdraft enclosure for handheld operation. (Courtesy of CenterLine (Windsor) Limited)

Fig. 6.18 Example of a fully enclosed acoustic spray booth for robotic cold spraying. (Courtesy of Able Engineering)

arm can be on the floor, ceiling, or wall inside the enclosure, depending on what may be best for optimum utilization of the work area. The robot electric motors and gearboxes as well as the indexing turntable all need to be protected from metal dust. The latter is often achieved by using a pressurized protective suit as shown in Fig. 6.12. All electrical equipment inside the enclosure must be rated for hazardous location Class II Division 1 Group E or Zone 21 as explained in the section "Explosion Hazards Associated with Metal Dust".

6.3 Concluding Remarks

Cold spray is a promising technology within the larger thermal spray family which, over the last decade, has seen a rapid evolution of equipment for both downstream and upstream injection methods. Each equipment configuration has its own advantages and limitations; the selection of the right cold spray equipment should be based on the specific considerations of the application, including whether it is a field or contained application, spray material requirements, desired coating properties, accessibility considerations, equipment reliability, and economics of the required procedure. Because of the presence of fugitive metal dust, the cold spray operation must be complemented with the utilization of safety auxiliary equipment such as cabinets, spray booths, and dust collection equipment. The cold spray process can be fully automated using a diversity of available automation options. Fully automated cold spray operations follow same industry standard practices for safety quite similar to thermal spraying.

References

ACGIH. 2013. *Industrial ventilation: A manual of recommended practice for design*. American Conference of Governmental Industrial Hygienists (ACGIH). Oxford: Elsevier.

CenterLine (Windsor) Limited. 2015. http://www.supersonicspray.com/. Accessed 23 Jan 2015.

Dymet Corporation. 2014. http://dymet.info/indexe.html. Accessed 7 July 2014.

Impact Innovations. 2014. http://www.impact-innovations.com/en/coldgas/cg_index_en.html. Accessed 29 July 2014.

Kashirin, A. I., O. F. Klyuev, and T. V. Buzdygar. 2002. Apparatus for gas-dynamic coating. US Patent 6,402,050, June 11.

Maev, R., and V. Leshchynsky. 2008. *Introduction to low pressure gas dynamic spray: Physics & technology*. Weinheim: Wiley.

Oerlikon Metco. 2014. http://www.oerlikon.com/metco/en/products-services/coating-equipment/thermal-spray/systems/cold-spray/. Accessed 29 July 2014.

Papyrin, A. 2001. Cold spray technology. *Advanced Materials & processes* 159 (9): 49–51.

Papyrin, A., V. Kosarev, S. Klinkov, A. Alkimov, and V. Fomin. 2007. *Cold spray technology*. Oxford: Elsevier.

Plasma Giken. 2015. http://www.plasmagiken.com/products/coldspray.html. Accessed 6 Jan 2014..

Villafuerte, J. 2010. Current and future applications of cold spray technology. *Metal Finishing* 108 (1): 37–39.

VRC. 2014. http://www.vrcmetalsystems.com/products.html. Accessed 2 Aug 2014.

Chapter 7
Laser-Assisted Cold Spray

D. Christoulis and C. Sarafoglou

7.1 State of the Art

In thermal spray technology, coatings' adhesion is a property of major concern because it is essential for the coating to adhere throughout the design life of the coating system (Davis 2004) under various and severe conditions. The same requirements of high coating adhesion hold true for cold spray coatings. In order to achieve advanced coatings of high adhesion, the substrate should be prepared correctly. Specific preparation steps of the substrates should be followed; otherwise, a total failure of the coating could occur (Davis 2004).

Cleaning and roughening of the substrates are the main steps which are preceded before the formation of a thermal-sprayed coating. Surface cleaning is carried out either by using organic solvents (methyl alcohol or acetone) or, in the case of degreasing large pieces, by employing hot pressured water or water vapour (Pawlowski 2008). After the surface degreasing, it follows the roughening of the substrate, which, in turn, promotes the mechanical anchoring of the coating onto the roughened substrate. The most common procedure to increase the roughness of the substrate is the dry abrasive grit-blasting method (Wigren 1998; Rosales and Camargo 2009; Sen et al. 2010; Bahbou et al. 2004). Grit blasting of the substrates can lead to an improvement of the adhesion strength of the thermal-sprayed coatings and also to an increase of their fatigue resistance. It has been shown experimentally that both adhesion strength (Vilemova et al. 2011; Makinen et al. 2007; Paredes et al. 2006) and fatigue resistance (Jiang et al. 2006; Multigner et al. 2009a, b) of thermal-sprayed coatings are enhanced with the modification of the substrate roughness by

D. Christoulis (✉)
CeraMetal Surface Engineering S.A., Koropi, Greece
e-mail: dimitris.christoulis@gmail.com

C. Sarafoglou
School of Naval Architecture and Marine Engineering, National Technical University of Athens, Athens, Greece
e-mail: chsaraf@mail.ntua.gr

J. Villafuerte (ed.), *Modern Cold Spray*, DOI 10.1007/978-3-319-16772-5_7

employing grit blasting. However, grit blasting may provoke contamination of the substrate by grit inclusions which can be catastrophic for some applications. More precisely, in the case of automotive sector (Barbezat 2005, 2006), for the production of engine blocks with thermally sprayed cylinder, other techniques should be used for the preparation of the substrate since the grit can remain within the many internal passages of the engine block. The grit can come loose during operation and later cause engine breakdowns (Schlaefer et al. 2008). Furthermore, the adhesive strength has been correlated with the grit inclusions (Maruyama et al. 2007), and also it has been shown that an extensive grit blasting could reduce the adhesive strength of thermal-sprayed coatings (Yang et al. 2006; Ichikawa et al. 2007). The fatigue resistance of thermal-sprayed coatings could also be decreased after the grit blasting of the substrates (Leinenbach and Eifler 2006; Multigner et al. 2009a, b).

Moreover, it should be noted that grit blasting is environmentally unfriendly, and also it can be unhealthy for the operators of grit-blasting apparatus since the grits have been correlated with serious diseases such as silicosis, aluminosis, lung scarring, pneumoconiosis, or emphysema (Petavratzi et al. 2005).

For the above-mentioned reasons, other conventional techniques have also been developed for the preparation of the substrate for the thermal-spraying processes. Such techniques are the waterjet, the chemical etching (Pawlowksi 2008), and the macroroughening (Davis 2004). Waterjet pretreatment has been used in the case of shrouded plasma-sprayed MCrAlY coatings onto nickel superalloys (i.e. Inconel 718, Rene 80 and Mar-M 509; Pawlowski 2008). The surface of the substrate is roughened by the waterjet, which produces a much finer surface than that of a sand-blasted one (Pawlowski 2008).

Advanced thermal-sprayed coatings can be produced by the combination of thermal spray and laser processing, which can be considered as half-brothers since they show many common features due to the use of a (more or less) high-energy source for both (Jeandin et al. 2010). Laser thermal spray hybrid processes have already been developed as a result of the successful combination of laser and thermal spray technologies. Various types of laser have been combined with the guns of the thermal spray process. Combining laser processing to thermal spray resulted in a major improvement for thermal spray in three subareas, that is, that of pretreatment (Garcia-Alonso et al. 2011; Zieris et al. 2003, 2004; Costil et al. 2004a, b; Danlos et al. 2011) that of posttreatment (Garcia-Alonso et al. 2011; Zieris et al. 2003, 2004; Pokhmurska et al. 2008; Li et al. 2010; Wang et al. 2010; Jeandin et al. 2003) and that of simulation of thermal and kinetic phenomena (Barradas et al. 2007; Guetta et al. 2009; Guipont et al. 2010; Fabre et al. 2011; Bégué et al. 2013).

Laser-assisted cold spray (LACS) combines the advantages of cold spray and laser technologies leading to the formation of advanced coatings. The term "laser-assisted cold spray" has been referred mainly for the laser pretreatment (ablation and texturing) of the substrates prior to the formation of cold-sprayed coatings (Bray et al. 2009; Christoulis et al. 2009; Olakanmi and Doyoyo 2014). However, post-treatment of cold-sprayed coatings by employing laser technology has also been carried out (Sova et al. 2013; Marrocco et al. 2011). This chapter focuses on the pretreatment case of the substrates.

7.2 Surface Pretreatment Techniques (Thermal Spray)

The adhesion of a coating layer on the substrate or on a previously deposited layer is essential for good cohesion of the coating to be achieved (Danlos et al. 2008). A critical factor that affects significantly the good performance of coating is the correct preparation of substrate's surface. Wrong practices on surface preparation prior to coating deposition can lead to a total coating failure. Thus, surface preparation is considered as an important financial and functional factor for thermal spraying (Davis 2004). Furthermore, surface pretreatments need to be performed on the substrate before thermal spraying in order to improve the adhesion and material properties. Preprocessing techniques are designed to remove grease and other contaminants from the surface, and also change the physicochemical properties and/or surface morphology.

7.2.1 *Conventional Pretreatment Techniques*

There are several steps on surface preparation, each of which alters the surface in a way so that a coating of good quality is produced (Pawlowski 2008). Cleaning, surface activation, roughening, and preheating are the most important steps preceding the spraying process. Among the conventional techniques, degreasing and sandblasting are used in most cases. The degreasing agent leads to chemical modifications of the surface while sandblasting modifies the surface morphology by creating a uniform roughness, thus providing a mechanical anchorage of the incoming particles to the substrate (Lamraoui et al. 2010). Described below are the most common techniques used for the implementation of these steps prior to thermal spraying, giving a picture of both the conventional and the most advanced and innovative methods.

Cleaning is a key parameter for the contact quality which depends on the droplet wetting and desorption of the pollutants adsorbed on the surface or on the underlying layer (Danlos et al. 2008). Degreasing substrates is usually realized either by using organic solvents (methyl alcohol, acetone) or by vapour or wet abrasive blasting.

Surface activation is the most important step of surface preparation prior to thermal spraying. Without activation, the coating would not adhere to the substrate surface. Activation by abrasive grit blasting, i.e. roughening, remains presently the most frequently applied process (Pawlowski 2008). A surface roughening is implemented in order to increase the surface area and produce a structure which facilitates the mechanical interlocking of the coating with the substrate and guarantees the adhesion of the coating. In principle, there are a number of processes available for the activation of the surface, such as abrasive grit blasting, waterjet treatment, chemical attack, and machining or microroughening.

7.2.1.1 Abrasive Grit Blasting

Dry abrasive grit blasting is the most commonly used surface roughening technique. Surface roughening by using the grit blast process, as a method to improve adhesion, is a debatable issue among researchers. Several studies state that thermal-sprayed coatings, without exception, have significantly higher bond strength to surfaces which have been previously grit blasted. For example, Gonzalez-Hermosilla et al. (2010) investigated the effect of substrate roughness on the fatigue behaviour of an SAE 1045 steel substrate coated with a WC–10Co–4Cr cermet by high velocity oxygen fuel (HVOF) thermal spraying. Their results indicate that fine-grinding impairs the mechanical bonding of the coating, giving rise to its delamination from the substrate at elevated maximum alternating stresses. Mohammadi et al. (2007) studied the influence of grit blasting parameters on the surface roughness of Ti–6Al–4V alloy as the substrate for plasma-sprayed hydroxyapatite (HA) coatings. Their findings led to the conclusion that the substrate surface topography significantly influences the adhesion properties of the coating at the interface. Staia et al. (2000) in their research showed that adhesive properties of WC–Co thermal-sprayed coatings are positively affected by roughness variations derived from alterations of the grit blasting pressure.

7.2.1.2 Waterjet Treatment

Waterjet treatment is another way of keeping the activated surface clean (Pawlowski 2008). This technique was applied to activate superalloys (Inconel 718, Rene 80 and Mar-M 509) with prior shrouded plasma spraying (SPS) of MCrAlY coatings. Furthermore, ultrahigh-pressure water jetting over 172 kPa has gain a lot of attention, due to its ability to remove high percentages of soluble salts from the steel surface. In addition, it has the advantage of not generating spent abrasive and not incurring the cost of abrasive disposal. At the higher pressures, lower volumes of water are needed, making thus the disposal costs lower compared to the traditional water blasting methods. Ultrahigh-pressure water jetting leaves a warm surface from which the residual water can easily dry, but does not generate heat that can cause thermal stress in the steel surface. Ultrahigh-pressure water jetting is an extremely versatile and effective method of removing paint and metal coatings, soluble salts, and other contaminants from steel surfaces. It is environmentally friendly, though at present is more expensive as compared to traditional blast cleaning methods.

7.2.1.3 Mechanical Activation (Machining)

Where it is not possible to clean by abrasive blasting, hand and power tool methods may be the only acceptable alternative methods. Macroroughening is usually accomplished by machining grooves or threads into the surface to be sprayed. Typically, rough machined surfaces are also grit blasted prior to spraying. Surfaces roughened to this magnitude are often used for thick coatings to restrict shrinkage stresses and to disrupt the lamellar pattern of particle deposition in order to break up the shear stresses parallel to the substrate surface (Davis 2004).

7.2.2 Laser Pretreatment Techniques

The conventional pretreatment processes exhibit certain drawbacks (Danlos et al. 2008). Chemical wastes from the cleaning process need to be carefully discarded as can be environmentally very harmful, while grit blasting can cause worsening of fatigue strength of the material and also issues concerning adhesion due to grit inclusions on the surface. In detail, surface degreasing is most often carried out using solvents such as carbon fluorochloride (CFC) or trichloroethylene. However, their use causes environment problems, recycling, and operator health protection issues. Moreover, the grit blasting process is not easily controlled with high precision. It can lead to substrate damage and subsequently failure on coating performance. In fact, this process can cause a modification of the surface mechanical properties and a decrease in resistance to fatigue for ductile materials such as aluminium and titanium due to the fragility by notch effect or sand encrustation when grit blasting air pressure and incidence angle are incorrectly used. Residues can be entrapped in the substrate; the more ductile the substrate material, the higher the amount of grit particles entrapped. Finally, very thin substrates can be easily deformed by conventional grit blasting as well (Wigren 1998; Coddet et al. 1999; Costil et al. 2005).

Hence, other cleaning processes have been developed to overcome these drawbacks and substitute the conventional techniques. Coupling of a high-power laser with thermal spraying is a technique developed for pretreatment of substrates prior to spraying (Jeandin et al. 2010).

The technique of pulsed-laser cleaning has emerged in the early 1970s as an attractive process surface preparation substituting traditional methods where chemical solvents are widely used (Tam et al. 1998). Later, it was effectively applied to remove small particles from semiconductor in micro-electrical fields. Most significant advantages of this process include:

- Quiet process and environmentally friendly compared to traditional processes
- High flexibility by using an improved laser delivery system
- Easily monitored and automated

In recent developments dedicated to industrial applications, a laser operated at short-pulse mode (~ns) can offer higher efficiency and therefore has been preferably employed. Laser preparation offers cleaning of various surface contaminants (grease, oils, contamination particles, inclusions, etc.) and prohibits the recontamination of surface during the deposition stage (Li et al. 2006).

7.2.2.1 Laser Ablation

Surface laser ablation also seems favourable to the liquid droplet spreading out onto the substrates, thus representing a very good wettability of splats on the substrate, which is one of the most important basic processes during thermal spraying. The study of the mechanisms of particle impact on substrate associated with splat morphology analyses attracts much attention since all the coating properties are connected to these processes. The particle impact and solidification depend on particle properties (kinetic energy, viscosity), as well as substrate properties (temperature, thermal conductivity, surface quality; Costil et al. 2005). In addition, Danlos et al. (2011) in their research, prior to cold spray deposition of aluminium coatings, they treated the substrates (aluminium alloy 2017) by conventional (degreasing and sandblasting) and laser processes. The linked adhesion measurements to the surface modifications were induced by the different preparations and it was found that laser processes improved the contact strength. Ablation laser cleaned the surfaces that promote intimate bonding between the coating and the substrate. This process also modified the surface structure and created morphology adapted to the specific characteristics of the cold spray particles. The addition of a heating laser improved the surface topography and improves significantly the coating adhesion. Lasers permit to prepare the surface with a high efficiency, thanks to the flexibility and the quickness of this process. Laser treatment can also be adapted to the characteristics of the material, so it is possible to create specific surface modifications. This technology can be used to texture substrate and create optimal surface topography to improve coating adhesion (Danlos et al. 2011).

7.2.2.2 Laser Cleaning and Heating

Laser surface cleaning has emerged as a very suitable tool for the substitution of wet cleaning techniques. The laser cleaning principle is based on specific interaction modes. The cleaning efficiency is a compromise between contaminants elimination and integrity of the substrate, which is very important when modifications are desirable. Verdiera et al. (2003) in their study clarified the laser–matter interaction mechanisms and effects on metallic materials for initial roughness of different surfaces and laser beam energy densities. They investigated both topographic and energetic modifications, roughness evolution, and wettability conditions. They found that increasing the laser beam energy density leads to a smoothing of rough surfaces and the formation of craters at high fluences, resulting from the removal of surface inclusions due to thermal effects. Contribution of direct vaporization and hydrodynamical sputtering were then pointed out. However, smoothing or crater formations do not have any significant effect on the surface roughness. The sessile drop technique allowed to define experimental conditions leading to best wetting conditions required for the development of the PROTAL® process in the field of thermal spraying (Verdiera et al. 2003).

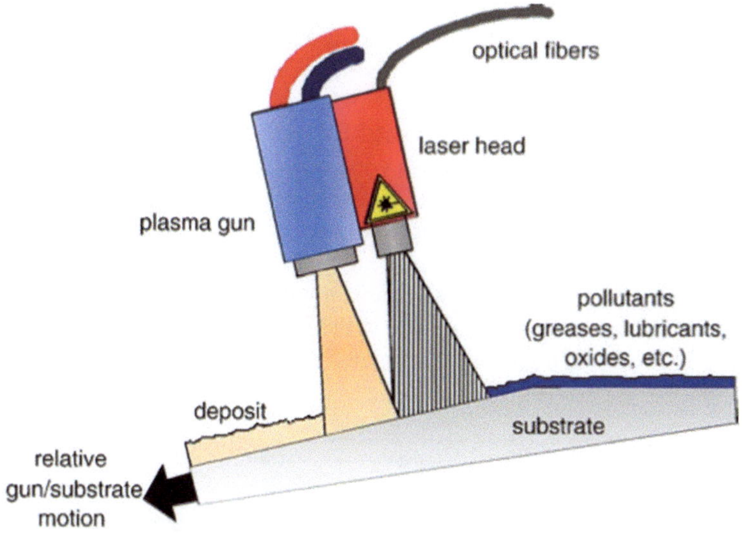

Fig. 7.1 Schematic presentation of the PROTAL® process. (Coddet 2006)

7.2.2.3 PROTAL® Process

In 1993, Coddet and Marchione proposed a treatment employed simultaneously to the thermal spraying process, called the PROTAL® (Quantel Lannion, France) process (French acronym for "PROjection Thermique Assisté par Laser", i.e laser-assisted thermal spray). This process combines the spraying operation and the surface preparation in a single step (Fig. 7.1).

Laser irradiation is applied in order to remove any contamination films and oxide layers from the substrate's surface, to generate a uniform surface condition enhancing the deposit adhesion and additionally to restrict the re-contamination of the deposited layers by condensed vapours. The PROTAL® process refers to a technique that allows simultaneous surface preparation and coating operation. This is obtained by the association of a spray gun (any kind) with a specific laser gun. The geometric arrangement between both guns is achieved in such a way that the laser treatment precedes immediately or even overlaps the thermal deposition stage. Hence, the molten particles impinge onto a surface free of oxides and pollutants. Eliminating the surface contamination layers simultaneously to the coating deposition not only allows the development of physical bonds between the coating and the substrate but also reduces the overall number of treatments necessary to coat a part and thus improves the effectiveness of the process (Coddet 2006).

Some early studies of PROTAL® noticed the laser-induced modification of surface morphology and surface energy, and then the subsequent works focused on the surface morphological evolutions as well as their correlations with the deposit adhesion. It was shown that this laser pretreatment permits only a small variation of the surface roughness but has a considerable influence on the deposit adhesion, the mean arithmetic roughness (R_a) varying at the level of less than one micrometre.

Therefore, it was suggested that such a laser irradiation may not only generate an efficient removal of the surface pollutants but also promote interfacial physicochemical bondings. Nevertheless, this aspect is still unclear and the mechanisms of laser-induced modifications have not been well understood yet (Li et al. 2006). Moreover, it has been seen that the nanosecond pulsed-laser irradiation effects depend strongly on the substrate nature and the surface conditions. For titanium substrate, two aspects dominate the surface modification as follows: (i) craters are formed due to a preferential laser ablation on surface defects, and (ii) rapid surface melting and cooling are caused by superficial superheating. The latter factor is essentially noted for titanium-based substrate in comparison with aluminium alloy, which is less prone to surface oxidation probably due to the lower surface temperature owing to its high reflectivity and high thermal conductivity (Li et al. 2006).

7.3 Laser-Assisted Cold Spray

The LACS process has been related with the substrate pretreatment (Bray et al. 2009; Christoulis et al. 2012; Olakanmi and Doyoyo 2014) and the simultaneous coatings build-up, in a single step, for the production of advanced protective coatings of high adhesion strength.

Parallel to the evolution of the LACS process, laser-assisted downstream injection cold spray (LALPCS) has also been developed (Kulmala and Vuoristo 2008). Since the current chapter is mainly focused to the LACS, the LALPCS is presented first (in brief) in Sect. 7.3.1.

7.3.1 Laser-Assisted Downstream Injection Cold Spray

In the downstream injection cold spray process, the coatings can be built-up by blending a ceramic powder (e.g., alumina) with a metallic powder; the ceramic powder activates the sprayed surfaces and hammers the substrate/sprayed layers by shot-peening (Hussain 2013). The downstream injection system developed in Laser Application Laboratory of the University of Tampere in Finland is presented in Fig. 7.2 (Kulmala and Vuoristo 2008). A DYMET 403K (Licenceintorg, Russia) downstrem injection cold spray system was coupled with a high-power-diode laser (Rofin

Fig. 7.2 Laser-assisted downstream injection cold spray, developed in University of Tampere, Finland. (Kulmala and Vuoristo 2008)

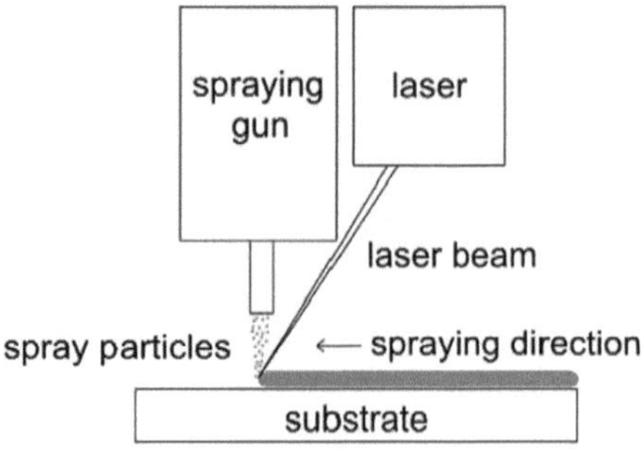

Table 7.1 Materials and sprayed parameters of the LALPCS. (Kulmala and Vuoristo 2008)

Sprayed material	$Cu+Al_2O_3$	$Cu+Al_2O_3$	$Ni+Al_2O_3$	$Ni+Al_2O_3$
Gas	Air	Air	Air	Air
Gas pressure (MPa)	0.6	0.6	0.6	0.6
Gas temperature (°C)	445	445	600	600
Standoff distance (mm)	30	30	30	30
Traverse speed of the gun (mm/s)	40	40	40	40
Passes	2, 5, 10	2, 5, 10	2, 5, 10	2, 5, 10
Pyrometer (°C)	0	650–800	0	650–800
Laser type	Diode laser (6 kW)	Diode laser (6 kW)	Diode laser (6 kW)	Diode laser (6 kW)
Laser power (kW)	0	1.8–2.4	0	1.8–2.4

LALPCS laser-assisted downstream injection cold spray

DL 060 H2 6 kW continuous wave laser, Hamburg, Germany). Laser irritated the sprayed particles and simultaneously provoked the substrate heating. The laser beam was rectangular (5.8×23.5 mm) while the cold spray spot was circular with a diameter of 5 mm. The parameters of LALPS as well as the materials which were sprayed by Kulmala and Vuoristo (2008) are presented in Table 7.1. As it shown in Table 7.1, coatings were also sprayed by employing only the conventional low cold-spray system without simultaneous laser irradiation of sprayed particles and substrate.

It should be mentioned that during the spraying, the laser power controlled by a pyrometer measured the surface temperature of the deposited coating. As soon as coating's temperature increased above the 650 °C, the laser power was decreased from 2.4 kW to about 1.8–2.0 kW.

For all the cases, 2, 5, or 10 gun passes were carried out for both $Cu+Al_2O_3$ and $Ni+Al_2O_3$ coatings. It was found that the coatings' deposition rate was increased by using the advanced LALPCS compared with the conventional LPCS. The higher deposition rates led to the formation of thicker coatings.

Also, sprayed $Cu+Al_2O_3$ coatings were denser compared to conventional downstream injection cold-sprayed coatings. Open porosity experiments (open cell potential measurements by exposing the coatings to 3.5 wt.% NaCl solution at room temperature for 11 days) proved that the laser-assisted downstream injected copper coatings did not show open porosity, but without the laser, copper coating was fully corroded (Kulmala and Vuoristo 2008).

7.3.2 *Laser-Assisted Cold Spray*

The LACS process has been developed with laser heads which provoke:

- Heating of the substrate (Sect. 7.3.2.1)
- Ablation of the substrate (Sect. 7.3.2.2)
- Both heating and ablation of the substrate by using several laser heads (Sect. 7.3.2.3)

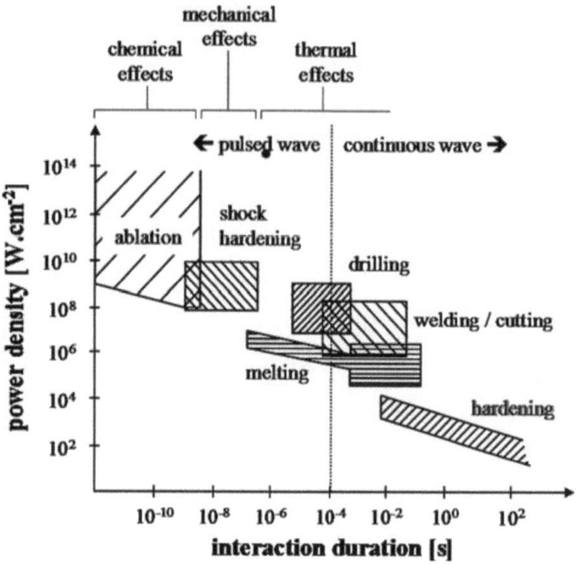

Fig. 7.3 Laser treatments on metallic substrates as a function of the power density and the interaction time. (Garcia-Alonso et al. 2011)

The pretreatment type of the substrate can be varied depending on the type of the material to be irradiated and the laser beam characteristics (Garcia-Alonso et al. 2011). Garcia-Alonso et al. demonstrated the effect of the laser treatments on metallic substrates as a function of the power density and the interaction time (Fig. 7.3).

7.3.2.1 Laser Heating

The first LACS system was presented in 2009. It was a LACS set-up which employed a diode laser that was demonstrated by Bray et al. (2009). Oxide-free titanium coatings of high density were deposited by this LACS system.

A nitrogen supplied cold spray system was used for spraying fine titanium powder (particle size <45 μm) onto mild steel substrates under the parameters which are presented in Table 7.2.

The cold spray gun was coupled with a laser diode of wavelength 980 nm (Fig. 7.4) in such a configuration which allowed the laser irradiation of both sprayed particles and an area of the substrate. The laser diode had a heating effect on the substrate as well as on the sprayed particles, which was beneficial for the formation of the titanium coating. More precisely, the heating provoked the reduction in strength of sprayed particles and substrate, allowing extensive particle deformation and thus bonding to occur (Bray et al. 2009). However, it should be mentioned that the titanium particles were deposited only to laser-treated areas of the substrate. The spot diameter of the laser was 4 mm while the powder beam had a diameter of 8 mm, and it was found that the titanium cold-sprayed particles were deposited only within the area of the substrate that was illuminated by the laser.

Table 7.2 Materials and sprayed parameters of the LACS. (Bray et al. 2009)

Sprayed material	Ti
Gas	Nitrogen
Gas pressure (MPa)	3.0
Gas temperature (°C)	Unheated (ambient temperature)
Standoff distance (mm)	50
Traverse speed of the gun (mm/s)	500
Passes	1
Laser type	Diode laser (wavelength 890 nm)
Laser power (kW)	≤1

A high-speed infrared pyrometer (Fig. 7.4) was used in order to control the temperature of the deposition sited during the coating build-up. The power of the diode laser was altered on-line by the measurements of the pyrometer in order to keep the substrate temperature above 550 °C.

It was observed that when the substrate temperature was measured with the pyrometer below 450 °C, only few cold-sprayed particles deposited on the substrate and thus coating was not formed. The increase of the laser power to 650–1000 W, which in turn provoked the increase of the substrate temperature from ~550 up to 900 °C, led on to the formation of dense titanium coatings.

The titanium coating produced using the LACS process were compared with titanium coatings produced by conventional cold spray equipment. It should be mentioned that for both LACSprayed coatings and cold-sprayed coatings similar materials and equipment were used. It was found that the LACSprayed coatings presented improved properties (Table 7.3) compared to the conventional cold-sprayed titanium coatings.

Furthermore, according to Bray et al. (2009), the deposition mechanism of cold-sprayed particles differs from those of LACS particles. In the case of conventional cold-sprayed coatings of high density, a lamellar structure is presented where the

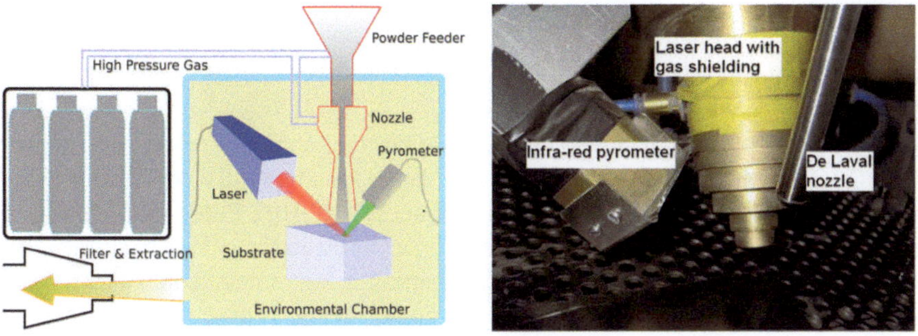

Fig. 7.4 Laser-assisted cold spray, developed in the University of Cambridge, UK, by Bray et al. (2009)

Table 7.3 Properties of titanium coatings. (Bray et al. 2009)

	Build rate (g/min)	Porosity (%)	Oxygen content (%)
LACSprayed titanium coatings	<45	<1	<0.6
Cold sprayed titanium coatings	<25	<5	<0.6

particles have elongated horizontal morphologies with random particle indentation, indicating little or no temperature gradient. On the other hand, in the case of LACS, the sprayed particles have indentations on their top side, indicating that a relatively cold, hard particle impacted a relatively hot particle.

The same set-up of LACS (Fig. 7.4) was used by Lupoi et al. (2011) for the formation of titanium coatings onto steel round bars. In order to achieve the formation of coatings onto round bars, the cold-spray gun, the diode laser head, and the pyrometer were stationary on a single mount, whilst the round bar was moved using a CNC X–Y system. Titanium coatings of 4 mm thickness were deposited on carbon steel tube. After the deposition, the coatings were machined on a lathe in order to reduce the coating's thickness at 3 mm. The LACS coatings presented machinability and ductility, as no cracks or detachment of coating were observed during the lathe operation. It should be mentioned that LACSprayed titanium coatings presented adhesion strength (pulloff test, ~77 MPa) almost four times higher than the adhesion strength of conventional cold-sprayed titanium coatings.

7.3.2.2 Laser Ablation

The LACS system has also been developed in combination with laser heads which provoke the ablation of the substrate. According to Garcia-Alonso et al. (2011), the laser wavelength and the pulse time duration are the key parameters of the ablation mechanism. Principally, the ablation results either from the thermal effect due to IR low-energy photons or from the photonic effect due to UV high-energy photons (Garcia-Alonso et al. 2011). Laser surface cleaning and laser ablation imply the removal of contaminants (oxides, oils, etc.) or matter, respectively, by a transition from their solid state to dispersed phases (Garcia-Alonso et al. 2011).

Aluminium-Based Coatings

In 2009, a pulsed Nd-YAG laser (PROTAL®, Quantel Lannion, France) was coupled with the cold-spray gun to result in the laser beam passing milliseconds prior to the cold-spray jet for deposition, (Christoulis et al. 2009, 2010). Pulsed Nd-YAG laser was operated only during the first pass in order to clean the substrate surface. By using the experimental set-up of Fig. 7.5, aluminium was sprayed onto an Al-based alloy (AISI 2017). The surface preparations and sprayings experiments were carried out in the McGill Aerospace Materials & Alloy Development Centre (MAMADC)

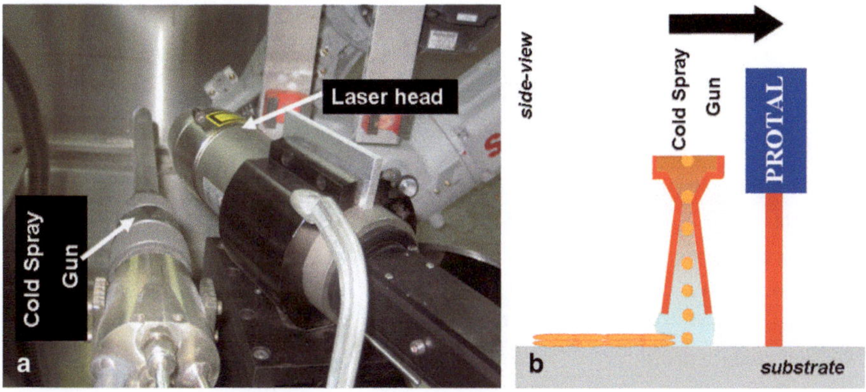

Fig. 7.5 **a** Experimental set-up, **b** side view of experimental set-up. (Christoulis et al. 2009)

cold spray laboratory located at the Industrial Materials Institute of the National Research Council of Canada.

In 2013, Olakanmi et al. (2013) also coupled Nd-YAG (ROFIN DY 044, Hamburg, Germany) with the cold-spray gun in order to spray Al-12Si powder onto grit-blasted stainless steel (304 L) substrates. This LACS equipment is located at the National Laser Centre/Council for Scientific and Industrial Research (NLC/CSIR), Pretoria, South Africa.

The main difference between the two set-ups, is that the Nd-YAG laser of Fig. 7.5 operates in pulsed mode while the Nd-YAG laser of Olakanmi's set-up operates on continuous mode.

The experimental conditions for both Al and Al-12Si powders are presented in Table 7.4.

For the set-up of Fig. 7.5, it is noted that the spraying conditions were selected by measuring the particles' mean velocity for various standoff distances. The particle velocity was measured by using the ColdSprayMeter® (Tecnar Automation Inc., St-Bruno, QC, Canada; Christoulis et al. 2009; Jeandin et al. 2010). In this case, the aluminium coatings were formed onto:

- Nd-YAG laser-pretreated substrates
- As-received substrates
- Grit-blasted substrates
- Mirror-polished substrates

The Nd-YAG laser was proceeding of the cold spray gun. The scanning of the substrate was realized in a specific way (Fig. 7.6) for all the substrates (conventionally prepared and laser-irradiated substrates), since experiments indicate that substrate temperature increases during spraying due to the heated propelling gas (Irissou et al. 2008).

Two different laser energy densities were examined: 1.0 and 2.2 J cm^{-2}. Depending on the laser energy density, the size of the laser spot was changed (Fig. 7.6b).

Table 7.4 Spraying and laser parameters

	Set-up of Christoulis et al. (2009)	Set-up of Olakanmi et al. (2013)
Materials		
Powder	Al	Al-12Si
Powder's granulometry	17–35 μm	45–90 μm
Substrate	AISI 2017	AISI 304 L
Spraying conditions		
Gas pressure (MPa)	3.0	1.25
Gas temperature (°C)	350	Ambient
Standoff distance (mm)	20	50
Gun traverse speed (mm/s)	100	10
Nozzle characteristics		
Type of the nozzle	PBI-33	DLV-180
Exit diameter (mm)	10	6
Throat diameter	2.7	2.0
Expansion ratio	13.7	9.0
Total length (mm)	220	210
Nd-YAG laser conditions		
Mode	Pulsed	Continuous
Wave length (μm)	1.064	1.06
Pulse frequency (Hz)	18.75, 37.5 and 150	–

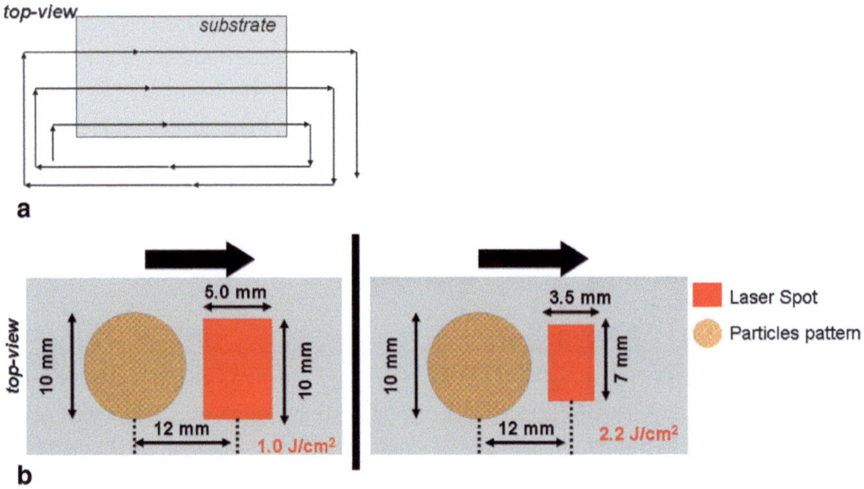

Fig. 7.6 **a** Movement geometry of coupled cold-spray gun-laser head and **b** top view of the pattern of both cold-sprayed particles and laser spot

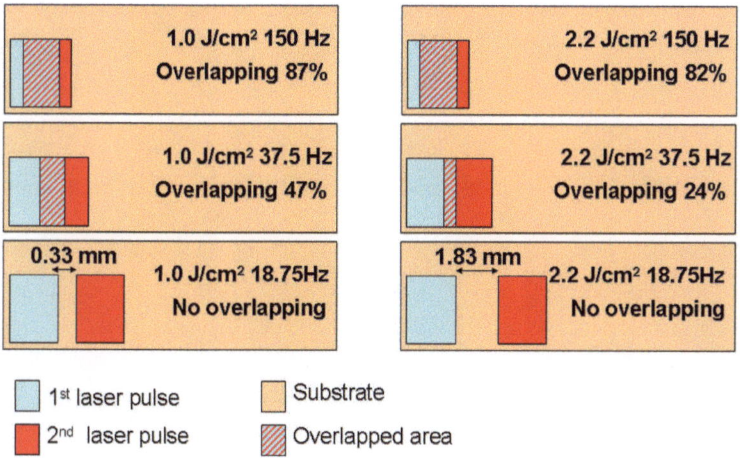

Fig. 7.7 Overlapping for two sequential Nd-YAG laser pulses

By taking into account the frequency of the laser, the size of the spot, and the movement parameters of the gun (Table 7.4), overlapping percentage was calculated. The overlapped area for two sequential laser pulses for different laser energy and frequency is shown in Fig. 7.7. In the case of lower laser frequency (18.75 Hz), inhomogeneous (treated and untreated) areas were created on the substrate (Fig. 7.7).

Thick coatings were deposited on the substrates for all the cases. The mean thickness and the percentage of cracked interfaces were determined by means of image analysis software (ImageJ Version 1.38x, Wayne Rasband, National Institute of Health, USA). The results are presented in Table 7.5 while representative images of cross-sections of the coatings are presented in Fig. 7.8.

Coatings' mean thicknesses were almost the same for conventionally prepared substrates and for low-energy (1.0 J cm^{-2}) irradiated substrates (Table 7.5). When the laser energy density was increased from 1.0 to 2.2 J cm^{-2}, the mean thickness

Table 7.5 Mean thickness of aluminium coatings

Substrate pretreatment	Mean thickness (μm)	Average deviation (μm)
As-received	340	±17
Mirror-polished	305	±19
Grit-blasted	324	±6
Nd-YAG laser (1.0 J cm^{-2}, 18.75 Hz)	300	±17
Nd-YAG laser (1.0 J cm^{-2}, 37.5 Hz)	310	±17
Nd-YAG laser (1.0 J cm^{-2}, 150 Hz)	323	±16
Nd-YAG laser (2.2 J cm^{-2}, 18.75 Hz)	304	±4
Nd-YAG laser (2.2 J cm^{-2}, 37.5 Hz)	393	±10
Nd-YAG laser (2.2 J cm^{-2}, 150 Hz)	410	±13

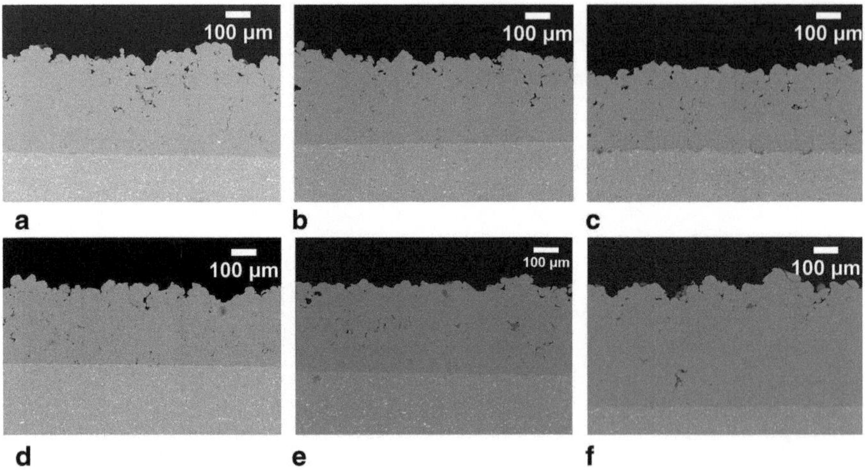

Fig. 7.8 Scanning electron microscope (SEM) images of cross-sections of cold-sprayed Al coatings formed onto **a** as-received, **b** mirror-polished substrate, **c** grit-blasted substrate, **d** laser-irradiated substrate (1.0 J cm^{-2}, 37.5 Hz), **e** laser-irradiated substrate (2.2 J cm^{-2}, 37.5 Hz), **f** laser-irradiated substrate (2.2 J cm^{-2}, 150 Hz)

of the coatings was also increased significantly µm. A similar behaviour was also observed by Olakanmi et al. (2013), where a continuous-mode Nd-YAG laser was used. It was found that the mean coating thickness of LACSsprayed Al-12S was increased from 48 µm (Fig. 7.9) to 847 µm (Fig. 7.9) as the laser power was increased from 1.0 to 3.5 kW.

Nd-YAG laser ablation also promoted a better interface with much less interfacial cracks for both pulsed-mode laser (Christoulis et al. 2009; Jeandin et al. 2010) and continuous-mode laser (Olakanmi et al. 2013). For the continuous Nd-YAG laser, at the highest laser power, it was found that that the coating was coherently

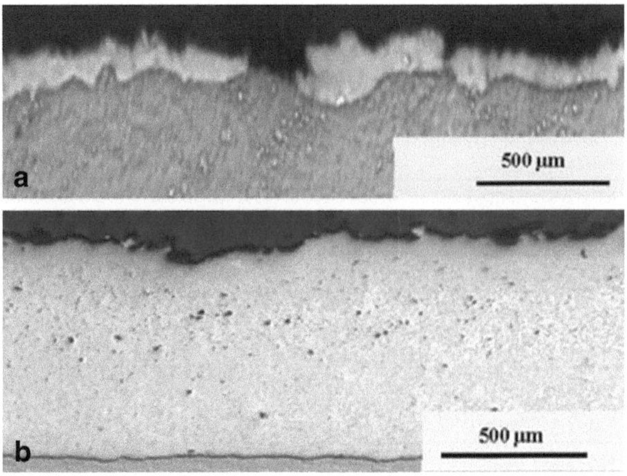

Fig. 7.9 Cross-sections of LACSpraed Al-12Si coatings. Laser power **a** 1.0 kW, **b** 3.5 kW. (Olakanmi et al. 2013)

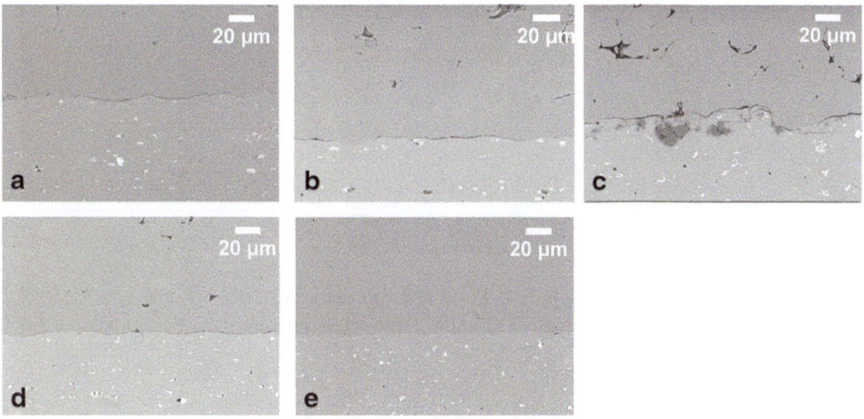

Fig. 7.10 Scanning electron microscope (SEM) images of cross-sections of cold-sprayed Al coatings formed onto **a** as-received substrate, **b** mirror-polished substrates, **c** grit-blasted substrate, **d** laser-irradiated substrate (1.0 J cm^{-2}, 150 Hz), **e** laser-irradiated substrate (2.2 J cm^{-2}, 150 Hz)

bonded at its interface with the substrate without the existence of cracks and porosities at the interface. Also, in the case of pulsed Nd-YAG laser, for the highest laser energy density (2.2 J cm^{-2}) and under the highest frequency (150 Hz), cracks could not be observed (Fig. 7.10; Jeandin et al. 2010).

The beneficial effect of the pulsed Nd-YAG laser ablation was also evaluated by transmission electron microscope observations. A typical oxide layer of about 100 nm in thickness could be observed at the coating–substrate interface in cold-sprayed "as-received" Al 2017 (Fig. 7.11). The energy-dispersive X-ray imaging (EDX) indicates that this oxide layer is in fact divided in two regions of distinct Al/O ratio (in gray in Fig. 7.11b). The layer at the substrate side has the stoichiometry of alumina indicating that it is the native oxide while the other one at the coating side is richer in oxygen (~35 wt.% Al, 65 wt.% O)

In contrast, for laser-processed Al 2017, no oxygen could be detected at the interface (Fig. 7.12) neither on scanning transmission electron microscope (STEM) images nor by EDX profile. This profile was obtained with a probe size of 1 nm

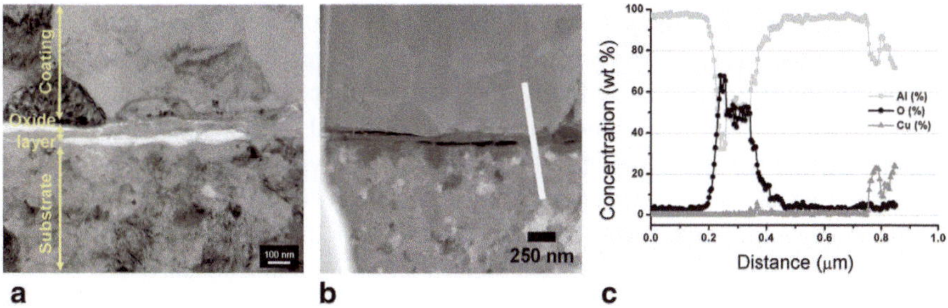

Fig. 7.11 TEM pictures of cold-sprayed Al on as-received AISI 2017 substrate. **a** Bright field (BF) image, **b** HAADF (high-angle annular dark field) image, **c** EDX profiles along the *white line*

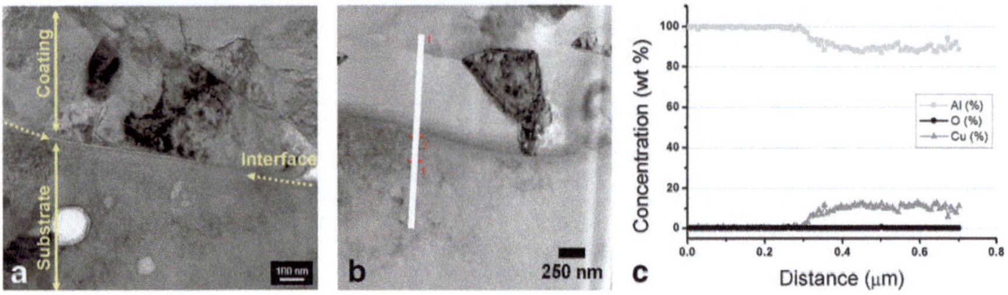

Fig. 7.12 TEM pictures of cold-sprayed Al on laser pretreated (2.2 J cm^{-2}, 150 Hz) AISI 2017 substrate. **a** Bright field (BF) image, **b** HAADF (high-angle annular dark field) image, **c** EDX profiles along the *white line*

(enlarged to 3 nm at the exit side of the thin foil), 10 nm between the consecutive analysis spots, and a limit of detection of 1 wt.% for O. It can therefore be inferred that the native layer was removed by the laser treatment, and if an oxide thinner layer was formed prior to the particle reached the substrate, the thickness of this layer would not exceed a few nanometres.

Finally, it should be noted that in the case of the grit-blasted substrates, on the one hand, the grit blasting contributed to the increase of the mean thickness of the coatings compared to the other conventional pretreatment. However, on the other hand, the grit-blasting method increased the percentage of cracked interface, and alumina particles entrapped onto the substrate were also seen (Fig. 7.10c).

Nickel-Based Coatings

Spherical powder of Ni-20Cr (Höganäs, 1616–09/PS) with particle size ranging from 20 to 53 μm was sprayed onto Inconel alloy 718 substrate by employing the set-up of Fig. 7.6 (Jeandin et al. 2010). The Ni-20Cr powder was sprayed onto:

- Nd-YAG laser pretreated substrates
- As-received substrates
- Grit-blasted substrates
- Mirror-polished substrates

Ni-20Cr powder was sprayed with KINETICS® 3000-M System (CGT-GmbH, Ampfing, Germany) and by using nitrogen as process gas. The spraying conditions are presented in Table 7.6, and they are selected after the measurements of particles' in-flight velocity. The so-called "MOC" (method of characteristics) circular standard nozzle of CGT-GmbH was used for the spraying experiments. The nozzle MOC has an inner diameter of 6.6 mm, an expansion ratio of 6.0, and a total length of 175 mm. Experiments were carried out in the Industrial Materials Institute (Boucherville, Quebec) of the National Research Council of Canada.

The laser conditions and the laser movement were the same with the case of LACSprayed aluminium powder (Table 7.4, and Fig. 7.6).

Table 7.6 Spraying conditions of Ni-20Cr

Process gas	100 % N_2
Gas pressure (MPa)	3.0
Gas temperature (°C)	660
Standoff distance (mm)	40
Nozzle traverse speed (mm/s)	100
Nozzle step (mm)	2
Passes	2

The mean thickness of the coatings was calculated via the cross-sections of the coatings. The mean thickness of the coatings was calculated by observing 12 scanning electron microscope (SEM) images (Fig. 7.13) in standard magnification of ×200 and by using image analysis software (ImageJ Version 1.38x, Wayne Rasband, National Institute of Health, USA). The mean thickness of the coatings for the different pretreatment methods of the substrate is presented in Table 7.7. Based on these measurements, it seems that the optimum ablation conditions for LAC-Sprayed Ni-20Cr coatings were: laser energy of 2.2 J cm^{-2} and laser frequency of 37.5 Hz. Under these laser conditions, it was also found (Jeandin et al. 2010) that the coating–substrate interface presented much less interfacial cracks compared to the case of as-received substrates or to conventional pretreated substrates (grit-blasted substrates and mirror-polished substrates). Figure 7.14 shows several interfaces between Ni-20Cr and Inconel 718 substrate, where it is obvious that in the case of grit-blasted substrates (Fig. 7.14c), alumina particles have provoked the contamination of the Inconel substrate.

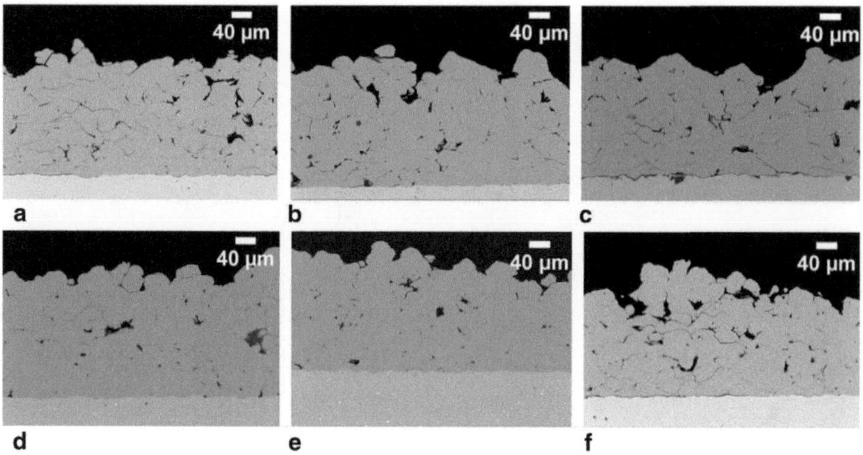

Fig. 7.13 Scanning electron microscope (SEM) images of cross-sections of Ni-20Cr cold-sprayed coatings formed onto **a** as-received substrate, **b** mirror-polished substrate, **c** grit-blasted substrate, **d** laser-irradiated substrate (1.0 J cm^{-2}, 37.5 Hz), **e** laser-irradiated substrate (2.2 J cm^{-2}, 37.5 Hz), **f** laser-irradiated substrate (2.2 J cm^{-2}, 150 Hz)

Table 7.7 Mean thickness of aluminium coatings

Substrate pretreatment	Mean thickness (μm)	Average deviation (μm)
As-received	225	±25
Mirror-polished	210	±10
Grit-blasted	196	±26
Nd-YAG laser (1.0 J cm^{-2}, 18.75 Hz)	228	±18
Nd-YAG laser (1.0 J cm^{-2}, 37.5 Hz)	227	±21
Nd-YAG laser (1. 0 J cm^{-2}, 150 Hz)	196	±26
Nd-YAG laser (2. 2J J cm^{-2}, 18.75 Hz)	218	±21
Nd-YAG laser (2. 2 J cm^{-2}, 37.5 Hz)	230	±21
Nd-YAG laser (2.2 J cm^{-2}, 150 Hz)	195	±22

The further increase of the laser frequency at 150 Hz for the highest laser energy (2.2 J cm^{-2}) provoked a decrease of the mean thickness of the coating and significant increase of the interfacial cracks (Jeandin et al. 2010). It seems that the increase of the laser frequency resulted in extensive melting of the substrate, which in turn could increase the pores and the cracks due to the change of the interaction between the sprayed particles and the substrate (Christoulis et al. 2012).

7.3.2.3 Combination of Laser Heating and Ablation

In 2010, Danlos et al. (2011) coupled the cold spray gun of a KINETICS® 3000 System (CGT-GmbH, Ampfing, Germany) with two type of lasers: an ablation laser

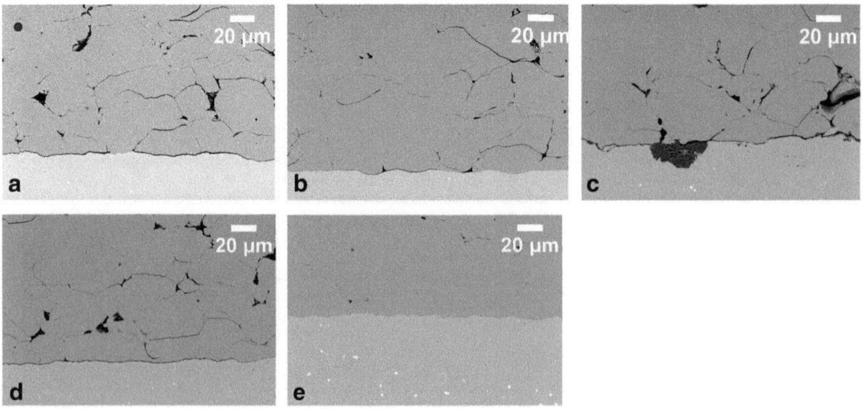

Fig. 7.14 Scanning electron microscope (SEM) images of cross-sections of cold-sprayed Al coatings formed onto **a** as-received substrate, **b** mirror-polished substrates, **c** grit-blasted substrate, **d** laser-irradiated substrate (1.0 J cm^{-2}, 18.75 Hz), **e** laser-irradiated substrate (2.2 J cm^{-2}, 37.5 Hz)

which was used to eliminate adsorbed pollutant molecules and a heating laser which was used to preheat the substrate and thus to improve the contact between substrate and sprayed particles.

More precisely, for the ablation of the substrates, a Q-switched Nd-YAG laser (wavelength 1064 nm, pulse duration 10 ns) of rectangular shape (6.5×3.6 mm^2) was used. The energy density of this laser was at 2.3 J cm^{-2}. On the other hand, the heating of the substrate was achieved by a second pulsed Nd-YAG laser (Cheval, Pirey, France; Danlos et al. 2011). The heating laser was a millisecond laser with a wavelength of 1064 nm. The laser beam has a circular shape (diameter: 10 mm) with a Gaussian energy distribution. The pulse duration was at 2 ms while the energy density was at 29.7 J cm^{-2}.

It should be noted that for both lasers, the pulse repetition rate was set at 60 Hz in order to allow the synchronization of the two lasers and by this way to control the laser treatment. However, the overlapping between two impulsions was different for the two lasers. For the ablation laser, an overlapping of 75 % was obtained, and for the heating laser, an overlapping of 85 % was obtained. The overlapping is different due to the different sizes of the laser spots.

By using this set-up, Danlos et al. (2011) sprayed aluminium 6061 powder (18.5–74.7 μm) onto four different pretreated AISI 2017 aluminium substrates:

- Substrates just degreased before the spraying.
- Grit-blasted substrates.
- Laser-ablated substrates. Only the ablation laser was operated.
- Substrates pretreated by both lasers: heating laser and ablation laser.

The spraying conditions of aluminium powder are presented in Table 7.8.

The pretreatment method did not have an effect on the thickness of the coatings. For all the cases, the mean thickness of the coatings was 450 μm. However, the pretreatment method had a significant effect on the adhesion of the coatings. The adhesion was measured according to ASTM C633–79 method. It was observed that the coating's adhesion on the degreased substrate was 28.09 MPa, and it was increased at 36.1 MPa in the case of grit-blasted substrates. Further improvement of the adhesion of coatings took place with the laser pretreatment of the substrate.

Table 7.8 Spraying conditions for the LACS set-ups of Danlos et al. (2011) and Perton et al. (2012)

	Aluminium powder (Danlos et al. 2011)	Ti6Al4V powder (Perton et al. 2012)
Process gas	100 % air	100 % N_2
Gas pressure (MPa)	2.8	4.0
Gas temperature (°C)	350	800
Standoff distance (mm)	20	40
Nozzle traverse speed (mm/s)	100	330
Nozzle step (mm)	2	2
Passes	2	3

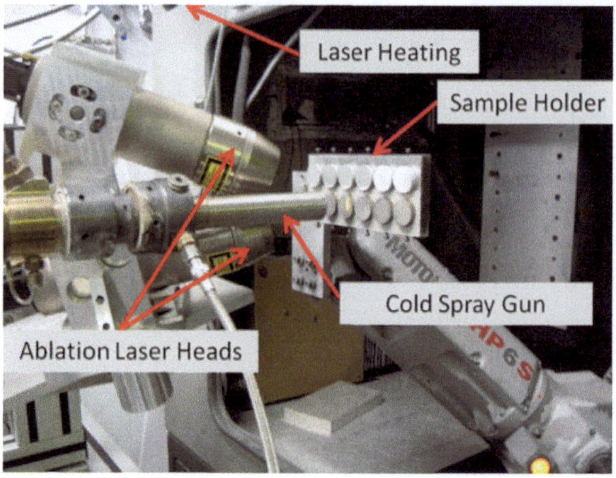

Fig. 7.15 LACS set-up composed by a gold spray gun, a heating laser, and two heads of ablation laser. (Perton et al. 2012)

Coating's adhesion was measured at 51.2 MPa, on laser-ablated substrate (only the Nd-YAG ablation laser was operated), and it was increased at 64.99 MPa in the case of the substrates which have been pretreated by both lasers.

Except from the improvement of the adhesion, the coating–substrate interface was slit and clean when the substrates were pretreated by the lasers, while in the case of grit-blasted substrates, grit inclusions, which can be harmful for the coating, were observed.

Perton et al. (2012) also combined two kinds of lasers with a cold spray technology. The set-up of Perton et al. (2012) is presented in Fig. 7.15, and it was consisted by a pulsed Nd-YAG ablation laser (PROTAL®) and an Nd-YAG continuous heating laser (CW 020 from Rofin Sinar, Hamburg, Germany). The main difference between the LACS system of Danlos et al. (2011) and Perton et al. (2012) is the heating lasers which were used; pulsed-mode laser and continuous-mode laser, respectively.

The laser beam of the heating laser had a circular shape (10 mm diameter) and two laser power levels were used: 750 and 1650 W. By using an infrared camera, it was found that for the laser power of 750 W, the surface temperature of the substrate was about 95°C. As soon as the laser power increased at 1650 W, the substrate temperature was also increased at 175°C.

The ablation of the substrates was carried out by four Q-switched Nd:YAG lasers which were combined in two laser heads. The laser beam of the ablation laser had rectangular shape ($4 \times 13.5\ mm^2$). The pulse duration of the laser beam was 10 ns and the frequency has been set at 150 Hz. Two pulse fluencies conditions were used: 1.3 and 2.2 J cm^{-2}.

Ti6Al4V powder (mean particle size of 30 μm) was sprayed by employing the LACS system of Fig. 7.15. The spraying parameters are presented in Table 7.8, and it is noted that the laser ablation was maintained throughout the build-up of the whole coating, for all the three passes.

The nozzle of the cold spray gun had a 5.3 mm diameter round outlet, an expansion ratio of 3.94 mm, and a divergent section of 120 mm. In Perton's set-up, the cold spray gun with the laser heads was stationary and the substrates were moved. The relative movement of LACS-substrates was accomplished in the same geometry of Fig. 7.6.

The Ti6Al4V coatings were formed onto Ti6Al4V substrates. As it is presented in Table 7.9, the substrates have been prepared under six different ways resulting in various levels of mean arithmetic roughness R_a.

The adhesion strength of the Ti6Al4V coatings was measured by using laser shock adhesion test (LASAT). The results are presented in Table 7.9. LASAT consists in irradiating the rear surface of a substrate with a laser to generate a shock wave. This shock wave propagates and reflects in the material into a release wave crossing the unloading wave, leading to tensile stresses. The principle of LASAT method has been described in detail in several articles (Barradas et al. 2005; Boustie et al. 2000; Bolis et al. 2007)

When the Ti6Al4V is sprayed without laser pretreatment (neither heating laser nor ablation laser), the higher adhesion strength was found on the mirror polished substrates (Table 7.9). In the case of mirror-polished substrate, the laser pretreatment of the substrate by the ablation laser only, provoked a slight increase of the adhesion strength.

The effect of laser treatment of the substrates by both lasers (heating and ablation) was examined in the case of Ti6Al4V which have been grounded with SiC papers (grit 400). In this case, the highest adhesion strength (~910 MPa Table 7.9) was found when the heating laser was operated at 1650 W and the laser energy density of the ablation laser was set at 1.3 J cm^{-2}.

7.4 Concluding Remarks

The LACS process can contribute to the improvement of cold-sprayed coatings in significant level, which in turn can expand the applications of cold-sprayed coatings.

Experimental studies have proved that the LACSprayed coatings presented advanced properties compared to the properties of conventional cold-sprayed coatings. The major improvements appear to the adhesion strength as well as to the cohesion strength of the coatings.

In order to improve the adhesion strength of conventional cold sprayed, several pretreatment methods are performed before applying the coating: grit blasting, high-pressure waterjet, machining, etc. Among these methods, the most common procedure is grit blasting. However, often time the grit blasting procedure results in the inclusion of small grit particles at the interface between the coating and substrate. This limits its use when fatigue properties are important. In addition, grit blasting produces waste and can be harmful to the operator as discussed earlier.

Table 7.9 Adhesion strength of LACSprayed Ti6AlV. (Perton et al. 2012)

		Mirror polished	Ground with 800 grit	Ground with 400 grit	As-machined	Grit blasted with alumina (grit 100)	Grit blasted with alumina (grit 24)
R_a (μm)		0.05	0.12	0.22	2.56	3.21	5.53
Adhesion strength (MPa)							
No laser pretreatment		900	613	~420 to ~600	70	242	371
Only ablation laser	1.3 J cm^{-2}			~500			
	2.2 J cm^{-2}	~920		~300	Debonding		
Only heating laser	750 W			~425			
	1650 W			~590			
Ablation + heating laser	1.3 J cm^{-2}and 750 W			~750			
	1.3 J cm^{-2}and 1650 W			~910			
	1.3 J cm^{-2}and 750 W			~425			
	1.3 J cm^{-2}and 750 W			~825			

In cold-sprayed coatings, thermal posttreatment method should be applied to enhance their cohesive strength. This adds another step in the production of a protective or functional coating but more importantly, in some cases, this procedure would hinder one of the advantages of the cold spray process which is to preserve the structure characteristics of the powder (i.e. nanostructure, phases, etc.).

So, the techniques for the improvement of adhesion and cohesion strengths of cold-sprayed coatings are a two-, three-, or even four-step process.

On the other hand, the LACS process permits the creation of dense coatings with high adhesion and cohesion strength in a mere one-step process. Furthermore, experiments have shown that further properties (porosity, hardness, etc.) of the coatings are also improved. Last but not least, the LACS process presents higher deposition efficiency compared to the conventional cold spray process.

References

Bahbou, M. F., et al. 2004. Effect of grit blasting and spraying angle on the adhesion strength of a plasma-sprayed coating. *Journal of Thermal Spray Technology* 6 (2): 508–514.

Barbezat, G. 2005. Advanced thermal spray technology and coating for lightweight engine blocks for the automotive industry. *Surface and Coatings Technology* 200:1990–1993.

Barbezat, G. 2006. Application of thermal spraying in the automobile industry. *Surface and Coatings Technology* 201:2028–2031.

Barradas, S., et al. 2005. Application of laser shock adhesion testing to the study of the interlamellar strength and coating–substrate adhesion in cold-sprayed copper coating of aluminum. *Surface and Coatings Technology* 197:18–27.

Barradas, S., et al. (2007) Laser shock flier impact simulation of particle-substrate interactions in cold spray. *Journal of Thermal Spray Technology* 16 (4): 548–556.

Bégué, G., et al. 2013. LAser Shock Adhesion Test (LASAT) of EB-PVD TBCs: towards an industrial application. *Surface and Coatings Technology* 237:305–312.

Bolis, C., et al. 2007. Physical approach to adhesion testing using laser-driven shock waves. *Journal of Physics D: Applied Physics* 40:3155–3163.

Boustie, M., E. Auroux, and J. P. Romain. 2000. Application of the laser spallation technique to the measurement of the adhesion strength of tungsten carbide coatings on superalloy substrates. *The European Physical Journal Applied Physics* 12:47–53.

Bray, M., et al. 2009. The laser-assisted cold spray process and deposit characterization. *Surface and Coatings Technology* 203:2851–2857.

Christoulis, D. K., et al. 2009. *Cold spraying combined to laser surface pre-treatment using PROTAL®*. Thermal Spray 2009: Proceedings of the International Thermal Spray Conference, ed. B. R. Marple et al., pp. 1151–1156.

Christoulis, D. K., et al. 2010. Cold-spraying coupled to nano-pulsed Nd-YaG laser surface pretreatment. *Journal of Thermal Spray Technology* 19 (5): 1062–1073.

Christoulis, D. K., et al. 2012. Laser-assisted cold spray (LACS). In *Nd YAG laser*, ed. Dr. Dan C. Dumitras, InTech Europe, Croatia, ISBN: 978–953-51–0105-5, InTech. doi:10.5772/36104. http://www.intechopen.com/books/Nd-YAG-laser/laser-assisted-cold-spray-lacs-.

Coddet, C. 2006. On the use of auxiliary systems during thermal spraying. *Surface and Coatings Technology* 201:1969–1974.

Coddet, C., et al. 1999. Surface preparation and thermal spray in a single step: The PROTAL process—example of application for an aluminum-base substrate. *Journal of Thermal Spray Technology* 8 (2): 235–242.

Costil, S., et al. 2004a. Role of laser surface activation during plasma spray coating of metallic materials. Proceedings of 18th International Conference on Surface Modification Technologies, ISBN 0871708337, Dijon, France, November 2004.
Costil, S., et al. 2004b. New developments in the PROTAL® process. Proceedings of Thermal Spray 2004: Advances in Technology and Application, ISBN 0-87170-809-4, Osaka, Japan, May 2004.
Costil, S., et al. 2005. Influence of surface laser cleaning combined with substrate preheating on the splat morphology. *Journal of Thermal Spray Technology* 14 (1): 31–38.
Danlos, Y., et al. 2008. Combining effects of ablation laser and laser preheating on metallic substrates before thermal spraying. *Surface and Coatings Technology* 202 (18): 4531–4537.
Danlos, Y., et al. 2011. Influence of Ti-6Al-4V and Al 2017 substrate morphology on Ni-Al coating adhesion—impacts of laser treatments. *Surface and Coatings Technology* 205 (8/9): 2702–2708.
Davis, J. 2004. *Handbook of thermal spray technology*. Materials Park: ASM International (ISBN 0-87170-795-0).
Ernst, P., and B. Distler. 2012. Optimizing the cylinder running surface/piston system of internal combustion engines towards lower emissions. SAE Technical Paper 2012-32-0092. doi:10.4271/2012–32-0092.
Fabre, G., et al. 2011. Laser shock adhesion test (LASAT) of electron beam physical vapor deposited thermal barrier coatings (EB-PVD TBCs). *Advanced Materials Research* 278:509–514.
Garcia-Alonso, D., et al. 2011. Pre-/during-/post-laser processes to enhance the adhesion and mechanical properties of thermal-sprayed coatings with a reduced environmental impact. *Journal of Thermal Spray Technology* 20 (11): 719–735.
Gonzalez-Hermosilla, W. A., et al. 2010. Effect of substrate roughness on the fatigue behavior of a SAE 1045 steel coated with a WC-10Co-4Cr cermet, deposited by HVOF thermal spray. *Materials Science and Engineering A* 527:6551–6561.
Guetta, S., et al. 2009. Influence of particle velocity on adhesion of cold-sprayed splats. *Journal of thermal spray technology* 18 (3): 331–342.
Guipont, V., et al. 2010. Bond strength determination of hydroxyapatite coatings on Ti-6Al-4V substrates using the LAser Shock Adhesion Test (LASAT). *Journal of Biomedical Materials Research Part A* 95 (4): 1096–1104.
Hussain, T. 2013. Cold spraying of titanium: A review of bonding mechanisms, microstructure and properties. *Key Engineering Materials* 33: 53–90.
Ichikawa, Y., et al. 2007. Evaluation of adhesive strength of thermal-sprayed hydroxyapatite coating using the LAser Shock Adhesion Test (LASAT). *Materials Transactions* 48 (4): 793–798.
Irissou, E., et al. 2008. How cold is cold spray? An experimental study of the heat transfer to the substrate in cold gas dynamic spraying. Proceedings of Thermal Spray 2008: Thermal Spray Crossing Borders, Maastricht, The Netherlands, June 2008. ISBN: 978-3-87155-979-2.
Jeandin, M., et al. 2003. Thermal spray and lasers. Proceeding of the 2nd International Conference on Materials Processing for Properties and Performance (MP3), Yokohama, Japan, October 2003.
Jeandin, M., et al. 2010. Lasers and thermal spray. *Materials Science Forum* 638–642:171–184.
Jiang, X. P., et al. 2006. Enhancement of fatigue and corrosion properties of pure Ti by sandblasting. *Materials Science and Engineering: A* 429 (1/2): 30–35.
Kulmala, M., and P. Vuoristo. 2008. Influence of process conditions in laser-assisted low pressure cold spraying. *Surface and Coatings Technology* 202:4503–4508.
Lamraoui, A., et al. 2010. Laser surface texturing LST treatment before thermal spraying—a new process to improve the substrate-coating adherence. *Surface and Coatings Technology* 205:S164–S167.
Leinenbach, C., and D. Eifler. 2006. Fatigue and cyclic deformation behaviour of surface-modified titanium alloys in simulated physiological media. *Biomaterials* 27 (8): 1200–1208.
Li, H., et al. 2006. Surface modifications induced by nanosecond pulsed NdYAG laser irradiation of metallic substrates. *Surface and Coatings Technology* 201:1383–1392.
Li, C., et al. 2010. Laser surface remelting of plasma-sprayed nanostructured Al_2O_3–13 wt%TiO2, coatings on magnesium alloy. *Journal of Alloys and Compound* 503:127–132.

Lupoi, R., et al. 2011. High speed titanium coatings by supersonic laser deposition. *Materials Letters* 65:3205–3207.

Makinen, H., et al. 2007. Adhesion of cold sprayed coatings: Effect of powder, substrate and heat treatment. Proceedings of Thermal Spray 2007: Global Coating Solutions, ISBN 0-87170-809-4, Beijing, China, May 2007.

Marrocco, T., et al. 2011. Corrosion performance of laser posttreated cold sprayed titanium coatings. *Journal of Thermal Spray Technology* 20 (4): 909–917.

Maruyama, T., et al. 2007. Effect of the blasting angle on the amount of the residual grit on blasted substrates. In *Thermal spray 2007: global coating solutions*, eds. B. R. Marple, M. M. Hyland, Y.-C. Lau, C.-J. Li, R. S. Lima, and G. Montavon. Materials Park: ASM International® (Copyright© 2007).

Mohammadi, Z., et al. 2007. Grit blasting of Ti-6Al-4 V alloy_Optimization and its effect on adhesion strength of plasma-sprayed hydroxyapatite coatings. *Journal of Materials Processing Technology* 194:15–23.

Multigner, M., et al. 2009a. Influence of the sandblasting on the subsurface microstructure of 316LVM stainless steel: Implications on the magnetic and mechanical properties. *Materials Science and Engineering: C* 29 (4): 1357–1360.

Multigner, M., et al. 2009b. Interrogations on the sub-surface strain hardening of grit blasted Ti -6Al-4 V alloy. *Surface and Coatings Technology* 203 (14): 2036–2040.

Olakanmi, E., and M. Doyoyo. 2014. Laser-assisted cold-sprayed corrosion and wear-resistant coatings: A review. *Journal of Thermal Spray Technology* 23 (5): 765–785.

Olakanmi, E. O., et al. 2013. Deposition mechanism and microstructure of laser-assisted cold-sprayed (LACS) Al–12 wt.%Si coatings: Effects of laser power. *Journal of Materials* 65 (6): 776–783.

Paredes, R. S. C., et al. 2006. The effect of roughness and pre-heating of the substrate on the morphology of aluminium coatings deposited by thermal spraying. *Surface and Coatings Technology* 200:3049–3055.

Pawlowski, L. 2008. *The science and engineering of thermal spray coatings*. 2nd ed. Hoboken: Wiley (ISBN: 978-0-471-49049-4).

Perton, M., et al. 2012. Effect of pulsed laser ablation and continuous laser heating on the adhesion and cohesion of cold sprayed Ti-6Al-4 V coatings. *Journal of Thermal Spray Technology* 21 (6): 1322–1333.

Petavratzi, E., et al. 2005. Particulates from mining operations: A review of sources, effects and regulations. *Minerals Engineering* 18 (12): 1183–1199.

Pokhmurska, H., et al. 2008. Post-treatment of thermal spray coatings on magnesium. *Surface and Coatings Technology* 202 (18): 4515–4524.

Rosales, M., and F. Camargo. 2009. Characterization of boron carbide thermal sprayed coatings for high wear resistance performance. Thermal Spray 2009: Proceedings of the International Thermal Spray Conference, 1175–1177.

Schlaefer, T. et al. 2008. Plasma transferred wire arc spraying of novel wire feedstock onto cylinder bore walls of AlSi engine blocks. Proceedings of Thermal Spray 2008: thermal Spray Crossing Borders, Maastricht, The Netherlands, June 2008. ISBN: 978-3-87155-979-2.

Sen, D., et al. 2010, Influence of grit blasting on the roughness and the bond strength of detonation sprayed coating. *Journal of Thermal Spray Technology* 19 (4): 805–815.

Sova, A., et al. 2013. Cold spray deposition of 316 L stainless steel coatings on aluminium surface with following laser post-treatment. *Surface and Coatings Technology* 235:283–289.

Staia, M. H., et al. 2000. Effect of substrate roughness induced by grit blasting upon adhesion of WC–17 % Co thermal sprayed coatings. *Thin Solid Films* 377–378:657–664.

Tam, A. C., et al. 1998. Laser cleaning of surface contaminants. *Applied Surface Science* 127–129:721–725.

Verdiera, M., et al. 2003. On the topographic and energetic surface modifications induced by laser treatment of metallic substrates before plasma spraying. *Applied Surface Science* 205 (1–4): 3–21.

Vilemova, M., et al. 2011. Effect of the grit blasting exposure time on the adhesion of Al_2O_3 and 316 L coatings. International Thermal Spray Conference & Exposition 2011.
Wang, Y., et al. 2010. Laser remelting of plasma sprayed nanostructured Al_2O_3-TiO_2 coatings at different laser power. *Surface and Coatings Technology* 204 (21/22): 3559–3566.
Wigren, J. 1998. Technical note: Grit blasting as surface preparation before plasma spraying. *Surface and Coatings Technology* 34 (1): 101–108.
Yang, H., et al. 2006. Influence of substrate roughness on adhesive strength. *International Thermal Spray Conference (2006): Building on 100 years of success*. Proceedings of the 2006 International Thermal Spray Conference, eds. B.R. Marple, M.M. Hyland, Y.C. Lau, R.S. Lima, and J. Voyer., Pub. ASM International, Materials Park, OH, USA, 15–18 May 2006, Seattle, Washington, CD-Rom. ISBN 0-87170-809-4.
Zieris, R., et al. 2003. Characterization of coatings deposited by laser-assisted atmospheric plasma spraying. Proceedings of Thermal Spray 2003: Advancing the Science & Applying the Technology, ISBN 0-87170-785-3, Orlando, United States of America, May 2003.
Zieris, R., et al. 2004. Investigation of AlSi coatings prepared by laser-assisted atmospheric plasma spraying of internal surfaces of tubes. Proceedings of Thermal Spray 2004: Advances in Technology and Application, ISBN 0-87170-809-4, Osaka, Japan, May 2004.

Chapter 8
Quality Assurance

L. Pouliot

8.1 Introduction

As emphasized in several chapters of this book, including Chap. 2, it is widely recognized that in the cold spray (CS) process, the particle velocity represents a key parameter governing the deposition process. This parameter relates specifically to the velocity at which spray particles are traveling at the time of impingement upon the substrate material. That topic is widely discussed in other numerous articles and books (Schmidt et al. 2009; Karthikeyan and Kay 2003; Tucker Jr. 2013; Champagne 2007; Karimi et al. 2013; Irissou et al. 2011). In practice, every material displays a material temperature-dependent "critical velocity" over which the material tends to stick to the substrate and form a coating. Below its critical velocity, the material most likely rebounces, becomes fugitive, and is collected as waste into the exhaust system.

Although particle velocity is the most critical parameter in CS, other factors do influence the ability to obtain a useful coating or deposition, namely cold sprayability. Some of these other factors include:

- Substrate surface preparation method (e.g., degreasing, grit blasting, other)
- Substrate temperature during spraying (Legoux et al. 2007)
- In-flight particle temperature
- Size distribution of the feedstock material
- Surface shape and texture of the particles
- Spatial distribution of the different particle sizes in the spray plume (after nozzle exit)
- Relative particle flow/flux

Many have emphasized on the importance of diagnostic tools to ensure reliability and repeatability in production processes that take advantage of the CS process.

L. Pouliot (✉)
Tecnar Automation Ltd., St-Bruno, QC, Canada
e-mail: lpouliot@tecnar.com

J. Villafuerte (ed.), *Modern Cold Spray*, DOI 10.1007/978-3-319-16772-5_8

Among other measurable process parameters, measuring the velocity of the particles for any given set of conditions has become a tool to establish a reference point or standard.

As an example, Dr. Eric Irissou from National Research Council (NRC), Canada, explains that:

> In the CS process, the particle velocity is the most important variable that influences its efficiency and the coating properties. Having the possibility to measure this parameter is INVALUABLE for conducting fundamental study… for establishing correlation between the velocity and the coating properties, for determining critical V and for validating models. Measuring particle velocity is also decisive when we optimize the process parameters in order to reach specific coating properties… it minimizes the number of samples that we need…. We also foresee the use of diagnostic tools in production as a process control. Most of the issues that we faced with the CS process translated in a decrease of particle velocity until the problem was fixed. Measuring particle velocity before and after spraying on a part therefore provides rapid confirmation that the process was or was not stable during all the coating procedure.

Similarly, Professor Bertrand Jodoin from the University of Ottawa explains:

> Sensors…allow us to validate numerical models (CFD) that can then be used to design cold spray nozzles that meet specific requirements in terms of particle velocity. Furthermore, they allow us to better understand the spray deposition windows of reactive materials to ensure that we avoid powder reaction during consolidation and also to maximize the reactivity of the consolidated powders. Particle velocity characterization also allows us to better evaluate the activation energy of these reactive materials and thus help us design new reactive materials.

In this chapter, we focus on practical methods of measuring particle velocity and particle size, including the main challenges associated with these methods. In particular, we explain the influence on the precision of the size measurement of the particles' shape and surface texture. The influence of the spatial distribution of the different particle sizes are briefly discussed. Finally, we demonstrate how relative flow measurements can be achieved.

8.2 Detecting Cold Particles

One of the main challenges for detection of CS particles is the fact that the particles are in effect cold, posing several challenges for their detection. That includes:

- First and foremost, any particle below 1000 °C would not emit enough infrared (IR) radiation to be "seen" or detected using Si or InGaAs-based photo detectors/cameras.
- Second, in the CS process, particles are traveling at very high speeds (typically between 350 and 1200 m/s), so the time available for characterization is very short.
- Third, particles can have very different shapes and surface textures.
- Last but not least, some particles are quite fine (in the 5–15 μm range);

8.2.1 Particle Detection Methods

Being able to "see" the particles is the first step towards particle characterization. In CS, particles exit the nozzle at a temperature ranging from 20 to about 750 °C. Unfortunately, in that temperature range, particles do not emit enough IR radiation to be seen or detected using conventional photo detectors. As a direct consequence, an illumination system must be used that shines enough light onto these particles. The scattered light is then collected by an optical system and transmitted to the detection module by means of an optical fiber. Two optical setups are typically used for light-scattering-based measurements, namely the "forward-" and "back-"scattering approaches, as illustrated in Fig. 8.1. Both setups offer advantages and drawbacks as listed in Table 8.1. In practice, the back-scattering approach is used most of the time, which avoids cumbersome and more fragile setups.

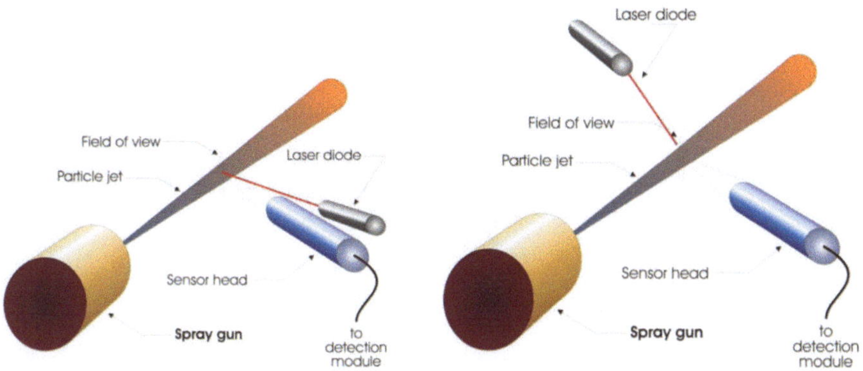

Fig. 8.1 Back-scattering versus forward-scattering setups. (Courtesy of TECNAR Automation Ltd.)

Table 8.1 Pros and cons of back scattering versus forward scattering

	Advantages	Drawbacks
Back scattering	Easy setup, not cumbersome Quite minimal clouding effect	Lower light gathering Higher sensitivity to particle surface finish, leading to lower precision on diameter measurement
Forward scattering	Much higher light gathering (Mie scattering theory) Less dependence on particle surface texture, leading to better precision on diameter measurement	More cumbersome sensor head and setup Loses detection effectiveness when particle density becomes high (clouding effect)

8.2.2 The Mie Scattering Theory

There exists a now well-known solution to Maxwell's equations called the "Mie solution" (Bohren and Huffman 1983), developed by German scientist Gustav Mie. It accurately describes the scattering of an electromagnetic plane wave by a homogeneous sphere. A polar representation of Mie's theory applied to a spherical 30-μm Ni particle illuminated with 900 nm monochromatic light is shown in Fig. 8.2. It clearly shows that much higher light gathering is expected with a forward-scattering approach (30 ° and less) than with a back one (between 150 and 180).

8.2.2.1 The Mie Effect in Cold Spray Particles

In the late 1990s, TECNAR and National Research Council (NRC) of Canada-jointly carried out experiments to confirm applicability of Mie scattering theory

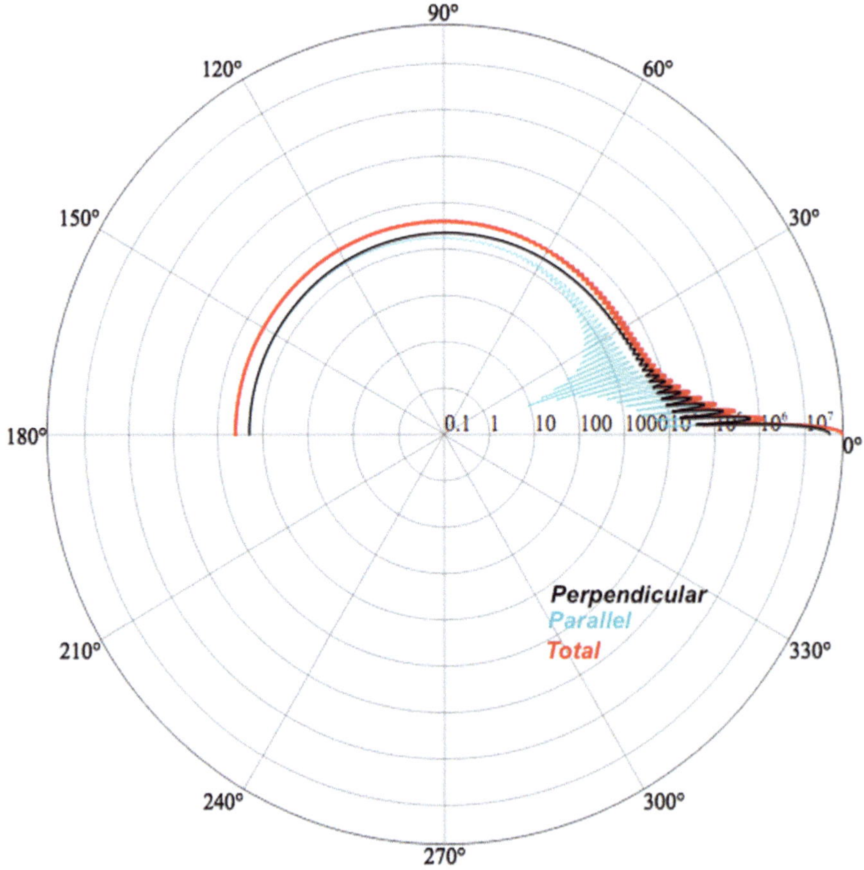

Fig. 8.2 Mie scattering polar diagram (30-μm Ni particle). (Courtesy of TECNAR Automation Ltd.)

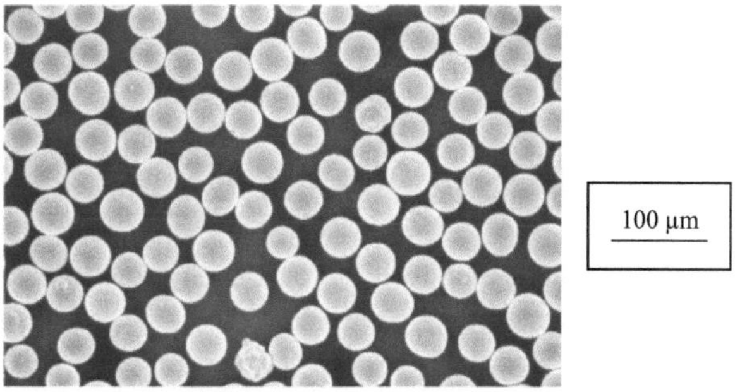

Fig. 8.3 Highly spherical Mo particles. (Courtesy of TECNAR Automation Ltd.)

to cold particles typically encountered in thermal and CS processes. Details of the experimental setup used for these measurements are described in Sect. 8.3.3. A narrow cut of highly spherical Mo particles as shown in Fig. 8.3 was used. Detailed scanning electron microscopy (SEM) analysis revealed a 36 μm average particle size with a standard deviation of ±3 μm. An experimental setup was then put together to allow for scattering measurements at 15 °, 90 °, and 165 °. With each setup, average light intensity as well as average diameter and standard deviation were measured. The results are summarized in Table 8.2. Those experimental results fully confirm that Mie scattering theory applies to our case. Indeed, forward scattering yielded six times more light than the back-scattering approach, and a much more precise diameter measurement (much lower standard deviation), as predicted by Mie (Table 8.2).

8.2.3 Effects of Particle Shape and Surface Texture

In order to quantify the influence of particle shape and surface texture on the size measurement, three different powders have been selected as shown in Fig. 8.4. As expected, measurements with the highly spherical (and reflective) powder yielded a very precise diameter value (within 3 % of the reference diameter as measured with the SEM). As we switched to the spherical powder containing aggregates, precision on the diameter measurement dropped to 7–10 %, most likely due to the nonspherical shapes of the aggregates. Finally, measurements with the

Table 8.2 Influence of incident light angle

Angle (°)	Average diameter (μm)	Standard deviation (μm)	Normalized light intensity
165 Back scattering	39	11	1
90	34	10	2
15 Forward scattering	36	3	6

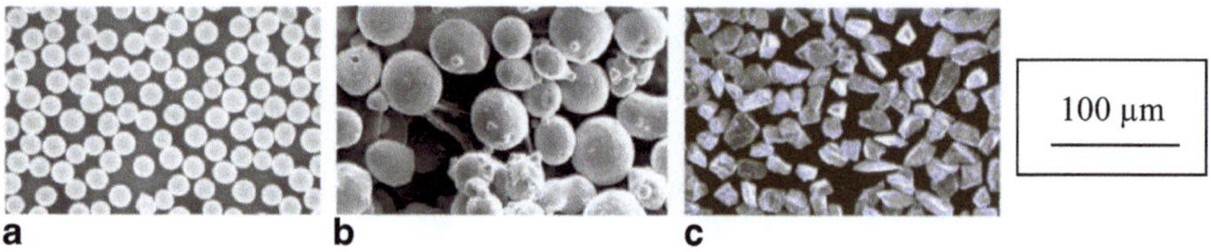

Fig. 8.4 **a** Highly spherical particles, **b** spherical particles with aggregates, **c** angular-shaped particles. (Courtesy of TECNAR Automation Ltd.)

angular-shaped powder yielded a very poor (25–50 %) precision on diameter. Indeed, with such "rice-grain" type of particles, the intensity of the scattered light will strongly depend on which face of the particle is illuminated at a given moment.

8.3 Measuring Particle Velocity and Diameter

Once the particles are "seen," it is possible to think about characterizing them. In the CS community, it is widely accepted that velocity is the most critical parameter that governs the coating formation and properties, yet some other parameters also have an impact as mentioned earlier.

8.3.1 The Particle Image Velocimetry Technique

Particle image velocimetry (PIV) has been used for more than two decades to characterize in-flight particles (or droplets), both hot and cold. There exist different implementations of PIV, but the basic principle is as follows:

1. A laser light sheet or spot is shined onto liquid or solid particles.
2. The light scattered by the particles is detected by a charge-coupled device (CCD) camera sensor equipped with a specific imaging lens.
3. In conventional PIV, the light scattered by the particles is recorded on two separate frames using a high-speed CCD camera. A cross-correlation image analysis algorithm is used to determine the mean displacement of the particles within the two frames. Then, knowing the time delay between the two frames and the optical magnification of the system, velocity of the particles can be determined.
4. Another approach to PIV consists of measuring the length of particle streaks within single video frames. Knowing the camera shutter speed and the optical magnification of the system, velocity can also be determined. Examples of the results obtained with such approach are shown in Fig. 8.5.

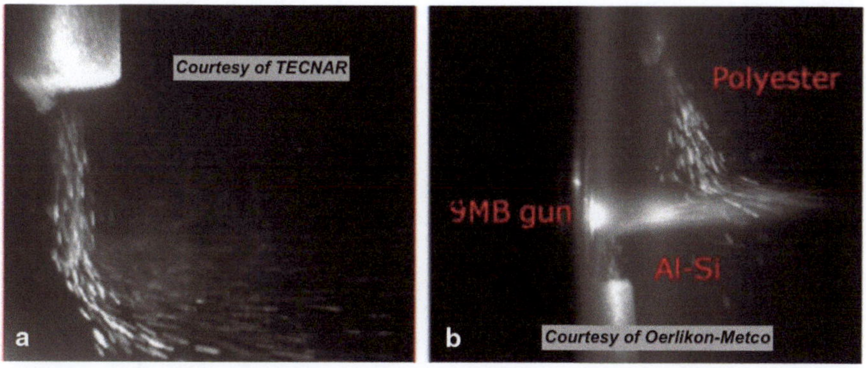

Fig. 8.5 **a** PIV on room temperature Cu particles, **b** PIV on Al–Si and polyester co-injected (plasma process)

PIV offers the undeniable advantage to be highly visual, in addition to allowing for quite precise (typically 2% or better) velocity measurements. However, the precision on the diameter measurement is limited by the pixel resolution, which can be an issue for size cuts typically used in CS. Also, measurements at high particle density are more difficult since it becomes very challenging to resolve the individual particle streaks.

8.3.2 The Single Particle Counting Technique

This approach also requires the use of a light source (usually a laser) to illuminate the cold particles. The optical scattered light collection system is designed in such a way that only one particle is detected in the measurement volume at a given point in time, which ensures single (or individual) particle's characterization.

An optical device, called the two-slit photomask (see Fig. 8.6), is placed in the image plane of the light collection system so that a particle crossing the measurement

Fig. 8.6 Two-slit photomask. (Courtesy of TECNAR Automation Ltd.)

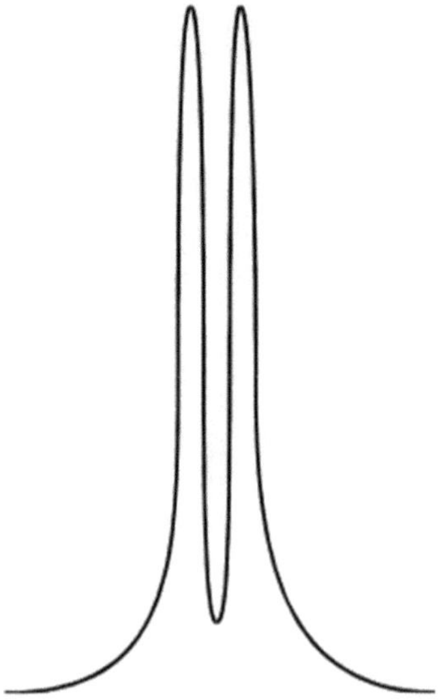

Fig. 8.7 Twin peak. (Courtesy of TECNAR Automation Ltd.)

volume of the system will automatically generate a highly symmetrical and very characteristic twin-peak signal (see Fig. 8.7) easy to distinguish from noise or any other signal. Since the photomask is manufactured by photolithography, the center-to-center distance d between the two slits is precisely known. Using a high-speed, high-precision digitizing board, we can then accurately measure the time of flight (TOF) between the two peaks. Finally, knowing the optical magnification (OM) of the system, particle's velocity V_p is very precisely calculated using the following formula:

$$V_p = \left(\frac{d}{\mathrm{TOF}}\right) * \mathrm{OM}\cdot$$

With that technique, velocity is measured at a precision of 1 % or better. Making the assumption that particles are spherical, we can also determine their diameter (Bisson and Moreau 2003). One can easily demonstrate that for spherical particles, the diameter D_p can be expressed as

$$D_p = \sqrt{E / K},$$

where E corresponds to the intensity of the scattered light (that we measure), and K is a constant that depends on the complex index of refraction of the particles, detection angle, solid angle of detection, sensor detectivity, etc. K needs to be experimentally determined. This approach is applicable if the intensity of the illuminating source is approximately uniform in the measurement volume.

Such approach can yield quite precise (2–3 % or better) diameter measurements if the shape of the cold particles used is almost spherical. However, if the

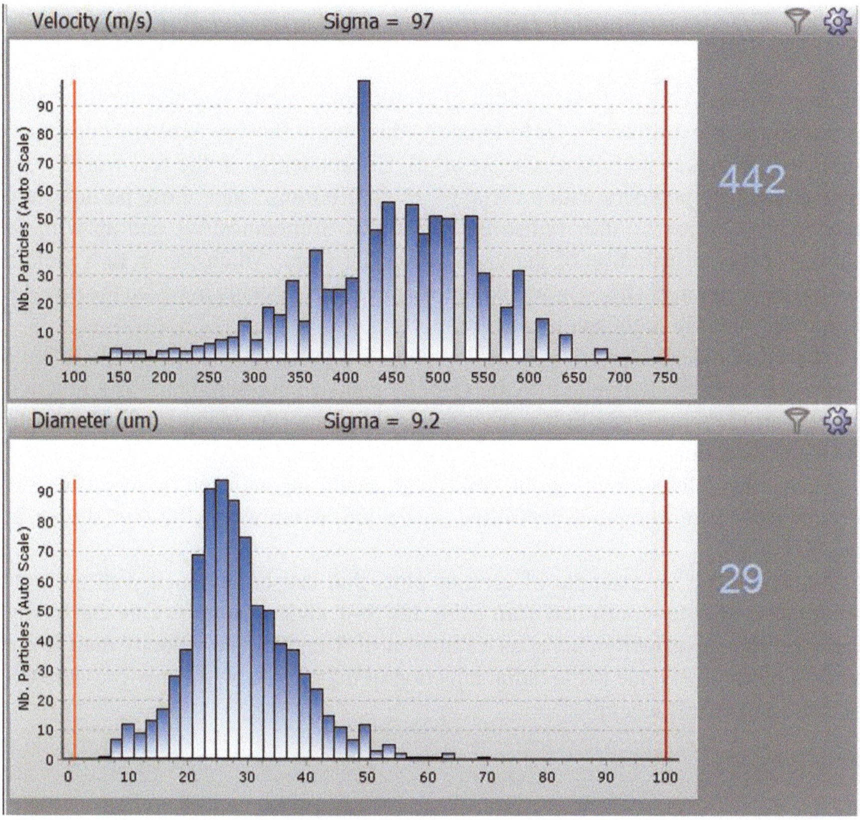

Fig. 8.8 Typical velocity and size distributions. (Courtesy of TECNAR Automation Ltd.)

feedstock material contains many aggregates, or if it is made of angular-shaped particles (as shown in Fig. 8.4c), the precision on diameter decreases rapidly (around 20 %) and can end up being very poor (50 % or less) as mentioned above. Figure 8.8 illustrates a typical example of the results obtained with a single-particle counting device:

8.3.3 The Laser Doppler Anemometry Technique

Clearly, the long-time known laser Doppler anemometry (LDA) technique could also be used to obtain a bulk-average velocity measurement in CS. However, such approach is rather costly and more complex to use than the previous two. As a consequence, it is rarely used for CS characterization (see, for example, Assadi et al. 2003) and, as such, we do not discuss it in detail in the present chapter.

8.3.4 Effects of Spatial Distribution of Multisized Particles

In a thermal spraying process, ideally one would spray mono-size powders to obtain the best possible uniformity. Unfortunately, this would be very expensive and such an approach makes no business sense at all. Consequently, in the real world, cold sprayers do use powders with a certain size distribution. Once those particles are exiting the CS nozzle, they redistribute spatially. Sometimes, we end up with an uneven spatial distribution of the various particle sizes, which can in turn yield a nonuniform coating. This is more likely to occur in CS systems in which the powder is injected radially in the main gas flow.

Therefore, when developing or optimizing a CS process, it is very useful to have diagnostic equipment able to provide velocity and size measurements as a function of the particle's positions in the cross-sectional plane (the plane perpendicular to the spray direction).

Such spray plume mapping capability can easily be achieved by mounting the sensor head on a computer-controlled *X–Y* slide assembly as the one shown in Fig. 8.9:

Figure 8.10 is an example of contour plots that can be obtained with a cross-sectional, computer-controlled scan using the *X–Y* slide assembly. One can easily visualize the iso-velocity curves as a function of *X* and *Y*. The velocity map is not axisymmetric likely due to the radial injection of the powder in the low-pressure CS system used in this example.

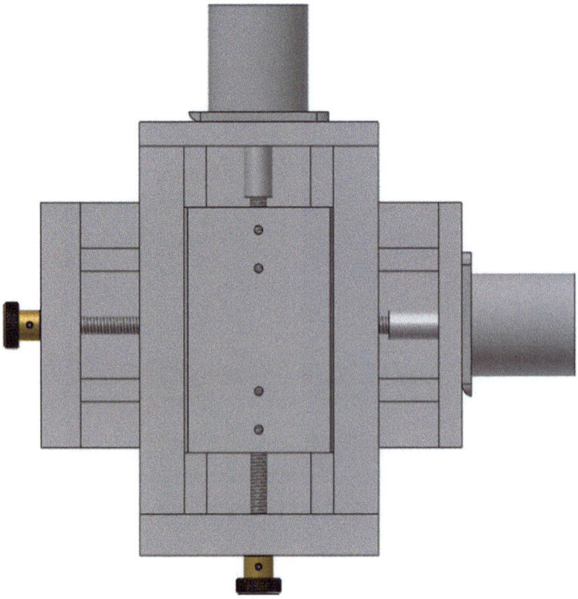

Fig. 8.9 Typical *X–Y* displacement unit for plume mapping. (Courtesy of TECNAR Automation Ltd.)

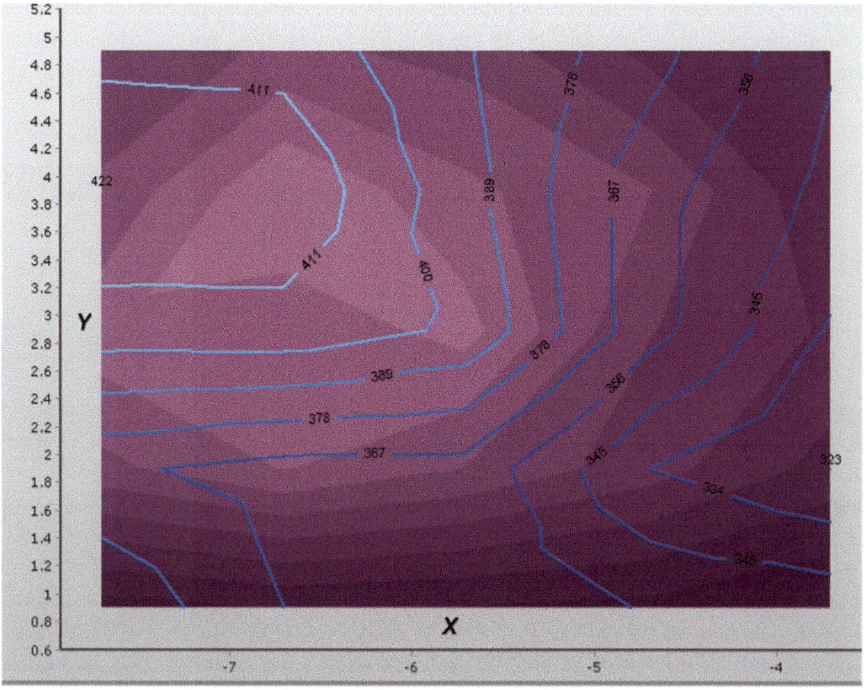

Fig. 8.10 Typical iso-velocity contour plots in a low-pressure cold spray process. (Courtesy of TECNAR Automation Ltd.)

8.4 Relative Particle Flow Measurement

With both the PIV and single-particle counting techniques, it is possible to achieve a decent relative particle flow measurement (RPFM). However, and as always, in both cases there are limitations. In the present section, we briefly discuss both approaches as well as their main limitations.

8.4.1 RPFM Using PIV

With the advent of high-resolution, high-speed digital cameras and the availability of fast computers at a reasonable price, it is now quite easy to analyze several video frames per second in order to provide real-time measurements. Moreover, most of the digital cameras are nowadays delivered with region of interest (ROI) capabilities that allow for detailed analysis of a preselected area of pixels.

One reasonable hypothesis is that the number of particle streaks in a given video frame will be somewhat proportional to the average powder feed rate. That hypothesis has been verified experimentally. However, such approach suffers from a fundamental limitation: as soon as the field of view is filled with particle streaks, the system becomes blind to any additional particle, whether it travels beyond or behind the object plane. In other words, there is a certain maximum powder feed rate (threshold) over which the RPFM will saturate. That saturation threshold will vary depending on powder size cut, the characteristics optical system (camera and lenses), and the nozzle geometry. Considering the typical CS conditions that are used in the field, the PIV approach has a quite limited range of application for flow measurement (saturation comes rapidly).

8.4.2 RPFM Using Single Particle Counting

With such an approach, a particle is detected following a trigger event. Consequently, it is reasonable to think that the number of trigger events per second is somewhat proportional to the average powder feed rate. Again, that hypothesis has been verified experimentally (both in thermal and cold spray). The main limitation of that approach is of the same nature as for PIV, that is, the RFPM is of good quality until a certain saturation threshold is reached. In this case, saturation occurs when the measurement volume (cross-sectional area x depth of field) of the system if completely filled with particles. Experience shows that although saturation can be reached for some current CS conditions, the range of powder feed rates that can be characterized using the single-particle counting approach is much larger than for PIV.

8.5 Concluding Remarks

Very clearly, particle velocity is the key factor governing CS coating formation. We have shown that it can be efficiently measured using various techniques, which involve the use of a laser source. The interest of measuring particle sizes and relative flow as well as how it can be achieved was also discussed. Finally, the influence of the particle shape and surface texture as well as of the illumination angle on the precision of the measurements has been explained. Commercial sensor systems are available in the market (www.tecnar.com, www.Oseir.com) to measure particle velocity, size, and flow rate for the CS process. The use of such systems is expected to become a standard practice for most applications of the CS technology. Figure 8.11 shows one of the commercially available sensors: the ColdSprayMeter (Fig. 8.12).

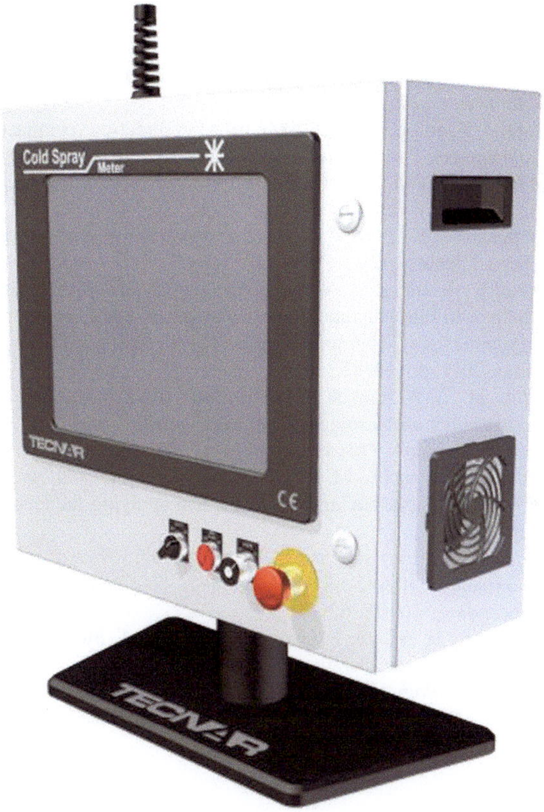

Fig. 8.11 The ColdSprayMeter system. (Courtesy of TECNAR Automation Ltd.)

Fig. 8.12 The TECNAR ColdSprayMeter sensor head and scanner in action. (Courtesy of National Research Council of Canada)

References

Assadi, H., F. Gartner, T. Stoltenhoff, and H. Kreye. Bonding mechanism in cold gas spraying. *Acta Materialia* 51:4379–4394.

Bisson, J. F., and C. Moreau. 2003. Effect of direct-current plasma fluctuations on in-flight particle parameters: Part II. *Journal of Thermal Spraying Technology* 12:258–264.

Bohren, C. F., and D. R. Huffman. 1983. *Absorption and scattering of light by small particles*. New York: Wiley.

Champagne V. K. 2007. *The cold spray materials deposition process: Fundamentals and applications*. Cambridge: CRC Press.

Irissou, E., F. Ilinca, W. Wong, J.-G. Legoux, and S. Yue. 2011. Investigation on the effect of helium-to-nitrogen ratio as propellant gas mixture on the processing of titanium using cold gas dynamic spray. Thermal Spray 2011: Proceedings of the International Thermal Spray Conference 2011, pp. 66–71.

Karimi, M., G. W. Rankin, B. Jodoin, et al. 2013. *Shock wave induced spraying process: Effect of parameters on coating performance*. Thermal Spray 2013: Innovative coating solutions for the global economy. Proceedings of ITSC 2013, pp. 178–183.

Karthikeyan, J., and C. M. Kay. 2003. *Cold spray technology: An industrial perspective*. Thermal Spray 2003: Advancing the science and applying the technology. Proceedings of ITSC 2003, pp. 117–121.

Legoux, J.-G., E. Irissou, and C. Moreau. 2007. Effect of substrate temperature on the formation mechanism of cold-sprayed aluminum, zinc and tin coatings. *Journal of Thermal Spraying Technology* 16:619–626.

Schmidt, T., H. Assadi, F. Gartner, H. Richter, T. Stoltenhoff, H. Kreye, and T. Klassen. 2009. From particle acceleration to impact and bonding in cold spraying. *Journal of Thermal Spraying Technology* 18:794–807.

Tucker Jr., R. C. 2013. Thermal spray technology. In *ASM handbook series,* vol. 5A, 54–59.

Chapter 9
Powder Reclamation Methods

J. Abelson

9.1 Basic Filtration Concepts

There are many filtration methods used to filter particulate from a fluid; mechanical filtration using a filter barrier (or media) is the preferred method for collection of thermal spray particulate since dry filtration allows easy disposal of captured particulate. Mechanical filtration is relatively unaffected by temperature or humidity, and the filtration efficiency increases over time. There are several filtration mechanisms (Fig. 9.1) involved in the mechanical filtration of particles:

- Sieving
- Inertial impaction
- Interception
- Diffusion

Sieving is the predominant filtration mechanism for larger particulate greater than 10 μm in size. Sieving occurs when the particle is physically too large to fit between two or more fibers. Sieving is the primary mechanism in liquid filtration and with very large dry particulate.

When collecting smaller particulate (less than 10 μm in size), the mechanisms of inertial impaction, interception, and diffusion work together to provide overall filtration efficiency. These three mechanisms use van der Waals intermolecular forces of attraction to hold a particle in place on a fiber after contact is made. The forces (intermolecular attraction) vary inversely as the seventh power of the interatomic space; therefore, weak attractive forces do hold smaller particles to the fibers. The mechanisms depend on the velocity of the airflow being suitable for the application and particulate size to allow adhesion to the fibers. In some cases, a very high media velocity will result in particles shedding as the van der Waals forces are overcome. This situation should be avoided in thermal spray powder recovery applications.

J. Abelson (✉)
Donaldson Company, Inc., 1400 W. 94th Street, Minneapolis, MN, USA
e-mail: jeff.abelson@donaldson.com

J. Villafuerte (ed.), *Modern Cold Spray*, DOI 10.1007/978-3-319-16772-5_9

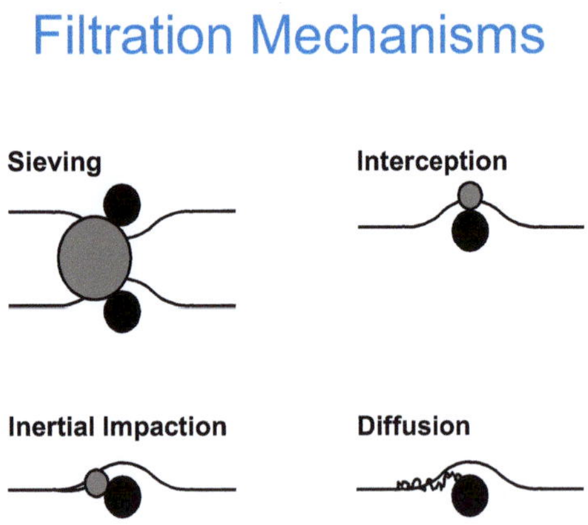

Fig. 9.1 Filtration mechanisms filtering the overspray powders

Inertial impaction is the filtration mechanism, which typically collects micron-sized particulate. Inertial impaction occurs as the airstream is displaced by the media fiber while the dust particle continues on its original path due to its mass and inertia. The particle collides with the fiber and "sticks."

Interception collects 0.2–1-μm-size particulate. Particles are collected here when the airflow moves the particle close enough to a fiber surface for adhesion to occur.

Diffusion is the predominant mechanism collecting very fine particulate of 0.1 μm size or less. Because these particles are so fine, they are influenced by atomic forces. The particles move in the general direction of airflow but are able to move independent of it (Brownian motion). Up to 30–50 % of the powders produced from thermal spray operations are in the 0.02–0.1-μm range. Filtration media produced for thermal spray industry applications should be designed to collect these very fine powders.

Figure 9.2 models efficiency versus particle size for each filtration mechanism and the overall filtration efficiency as the particle size increases. The top blue line is the total efficiency curve while the orange line shows the contribution of diffusion. Note diffusion has a significant contribution (> 50 %) when particles are very small (less than 0.2 μm). The yellow line reflects the contribution of inertial impaction and which is most significant when particles are greater than 0.4 μm. The green line shows the contribution of interception, which is the most significant mechanism when particle sizes are more than 0.2 μm.

Finally, the brown line shows contribution for sieving which has no major significance with submicron particles.

Due to the high degree of variability between applications and the differences in powder characteristics, we cannot count on one filtration mechanism to provide effective filtration for all dusts. We need a filter media to provide filtration using all of the filtration mechanisms.

The overall efficiency for a filter media can be broken down to show where certain filtration mechanisms are predominant in specific particle size ranges.

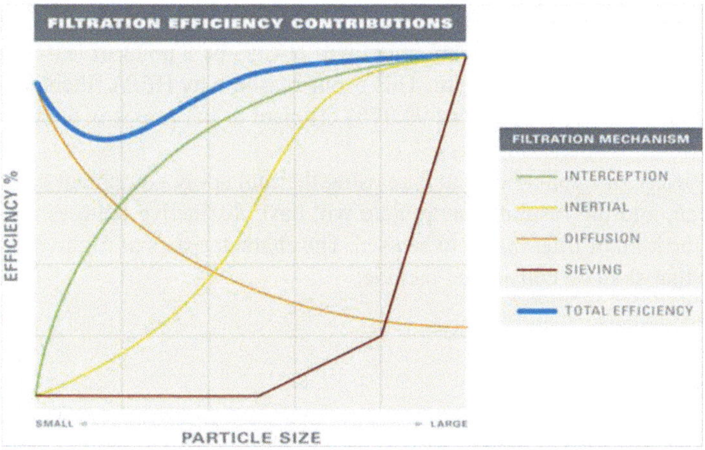

Fig. 9.2 The contribution of each filtration mechanism to overall efficiency dependent on powder size

This overall efficiency curve (Fig. 9.3) is for generic filter media. Note how diffusion is the dominant filtration mechanism for particulate less than 0.1 μm in size, diffusion and interception are the dominant filtration mechanisms for 0.1 to 0.5-μm particulate, and inertial impaction with interception are the dominant filtration mechanisms for greater than 0.5 μm.

Every mechanical filtration media—from human lungs to furnace filters, high-efficiency particulate air (HEPA) filters, or bags and cartridges—experience this

Fig. 9.3 Filtration mechanisms' impact on different-sized powders

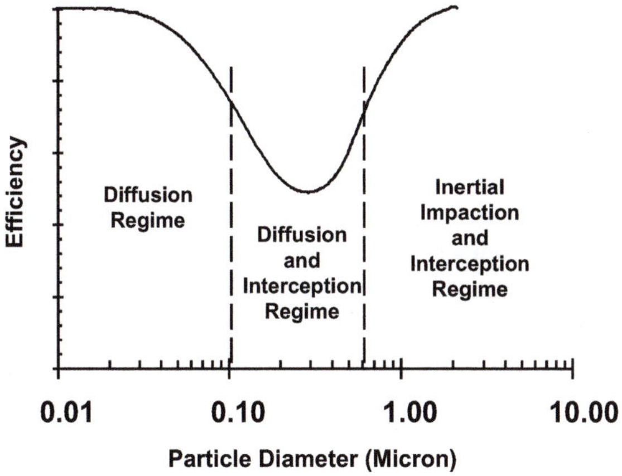

same general type of total efficiency curve. Although the efficiencies of each mechanical filtration media will vary, there will always be a point of least efficiency, generally between 0.1 and 0.4 μm. This is one reason why HEPA filters in the USA are rated on their performance of 99.97 % efficient at 0.3 μm; it is their lowest efficiency point of operation.

The variety of applications anyone using thermal spray equipment can encounter is large, and the resulting particulate will have distinctive features which help identify the kind of filter media to be used. The characteristics of fugitive cold spray powders that shall be considered include:

- Hygroscopicity
- Abrasiveness
- Corrosiveness
- Toxicity
- Explosiveness
- Fire hazard

Hygroscopic powders tend to take up and retain moisture, which then changes the physical characteristics of the powder layer on the filter media. There are different degrees of moisture absorption from slightly hygroscopic powders where moisture is absorbed slowly over an extended time to highly hygroscopic powders, which quickly absorb moisture from the air contacting the powder surface. If powders are hygroscopic, they tend to form dust layers on the filter media which are structurally locked to the media or become a hard "cake" on the surface of the media which reduces the ability of the air to flow through the media. This result requires more energy usage to maintain sufficient airflow in the system to continue capturing overspray powder at the collection point.

Abrasiveness is a characteristic of collected powder overspray which tends to rub away, or abrade, filter media. Examples of abrasive powders might include molybdenum, nickel chromium, or other nickel alloys where sprayed powders have very hard jagged edges. To protect filter media from wear, special care is often taken to control the inlet velocity and airflow patterns inside the collector to reduce the overall effects.

Corrosive and toxic powders are those which either break down materials they come in contact with or you should have minimal contact with. Corrosive powders are often hygroscopic as well, and when exposed to moisture, they break down filter media (if improperly chosen) more rapidly. Special coatings are sometimes necessary to protect the collector from corrosion (Abelson 2004).

Many processes create very small powders, which may become airborne, and they can settle on surfaces and in crevices throughout the plant. Eventually, these powders not only create a housekeeping issue but if the powders are combustible, they can also create a potentially explosive dust cloud if disturbed. It is essential for plant leaders to implement combustible dust control strategies by interfacing with experts on fire and explosion protection equipment and solutions. Combustible dusts often present both fire and explosion risks so it may help to consider the management of these risks separately.

The material safety data sheet (MSDS) is often a source for information on how to handle powders including a spill or fire. If the MSDS sheet is not available, or does not provide guidance on fire or explosion risks, services are available to provide explosiveness testing for a fee.

Dust collection companies generally offer assistance in determining some of the powder properties described above as they are related to the selection and operation of dust collector. Some of the characterization techniques include scanning electron microscopy (SEM) and particle size analysis.

SEM images can:

- Provide insight into the agglomerative or abrasive nature of the powder.
- Display the particle shape, which provides insights on how powder will deposit on a filter media, which then facilitates dust collector sizing.

Particle size analysis can:

- Clarify the general powder sizes to be filtered.
- Clarify how respirable powder will be with most respirable dusts ranging from submicron to 50 μm in size.
- Provide insights into how densely the powder will deposit on the filter media. The smaller and more uniform or "narrow" the distribution of powder sizes, the denser the powder cake deposited on the filter media will be and the more frequently the filter media will need to be cleaned to retain sufficient airflow capacity.
- Help companies understand the level of risk for combustible powders. In general, the smaller the particle size the more hazardous the combustion risks for a powder.

Figure 9.4 illustrates the particle size distribution of an aluminum–aluminum oxide powder blend before and after it was sprayed with a downstream injection cold spray system. Powder was sprayed using 12 bar and 350 °C gas temperature. The cold spray process, in this case, appears to have little effect on the overall size distribution of the overspray. This distribution represents a "normal distribution" curve.

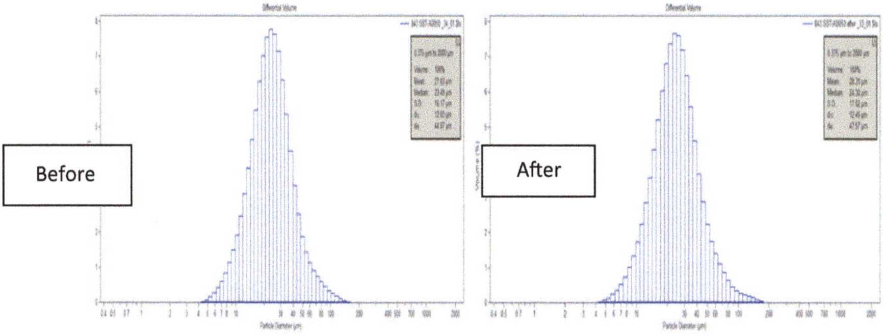

Fig. 9.4 The particle size distribution before and after the powder was sprayed using a downstream injection cold spray booth. (Courtesy of Centerline Windsor Ltd.)

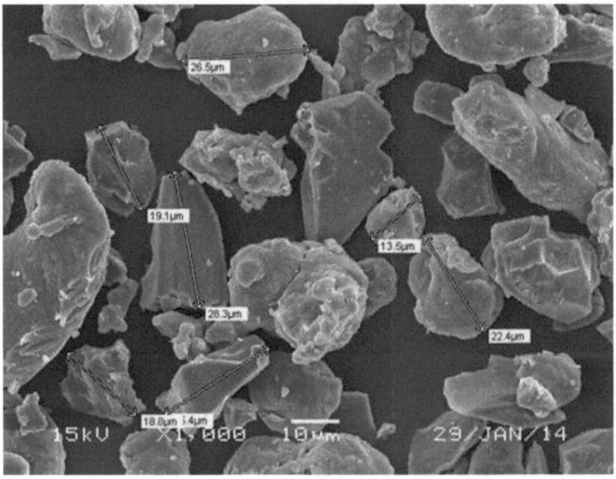

Fig. 9.5 Cold-sprayed aluminum magnified at 1000×

Such distribution is broad enough to form layer on the filter media that allows sufficient permeability; that is, allowing air to freely flow through as the particles will not pack tightly enough to restrict airflow. Sizing of dust collectors varies based on narrow or wide particle size distributions and how permeable the powder layer is expected to be.

While particle size analyses show size distribution of the powder, SEM pictures reveal the shape of small particles as they approach the filter media. Very tiny particles seen on SEM images may not show up on the size distribution curve because by mass they are insignificant. By population, they can represent a significant number of particles challenging the filter media.

Figure 9.5 shows a cold-sprayed aluminum–alumina powder blend at 1000×magnification. Note in this picture the presence of jagged (alumina) and rounded particles (aluminum). The sharp edges of jagged particles (alumina) would be destructive to the media, especially if the particle velocity is over 20 m/s. Even short periods of time can fatigue or wear away components. The abrasive characteristic of the particles may influence the size of dust collector required to provide more open area (larger footprint) for a dropout zone to allow particles to slow down as much as possible.

9.2 Filtration Media Formulation

The formulation of a filtration system must take into account several factors including:

- "What level of filtration is needed? For example is the filtered air being returned to an occupied space, or discharged to atmosphere?
- Is there a particular efficiency, powder loading capacity or pressure drop requirement?

- What are the sources for the particles in the air? What material do the particles consist of? Are the particles toxic or hazardous in nature?
- Are their vapors or gases present and do they involve hazards?
- How often will the process be running? Will it run constantly (24/7) or intermittently (for a few hours per day)?
- How much space is available for collector and where is that space?
- Are there codes or standards discovered in the risk assessment to be considered?
- What is the budget impact for dust collection equipment?
- What are the environmental conditions this filter will be subjected to? Will it be hot or cold? Will it be subjected to moist conditions? Will the collector be located indoors or outdoors?
- How will the filters be handled and disposed of after the end of life?" (Spengler et al. 2001)

Once the application requirements are understood, the filtration company should review existing filter media to see if there is a match meeting their application requirements. If there is not a match, a modified or new media may have to be developed.

Modification of an existing media may involve changes in the media structure such as different fibers, media resins, or chemical treatments (e.g., oil resistant coatings). Other adjustments may include the addition of surface loading technology such as nanofibers or polytetrafluoroethylene (PTFE) membrane. Performing any of these developments efficiently requires a process utilizing computer modeling. Computer modeling can simulate the media characteristics, airflow, and particles to determine if the new media design can meet requirements. The modeling should take into account all the filtration mechanisms for the size of the powder particles, the velocity of air, and the expected powder loading on the filter media. Assuming positive results from the computer model, representative samples of the media can be produced for evaluation of physical properties. Typical properties tested include:

- Thickness—important because it may limit the amount of media packaged into a finished filter. A thick media can also increase energy required to pull the air through it, which can negatively affect expected performance. This can increase the cost of running the collector.
- Pore size—the measured width of the openings in the filter media. The more porous the media, the less resistance it has and air will flow through it more easily. This also means it may be less efficient because large pore sizes allow larger particles through. Depending on the application, this could be negative or positive.
- Loose fibers—fiber shedding simply meaning the fibers used to form the filter media come off during use either due to the velocity of the air through the media or from cleaning actions on the filters. Some applications cannot allow loose fibers in the process. An example of where this may be significant is food applications if the filtered powder is reused in the process.
- Tensile testing—evaluates the tensile strength and elongation properties of the media. These properties are especially important in packaging a filter media that

may be pulsed for cleaning where a filter media could see up to 2 psi of pressure instantaneously.
- SEM—looks deep into the structure of the media to view the fiber arrangement pattern from above and from a cutaway side view. This image capability gives filter engineers information on how fibers and media resins interact. This also helps the media engineers to troubleshoot media problems if necessary.

9.2.1 Filtration Media Selection

Choosing the appropriate filter media is important for the overall effectiveness and efficiency of the packaged filter. Typically, the best filter media for cold spray applications are one of the following:

- Cellulose with nanofibers
- Cellulose with nanofibers and a flame retardant additive

Cellulose filter media generally are made from naturally occurring materials. While the resultant media is similar to paper, understand that the media is highly engineered for the purpose of filtration, not for printing. Cellulose filter media can also have flame retardant characteristics added if necessary for an application. It is important to understand flame retardant media are not flameproof, nor will they change the properties of the powder deposited on them. The flame-retardant media typically has an additive to reduce the ability of the media to support combustion by itself. So if a hot spark hits the clean media with no powder on it, when the spark extinguishes, the flame-retardant media will not continue to burn. Flame-retardant media can be damaged and will burn if they have a constant heat source applied to them. This means if combustible powder on the surface of the media caught fire, the filter will burn.

Cellulose filter media provides a typical level of filtration efficiency. To improve that efficiency, some suppliers add nanofiber layer to the media. Nanofibers are durable synthetic fibers and polymer with a mean fiber diameter of 0.2 μm. They increase efficiency of media by promoting the rapid development of a dust cake layer on the surface of the cellulose media. Earlier in the chapter it was stated the use of all of the filtration mechanisms facilitated the development of a powder buildup which increased the efficiency of the filter because the filtration mechanisms for dry dust collection are enhanced by the increase in active surface offered by a porous filter cake. Looking at a 10-μm-size particle, you can see the comparison of a media that has a nanofiber layer and a cellulose media that does not (Fig. 9.6).

One method of producing a nanofiber layer is an electro-spinning process to create a very fine, continuous, resilient fiber of 0.2–0.3 μm in diameter, which is then applied to a filter media substrate material. The nanofibers form a permanent web with very fine interstitial spaces on the surface of the substrate. The web collects dust, dirt, and contaminants on the surface of the filter; offering many benefits

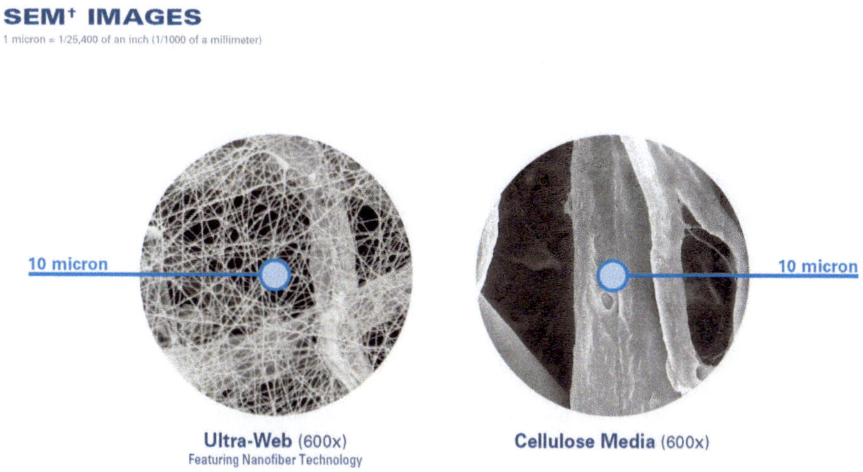

Fig. 9.6 Images with (*left*) and without (*right*) nanofibers. Note the difficulty the 10-μm powder has passing through the nanofiber layer. (Courtesy of the Donaldson Company, Inc. 2011)

over conventional filters built with cellulose, cellulose/synthetic, spunbond, or even meltblown commodity filtration media.

Although the cost to purchase premium nanofiber performance filters for a collector can initially be higher than commodity media filters, an argument can be made that nanofiber layer filters offer the following benefits:

- *Higher initial and ongoing efficiency.* A primary function of a collector is to control and minimize emissions from a manufacturing process. Clean, new filters typically offer lower efficiencies (and, therefore, higher emissions) compared to filters with a buildup of powder on them. A nanofiber layer on the surface of the filtration media captures powder particles better than filters without a nanofiber layer. This performance is made possible through previously discussed filtering mechanisms including interception, diffusion, and impaction.
- *Fewer potential emissions.* Most industrial collectors utilize filter cleaning to manage the buildup of powder on the filter surface. Each time a filter is cleaned (typically by a reverse pulse of compressed air), the powder on the filter media is disrupted, and an emission-producing event can occur. With a nanofiber layer, collected powder accumulates on the surface of the filtration media rather than within the media structure and is cleaned off with fewer pulses. Fewer pulses results in fewer potential emission-producing events.
- *Lower pressure drop across the filtration media saves energy.* Collectors rely on a fan to draw powder-laden air from a contaminant source to the collector and through the filter media. The energy (static pressure) required to move that air through the system defines the necessary fan size and horsepower, therefore, the effective energy required to operate the system. The restriction created by the filter media and particulate can add significantly to the overall fan energy requirement of a system. In commodity media filters, filtered powder can be driven

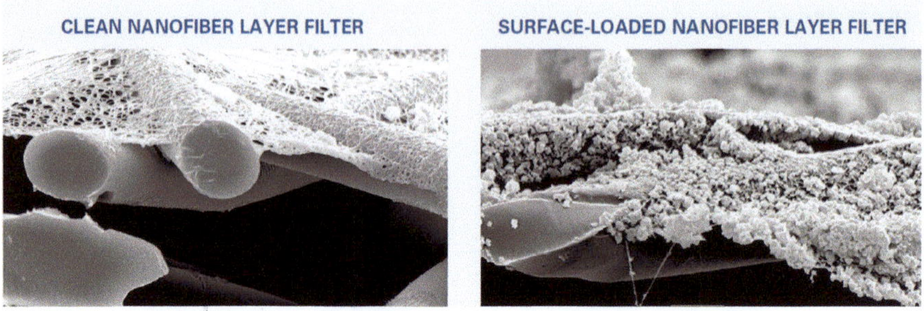

Fig. 9.7 Nanofiber media is loaded with ISO fine dust. Dust particles collect on the surface of the media and clean off easily while the substrate stays clean. A depth-loading filter would allow dust particles to penetrate deeply into the substrate where they build up and choke off the airflow. (Courtesy of the Donaldson Company, Inc. 2014)

deep into the media's pores where it cannot be cleaned. When the captured powder cannot be cleaned from the media depth, higher stable pressure differentials are created across the media and energy demand goes up. Because the nanofiber media captures powder on the surface, it cleans more completely and operates at a lower pressure differential across the media reducing energy demands. Even greater energy savings are possible if the system fan is configured with a variable frequency drive (VFD) control system.

- *Less compressed air consumption required for pulse-jet-cleaned dust collection systems.* As mentioned earlier, powder captured on the surface of a filtration media requires fewer compressed air pulse cycles to clean than a depth-loaded media. Fewer compressed air pulse cycles results in lower overall compressed air consumption, which in turn, reduces the energy demands on the compressor and the costs for compressor operation (Fig. 9.7).
- *Longer filter life.* Most filters reach "end of life" when the media is fully depth loaded and can no longer be cleaned allow design airflow with the available fan in the system. Because of the nanofiber layer efficiency and surface loading characteristics, nanofiber filters can last significantly longer than traditional commodity filter elements. Longer filter life means buying new filters less frequently, saving considerable money over time. Additionally, longer filter life reduces costly operation downtime for filter maintenance activities.
- *Flexibility in filter configuration to help solve problems.* Nanofiber performance layer media can be produced on a variety of substrate materials and in different filter configurations. A nanofiber layer on cellulose, synthetic, or spunbond media substrates improves the performance of each of these media. The substrates can be selected for antistatic, temperature, or moisture-resistant properties while still obtaining the advantages of nanofiber performance layer. Nanofiber performance layer filters have been available as cartridge filters for years, but manufacturers are now offering these media in pleated bag and fluted filter configurations.

The expansion of filter configurations and the availability of premium nanofiber performance layer filters for a variety of applications mean more collector operators can make the switch from commodity filters, reducing their emissions, saving energy, and improving their bottom line (Woolever 2013).

9.3 Dust Collection Technologies

When designing a collector for overspray from a cold spray process, three things should be considered by the design engineer:

- Managing the airflow within the collector to minimize operating cost. Consider the cost per cubic foot of air moved to collect overspray powder.
- Minimizing the load of powder sent to the filter to reduce pressure drop and extend filter life.
- Develop an effective filter cleaning system.

Managing airflow in the collector is important, and typically collectors have three ways air can enter the collector: from the top, from the side, or from the bottom.

Of the three inlet configurations, air entering from the top of the collector has been proven to provide superior results. The Environmental Protection Agency (EPA)-sponsored research has shown superior performance results from downdraft flow. This downdraft airflow reduces redeposition since it aids gravity in moving dust particles toward the hopper (*Industrial ventilation—A manual of recommended practice for design,* ACGIH 2013). The downdraft airflow guides dust into the storage device under the collector for disposal. Figure 9.8 helps clarify this action.

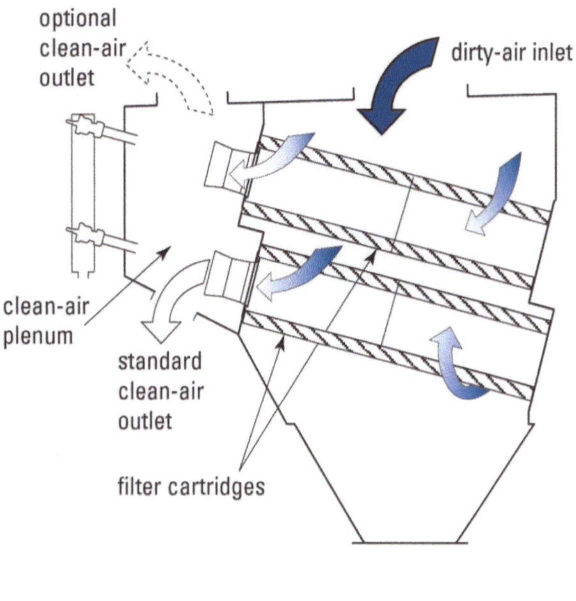

Fig. 9.8 Cutaway view of a dust collector showing how air flows through the collector. Air enters the top, flows on the outside of the filters through the middle of the filter, and exits through the clean air outlet. (Courtesy of the Donaldson Company, Inc. 2014)

The overspray powder is collected and carried through a round piece of duct at a relatively high velocity. This duct connects to the dirty air inlet on the top of the collector and then flows through the collector. The majority of the powder goes straight to the hopper in this design. Any powder that does not naturally fall out of the airstream will be directed to the filters where it collects on the outside of the media, the cleaned air then passes out through the clean air outlet.

As an alternative to the top entry, side-entry inlets can also be utilized. If a collector is configured with specially packaged filters, air patterns can be managed by having the air enter from the side. One advantage of this design is a heavy dust can drop directly into the hopper without passing by filters at all. As air turns into the collector, it slows down and the momentum of the particles carries them in the original direction out of the general air stream. Another advantage of a side entry is the collector can fit into tight spaces where low headroom is a requirement. Examples include situations where companies have overhead cranes or other equipment to be negotiated when installing equipment. Figure 9.9 is an example of a side-entry collector.

The last inlet configuration for a collector is bottom entry. Typically, bottom entry occurs in the hopper where many designs include a perforated strike plate. This plate should knock the heavy materials down and force air to turn to drop out as much powder as possible before the air rises to the filter media.

Managing the powder loading to the filters is the next consideration. Some strategies include:

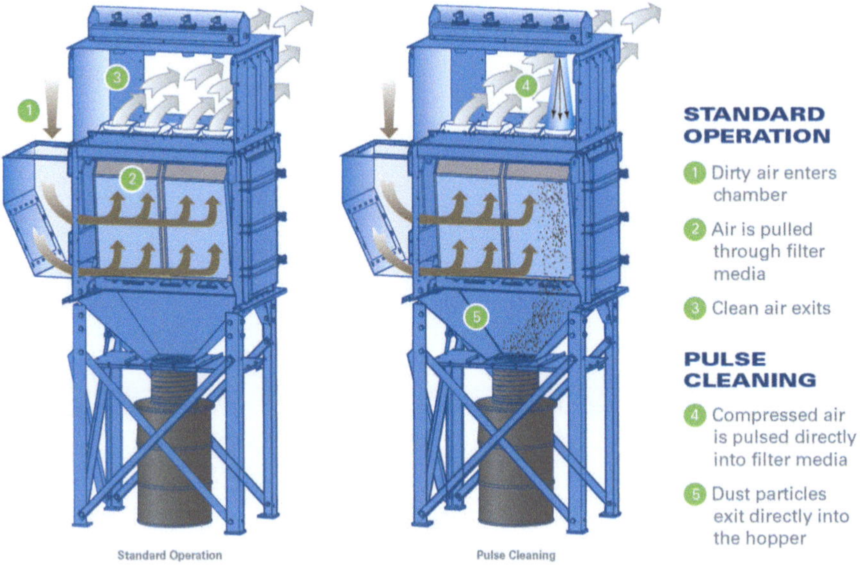

Fig. 9.9 Cutaway of a side-entry collector showing how the air flows through the collector. Also shows what happens when the collector pulse cleans the filters with compressed air. (Courtesy of the Donaldson Company, Inc. 2014)

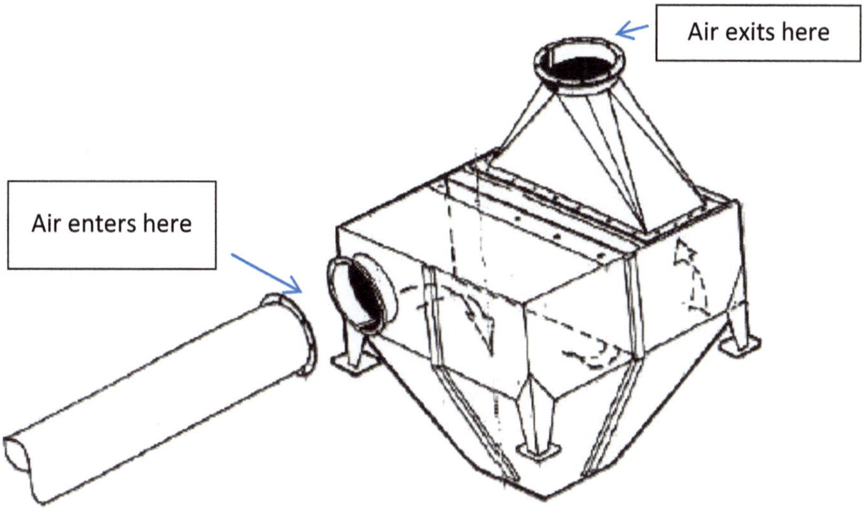

Fig. 9.10 Example of a simple dropout box design. (Courtesy of the Donaldson Company, Inc. 2014)

- Dropout boxes before the dust collector
- Dropout chambers as part of the collector
- Cyclones before the collector
- Increase in space inside the collector by modifying the filter shape or the collector size

A dropout box (see Fig. 9.10) is simply a box with an inlet and an outlet arranged to force air to turn inside the box. Each time the air turns, it gives the powder an opportunity to drop out of the air stream and collect as it tends to slow down. The advantage of a dropout box is their relatively inexpensive cost to construct. The challenge is if the box is improperly designed or if the powder is not not properly emptying out of box, the box will have little to no effect in reducing powder loading to the collector.

A dropout box can be built into the side of the collector itself or as an additional module with no filters. The addition of a dropout box on the side of the collector is it offers a compact design for many end users who have space limitations. The air usually enters at the bottom and has plates welded on the inside where the air is put through a tortuous path and drops as much powder as possible. The other option is to design in a dropout chamber as a part of the dirty air chamber or as a separate bolt on chamber, each not having filters in it. The advantage of these chambers is that the square feet of area is larger than the incoming ductwork carrying the overspray and simply opening up that area slows the air down and drops particulate where the particulate may go directly to the collection device never even touching the filters. Also in each case if there are no filters in it, there is no risk of having a filter be subjected to high velocities compromising the integrity of the filters.

The next strategy to consider is a cyclone (see Fig. 9.11) placed before the collector. Air enters near the top and centrifugal force and friction cause particles to drop out at the bottom of the vortex. The advantage of the cyclone is that the loading to the collector can be decreased significantly, and powder size can be shifted from a narrow to a more normal size distribution because particles typically agglomerate. This change in particulate size distribution may promote a more permeable dust cake, allowing air to pass more easily. The argument can then be made for a collector having less filter area because thicker powder overspray deposits can develop without significant pressure drop increases. The other advantage of a cyclone is that residence time in the cyclone allows the powder a chance to cool before entering the collector. Filtering cool powder is much easier. Lastly, a cyclone offers the opportunity external ignition sources to extinguish before they enter the dust collector, reducing the probability of a fire in the collector.

The last strategy a designer may consider to reduce the powder load to the filters is changing the shape of the filters. If the filters are changed from round to an oval shape, the space between the filters can be increased by as much as 20–25 % while keeping the box the same size. This allows 20–25 % more space for the air to travel, slow down and drop powder out of the air stream. If the designer chooses to change

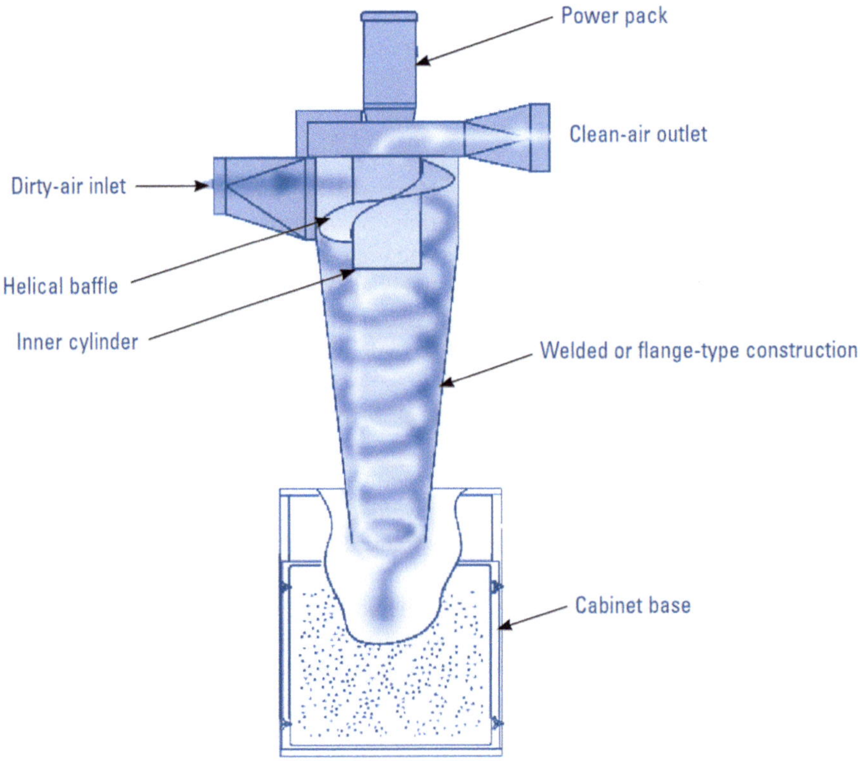

Fig. 9.11 Cyclone style collectors. (Courtesy of the Donaldson Company, Inc. 2014)

Table 9.1 Summary of different options to reduce particulate loading

Strategy	Benefit	Challenges
Dropout box before the collector	Easy to make on own Easy to install as an add-on	Must maintain regularly and may increase fan HP required to clean air
Cyclone	Changes powder sizes potentially reducing collector size Reduces possibility of an external ignition source entering collector	Increases fan HP required increasing energy required to clean air
Dropout chamber on collector	Space saving design Easy to maintain	Cannot be easily added after the fact
Change the space in collector	Space-saving design Easy to maintain Handle high loads and abrasive dusts	Cannot be easily added after the fact

HP horse power

the shape and adjusts the collector box, the collector has even more space to allow enhanced dropout of powder.

A summary of these different options the benefits and challenges are compared in the Table 9.1.

An effective cleaning system is extremely important to collector design. The performance of the cleaning system drives how much overspray powder the collector can handle and how much energy the collector consumes when cleaning the air. An effective cleaning system will:

- Properly disturb powder buildup on the dirty side of the filter. Disturbed powder will leave the filter but some amount may return to the surface of the filter, which will enhance efficiency. A side benefit of maintaining proper powder cake on the filters is assuring airflow at the point of overspray generation is maintained at a desired rate.
- Evenly distribute pulsing energy along the full length of the filter. If pressure is uneven, powder may build up on one end of the filter causing air to take the path of least resistance in the areas getting more pulse energy. The effect of higher velocity going through a small amount of media could be more emissions, premature plugging of media.
- Have as few turns as possible as compressed air leaves the diaphragm valve on its way to the clean side of the filter media. Each turn in direction the pulsed air is required to make consumes some of the energy the pulse will have when it finally reaches the dirty side of the filter media.
- Move deposited powder from the filter surface to the collection device at each pulse. The faster the powder is moved toward the collection device, the more quickly powder can be disposed of.
- Have proven performance by testing in the laboratory and the field.

9.4 Guidelines for Ventilation of Thermal Spray

The variety of thermal sprayed part geometries creates many unique opportunities and challenges for collecting overspray. Depending on the part geometry, the designer of the overspray ventilation system may have to review sound attenuation needs, as well as considering any gases produced from the thermal spray processes, or other general safety issues. One such issue may be the combustible nature of thermal spray powders; "many process requirements may make elimination of combustible dust impractical. However, it may still be possible to manage the dispersion of dust within the plant by using an appropriate and effective industrial ventilation system including dust collection. A well-designed, maintained, and operated industrial ventilation system including good hoods or booths, proper doct sizes and properly selected collection equipment can provide effective powder control and can therefore help manage the presence of dispersed powder" (Dust Collectors and Combustible Dust Strategies 2011) from the powder spraying process.

When designers consider methods to ventilate overspray, typically a local exhaust hood or booth is chosen.

Local exhaust hood strategies generally provide airflows to capture overspray with less total air volume than a large booth. Local exhaust hoods (example Fig. 9.12) are also nice when they can follow the gun as the operator or robot sprays a part. A local exhaust hood is also used when part geometries are simple, and/or require very little movement of the gun. A typical application where a simple local exhaust hood would be used is coating of a round roller for the printing industry. The local exhaust hood is constructed so it follows the contour of the roller and typically has 1–1.5 m/s capture velocity with capture velocities below 1.5 m/s to avoid affecting the actual coating process.

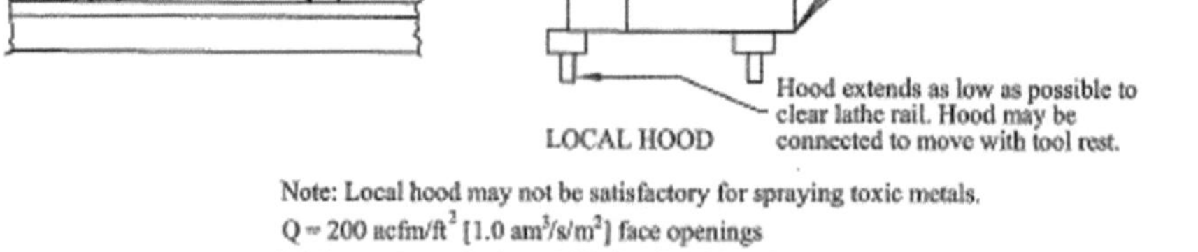

Note: Local hood may not be satisfactory for spraying toxic metals.
$Q = 200$ acfm/ft^2 [1.0 am^3/s/m^2] face openings
Minimum duct velocity = 3500 fpm [17.50 m/s]
$h_e = 0.25\ VP_d$

Fig. 9.12 Typical hooding arrangement for a local exhaust hood that remains stationary while the part being sprayed moves. (Used with permission from ACGIH 2013)

The other common option is a spray booth. Spray booths offer several advantages including the ability to handle complicated part geometries when the overspray can occur in many different directions making it difficult to locate a smaller local exhaust hood. The powder is captured in the booth and then drawn into a duct for transport to the collector where it is separated from the air and the filtered air leaves the collector. There are many considerations in designing a booth including how loud some thermal spray processes are, so sound dampening is something to be considered. The amount of air necessary to moderate temperature inside the booth may also be a consideration. Putting too little air can elevate booth temperature, which can affect coatings and spray equipment while putting too much air in can be a waste of significant energy for very little benefit. In order to pull overspray and maximize collection, it is recommended the cross ventilation velocity in the booth be typically 0.4–1 m/s. Figure 9.13 displays similar recommendations from American Conference of Governmental Industrial Hygienists (ACGIH). Note the toxic nature of powder is noted on the diagram. Special care must be exercised in the use of potentially toxic spray powders.

"From that starting point of a booth or hood, a proper dust collection system can be designed. This will usually include a duct to transport the powder, a filter to remove the powder from the air, and a fan to provide the energy to generate the airflow. The exhaust fan for a 10,000-cubic-feet-per-minute (cfm) system might require 30–40 hp. The airflow requirement is generally fixed and should not change unless the

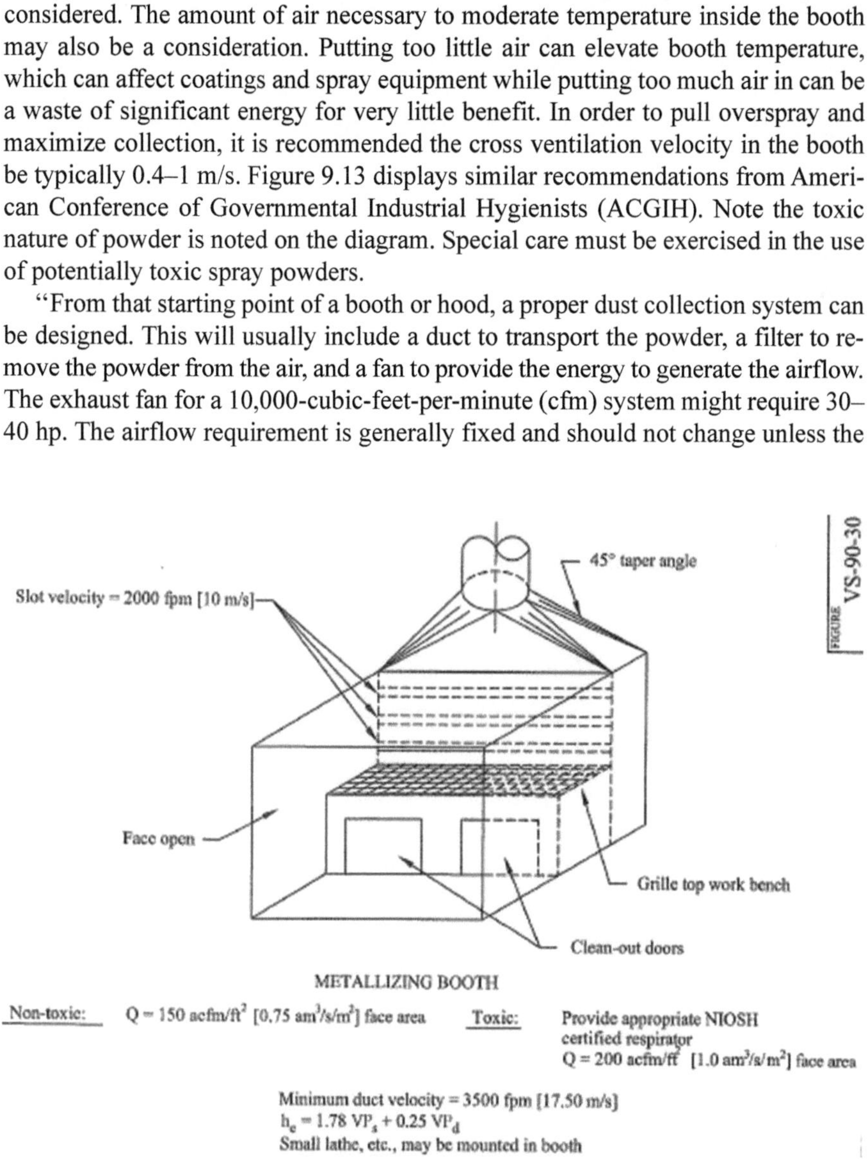

Fig. 9.13 A typical metallizing booth arrangement with guideline for ventilation. (Used with permission from ACGIH 2013)

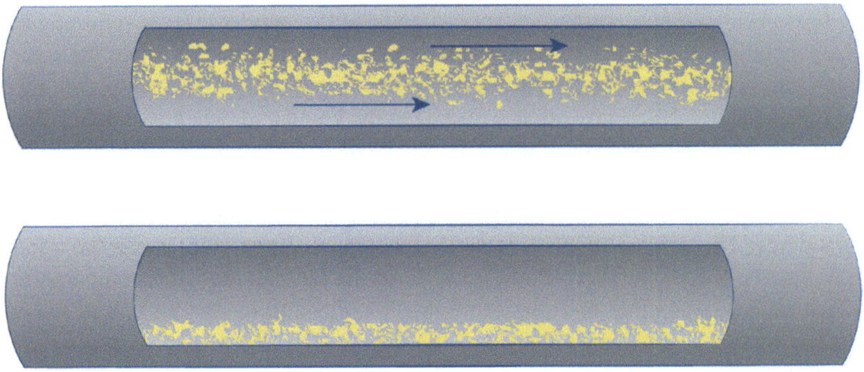

Fig. 9.14 The *top image* shows that dust particles are suspended when air velocity is greater than 3500 fpm, and the *bottom image* shows how particles can build up when air velocity is less than 3500 fpm

enclosure is redesigned. It is commonly accepted that a velocity of 3500–4000 feet per minute (fpm) is optimal to transport powder in a round duct 1. Moving the air more slowly will allow the powder particulate to drop out and settle in the bottom of the duct—creating a fire hazard and potentially blocking the duct. Moving air more quickly wastes fan energy and creates unnecessary wear and tear on the ducts. For our example, moving 10,000 cfm at a velocity of 3500–4000 fpm requires the selection of a round duct 22″ in diameter. This selection has a cross sectional area of 2.6398 square feet, resulting in a velocity of 3,788 fpm" (Richard 2011; Fig. 9.14).

9.4.1 VFDs and Airflow Control Systems

A better way to control the fan and maintain constant airflow in the system is with a VFD. A VFD operates the fan motor at a specific rotational speed based on adjusting the hertz frequency. Whereas normal three-phase power in North America usually runs on 60 Hz frequency, a VFD allows the operator to select a specific frequency slowing or accelerating the rotational speed of the fan. In an ideal system, the system would run at full speed only when the static pressure load of dirty filters required it. The rest of the time, the fan would be running at a slower speed to generate exactly the amount of static required. This method of operation offers benefits in cost savings. Compared to operators who use the "set and forget" approach where they always run their dust collection systems at speeds faster than necessary (to ensure complete ventilation of the thermal spray enclosure), the VFD approach uses a smart system running at exactly the required airflow speed saving energy.

Mathematical models can demonstrate savings, using a few simple assumptions and some system variables. Generally, an upgrade to VFD and airflow control system can pay for itself in less than 2 years, and more importantly, the dust collection system will be running at the right design air volume. This saves wear and tear on the system, and especially on the high-end, surface-loading filters required for thermal spray. Any decision to upgrade to a VFD and airflow control system should include the following savings as factors:

- Filter costs
- Labor costs
- Disposal costs
- Inventory costs
- Shipping costs (for new filters and disposal of old filters)
- Quality process
- Operational stability of the system and proper maintenance of airflow in the system

9.4.1.1 Controlling the VFD

Once the decision to use a VFD has been made, the next step is to determine the method of providing operational input. The objective is to maintain a desired air volume regardless of changes in system static pressure. An airflow-measuring device in the duct system can send a signal to the controller to adjust fan speed as airflow changes. These instruments are suited for clean air environments, and therefore are usually installed in ducts after the air is filtered. This could be a duct on the outlet of the fan where the length is sufficient to provide a smooth, reliable indication of the total airflow moving through the duct.

An alternative method is to measure the system static pressure, rather than the actual airflow at a point in the duct system just before air enters the dust collector. At the design airflow, the amount of static necessary becomes a function of factors which should remain unchanged as long as the system is not modified mechanically. Filters will get dirty and be pulsed clean, but static pressure resistance at the inlet of the dust collector should stay the same if the system is not modified and is running at design airflow. A controller maintaining static is a simple way to effectively control a VFD in a dust collection system. As filters build up resistance, the airflow delivered by the fan drops. This reduction in airflow produces a lower static resistance in the duct in front of the collector, so the controller will adjust the VFD speed to increase airflow returning the system to design flow and static resistance. Conversely, as filters are pulsed clean, the resistance across the filters falls and flow through the duct increases with an increase in system static. The VFD will decrease power to return to the system to design flow and the same level of static at design conditions. The result is a smooth consistent airflow with the associated benefits and savings that come with it.

9.4.1.2 Considerations

There are situations where a VFD airflow control system will have limited benefits. An example is when a dust collector (and fan) is used intermittently to service one or two cells at a time in a multiple thermal spray cell system. Since the duct system is not mechanically stable, there is no simple way to utilize the variable static of the system to adjust power of the fan effectively. This is a limitation of the duct system—not a limitation of the VFD (Richard 2011).

9.4.2 Components of a Dust Collector to Consider When Contaminants May Be Harmful to Health

Depending on the material sprayed for the application, limiting exposure to the powder overspray may be appropriate. Caution is advised when operators change filters or empty collection bins when the powder has harmful properties. There are options available to reduce operator exposure to contaminants and general exposure of contaminants to atmosphere.

The bag-in/bag-out filter removal option allows installation and removal of filters and collected powder with reduced operator and environmental exposure to contaminants by combining a bag collar mounted around the filter access cover. Filter cartridges are then removed and replaced using a sealed bag (Fig. 9.15).

Bag-in/bag-out features are also available for the hopper discharge. This features a drum or pail, liner, and elastic top bags secure to a hopper adapter collar. Collected

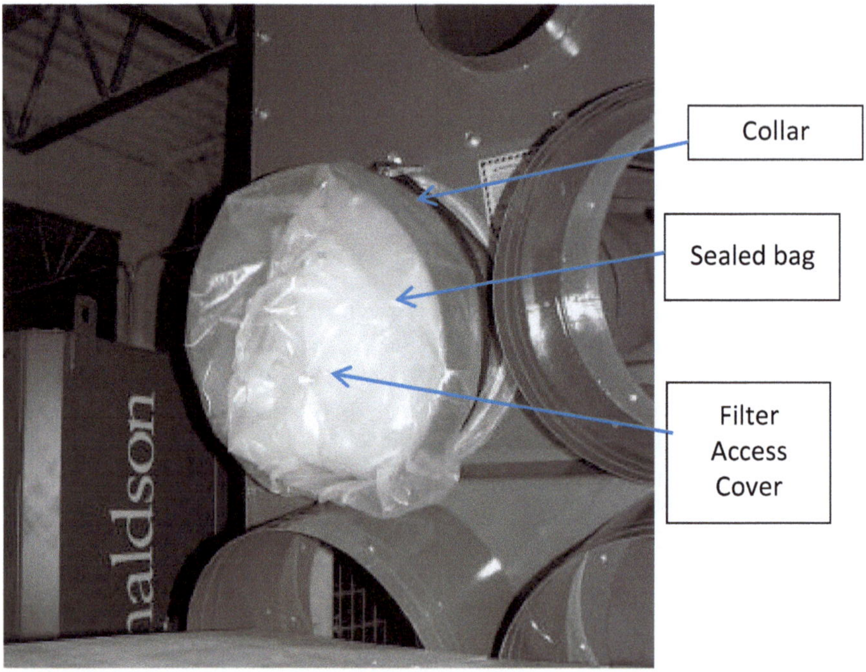

Fig. 9.15 A bag-in/bag-out assembly used for filter removal

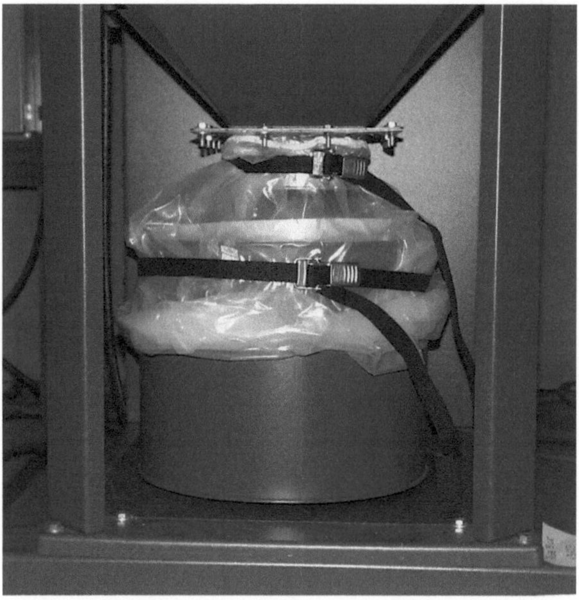

Fig. 9.16 A bag-in/bag-out assembly used for the hopper discharge. (Courtesy of the Donaldson Company, Inc. 2014)

dust is removed and disposed of in a sealed bag. An example of what the assembly looks like is here (Fig. 9.16).

The above components show what can be applied to a collector when the contaminant is hazardous. Remember that when a thermal spray process is used, it will create very small particles of powder, which may easily become airborne, eventually settling on surfaces and in crevices throughout the plant. These particles not only create a housekeeping issue, and if they are considered combustible, they could also represent a potentially explosive dust cloud if disturbed.

"It is essential for plant leaders to understand the risks of combustible dust and ensure they manage combustible powders in their facilities. Dust collector manufacturers can supply dust collectors which may assist in supporting combustible dust control strategies in a plant by interfacing with the enduser and their experts in fire and explosion protection equipment and solutions. Combustible dusts generally present both fire and explosion risks so endusers should consider both risks separately" (Dust Collectors and Combustible Dust Strategies 2011). In Table 9.2 are different collector options for strategies to mitigate risks identified by plant leaders.

Risks or consequences of events are typically identified through detailed process hazard assessments. A detailed process assessment should identify all of the processes that generate risks for the end user and the conditions that may lead to those risks. Each risk is then evaluated and decisions taken on how or if the risk should be eliminated or reduced. If a risk cannot be eliminated, contingency plans may be put in place to minimize the consequences should an event still occur. This process hazard analysis is mandated for review on a regular basis and when any changes are made in the processes or materials.

Table 9.2 Collector options which may assist in supporting combustible dust strategies in a plant. (Table courtesy of Donaldson Company, Inc. 2012)

		Risk Mitigation	
		Explosion	Fire
Fire protection			
	Fire extinguisher/fire suppression/sprinkler coupling		✓
	Overflow drain		✓
	Automatic fast acting abort gate		✓
	In-line spark abatement		✓
	Spark detection and extinguishing system		✓
Explosion protection			
Mechanical	Explosion relief panels	✓	
Chemical	Chemical suppressant delivery device	✓	
	Suppression system control panel	✓	
	Detector/sensor—chemical or actuated system devices	✓	
Dust collector isolation			
Inlet			
Mechanical	Actuated knife gate—inlet	✓	
	Flow actuated isolation valve—inlet	✓	
Chemical	Chemical isolation device—inlet	✓	
	Suppression system control panel	✓	
	Detector/sensor—chemical or actuated system devices	✓	
Hopper			
	Rotary valve/airlock	✓	✓
Outlet (if air is returned to the building)			
Mechanical	Actuated knife gate—outlet	✓	
	Flow actuated isolation valve—outlet	✓	
Chemical	Chemical isolation device—outlet	✓	
	Suppression system control panel	✓	
	Detector/sensor—chemical or actuated system devices	✓	

9.5 Disposal

After the powder overspray has been removed from the airstream and cleaned from the filter either by mechanical or pulsing style methods, the particles should be collected in some type of storage device. Some examples include drums, square containers, and large fabric sacks. When the storage device is full, the company must decide what to do with the waste. The options available typically include recycling, or disposal such as landfilling. Deciding what to do with a collected powder starts with evaluating the properties and risks of the powder before disposal. Generally, companies look to recycling to reduce residual waste going to a landfill. In some cases, companies get credit for how well the recycling programs perform. If recycling is a good option for the waste in question, the first factor a recycling company will look at is the material particles are made from. For material collected from the cold spray process is zinc or aluminum, there is a good chance the recycler will take your waste. Some companies put exactly what they recycle on their website. Sending a sample to the recycler is a typical way for the recycler to understand the material they are going to receive to determine the value of the material. Anytime a sample is sent, it is highly recommended an MSDS sheet accompany the sample. The MSDS sheet should be accessible to the shipper in the event it is required which suggests the MSDS sheet be included with the shipping paperwork. The next thing the recycler looks at is the volume generated and considered. The more volume generated, the more attractive the waste will be. Companies generating lower volumes tend to choose to store the material until enough has been collected to save on shipping costs. The above factors apply to recyclers who take the waste, recover the metals for reuse, and pay the company for the material.

Another source to help decide what to do with collected material can be the EPA for the USA or equivalent for other countries. Their website has definitions of the different types of wastes and regulations associated with hazardous wastes. MSDS sheets may also have a section denoting whether the powder in the shipped state is considered hazardous. It is worth noting some states have their own regulations in addition to federal regulations, and while most states' hazardous waste regulations are based on federal requirements, some states have developed more strict regulations. With this in mind, carefully evaluating the rules that apply is very important.

If recycling is not an option, disposal and possible landfilling may be the next choice. Contact an environmental consulting company familiar with the code requirements. They will likely request information related to the waste like:

- What powders are used in the process?
- Does an MSDS sheet exist for this material?
- What was the process used to generate this waste?
- How much of this waste is generated per week/month/year?

The consulting company should be able to help identify next steps and guide the company through the process including proper documentation to comply with the local regulations. Services like these are not usually free so this will be a consideration for some companies.

Lastly, the local waste management company can offer assistance with determining next steps when all other options have been exhausted.

The key to remember through all of this is that it is the waste generators' responsibility to comply with all national (country), state, and local codes required for waste disposal.

References

Abelson, J. 2004. Basic filtration concepts. *ITSA Spraytime* 11 (4): 2004.

ACGIH. 2013. *Industrial ventilation: A manual of recommended practice for design*. 28th ed. Cincinnati: Kemper Woods Center.

Donaldson Company, Inc. 2012. http://www2.donaldson.com/torit/en-us/pages/technicalinformation/efficientcontrolthermalspray.aspx. Accessed 30 Aug 2011.

Dust Collectors and Combustible Dust Strategies. 2012. http://www2.donaldson.com/torit/en-us/technicaldocuments/dustcollectorsandcombustibleduststrategies.pdf. Accessed 26 Dec 2012.

Richard, P. 2011. Efficient control of thermal spray dust collectors. *ITSA Spraytime* 18 (2): 2011.

Spengler, J. D., J. M. Samat, and J. F. McCarthy. 2001. *Indoor air quality handbook*. New York: McGraw-Hill.

Woolever, J. 2013. Top 5 reasons a nanofiber performance layer filter is worth it. Donaldson Company, Inc. 2013. http://www2.donaldson.com/torit/en-us/pages/technicalinformation/top-5-reasons-a-nanofiber-performance-layer-filter-is-worth-it.aspx. Accessed 11 June 2013.

Chapter 10
Applications

V. K. Champagne, P.K. Koh, T. J. Eden, D. E. Wolfe, J. Villafuerte and Dennis Helfritch

10.1 Repair and Restoration in Aerospace

10.1.1 Introduction

The aerospace industry is rather conservative when it comes to adopting new technologies; one of the main reasons is the stringent criteria used to qualify new processes for use on flight-safe critical components. If a new technology has not been adequately tested and evaluated for a specific application, the consequences could be catastrophic, resulting in the loss of human lives and expensive aircraft. Traditionally, this process takes upwards of 20 years because of the vast amount of testing and verification required. A significant investment must be made by numerous entities involved, including research, engineering, manufacturing, quality control,

V. K. Champagne (✉)
U.S. Army Reasearch Laboratory, Aberdeen Proving Ground, MD, USA
e-mail: victor.k.champagne.civ@mail.mil

P. K. Koh
SIM University, 461 Clementi Road, Singapore, Singapore
e-mail: pkkoh@unisim.edu.sg

T. J. Eden · D. E. Wolfe
The Applied Research Laboratory, The Pennsylvania State University, Reston, VA, USA
e-mail: tje1@psu.edu

D. E. Wolfe
e-mail: dew125@psu.edu

J. Villafuerte
Corporate —Supersonic Spray Technologies, CenterLine Windsor Ltd., Windsor, ON, Canada
e-mail: julio.villafuerte@cntrline.com

D. Helfritch
TKC Global, Hearndon, VA, USA
e-mail: dennis.helfritch@tkcglobal.com

J. Villafuerte (ed.), *Modern Cold Spray,* DOI 10.1007/978-3-319-16772-5_10

inspection, logistics, and acquisition. The process involves extensive testing and evaluation of laboratory samples, as well as subscale or full-scale components and/or assemblies, some of which require flight testing for flight safety. The adoption process progresses to feasibility assessment and finally demonstration and validation, before it can be applied to production and/or field use.

The intent of this section of this chapter is to present some of the applications that have been transitioned into the aerospace industry by the US Army Research Laboratory (ARL) and to present several case studies showing the tremendous impact that cold spray (CS) has made for a few select applications, while attempting to include some technical data substantiating the advantages and benefits gained by transition of the process. The focus application will involve the restoration of magnesium aerospace components where much data have been generated and have served as the starting point for the implementation of CS into the aerospace industry.

As explained in previous chapters, the uniqueness of CS for aerospace lies in its ability to produce a coating, provide dimensional restoration, or even produce a near-net-shaped part at temperatures well below the melting point of the powders being applied, thereby avoiding or minimizing many deleterious high-temperature reactions, which are typical of thermal spray processes. It is this characteristic of CS that makes it attractive as a method for coatings or for dimensional restoration, while retaining their own unique material properties.

10.1.2 Pioneering Work

The US ARL has led the development of CS technology over the past decade and has introduced a process to reclaim aerospace parts, especially those produced from magnesium, that shows significant improvement over existing methods. ARL has demonstrated and validated a CS process using aluminum and/or Al alloys as a cost-effective, environmentally acceptable technology to provide surface protection and a repair/rebuild methodology to a variety of magnesium and aluminum aerospace components for use on army and navy helicopters and advanced fixed-wing aircraft, which has been adopted worldwide and is being incorporated not only in the US Department of Defense (DoD) but also in private industry.

The development and qualification of the CS process to deposit aluminum and aluminum alloys was originally proposed by the Center for Cold Spray Technology at ARL in 2003, for providing dimensional restoration and protection to magnesium components, primarily targeting the aerospace and automotive industries. Initially, ARL concentrated on developing a "nonstructural" repair process that could be qualified quicker and less costly than a "structural" repair process that would require extensive component testing.

Soon multimillion dollar programs were established and executed by ARL, including the Environmental Security Technology Certification Program (ESTCP) that extended from 2005 to 2011 and culminated with the qualification of the CS process for use on the UH-60 Blackhawk, in collaboration with Sikorsky Aircraft Company and the establishment of the first dedicated CS repair facility at the Navy Fleet Readiness Center (FRC-East), Cherry Point, North Carolina. This program served as an international benchmark for the adaptation of CS for the aerospace industry. The CS process is now viewed as the best possible method for depositing the aluminum alloys to provide dimensional restoration to magnesium components, significantly improving performance, and reducing life cycle costs.

Subsequently, other military services and the private sector followed suit, and a multitude of efforts were undertaken to develop similar repair procedures. Other nationally recognized programs were the National Center for Manufacturing Sciences (NCMS) whose objective was to exploit the use of CS for corrosion control of magnesium. The project participants included all the branches of the armed services: The army ARL, navy, air force and marines, and the cross-industry companies: The Boeing Company, Delphi Corporation, Ford Motor Company, CenterLine (Windsor) Ltd., and Solidica. Historically, the advancement in thermal spray technology has been driven by the need to fulfill a demand in the aerospace industry. Rapid growth in revenue for the thermal spray industry was observed during the period between 1960 and 1990 when the application of the technology for aircraft engines grew simultaneously (Fig. 10.1). The key contributors of this growth were the commercial introduction of plasma spray, the detonation gun (or D-gun) and high-velocity oxy-fuel (HVOF) spray, and new and improved process control equipment and materials. All these coincided with the thermal spray application for advanced gas turbine engine components such as compressor blades, cases, stator vanes, bearing housings, and labyrinth seals. Many of these applications were subsequently leveraged by other industries such as the marine, oil and gas, medical, and even nuclear energy. It is no surprise that the application of CS in the aerospace industry is keenly watched by many thermal spray practitioners.

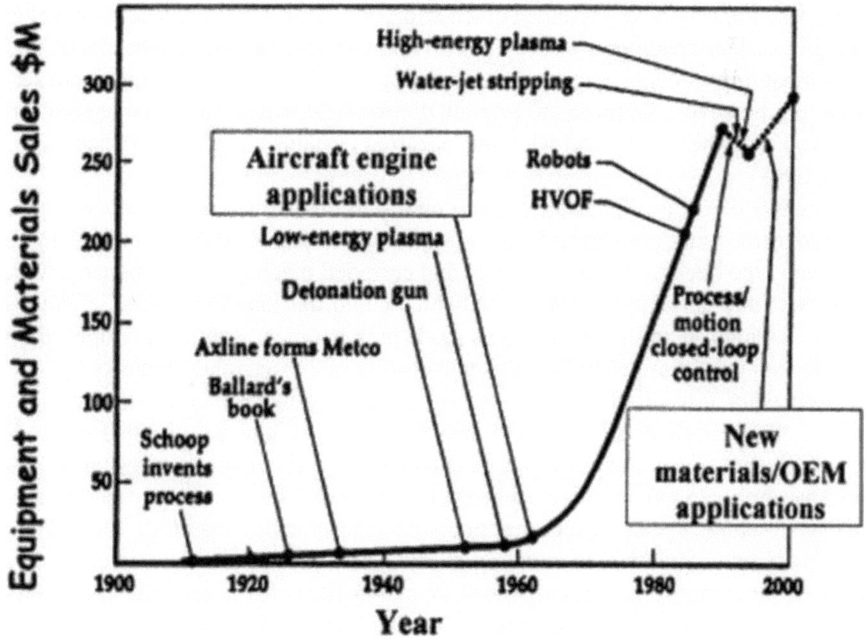

Fig. 10.1 Timeline of significant developments during growth of the thermal spray industry. *OEM* original equipment manufacturer. (Handbook of Thermal Spray Technology)

Plasma and HVOF thermal spray have been investigated for deposition of aluminum, but the results have generally been unsatisfactory due to inconsistent coating integrity. Poor adhesion and delamination of the coatings are typically the cause for high rejection rates. Both of these processes involve the use of high thermal energy to melt or partially melt the coating material before it is accelerated onto the surface of the substrate. The tiny molten or partially melted particles rapidly solidify upon impact with the substrate and contract forming tensile residual stresses in plasma spray coatings. This is not always the case with the HVOF process because the particles are accelerated at high velocity and have been known to form coatings that are in compression. Regardless, failures occur because the plasma and HVOF processes can generate excessive heat causing the formation of an oxide on the magnesium that is detrimental to adhesion. Besides, the thermal spray pattern is very wide so that it would be difficult to apply the coatings to localized areas requiring repair or significant dimensional restoration.

10.1.3 Cold Spray Repair for Magnesium Aerospace Parts

Many of the corrosion problems associated with magnesium helicopter components occur at the contact points between inserts or mating parts, where ferrous metals are located, creating galvanic couples (Vlcek et al. 2005). In addition, magnesium alloys are also very susceptible to surface damage due to the impact, which occurs frequently during manufacture and/or overhaul and repair. Scratches from improper handling or tool marks can result in preferential corrosion sites. The DoD and the aerospace industry have expended much effort over the past two decades to develop specific surface treatments to prevent corrosion, to increase surface hardness, and to combat impact damage for magnesium alloys in order to prolong equipment service life; however, the means to provide dimensional restoration to large areas on components where deep corrosion has occurred remains a challenge (Champagne et al. 2008).

The use of CS as a possible coating option in the aerospace industry was pioneered by the work performed by the US ARL (Champagne 2008). The project was driven by the need to restore damaged and corroded magnesium components back to serviceable condition in rotorcrafts which could translate to millions of dollars in cost savings annually. The progress made in this project significantly opened up application opportunities both within and outside of the aerospace industry.

Helicopter transmission gearboxes are commonly made of magnesium alloys due to their excellent stiffness and damping capacity, low density, thermal conductivity, good machinability, and wide availability. However, magnesium alloys are susceptible to galvanic corrosion as they are one of the most electrochemically active structural metals. Anode reaction with other metallic materials is common, especially when they are operating in a saltwater environment. Figure 10.2 shows the locations where galvanic corrosions are mostly likely to occur on the main transmission gearbox housings (Champagne 2008). Figure 10.3 shows the extent of corrosion damage on a H-53 tail gearbox housing (Leyman and Champagne 2008).

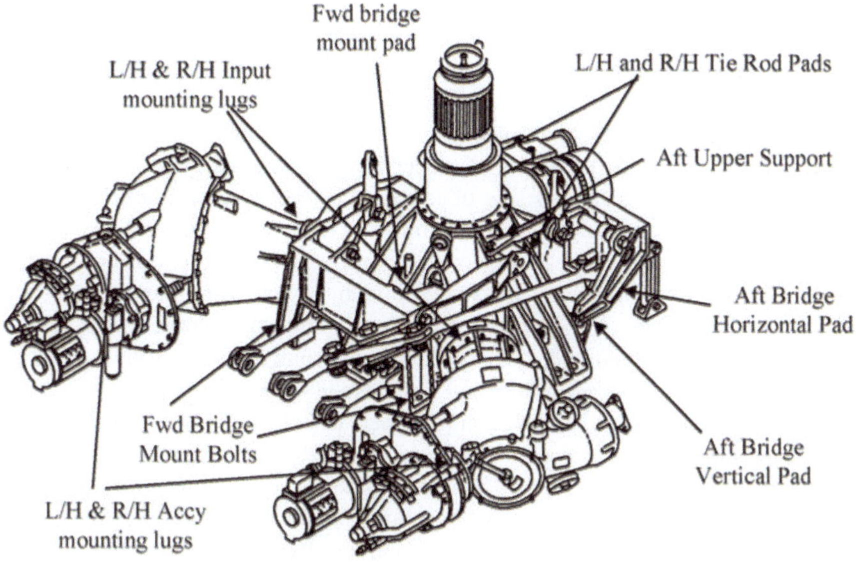

Fig. 10.2 Areas that are most susceptible to corrosion in UH-60 main transmission housing. *L/H* left hand, *R/H* right hand. (Champagne 2008)

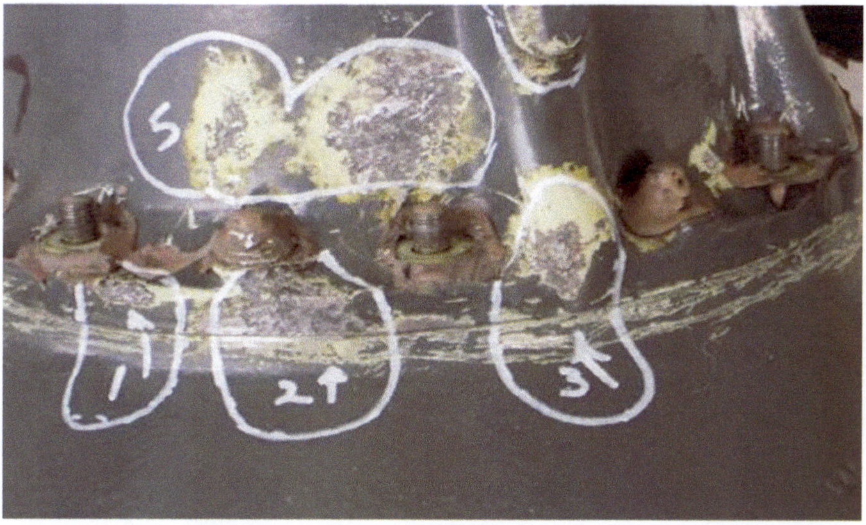

Fig. 10.3 Corrosion locations on a H-53 tail gearbox housing. (Leyman and Champagne 2008)

To improve on the corrosion resistance of magnesium alloys, pretreatments such as hard anodizing, chromate or phosphate coatings, and epoxy paint are usually utilized. However, all these pretreatment processes pose serious health and environment problems. Furthermore, magnesium alloys still continue to suffer severe degradation during service despite these surface treatments. Once the corrosion on gearboxes becomes too significant, the whole gearbox is required to be replaced, incurring high costs of replacement and significant impact on operation readiness.

Cold spraying of commercially pure (CP) aluminum or 6061 aluminum alloy (Al-6061) as an alternate repair was explored. The deposition of aluminum and aluminum alloys, deposited by CS, is now becoming an accepted practice to provide dimensional restoration and a degree of corrosion protection to magnesium components. It is known that the addition of aluminum to magnesium promotes the formation of better passive films than unalloyed magnesium. Therefore, it is not surprising that the application of cold-sprayed aluminum to magnesium and magnesium alloys constitutes a method to inhibit corrosion in aqueous media. The protective capability of pure aluminum cold sprayed on magnesium has been demonstrated by others (Zheng et al. 2006; McCune and Ricketts 2004; Gärtner et al. 2006; Balani et al. 2005). In all cases, the corrosion potentials of cold-sprayed magnesium coupons approached those of CP aluminum. Such polarization behaviors are promising since there is no galvanic protection strategy that is reasonable for magnesium, given magnesium's strong thermodynamic potential for oxidation. In galvanic corrosion, only small areas surrounding the dissimilar interface require protection, for which CS represents an innovative alternative to the use of washers and insulating bushings.

The final approved procedure of using CS aluminum to repair magnesium substrates involved the surface preparation of the affected area (cleaning surface oils and debris, oxide removal by Scotch-Brite pads or discs), masking, introduction of

Fig. 10.4 Before and after photos of cold-sprayed restored UH-60 main rotor transmission housing. (Champagne and Barnett 2012)

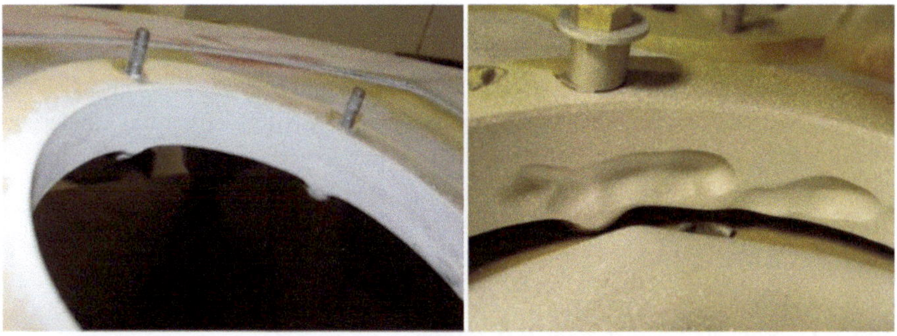

Fig. 10.5 Magnesium helicopter gearbox restored by cold spray using CP aluminum and 6061 aluminum alloy powders. (Champagne and Barnett 2012)

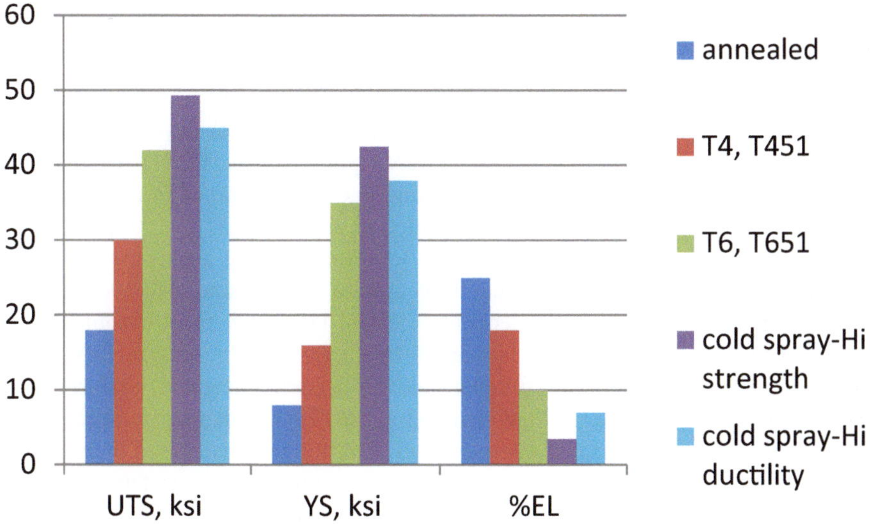

Fig. 10.6 Tension test results of cold spray 6061 Al compared to wrought aluminum. (Champagne and Barnett 2012)

a CS deposit, final machining, dimensional measurements, and visual inspection before final acceptance test (nondestructive testing and/or specific requirements).

Several magnesium components were successfully restored under the program. For example, the severely corroded UH-60 main rotor transmission housing was repaired using cold spray (Fig. 10.4), and the damaged helicopter gearbox was dimensionally restored with CP aluminum and 6061 aluminum alloy CS coatings (Fig. 10.5; Champagne and Barnett 2012).

The results listed in Fig. 10.6 clearly indicate the potential for CS 6061Al as a "structural" material that can be considered for use on other substrate materials than ZE41 A-T5 Mg (Champagne and Barnett 2012). The ultimate tensile strength (UTS) of the as-cold-sprayed 6061AL was as high as 50 ksi with a corresponding yield strength (YS) of 42 ksi and elongation (EL) of 3 %. A post-processing procedure of feedstock 6061 Al powder has been developed and can yield CS material with a minimum UTS of 45 ksi, YS of 38 ksi and as much as 7 % EL. This has been a major accomplishment for the implementation of CS for "structural" materials repair and for additive manufacturing.

10.1.4 Case Studies

10.1.4.1 Forward Equipment Bay Panels

A common problem on military aircraft is wear due to chafing around fastener holes in skin panels. This chafing results in skin panels that exceed fit tolerances at the fastener locations. The South Dakota School of Mines and Technology (SDSM&T) Repair Refurbish and Return to Service (R3S) Research Center and the ARL Center for CS in cooperation with the 28th Bomb Wing at Ellsworth Air Force Base (AFB), SD, USA; Air Force Engineering and Technical Services (AFETS); Oklahoma City Air Logistics Center; and H. F. Webster Engineering Services developed a repair process for a B1 Bomber left upper aft forward equipment bay (FEB) panel that had chafing damage to the fastener holes resulting in out-of-tolerance fit for the panel. This type of repair can be applied to other aircraft and serves as an example of how CS can be used as a repair technique. The locations of these panels are shown in Fig. 10.7 on the B-1. This FEB panel is secured to the airframe with 100° tapered flat head TRIDAIR fasteners (Widener et al. 2013). The fasteners are designed to be installed flush with the panel for laminar airflow over the skin surface. In service, the chamfer wears causing the fastener holes to become elongated, rendering the panels unserviceable (Fig. 10.8; Widener et al. 2013). The damage is accelerated by air turbulence on elongated fastener holes. Chafing wear is caused by repeated opening and closing of the panel. Steel fasteners are used to secure the aluminum panes. The panel is made from 2024-T6 aluminum. The CS repair utilized the VRC Gen III ARL system and 6061 Al powder sprayed normal to the chamfered surface of the panel (Widener et al. 2013). Load transfer, fatigue, and tensile tests were ac-

Fig. 10.7 FEB panels on B-1. There are eight panels per aircraft, four panels per left and right sides

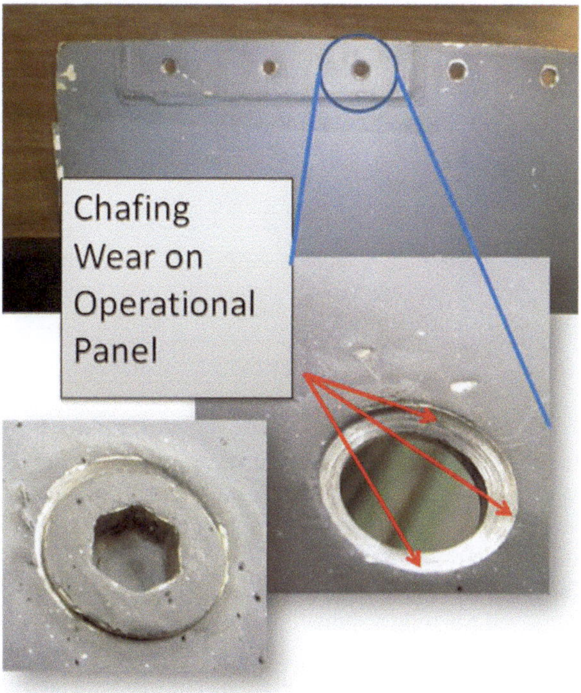

Fig. 10.8 Wear sites under fastener heads on FEB panels

complished, along with 3-lug shear testing and metallography to characterize the repair. The results demonstrated the capability of CS to provide a permanent repair for this application, restoring the full capability of the panel. Cold-sprayed coupons met or exceeded the required bearing loads for the parent material and fastener type for this application. Even when tested to failure (greater than 1.5 × bearing yield) the CS material did not separate from the panel (Fig. 10.9; Widener et al. 2013). Fatigue test results revealed that at 15 ksi tensile stress (typical upper end

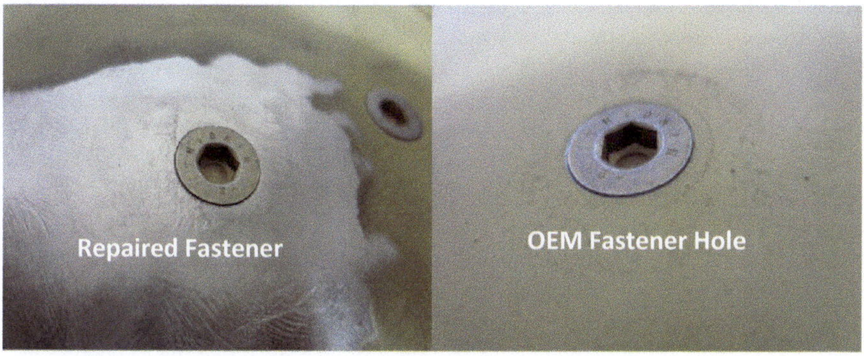

Fig. 10.9 Cold-spray-repaired fastener hole as compared to a fastener hole on a new part

Fig. 10.10 Even when tested to failure at 5600 lbs, the cold spray material did not separate

for aircraft skin loads) coupon lasted approximately 500,000 cycles (Fig. 10.10; Widener et al. 2013). Replacement panels cost as much as US$ 225,000. CS technology has recently been approved as a low-cost (and high return on investment) solution to repair FEB panels. The first panel was repaired and installed on a B-1B for flight-testing in August 2012. Cost savings to the B-1 program if this repair is approved for refurbishment of the remaining forward equipment bays is estimated at US$ 9.6 million annually. Further applications of this technology can be applied to all major design series (MDS) aircraft and are an example of infusing new technologies into current DoD maintenance processes to reduce sustainment cost while maintaining the viability of older weapon systems. If applied to other MDS across the DoD, cost savings could reach US$ 100 million annually (Widener et al. 2013).

10.1.4.2 Hydraulic Line Repair

Chafing of titanium (Ti) hydraulic tubing, Haynes AMS 4944 (Ti3Al2.5V), as a result of vibration and abrasive action is a major maintenance problem on the B-1B aircraft in terms of maintenance man-hours (Figs. 10.11 and 10.12; Leyman et al. 2012). A technological solution that reduces the frequency of hydraulic tubing chafing would have broad applicability across the DOD and could be incorporated for use on similar commercial components. CS preventative maintenance of hydraulic lines has been proven to be fairly simple and effective from both an economic and technical standpoint. A feasibility study was accomplished in 2009 which demonstrated the effectiveness of the CS process in preventing hydraulic tube chafing by the application of a CP Ti coating providing a wear surface in areas known to experience chafing problems. This preventative measure can be performed during programmed depot maintenance (PDM) or during the high-velocity maintenance (HVM) process to prevent or reduce occurrences of hydraulic tubing chafing in the field. A characterization study was accomplished which showed that a Ti coating could be successfully applied to Ti tubing providing an additional "sacrificial" wear surface (Fig. 10.13; Leyman et al. 2012). The study showed that the Ti coating had

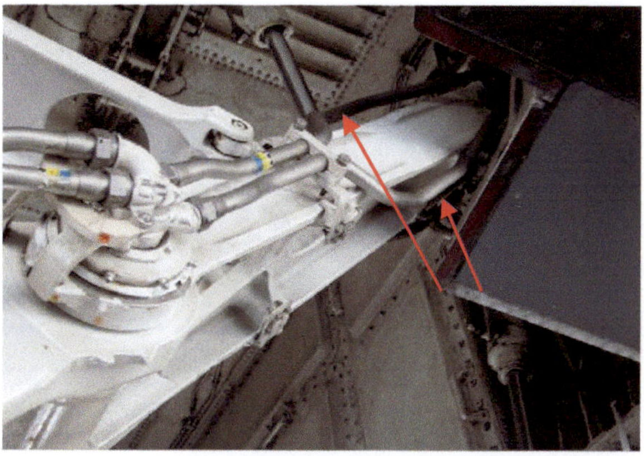

Fig. 10.11 Wheel well titanium hydraulic lines that chafe during flight

Fig. 10.12 B-1 spoiler actuator lines that were coated by the cold spray process

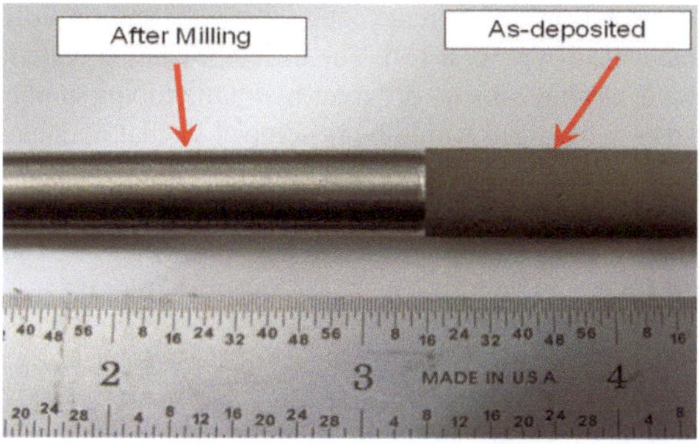

Fig. 10.13 As-sprayed CP Ti on the *right* and machined on the *left*

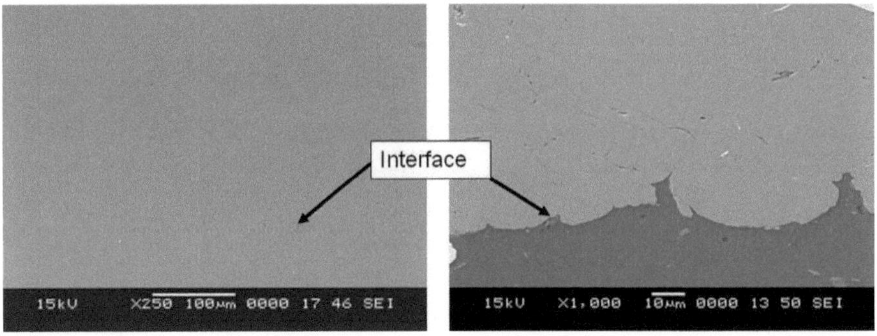

Fig. 10.14 Scanning electron microscopy of the interface between the cold spray CP Ti coating and the Ti3Al2.5V hydraulic line showing high density and mechanical mixing

adequate deposition efficiency (~70 %), bond strength (>12 ksi), density (~99 %), and hardness (93HRB). Additionally, the CS deposit was subjected to burst testing >16,000 psi and complete optical and electron microscopy (Fig. 10.14) of the microstructure, bond line integrity and porosity (<1 %, Leyman et al. 2012). An operational wear test was also accomplished to verify the results of the initial feasibility study. The test involved two B-1 aircraft from Ellsworth AFB, SD, USA. They had two previously identified hydraulic lines (per aircraft) that have a high incidence of chafing. The hydraulic lines of the main landing gear wheel well (behind the follow-up door) and wing spoiler actuator were sprayed with CP Ti to prevent the chafing.

Once the coated lines were installed, an operational wear test was conducted for approximately 5 months. The test involved performing weekly dimensional measurements of the CS coating to determine the effectiveness of the coating, as it related to chafing prevention. The coating was applied to a B-1 nose landing gear (NLG) accumulator hydraulic line (commonly referred to as a curly-Q line), installed, and has been flown for 4 years with no adverse effects and no observed chafing beyond limits and have several thousand flight hours to date.

Since the conclusion of the wear test, proposed logistical and maintenance processes to implement the new process were developed, including the research and development of different coating materials and application parameters based on the outcome of the wear test. The different coating materials will account for varying wear mechanisms on different hydraulic tubing applications. Finally, an indicator layer/mechanism would be developed so that maintenance personnel can identify when the sacrificial layer has been worn through.

10.1.4.3 AH-64 Apache Mast Support Repair

Corrosion and mechanical damage has rendered a number of AH-64 mast supports non-serviceable for continued use on the AH-64 Apache helicopter (Fig. 10.15; Leyman and Champagne 2009). The US ARL developed a portable CS repair that has been shown to have superior performance in the qualification tests conducted, is inex-

Fig. 10.15 AH-64 Apache helicopter mast support

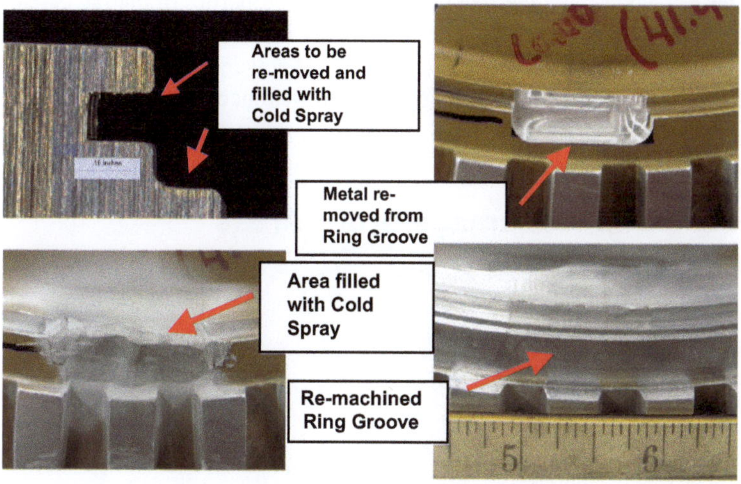

Fig. 10.16 Step-by-step process to repair the snap ring groove of the AH-64 by cold spray

pensive, can be incorporated into production, and has been modified for field repair, making it a feasible alternative over competing technologies. The goal of this effort was to repair both the corrosion and mechanical damage by blending and machining damaged areas, rebuilding lost material using CS with aluminum powder and finish blending and machining to the original dimensions (Fig. 10.16; Leyman and Champagne 2009). The mast support is fabricated from aluminum alloy 7149, and therefore protective finishes such as a conversion coating, primer, and topcoat may be applied to the repaired areas after the CS coating. While it is possible to perform a remediation without rebuilding lost material, the number of times this type of repair could be performed would be limited. With the use of CS to rebuild lost material, the component could be remediated as many times as necessary until its safe-life has been reached.

10.1.4.4 Aluminum Fan Case Corrosion Repair

Corrosion in the form of pitting is common on the fan cases of gas turbine engines due to the retention of water after flight operations. This is particularly pronounced between the 4 o'clock and 8 o'clock region where stagnant water usually resides (Fig. 10.17; Koh et al. 2012). The existing repair procedure allows the pits to be blended to a certain depth. However, beyond certain blending limits, the wall thickness of the fan cases falls below the structurally acceptable margins, deeming thc fan cases unserviceable.

Current repair methods may allow a variety of techniques such as plasma spray, HVOF, and epoxy bonds for the dimensional restoration of these aluminum fan cases. However, these repair repairs offer no structural advantage when applied to the affected area. Although the cases may be restored dimensionally, these existing repair methods are of no use as the underlying structure still violates minimum dimension conditions. Fusion welding processes, while capable of producing structural repairs, often result in unacceptable distortion due to the thermal stresses the fan cases are subjected to. In addition, the high temperature which fusion welding generates has a detrimental effect on the material properties of the fan cases.

The use of CS as a repair method was explored as an alternative (Koh et al. 2012). Spherical 6061 aluminum alloy (Al-6061) powder (Valimat, Fig. 10.18; Koh et al. 2012) with a size ranging from 5 to 50 μm was successfully coated on Al-6061 substrates and evaluated.

Fig. 10.17 Area on gas turbine engine fan case which is prone to corrosion due to retention of water

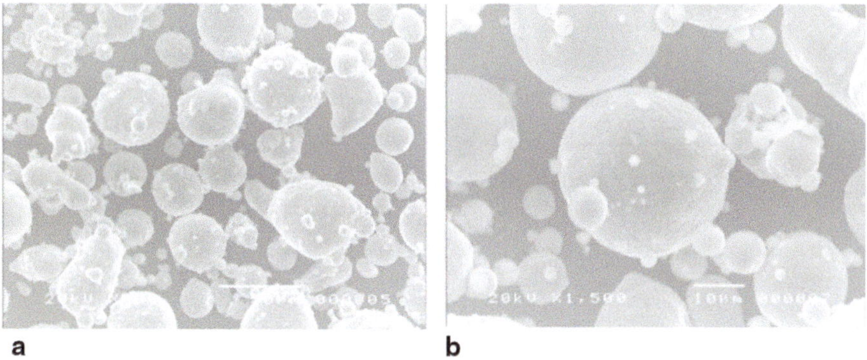

Fig. 10.18 Scanning electron microscopy (SEM) images of the Al-6061 feedstock powder at **a** 500× and **b** 1500×

The microstructure of the aluminum coating was analyzed and the scanning electron microscopy (SEM) pictures of the Al-6061 coating at different magnification are shown in Fig. 10.19 (Koh et al. 2012). The coating thickness was between 500 and 600 μm. The images demonstrated that there is very low porosity in the coating, consistent with typical coating obtained using a low-temperature CS process. This result is yet another validation of the ability of the CS system to produce high-density coating with little or no oxidation. The well-defined interface between the coating and substrate also suggested good adhesion of the coating to the base material.

Microhardness measurements were performed using a 100-g load ($HV_{100\ g}$) and a dwell time of 15 s. The microhardness results are the average of five measurements. An average microhardness value of 104.7 $HV_{100\ g}$ was obtained for the Al-6061 coatings. This is in comparison to the 103.8 $HV_{100\ g}$ mean value obtained for the base Al6061 material. The increased microhardness of the coating is likely to be the result of a gradual compaction due to the constant and repeated impact of the cold-sprayed impinging particles. The lack of porosity in the underlying layers observed by SEM in the deposited material as shown in Fig. 10.19 (Koh et al. 2012) is evidence of the coating coherency.

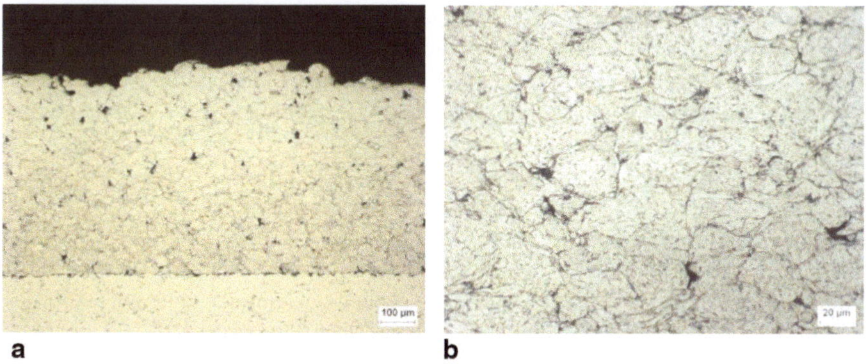

Fig. 10.19 SEM micrographs of a cross-section of the Al6061 coating at **a** 100× magnification and **b** 500× magnification

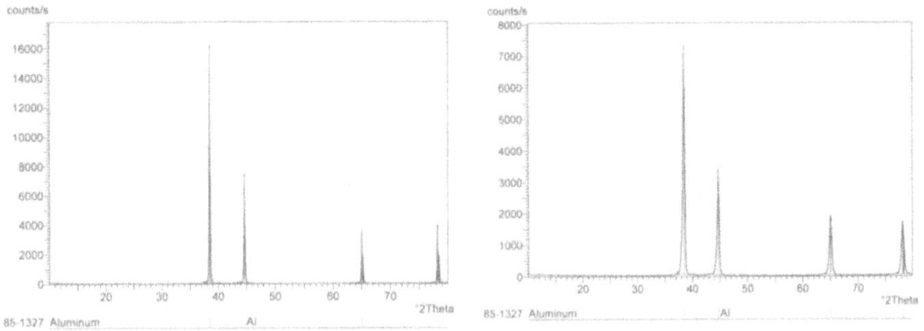

Fig. 10.20 XRD patterns for **a** Al-6061 powder and **b** coatings produced using helium gas as carrier gases

Tensile bond strength testing was carried out in accordance with ASTM C 633 standards. Circular test studs measuring 25.4 mm in diameter and 38.1 mm in overall length were first grit-blasted prior to the CS coating. An average of coating thickness of about 200 μm was obtained for all the test samples. The cold-sprayed coatings failed at 34±6 MPa, partly in the coating–substrate interface and partly in glue-pull off bar interface. Comparatively, the bond strength results for plasma-sprayed coatings performed under identical standards yielded much lower average adhesion strength of 20 MPa, indicating significantly stronger tensile bond strength for the cold-sprayed Al-6061 coatings.

X-ray diffraction (XRD) measurements were carried out using the Philips X'Pert X-ray generator diffractometer with a Cu K_{α} operated at 40 kV, 40 mA radiation at 2-theta-step increments of 0.02°, count rate of 1 °/min, fixed divergence angle of 1°, receiving slit width of 1°. Figure 10.20 (Koh et al. 2012) shows the XRD patterns for the Al-6061 powders together with the coating obtained using helium as the carrier gas. The coatings deposited were about 0.5 mm in thickness. The results show minimal differences between the starting powder and the coatings, implying that no microstructural changes took place during the spraying process, thus verifying the ability of the CS process to retain the feedstock properties in the coating.

The CS technology presents immense opportunity in providing a new platform to address corrosion issues for the gas turbine engine fan case. The results obtained indicated that the cold-sprayed Al-6061 coatings were almost identical in content from the feedstock powder, exhibiting no oxidation and very low porosity. They also exhibited superior adhesion strength and increased microhardness properties. The study validated the possibility of using the CS technology as a viable alternative repair.

10.1.4.5 Landing Gear

High-strength nickel alloys (Inconel) are commonly used in the aerospace industry due to their superior mechanical properties coupled with high chemical and ther-

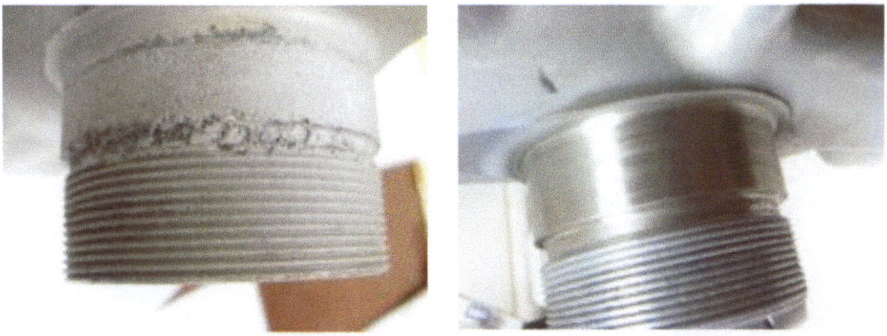

Before After

Fig. 10.21 Before and after photos of cold-sprayed repaired B737 nosewheel steering actuator barrel. (MOOG Aircraft Group 2012)

mal resistance. However, nickel alloy components are usually subjected to high thermal and mechanical stresses during flight and may result in wear and damages which deemed them unserviceable. One such component is the nosewheel steering actuator barrel which is located on the front of the aircraft. Due to its location, it is regularly subjected to moisture and dirt which work themselves into the joints when the landing gear is engaged, causing it to corrode.

Corroded nosewheel steering actuator barrel can be successfully repaired using cold-sprayed nickel alloy coatings. The process involves firstly the removal of the corroded area by pre-machining the barrel. The barrel is then restored using cold-sprayed nickel metallic powder. The component is subsequently machined to the original part dimensions. The repair not only brought in significant cost savings by eliminating the need to purchase replacement parts but also provided greater corrosion protection (Fig. 10.21; MOOG Aircraft Group 2012).

The use of ion vapor deposited (IVD) aluminum coating as an alternative to cadmium plating for high-strength steel landing gears was started in recent years. The IVD aluminum coatings however are susceptible to corrosion once they are damaged during flight operation. CS aluminum process has been established as a repair procedure to address the corrosion issue. The cold-sprayed aluminum has been shown to be non-embrittling to the high-strength steel as well as superior in terms of meeting corrosion requirements of MIL-TL-83488 specifications. One of its greatest advantages has to be its ability to be performed in the field with the use of a portable CS system (Fig. 10.22; Birtch et al. 2008). The costs associated with the removal of the landing gears from operation for repair is thus minimized.

Fig. 10.22 On-site cold spray aluminum repair of IVD aluminum coating using a portable cold spray system. (Birtch et al. 2008)

10.2 Antimicrobial Copper Coatings

10.2.1 Introduction

Bacterial growth on surfaces is a cause of concern in many medical-based industries due to the possibility of increased risk of bacterial infection (Page et al. 2009). The bacterial contamination of hospital surfaces, including patient rooms, nurse stations, and kitchens has been extensively documented (Rutala et al. 1983; Bernard et al. 1999; White et al. 2007; White 2006). Infections acquired during US hospital stays kill more people than breast cancer, auto accidents, and AIDS combined, with total costs of US$ 35–45 billion per year (White 2006).

10.2.2 Antimicrobial Copper

In the past few decades, work has been done on the antimicrobial properties of copper and its alloys against a range of microorganisms threatening public health in food processing and health-care applications (Faúndez et al. 2004; Grass et al. 2011). The use of copper and copper alloys for frequently touched surfaces such as door and furniture hardware, bed rails, IV poles, dispensers, taps, light switches, and cooking and food preparation surfaces can help reduce the amount of disease-causing microbes in hospitals and food dispensing organizations. Michels et al. 2005) showed that increasing copper content of alloys increased antimicrobial effectiveness against methicillin-resistant Staphylococcus aureus (MRSA), and copper-containing alloys were significantly more effective than stainless steel, see Fig. 10.23.

The US Environmental Protection Agency (EPA) registers 355 copper alloys with public health claim (Handbook of Thermal Spray Technology). All of the al-

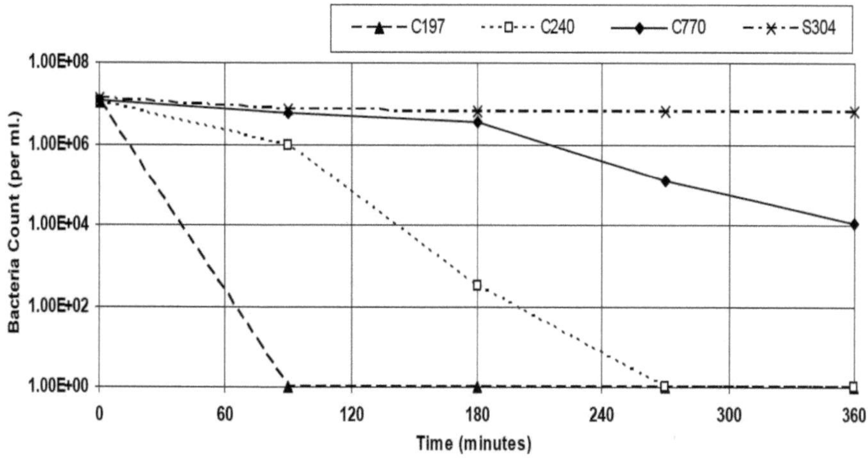

Fig. 10.23 MRSA viability on copper alloys and stainless steel surfaces at 20 °C

loys have minimum nominal copper concentrations of 60 %. The registration of copper and certain copper alloys such as brass and bronze means that the EPA recognizes these solid materials' antimicrobial properties. Products made from any of the 282 registered alloys are legally permitted to make public health claims relating to the control of organisms that pose a threat to human health. Laboratory studies conducted under EPA-approved protocols have proven copper's ability to kill, within 2 h of contact time, more than 99.9 % of the following disease-causing bacteria: *Staphylococcus aureus, Enterobacter aerogenes, Escherichia coli* O157:H7, *Pseudomonas aeruginosa, Vancomycin-resistant Enterococcus faecalis* (VRE), and MRSA.

10.2.3 Effect of Copper Characteristics

In order to make use of the antimicrobial ability of copper, surfaces that contact skin and foods should be composed of pure copper or copper alloy. This can be accomplished by either making equipment out of copper or coating the surface of equipment with copper. In general, cost considerations favor copper coatings over solid structural copper, and various metal spray techniques are available for the purpose of depositing copper on surface prone to transmission of harmful microorganisms.

The method by which copper destroys bacteria is not well understood, but most theories propose bacterial destruction by copper ions (Santo et al. 2011). Copper ions have the capacity to kill bacteria by destroying their cell walls and membranes. Cu^{2+} ions can penetrate the cell membrane through opening or closing of membrane channels. This alters the permeability of cellular membranes, which causes leakage of intracellular ions and low molecular weight metabolites. At the same time, Cu^{2+}

ions entering cells combine strongly with intracellular amino acids and proteases, resulting in degeneration that leads ultimately to the denaturation of proteins.

Several thermal spray technologies are capable of depositing a copper coating; however, the characteristics of CS deposits are quite unique, having significant advantages over other thermal spray methods. The process does not use thermal energy to melt the particles to be deposited but instead relies upon the supersonic impact of the particles on the substrate. Copper crystalline structure is not changed through melting. High impact forces produce a dense, nonporous deposit. The resulting electrical conductivity of the deposit is near that of wrought copper. High electrical conductivity yields high release of Cu^{2+} ions, yielding superior antimicrobial performance.

10.2.4 Deposition Method Comparison

In order to evaluate their antimicrobial efficiency, copper coatings were applied to coupons by three thermal sprays—plasma spray, wire arc, and CS (Champagne and Helfritch 2013). Approximately, 1-mm-thick coatings were applied to aluminum substrates. The coatings completely covered the metal substrates with an impervious seal. The coated coupons were inoculated with MRSA, and the plated samples were then held at room temperature for 2 h, after which survivors were resuspended and cultured, per standard EPA protocol, "Test Method for Efficacy of Copper Alloy Surfaces as a Sanitizer" (http://epa.gov/oppad001/pdf_files/test_meth_residual_surfaces.pdf). Two hours of MRSA contact with the abovementioned coatings resulted in the survival rates shown in Fig. 10.24. The spray method, hardness achieved, and microscopic cross section are identified for each test. The CS coatings are seen to yield a decrease in microbial survivorship by more than five orders of magnitude, while the other spray methods produced decreases less than two orders of magnitude. The material properties of the cold-sprayed coating are characterized by high hardness and low porosity.

The results show a greater-than-three order of magnitude difference in kill efficiency between the plasma and wire arc methods and the CS method of copper deposition. This large difference in antimicrobial effectiveness between copper spray deposition methods requires an examination of how the deposition mechanism affects the nature of the copper. The plasma and wire arc methods deposit molten particles at relatively low velocity (<200 m/s). The CS method deposits solid particles at high velocity (>600 m/s). It is known that the high-velocity impacts of cold-sprayed particles leads to extreme work hardening and correspondingly high dislocation density within the deposit, as explained in Chap. 4. It is also known that ion diffusion is augmented by the presence of dislocations through "pipe diffusion" and that the ionic diffusion pathway is principally through these dislocations. Dislocation density is proportional to the square of Vickers hardness, and ion diffusivity is directly proportional to displacement density. Ion diffusivity, thus, varies as the square of hardness. The diffusion of copper ions can therefore be significantly

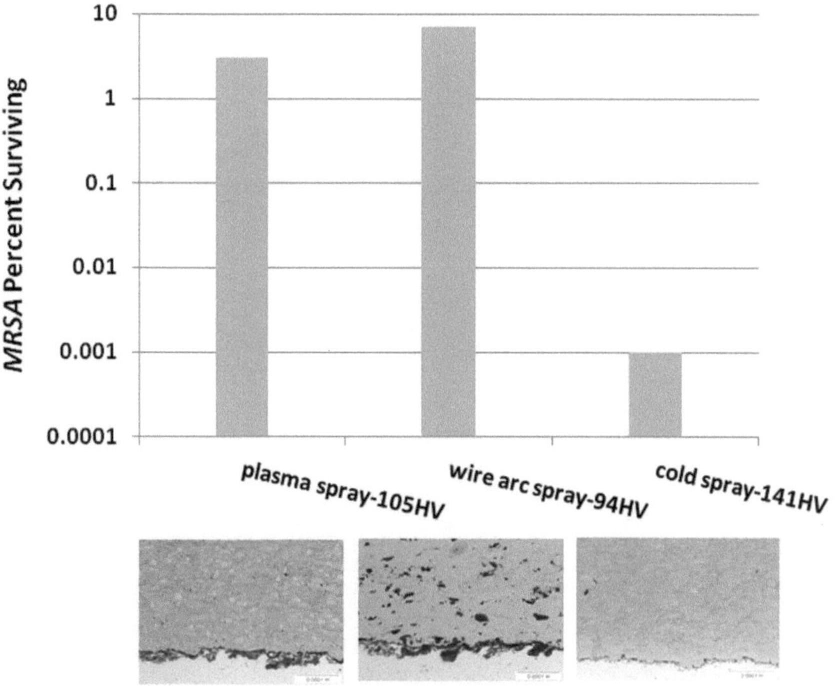

Fig. 10.24 Sterilization by various surfaces. *MRSA methicillin-resistant Staphylococcus aureus*

increased through the hardness increase produced by the CS process, which serves to enhance the flow of Cu^{2+} ions needed for microbial destruction.

The significant microbiologic differences between coatings produced by different spray techniques demonstrate the importance of copper application technique. The CS method shows superior antimicrobial effectiveness resulting from the high impact velocity imparted to the sprayed particles, which subsequently results in high dislocation density and high ionic diffusivity. The CS process can readily apply copper coatings onto touch surfaces. Figure 10.25 illustrates a hospital tray and the entire metal support structure that have been both coated with pure copper by CS.

10.3 Bactericide Coatings

10.3.1 Introduction

Conventional organic agents have traditionally been used as antibacterial tools. However, these agents usually contain noxious ingredients which can be harmful to humans. The use of ceramic powders as an alternative substitute has been explored by Zhang et al. (2007); some of these powders have been shown to exhibit

Fig. 10.25 Hospital tray copper coated by cold spray

significant antibacterial activity without the presence of light. Moreover, the use of these ceramics has the advantage of containing mineral elements that are essential to the human body (Yamamoto et al. 2001). One of these materials is ZnO, as explained in the next section.

10.3.2 ZnO–Ti Composite Antibacterial Coatings

ZnO–Ti material shows limited sprayability when deposited using conventional thermal spray processes; as explained in other chapters, conventional thermal spray processes tend to modify the chemistry of the deposited materials. Because of its low-temperature attributes, CS appears as a more viable alternative. However, in the absence of heating, the coating of brittle ceramic materials can be challenging. Therefore, in order to circumvent this obstacle, the ceramic material must be deposited as a composite containing a ductile phase, such as Ti. This has already been demonstrated by the successful coating of zinc oxide–titanium (ZnO–Ti; Sanpo et al. 2011) composite powder. In this study, ZnO–Ti coatings were deposited on Al-6061 substrates using CS parameters of 13–15 bars of helium gas at a temperature of between 300 and 400 °C. Three different ZnO–Ti composite powders comprising ZnO to Ti weight ratios of 20:80, 50:50, and 80:20 were prepared and coated. Energy dispersive X-ray (EDX) analysis of the coatings verified that the composi-

Table 10.1 ZnO–Ti contents (wt.%) in the powders and coatings

Powder composition	Powder feedstock		Coating	
	ZnO	Ti	ZnO	Ti
ZnO 20/Ti 80	20	80	9.45	90.55
ZnO 50/Ti 50	50	50	33.41	66.59
ZnO 80/Ti 20	80	20	53.78	46.22

tion of the Ti was higher in the coating than that of the feedstock (Table 10.1). The quantitative results of the coating compositions performed on the deposition seem to suggest the binding work of ductile material in holding the brittle materials together as the composite materials impact onto the substrate.

The coatings were subsequently put through an antimicrobial experiment. A bacterial qualitative test was performed by introducing *E. coli* onto luria broth (LB) agar surface which was poured onto the Petri dishes. The cold-sprayed samples were previously placed perpendicular to the base of the petri dish. The results presented in Fig. 10.26 show that all cold-sprayed samples have a distinctive killing effect on *E. coli*.

A lack of *E. coli* colonies growing in the region surrounding the ZnO–Ti-coated samples was observed and appeared as clearance zone. The area of the clearance zone increased with increasing ZnO powder concentration in the composite powder feedstock and cold-sprayed coating, showing that the killing effect on *E. coli* increased with increasing amount of ZnO.

To perform bacterial quantitative test, the *E. coli* stock was stored inside microcentrifuge tubes at −80 °C. Using a sterile wire loop, a single colony was extracted and streaked onto another agar plate. The plates were further incubated upside-down at 37 °C for another 24 h. Using a sterile wire loop, an isolated colony was placed into a test tube containing 10 ml of broth, and vortexed for 60 s. 100 μl of solution was plated onto agar dishes using a spreader. The plates were again incubated at 37 °C (upside down) for 24 h. Choose and count the number of colonies in plates containing 30–300 colonies per plate. One colony represents a colony-forming unit (CFU).

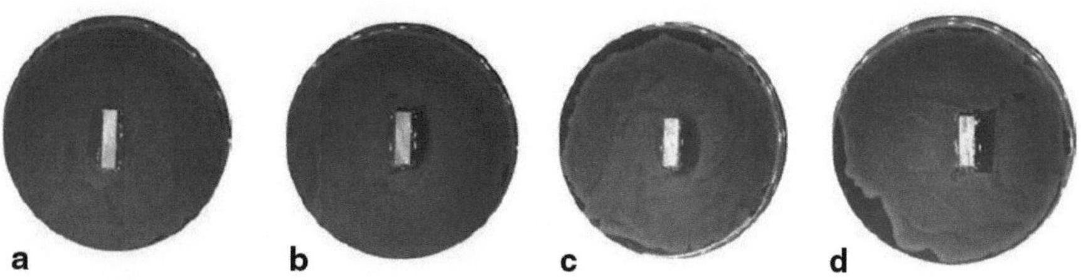

Fig. 10.26 Qualitative analysis on the antibacterial properties of ZnO–Ti coatings. A clearance zone (absence of *E. coli* colonies) surrounding the coated samples was observed on samples **b** ZnO 20/Ti 80, **c** ZnO 50/Ti 50, and **d** ZnO 80/Ti20 in an increasing manner compared with Al 6061 without coating (**a**)

Determine the geometric mean of the number of *E. Coli* from the triplicate incubation period control and incubation period treated samples by the following equation:

$$\text{Geometric mean} = (\log 10X1 + \log 10X2 + \log 10X3)/3\,,$$

where X is the number of organisms recovered from the incubation period control (substrate without coating) or incubation period treated samples (substrate with coating).

Use the following equation to calculate the percentage reduction of *E. coli:*

$$\%E.Coli \text{ reduction} = [(a-b) \times 100]/a,$$

where a and b are the antilog of the geometric mean of the number of *E. Coli* from the incubation period control and treated samples, respectively.

Figure 10.27 shows that all cold-sprayed samples displayed an even more pronounced killing effect on *E. coli*. The killing rate increased with increasing ZnO powder concentration in the composite powder feedstock and cold-sprayed coating.

Table 10.2 presents the percentage reduction of *E. coli* when placed on coatings of various ZnO–Ti composition. The results show that the reduction was 13.34, 25.38, and 32.06% for ZnO 20/Ti 80, ZnO 50/Ti 50, and ZnO 80/Ti 20 coatings, respectively, indicating a direct proportional relationship between concentration of ZnO and the percentage reduction of *E. coli*.

10.3.3 HA-Ag/PEEK Antibacterial Coatings

The antibacterial property of cold-sprayed silver-doped hydroxyapatite/poly-ether-ether-ketone (HA-Ag/PEEK) has also been investigated (Sanpo et al. 2009). The study was built on previous work which demonstrated the antibacterial effects of

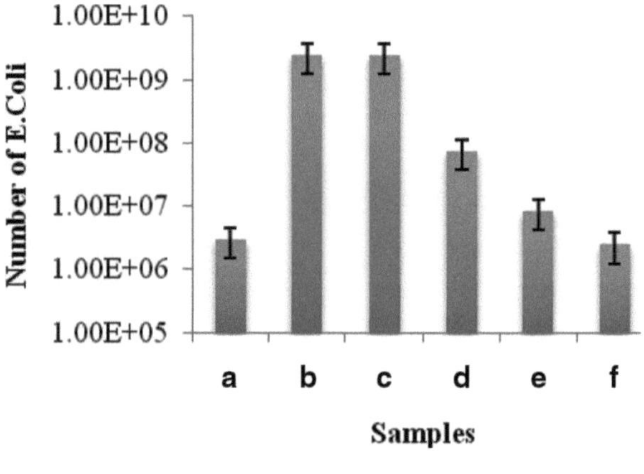

Fig. 10.27 Quantitative analysis of antibacterial properties of ZnO–Ti coatings: *a E. Coli* at 0 h, *b E. Coli* at 24 h, *c* Al6061 substrate, *d* ZnO 20/Ti 80, *e* ZnO 50/Ti 50, and *f* ZnO 80/Ti 20

Table 10.2 *E. coli* reduction calculation for various ZnO–Ti composition coatings

Samples	Calculation	
	Geometric mean (*X*)	Percent reduction of *E. Coli* (%)
Control sample (without coating)	9.132	–
ZnO 20/Ti 80	7.914	13.34
ZnO 50/Ti 50	6.814	25.38
ZnO 80/Ti 20	6.204	32.06

silver-doped hydroxyapatite coatings on implants (Feng et al. 1998). In addition, the incorporation of Ag+ ions into microporous hydroxyapatite (HA) coatings was shown to be an effective bioactive delivery system for the slow release of antibiotics (Shirkhanzadeh et al. 1995). PEEK was also considered because of its excellent thermal stability, friction reduction, and wear resistance (Yin et al. 2008) and its possible application to lower friction and wear of metallic substrates (Liao et al. 2001).

In this study, HA-Ag and PEEK nanopowders were mixed in the ratios of 80:20, 60:40, 40:60, and 20:80 (wt.%), blended in a ball mill for 24 h. PEEK was added as the ductile binder for the more brittle HA-Ag ceramic powder. The composite nanopowders were subsequently cold sprayed using parameters of 11–12 bar of compressed air at preheated temperature of 150 and 160 °C onto glass substrates (Fig. 10.28) and tested for bacteriostatic activity against *E. coli*. An average coating thickness of between 30 and 40 μm was obtained for all coatings.

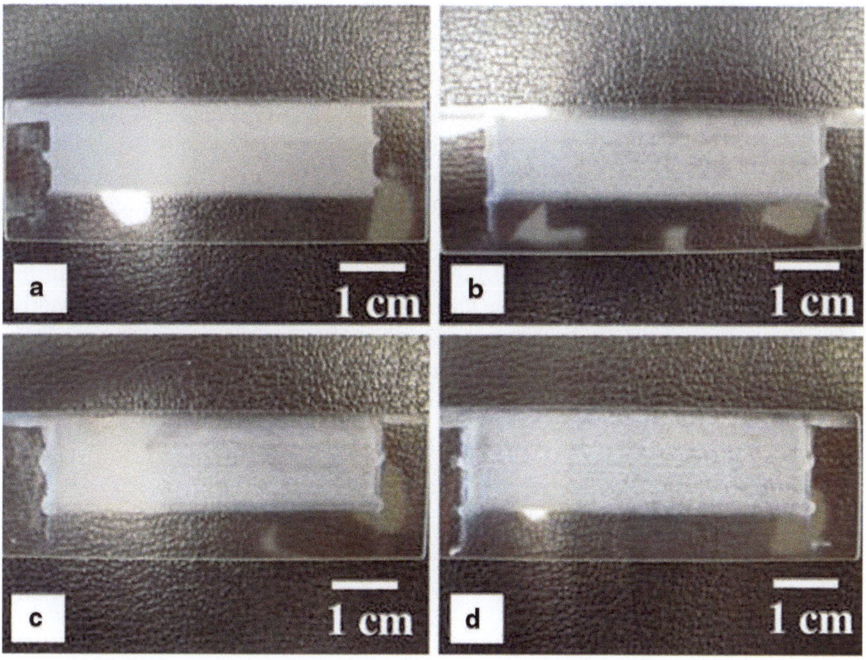

Fig. 10.28 Cold-sprayed coating samples of **a** HA-Ag 20/PEEK 80, **b** HA-Ag 40/PEEK 60, **c** HA-Ag 60/PEEK 40, and **d** HA-Ag 80/PEEK 20

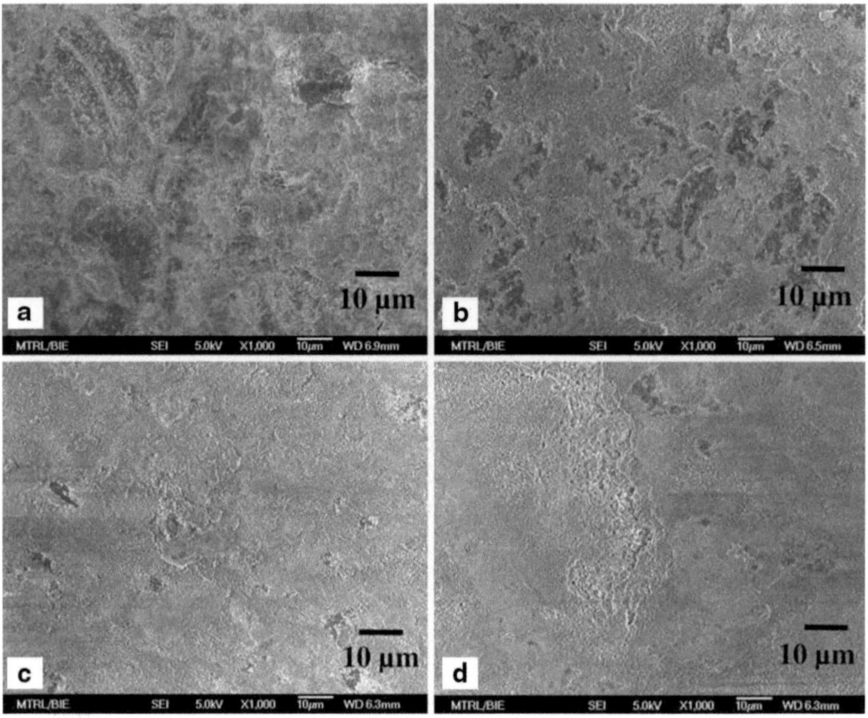

Fig. 10.29 SEM images of **a** HA-Ag 20/PEEK 80, **b** HA-Ag 40/PEEK 60, **c** HA-Ag 60/PEEK 40, and **d** HA-Ag 80/PEEK 20 coatings

The SEM images of the coating (Fig. 10.29) revealed that the surface of the cold-sprayed coating consisted of HA-Ag powder embedded in a continuous PEEK matrix. EDX analysis verified comparable HA-Ag/PEEK contents in the starting powders and as-sprayed coatings, inferring that the phase composition and ratio of the powder characteristics remained unchanged during the deposition process. In the bacterial quantitative test, the cold-sprayed HA-Ag/PEEK samples demonstrated pronounced killing effect on the *E. coli* (Fig. 10.30). The antibacterial activity increased with the increasing HA-Ag/PEEK nanopowder concentration in the coating. This study demonstrated the ability of CS in depositing a ceramic material (HA-Ag), a nanophase, and composite powder (HA-Ag/PEEK) which can retain and elicit a coating functionality (antibacterial) similar to the starting material.

10.4 Tribological Coatings

10.4.1 Introduction

Wear of metallic components is a major issue in a number of different industries such as automotive, aerospace, tooling, agriculture, and paper production. Several

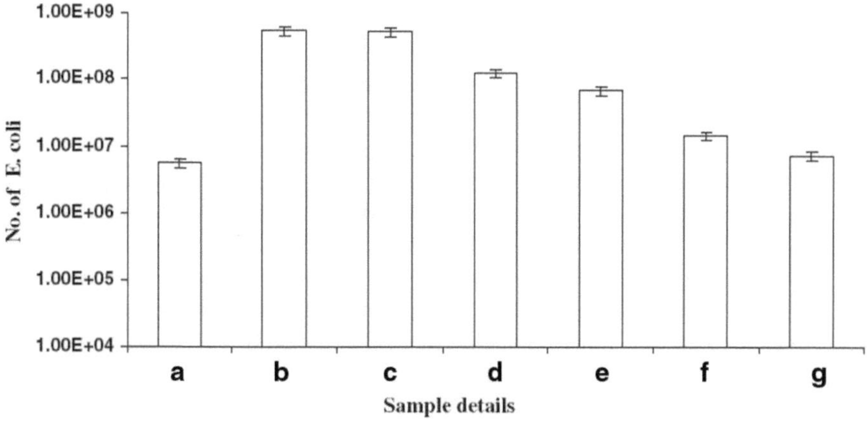

Fig. 10.30 Quantitative analysis on the antibacterial properties of HA-Ag/PEEK coatings: *a E. coli* at 0 h, *b E. coli* at 24 h, *c* pure glass, *d* HA-Ag 20/PEEK 80, *e* HA-Ag 40/PEEK 60, *f* HA-Ag 60/PEEK 40, and *g* HA-Ag 80/PEEK 20

technologies and approaches have been developed to mitigate wear damage. One of the most common is the application of wear-resistant coatings. The type of wear resistant coating and application method are determined by the material system, the operating environment (type of wear, lubrication, temperature, etc.), cost and the safety/reliability requirements. Improper selection of the wear coating or application method can lead to increased wear, damage to critical components and even complete failure. The continuing improvement in equipment, powders, and process development has greatly increased the use of CS for depositing wear resistant coatings.

10.4.2 Wear Modes

The three main classes of wear are mechanical, chemical, and thermal (Kato 2002; ASTM 1987). Mechanical wear is one of the wear modes,CS technology can effectively address. Mechanical wear is mainly controlled by deformation and fracturing of the material and can be divided into subcategories including adhesive (sliding), abrasive, fretting, erosive (impact), and other types. The first subcategory is adhesive wear which occurs when two bodies are in sliding contact and there is material transfer between the two surfaces. The material transfer causes plastic deformation and damage to the surface. The second category is abrasive wear which occurs when there is a progressive loss of material due to relative motion between that surface and a contacting substance or substances. The third subcategory is fretting wear which occurs when there is a very small oscillatory movement between two solids which results in a continuous cycle of microwelding and breaking the microwelds which results in cracks and loss of material. The mechanisms of erosive and impact wear due to solid particle impingement are very complex and poorly understood. No single coating material exists which offers erosion resistance properties at both high angle (60–90°), as well as low angle (20–40°) hard particle impingement. In general, high hardness (brittle) materials (carbides, nitrides and borides) provide excellent erosion resistance against particles with an impingement angle of less than 45°. However, erosion of the hard brittle materials increases at the higher angles of hard particle impingement, that is, 90°. In comparison, metallic (ductile) materials such as nickel-based alloys including Stellite 6 and Nucalloy45 show increased erosion resistance at the higher angles (90°), but poor resistance at the oblique (20–35°) angles of hard particle impingement. Therefore, an intermediate coating system for mitigating erosion is often WC-Co, WC-Ni, Cr, or NiCr-Cr_3C_2 coatings which can be applied by CS techniques. These coatings often provide a good balance against erosion under a variety of hard particle impingement angles.

10.4.3 Wear-Resistant Coatings

Wear can be reduced by increasing the surface hardness of the material that is wearing, reducing the friction between surfaces or a combination of both. The selection of the wear coating will depend on the working environment, substrate material, the coating material and the CS equipment and processing parameters. From the late 1990s to late 2000s, several attempts were made at developing methods to produce wear-resistant coatings for hard-face materials (Wolfe et al. 2006). A number of the various powder systems used for wear resistant coatings in the thermal spray industry have also been evaluated for deposition by CS. These included the cermets such as tungsten carbide cobalt (WC-Co), Nickel chrome-chrome carbide (NiCr-Cr_3C_2), and stellites. Attempts were made to apply these coatings to hard surfaces such as roller and drive shafts. These types of applications proved to be challenging for CS deposition as they required the use of high hardness powders which lacked the ductility necessary for successful bonding and coating build up. If the coatings did build up, they typically had poor adhesion strength and high amounts of porosity. Therefore, in order of depositing higher hardness ceramic materials, additional ductile material needed to be added to increase the adhesion strength and reduce the porosity, which unfortunately resulted in lower hardness and subsequently reduced wear resistance. Most recently, improvements in CS equipment (higher gas pressure and temperature) and powder availability have made it possible to deposit harder materials.

10.4.3.1 Stellite 6 on Carbon Steels

Stellite® cobalt alloys have complex carbides in a CoCr-based alloy matrix. They are commonly used to improve resistance to wear, corrosion, and galling during exposures to high temperatures. Applications include hard-facing of valves, valve seats, spindles, shafts, and rotating parts. Stellite 6 is the most used of the Stellite alloys and is frequently deposited using thermal spray methods. High-temperature deposition methods can lead to large thermal stresses and dissolution of the coating in the base material. These can reduce the properties of the coating.

Some recent work indicates the possibility of successfully depositing Stellite 6 on low carbon steel substrates using upstream injection CS equipment with nitrogen as the carrier gas at 800 °C and 38 bar (Cinca and Guilemany 2013). Results showed very low levels of porosity, high hardness (600 VHN) and abrasive wear rates measured following ASTM-G65-00 were on the same order as coatings applied by HVOF. The CS process reduces the residual stresses and eliminates any dissolution. However, additional work is needed improve deposition efficiency and characterize the adhesion, corrosion, and wear resistance before the process can be applied commercially.

10.4.3.2 WC-Co on Carbon Steels

WC-Co is a very common wear resistant coating that can be applied by a number of different thermal spray processes, such as HVOF, flame spray and plasma spray. It has high hardness, excellent wear resistance, and strength. WC-Co coatings are used extensively in several different industries such as oil and gas, petrochemical, and mining.

Current work has shown that there are applications where the application of WC-Co with CS is possible. Several researchers have published articles on the deposition of WC-Co coatings on steel and aluminum (Dosta et al. 2013; Kim et al. 2005a, b; Couto et al. 2013). During deposition WC-Co with traditional thermal spray methods, such as HVOF, the inherent thermal cycle, and exposure to combustion gases can result in phase transformation, decarburization, porosity, and oxidation. CS offers an alternative method which may preserve the good qualities of the feedstock while providing dense and high cohesive and adhesive bonding strength. One of the stumbling blocks for the implementation of CS in this type of application is the ability of CS to economically compete with the established HVOF as, in many industrial circles, the latter produces coatings that are acceptable. Recall that no powder melting occurs during the CS process so that unique alloy compositions can be maintained as well as prevent unwanted chemical reactions, especially at the interface that could reduce adhesion. Other areas of focus are varying the amount of Co binder as well as the size of the WC powder to improve deposition efficiency, hardness, and coating performance. Yet, other possibilities include the blending and/or encapsulation of WC-Co particles with more ductile materials, such as nickel, aluminum, or copper, to enhance sprayability and produce a composite material (Wang and Villafuerte 2009).

10.4.4 Self-Lubricating Coatings

Application of self-lubricating coating materials on different metallic substrates has been studied by many researchers (Segall et al. 1998; Culliton et al. 2013; Olakanmi and Doyoyo 2014; Manoj and Grossen 2011; Pitchuka et al. 2014; Bakshi et al. 2009; Smid et al. 2012). It has been generally shown that a thin lubricating film could largely protect the substrates from damage by significantly increasing wear resistance and reducing surface friction (Stark 2010; Walia 2006). The following summarizes the studies that investigated application of self-lubricating coatings on different substrates, using CS.

Walia (Hu et al. 2009) investigated application of a molybdenum disulfide (MoS_2) lubricant film on turbine blade dovetail joints to increase their wear resistance. Given the fact that MoS_2 particles are vulnerable at high temperatures, CS was used for deposition. The thin lubricant film of MoS_2 increased wear resistance and significantly reduced the coefficient of friction of coated surfaces on the dovetail joints.

In a more recent study, Stark (Walia 2006) reported the performance of hexagonal boron nitride (hBN) lubricant coatings on aluminum 6061 substrates. This material was selected for its lubricious property at high pressures and temperatures. In spite of good lubricating properties, bare hBN particles are considered a poor candidate for CS coating due to their nature, especially in the impact zone. In order to achieve the deformation required and create a bond between the particles and the substrate, hBN particles need to be encapsulated with nickel or other ductile material. The encapsulating nickel provides a ductile medium, which leads to a metallurgical bond between the particles and the substrate, allowing the coating to build up. In general, the results from this study demonstrated promising attributes for the self-lubricating coating material with high bond strength between the coating and substrate, low friction, and high wear resistance. For developing novel multifunctional CS coatings for tribological applications, powder preparation becomes critical in order to ensure that the hybrid coatings designed provide both increased wear resistance and solid lubrication while maintaining durability. For example, the preparation of BN-coated powders requires commercially available hBN powders to be characterized for shape and size distribution before being coating with nickel in order to ensure the optimized particles size and distribution are achieved for the CS process. Depending on the powder size and distribution, further sizing may be required in order to obtain the desired CS powder size and distribution after applying a ductile metallic layer. Other commonly used encapsulation methods include electrophoretic deposition, vapor phase deposition, electroplating, and electroless plating. Hu et al. (Neshastehriz 2014) indicated that electroless plating demonstrated higher corrosion and wear resistance than the other encapsulation methods, but further investigation is required. For encapsulation, catalytic metals are deposited onto particles in a metal salt solution. The success of deposition depends on different parameters such as the plating bath composition, temperature and pH. For example, electroless nickel plating was used to encapsulate micrometer-sized hBN particles (approximately 7 μm in size, Stark et al. 2012). In short, the hBN powders were cleaned in a nitric acid solution in order to improve the nickel bonding to the BN powders. Next, the surface of hBN particles were activated with tin chloride ($SnCl_2$) followed by palladium chloride ($PdCl_2$) solutions. Catalytic layers of tin and palladium ions on the surface of hBN particles provide sites for nickel deposi-

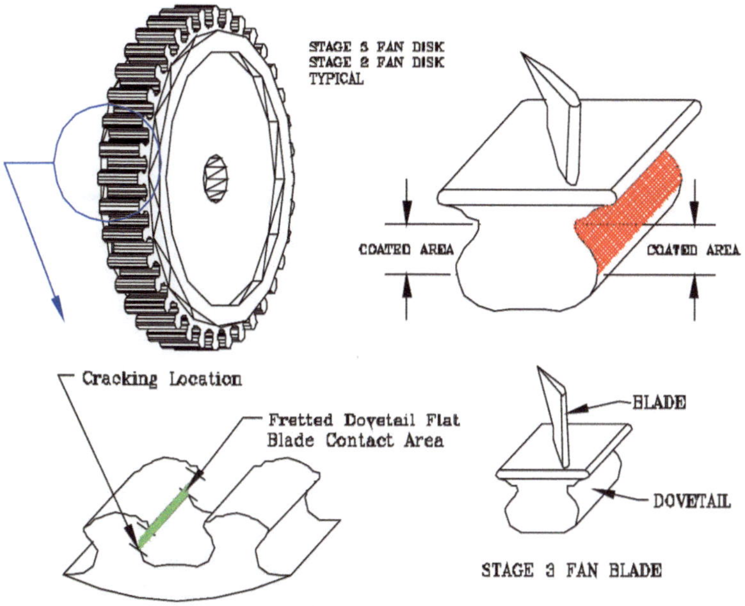

Fig. 10.31 Schematic of a blade and disk showing where fretting occurs. (Hager 2002)

tion. Nickel deposition is then completed in two steps: In the first step, a thin layer of nickel is deposited around the Sn–Pd-activated hBN particles, followed by the second step in which nickel deposition is completed to increase the thickness of the encapsulation layer around hBN particles.

An area where a self-lubricated coating is of great value is the matting surface of compressor or turbine blades in the dovetail joint where the vanes fit in the disk. The self-lubricating coatings can eliminate fretting fatigue in the vanes. The area of the blade that is coated is shown in Fig. 10.31 (Hager 2002).

10.4.5 Fretting Fatigue Resistant Coatings

Hager (2002) applied nickel (Ni), molybdenum (Mo), cobalt (Co), Amdry 9951 (40Co-32Ni-20Cr_3C_2), nickel chrome-chrome carbide (NiCr-Cr_3C_2) coatings to Ti alloy substrates. The coatings were evaluated for resistance to fretting fatigue. A schematic of the wear test is shown in Fig. 10.32. The Amdry 9951 and the nickel chrome-chrome carbide provided excellent wear results but were too abrasive on the mating material. Mo and Ni coating were evaluated and performed well in simulate engine tests. A solid lubricant was applied to the surface of the coatings prior to the engine test. A wear track for a cold-sprayed Mo coating after 3,000,000 cycles is shown Fig. 10.33.

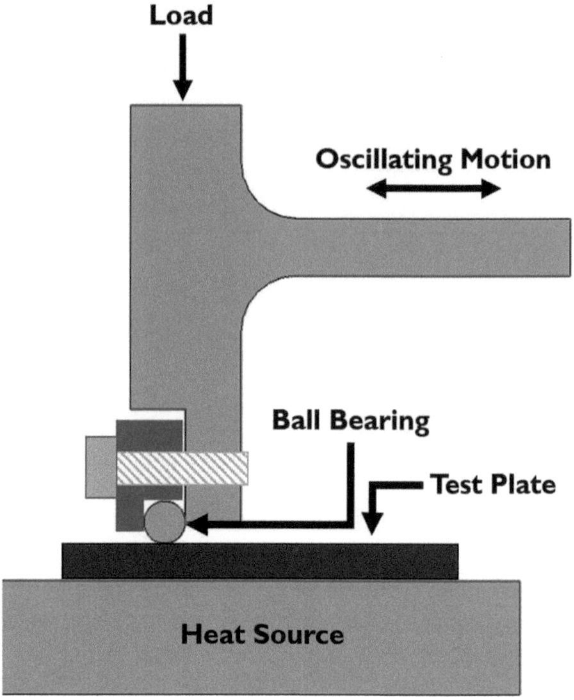

Fig. 10.32 TE77 Plint fretting/reciprocating wear tester utilizing a cylinder on flat configuration. (Hager 2002)

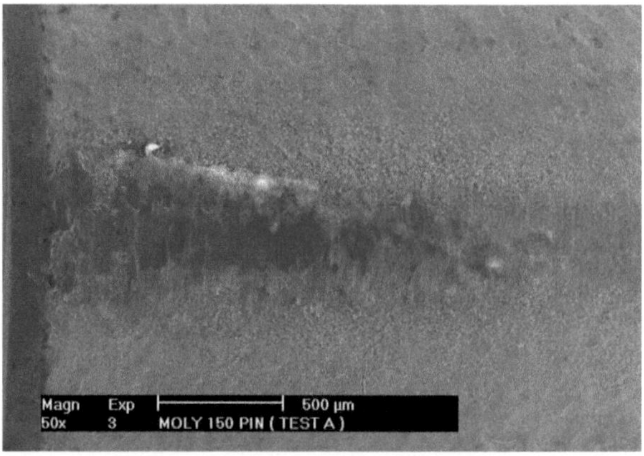

Fig. 10.33 Wear track of a cold-sprayed molybdenum coating after three million cycles of reciprocating wear. (Hager 2002)

10.4.6 Tribology

Wolfe et al. (2006) performed an extensive tribology study on applying wear resistant coatings to 4140 alloy substrate. The CS Cr_3C_2-Ni wear resistant coatings and uncoated 4140 alloys were evaluated against 100Cr6 steel (mating material) for the tribology testing under dry conditions in they reported the average weight loss of the coating, mating material, and friction coefficient of the various CS coatings. The average mean friction coefficient of the various coatings evaluated under dry conditions suggests that the polysulfone (PSU)-blend coating (tailored) had the lowest mean friction values for the coatings evaluated. However, in general, the friction increased with increasing time for all coatings tested under dry conditions. The authors attempted to measure the wear volume of the material estimated from a composite machining center (CMC) six-axis machine which measured the wear profiles. A minimum of four measurements were taken 90° apart were made in order to obtain a statistically weighted average loss of the samples. Most of the CS Cr_3C_2-based coated samples did not have a smooth uniform surface finish, and it was very difficult to accurately determine the amount of wear for the coating. As a result, the degree of wear from the tribology test was based upon the weight loss of the coating using a high precision balance. Overall, the chromium carbide coatings showed significant improvements in the wear rates as compared to the uncoated 4140 alloy, but additional testing was required to better understand the wear rates and friction coefficients. In addition, measurements and calculations were performed on the mating material (100Cr6 6 mm diameter ball) to determine the wear of the mating material against the various coatings. Additional sets of trial experiments were performed with a higher load of 10N and under sand-oil lubricated condition. The type of lubrication used was DTE hydraulic oil with 20 wt.% Iraq sand (more aggressive environment). The tribology results appeared to show mixed results as to which coating performed the best under lubricated conditions. It appeared that some of the coatings gained weight due to the lubricant being incorporated into the porous coatings. The differences in the wear results from the 5 and 10N loads were believed to be the result from localized changes in friction and coating microstructure. However, depending on the wear application it is not adequate to only look at the amount of wear associated with the coating material. Often it is important to consider both the coating and the mating material wear rates. The PSU blend #2 coating showed similar wear results for both the Cr_3C_2-based coating and the 100Cr6 mating material. Similarly, the TAFA 1375V CS coating showed the least amount of wear, but increased amount of weight loss for the mating material. Therefore, it is important to consider the results of each test and consider the results based on the desired application. For the purpose of their study (Wolfe et al. 2006), the authors were looking for a coating material that had uniform wear against 100Cr6 hardened alloy steel (1000VHN0.300).

10.4.7 Concluding Remarks

In summary, CS has proven multifunctional practicability in various applications from dimensional restoration of aluminum and magnesium castings and dimensional restoration of other casting components. It has also shown great possibilities with low temperature deposition of antimicrobial and bactericide materials as well as deposition of multifunctional wear resistant and self-lubricating coatings. By applying multifunctional tailored CS deposits, the surface properties of metallic and nonmetallic components can be enhanced for improved performance and increased life.

References

ASTM. 1987. *Standard terminology relating to wear and Erosion*. 3.2 vols. s.l: Annual Book of Standards.

Bakshi Srinivasa R., Di Wang, Timothy Price, Deen Zhang, Anup K. Keshri, Yao Chen, D. Graham McCartney, Philip H. Shipway, and Arvind Agarwal. 2009. Microstructure and wear properties of aluminum/aluminum–silicon composite coatings prepared by cold spraying. *Surface and Coatings Technology* 204 (4): 503–510.

Balani, K., T. Laha, A. Agarwal, J. Karthikeyan, and N. Munroe. 2005. Effect of carrier gases on microstructural and electrochemical behavior of cold sprayed 1100 aluminum coating. *Surface and Coatings Technology* 195 (2–3): 272–279.

Bernard, L., A. Kereveur, D. Durand, J. Gonot, F. Goldstein, J. L. Mainardi, J. Acar, and J. Carlet. 1999. Bacterial contamination of hospital physicians' stethoscopes. *Infection Control and Hospital Epidemiology* 20 (9): 626–628.

Birtch, W., G. Russell, and S. E. Hale. 2008. *Kinetic spray for corrosion protection and metal part restoration*. Baltimore: Commercial Technologies for Maintenance Activities (CTMA) Symposium.

Champagne, V. K. 2008. The repair of magnesium rotorcraft components by cold spray. *Journal of Failure Analysis and Prevention* 8 (2): 164–175.

Champagne, V. K., and B. Barnett. 2012. Cold spray technology for DOD applications, ASETS defense 2012: Workshop on sustainable surface engineering for aerospace and defense, San Diego, CA, Aug 27–30, 2012.

Champagne, V. K., and D. Helfritch. 2013. A demonstration of the antimicrobial effectiveness of various copper surfaces. *Journal of Biological Engineering* 7:8.

Champagne, V. K., P. F. Leyman, and D. J. Helfritch. 2008. *Magnesium Repair by Cold Spray*, ARL Technical Report ARL-TR-4438. 34.

Cinca, N., and J. M. Guilemany. 2013. Cold gas sprayed stellite-6 coatings and their wear resistance. 2013. *Journal of Material Science & Engineering* 2 (2): 1000122. http://dx.doi.org/10.4172/2169-0022.1000122.

Couto M., S. Dosta, M. Torrell, J. Fernández, and J. M. Guilemany. 2013. Cold spray deposition of WC–17 and 12Co cermets onto aluminum. *Surface and Coatings Technology* 235 (2013): 54–61.

Culliton, D., Anthony Betts, Sandra Carvalho, and David Kennedy. 2013. Improving tribological properties of cast Al-Si alloys through application of wear-resistant thermal spray coatings. *Journal of thermal spray technology* 22 (4): 491–501.

Dosta, S., M. Couto, and J. M. Guilemeny. 2013. Cold spray deposition of a WC-25Co cermet onto Al7075-T6 and carbon steel substrates. *Acta Materialia* 61:643–652.

EPA registers copper-containing alloy products. 2008. www.epa.gov/pesticides/factsheets/copper-alloy-products.htm.

Faúndez, G., M. Troncoso, P. Navarrete, and G. Figueroa. 2004. Antimicrobial activity of copper surfaces against suspensions of *Salmonella enterica* and *Campylobacter jejuni*. *BMC Microbiology* 4:19.

Feng Q. L., T. N. Kim, J. Wu, E. S. Park, J. O. Kim, D. Y. Lim, and F. Z. Cui. 1998. Ag-Hap thin film on alumina substrate and its antibacterial effects. *Thin Solid Film* 335:214–219.

Gärtner, F., T. Stoltenhoff, T. Schmidt, and H. Kreye. 2006. The cold spray process and its potential for industrial applications. *Journal Thermal Spray Technology* 15 (2): 223–232.

Grass, G., C. Rensing, and M. Solioz. 2011. Metallic copper as an antimicrobial surface. *Applied and Environmental Microbiology* 77 (5): 1541–1547.
Hager, C.H., Jr. 2002. Evaluation of coatings and coating processes for the fretting amelioration of titanium bladed disk assemblies found in the high temperature compressor of jet turbine engines. Master Thesis, The Pennsylvania State University.
Handbook of Thermal Spray Technology (#06994G), Introduction to Thermal Spray Processing, 2004 ASM International.
Hu, X., P. Jiang, J. Wan, Y. Xu, and X. Sun. 2009. Study of corrosion and friction reduction of electroless Ni-P coating with molybdenum disulfide nanoparticles. *Journal of Coating Technology* 6:275–281.
Kato, K. Sendai. 2002. Classification of wear mechanisms/models. Proceedings of the Institution Mechanical Engineers 216.
Kim Hyung-Jun, Chang-Hee Lee, and Soon-Young Hwang. 2005a. Fabrication of WC–Co coatings by cold spray deposition. *Surface and Coatings Technology* 191 (2): 335–340.
Kim Hyung-Jun, Chang-Hee Lee, and Soon-Young Hwang. 2005b. Superhard nano WC–12 % Co coating by cold spray deposition. *Materials Science and Engineering: A* 391 (1): 243–248.
Koh, P.K., K. Loke, P. Cheang, and C. T. Lee. 2012. Cold spray repair of gas turbine engine fan cases. Singapore Aerospace Technology & Engineering Conference, Singapore, Feb 13, 2012.
Leyman, P. F., and V. Champagne. 2008. Cold spray aluminium for magnesium gearbox repair. US Army Research Laboratory Weapons & Materials Research Directorate, Feb 26–28, 2008.
Leyman, P. F., and V. K. Champagne. 2009. Cold spray process development for the reclamation of the apache helicopter mast support. ARL-TR-4922, August 2009.
Leyman, P. F., V. K. Champagne, et al. 2012. *Titanium coatings using cold spray*. TMS 2012.
Liao, H., C. Coddet, and L. Simonin. 2001. Mechanical properties of thermal spray PEEK coatings, Thermals spray 2001: New surfaces for a new Millennium. In *ASM International,* ed. C. C. Berndt, 315. Materials Park, OH.
Manoj, V., and David, Grossen. 2011. Wear resistant thermal spray coatings for wind-turbine components. *Advanced Materials & Processes* 169 (11): 59–62.
McCune, R., and M. Ricketts. 2004. *"Selective galvanizing by cold spray processing" in "Cold Spray 2004"*. Akron: ASM International—TSS.
Michels, H., S. Wilks, J. Noyce, and C. Keevil. 2005. Copper alloys for human infectious disease control. Presented at Materials Science and Technology Conference, September 25–28, 2005, Pittsburgh, PA.
MOOG Aircraft Group. 2012. Advanced surface repair capabilities.
Neshastehriz, M. 2014. Influence of hardenability of Nickel encapsulated, Hexagonal boron nitride particles on coating bond strength via cold spray. M M.S. Thesis, The Pennsylvania State University, University Park, PA.
Olakanmi, E. O., and M. Doyoyo. 2014. Laser-assisted cold-sprayed corrosion-and wear-resistant coatings: A review. *Journal of Thermal Spray Technology* 23 (5): 765–785.
Page, K., M. Wilson, and I. Parkin. 2009. Antimicrobial surfaces and their potential in reducing the role of the inanimate environment in the incidence of hospital-acquired infections. *Journal of Materials Chemistry* 19:3819–3831.
Pitchuka, Suresh Babu, Benjamin Boesl, Cheng Zhang, Debrupa Lahiri, Andy Nieto, G. Sundararajan, and Arvind Agarwal. 2014. Dry sliding wear behavior of cold sprayed aluminum amorphous/nanocrystalline alloy coatings. *Surface and Coatings Technology* 238:118–125.
Rutala, W. A., E. B. S. Katz, R. J. Sherertz, and F. A. Sarubbi. 1983. Environmental-study of a methicillin-resistant staphylococcus -aureus epidemic in a burn unit. *Journal of Clinical Microbiology* 18 (3): 683–688.
Sanpo N., M. L. Tan, P. Cheang, and K. A. Khor. 2009. Antibacterial property of cold-sprayed HA-Ag/PEEK coating. *Journal Thermal Spray Technology* 18 (1): 10–15.
Sanpo, N., H. Chen, K. Loke, P. K. Koh, P. Cheang, C. C. Berndt, and K. A. Khor. 2011. "Biocompatibility and Antibacterial property of Cold Sprayed ZnO/Titanium Composite Coating, Science and Technology Against Microbial Pathogens", 2011, p 140–144.

Santo, E., E. W. Lam, C. G. Elowsky, D. Quaranta, D. W. Domaille, C. J. Chang, and G. Grass. 2011. Bacterial killing by dry metallic copper surfaces. *Applied and Environmental Microbiology* 77:794–802.

Segall A. E., Anatoli N. Papyrin, Joseph C. Conway, and Daniel, Shapiro. 1998. A cold-gas spray coating process for enhancing titanium. *JOM Journal of the Minerals, Metals and Materials Society* 50 (9): 52–54.

Shirkhanzadeh, M., M. Azadegan, and G. Q. Liu. 1995. Bioactive delivery systems for the slow release of antibiotics: Incorporation of Ag + Ions into micro-porous hydroxyapatite coatings. *Materials Letters* 24:7–12.

Smid, I., A. E. Segall, P. Walia, G. Aggarwal, T. J. Eden, and J. K. Potter. 2012. Cold- sprayed ni-hbn self-lubricating coatings. *Tribology Transactions* 55:599–605.

Stark, L. 2010. Engineered self-lubricating coatings utilizing cold spray technology. M.S. Thesis, The Graduate School Engineering Science and Mechanics, Penn State University.

Stark, L., I. Smid, A. Segall, T. Eden, and J. Potter. 2012. Self-lubricating cold-spray coatings utilizing microscale nickel-encapsulated hexagonal boron nitride. *Tribology Transactions* 55 (5): 624–630.

Vlcek, J., L. Gimeno, H. Huber, and E. Lugscheider. 2005. A systematic approach to material eligibility for the cold spray process. *Journal of Thermal Spray Technology* 14:125–133.

Walia, P. 2006. Development of Ni-Based self-lubricating composite coatings for TI-6Al-4V dovetail Joints using the cold spray process. M.S. Thesis, The Pennsylvania State University, University Park, PA.

Wang, J., and J. Villafuerte. 2009. Low pressure cold spraying of tungsten carbide composite coatings. *Advanced Materials and Processes ASM International* 167 (2): 54–56.

White, D. 2006. New report sheds sobering light on hospital infections. *ABC News*, 12/1/2006.

White, L. F., S. J. Dancer, and C. Robertson. 2007. A microbiological evaluation of hospital cleaning methods. *International Journal of Environmental Health Research* 17 (4): 285–295.

Widener, C., R. Hrabe, B. James, and V. Champagne. 2013. "B1 Bomber-FEB Panel Repair by Cold Spray", Cold Spray Action Team (CSAT) Meeting 2013, Worcester Polytechnic Institute, MA, 18 June 2013.

Wolfe D. E., Timothy J. Eden, John K. Potter, and Adam P. Jaroh. 2006. Investigation and characterization of Cr_3C_2-based wear-resistant coatings applied by the cold spray process. *Journal of Thermal Spray Technology* 15 (3): 400–412.

Yamamoto, O., K. Nakakoshi, T. Sasamoto, H. Nakagawa, and K. Miura. 2001. Absorption and growth inhibition of bacteria on carbon materials containing Zinc Oxide, *Carbon* 39:1643–1651.

Yin, J., A. Zhang, K. Y. Liew, and L. Wu. 2008. Synthesis of poly(Ether Ether Ketone) assisted by Mirowaves irradiation and its characterization. *Polymer Bulletin* 61:157–163.

Zhang, L., Y. Jiang, Y. Ding, M. Povey, and D. York. 2007. Investigation into the antibacterial behaviour of suspensions of ZnO nanoparticles (ZnO Nanofluids). *Journal of Nanoparticle Research* 9:479–489.

Zheng, W., C. Derushie, J. Lo, and E. Essadigi. 2006. Corrosion protection of joining areas in magnesium die cast and sheet products. *Materials Science Forum* 546–549:523–528.

Chapter 11
Cold Spray Economics

D. Helfritch, O. Stier and J. Villafuerte

11.1 Introduction

Cold spray (CS) is employed for scores of applications, from the formation of thin metallic coatings to the production of free-standing shapes. The costs from application to application can vary significantly, but the cost for each is dependent upon the same set of operating parameters. Once defined, these parameters can be used to accurately calculate the cost of a finished product or can be used to predict the cost of a potential product. The methods presented here assume that a complete product has not yet been fabricated, and cost estimates will be based upon the characteristics of the desired product. The methods presented can thus be used to calculate quotation prices for prospective customers, for example, by spray shops.

From an original equipment manufacturer (OEM) perspective, economic affordability of a proposed CS-based manufacturing process is prerequisite even for the release of development budget. In the run-up to a technical deep exploration, future costs need to be estimated largely in absence of knowledge about the final manufacturing process. When CS is considered to replace a prevailing thermal spray process, a simplified cost comparison may be appropriate. Benchmarks against entirely different manufacturing or repair routes like, for example, galvanization, casting, milling, cupping, sintering, extrusion, brazing, welding, or cladding require, however, the estimation of the total CS costs comprising consumables, investment, and labor. Those will have to be estimated reliably, using a minimum of speculative input about the CS application not yet developed.

D. Helfritch (✉)
TKC Global, Hearndon, VA, USA
e-mail: dennis.helfritch@tkcglobal.com

O. Stier
Corporate Technology, Siemens AG, Berlin, Germany
e-mail: oliver.stier@siemens.com

J. Villafuerte
Corporate —Supersonic Spray Technologies, CenterLine Windsor Ltd., Windsor, ON, Canada
e-mail: julio.villafuerte@cntrline.com

J. Villafuerte (ed.), *Modern Cold Spray,* DOI 10.1007/978-3-319-16772-5_11

Several approaches have been taken in the past to count the costs of CS (Papyrin 2002; Gabel 2004; Karthikeyan 2005; Pattison et al. 2007; Champagne 2007; Helfritch and Trexler 2011), from general considerations to full process simulations as featured in the web-based software by Kinetic Spray Solutions (KSS), Buchholz, Germany (http://kinetic-spray-solutions.com/, 2013). These studies have in common that they assume particular cases for the cost modeling, or use more process data than required for actual cost prediction. Therefore, a general analysis of the cost structure of CS is presented in this chapter which is apt to estimate manufacturing costs *from the minimum of input required.* It allows comparison of CS with different manufacturing processes and has proven useful for assessing the economic viability of technically interesting CS application ideas.

11.2 Basic Framework

CS accelerates powder particles suspended in a gas. The gas–particle suspension is accelerated by expansion through a supersonic nozzle. The resulting high-velocity particles impact upon a substrate to create a deposition. From this simple description, one can conclude that powder and gas costs and their rates of usage are major contributors to the overall product cost.

Manufacturing costs all reside within three classic categories:

Materials costs
Direct labor costs
Overhead costs

Gas and powder are materials costs. The salaries paid to workers while engaged directly in manufacturing a specific product are direct labor costs. All other costs such as utilities, depreciation, maintenance, etc. are indirect overhead. These costs are interrelated and are subject to task difficulty. In the sections below, we consider each cost category in detail.

11.2.1 Materials Costs

Materials costs can be determined based upon the dimensions of the deposited product and the efficiency of deposition. The product will contain a known volume of deposited powder. Assuming negligible porosity and known density, this volume gives the mass of deposit. The mass of powder needed for the product is therefore that of the deposit plus estimated overspray, divided by the fractional deposition efficiency (DE). The DE can be obtained from a trial spray of the powder, can be estimated from similar applications, or can be iteratively calculated. Once the mass of powder is known, the mass of gas needed for powder acceleration can be simply calculated by dividing the powder mass by the ratio of powder to gas, typically 0.05. Given the masses of powder and gas, the costs of these materials are simply the masses multiplied by the cost per mass, such as dollars per kilogram. In addition, the time needed for a single part fabrication can be calculated from the mass of the gas needed, the gas pressure and temperature, and the nozzle throat diameter. While this time is not needed for material cost calculations, it is needed for labor costs below.

11.2.2 Direct Labor

Labor rates, dollars per man-hour, are known, and the hours needed for product completion are known from the calculation described above. Time must be added to take into account initial planning and setup. This time includes onetime initial fixture assembly, robot programming, and operating parameter determination. If multiple pieces are manufactured, then this onetime cost is shared by piece. While rates may vary among workers, typically an average rate for all workers is assumed. The cost of direct labor as described here assumes that the worker is employed in other activities when not operating the CS system. If the worker is paid a salary regardless his activity, then this labor cost must be part of fixed overhead and must be subjected to utilization considerations, as described below.

11.2.3 Overhead

This cost category can be further divided into two sub categories:

Variable overhead, such as utilities for direct production, which changes as production changes

Fixed overhead, such as administration, rent, heating and lighting, maintenance, and capital recovery, which remains independent of production

The cost of direct use of electricity for the production of a product by CS can be easily determined. Electricity is used to heat and sometimes to compress the gas used. The usages can be straightforwardly calculated from the temperature, pressure, and flow rate of the gas. Other electrical usages are for robot motion and control systems, but these are negligible in comparison with gas treatment. Once the rate of electricity usage is determined, the total usage in kilowatt-hour can be calculated by multiplying by the production time. The cost is then obtained by multiplying by the purchased cost of electricity in dollars per kilowatt-hour.

Fixed overhead is apportioned to individual jobs depending on the amount of time that particular job requires in relation to all other jobs, and this is where the concept of utilization must be introduced. Utilization is the percentage of time the CS system is used with respect to the total time that is available. Utilization is a measure of how effectively the available resources are being put to use. So the amount of fixed overhead charged to an individual job is equal to (total fixed cost per year/available operating time per year) × (total job time/fractional utilization). It will be seen that utilization is a major cost factor.

The capital recovery factor (CRF) method for calculation of depreciation takes into account the declining cost of the equipment, as well as the cost lost to interest payments if the purchase money were used instead as a loan. This is similar to a mortgage payment, which consists of principal and interest. This calculation yields a higher fixed cost than straight-line depreciation, unless interest rates are zero. The yearly depreciation cost, calculated by the CRF method is given by:

$$\text{Yearly depreciation cost} = \text{CRF} \times (\text{capital cost} - \text{salvage cost})$$

$$\text{CRF} = \frac{i(1+i)^n}{(1+i)^n - 1} \tag{11.1}$$

where i is the fractional interest rate, for example, 5 % = 0.05, and n is the years of ownership.

Yearly maintenance is generally estimated as a percentage of capital cost, for example, 5 %. Other fixed overhead costs, such as rent and administration are straightforward, and should be apportioned with respect to fraction of total floor space used and fraction of administrative time devoted to CS.

11.2.4 Combined Costs

The cost determination steps described above are easily assembled and carried out in a spreadsheet program. An example of a typical spreadsheet is shown in Fig. 11.1. Values that must be input, such as gas used and labor rate, are shown in italic. The remaining values such as flow rate and time for completion are calculated by the spreadsheet. The costs by category are then also calculated. A pie chart allows for quick assessment of the importance of various cost drivers. For this spreadsheet, the gas flow is calculated from knowledge of the nozzle throat diameter and the gas conditions upstream of the throat. Time for completion, which is needed for the determination of most cost contributions, is simply calculated by dividing total powder mass used by the powder feed rate. The powder mass needed is affected by the volume of the part, the DE, and the overspray fraction. For example, the time needed to complete a single part is given by

$$\begin{aligned}\text{Powder mass needed} &= \left[(\text{part volume})(\text{metal density})/(\text{fractional DE})\right] \times \\ &\quad \left[1 + \text{fractional overspray}\right]\\ \text{Time for completion} &= \text{powder mass used} / \text{powder feed rate}\end{aligned}$$

The electricity needed to produce a part is calculated by adding the usages of the gas heater, the gas compressor, and the exhaust fan and then multiplying by the time needed to complete a part. Electricity usage of the compressor and heater can be calculated from the known flow rate and conventional power equations. The ventilation fan power is assumed to be 15 kW for this spreadsheet.

INPUT DATA

nitrogen (0) helium (1)	0	
throat diameter	2.7	mm
feed rate	5.5	kg/hour
gas temp after heating	500	degree C
compressed gas pressure	4	MPa
deposit volume	125	cu cm
number of pieces	100	
overspray	10	%
material density	9	gm/cc
deposition efficiency	70	%
unit powder cost	100	$/kg
unit nitrogen cost	0.14	$/kg
unit helium cost	30	$/kg
unit electricity cost	0.15	$/kWhr
pre-spray set up time	8	hours
set-up time per piece	0.20	hours
hourly labor rate	75	$/hr
equipment capital cost	900,000	$
equipment life	15	years
salvage cost	300,000	$
interest rate	3	%
system utilization rate	75	%
maintenance, % of capital	5	%
yearly administration	150,000	$
yearly rent, utilities	100,000	$

CALCULATED VALUES

gas flow, NCMH	94.1
powder to gas, %	4.7
powder per piece, kg	1.77
time per piece, hr	0.32
elect per piece, kWhr	11.9
powder cost	$17,679
gas cost	$530
labor cost	$4,511
administration	$6,014
utilities & rent	$4,010
maintenance	$1,804
depreciation	$2,015
production electricity	$179
Cost per Piece	$366
Total Job Cost	$36,562

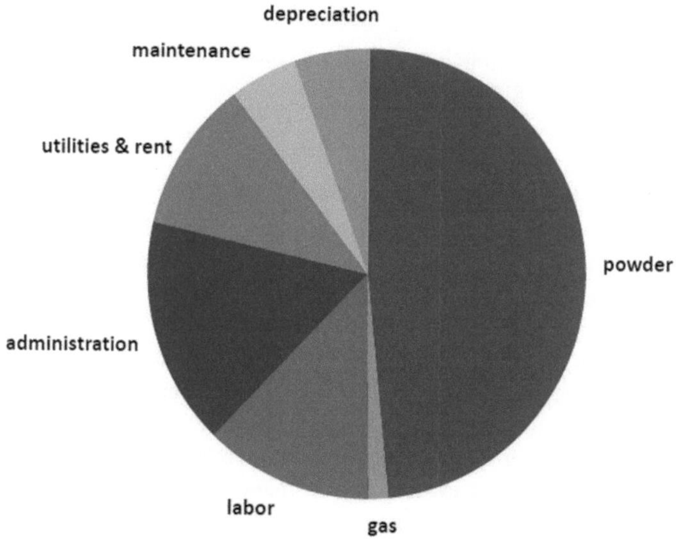

Fig. 11.1 The cost calculation spreadsheet. *NCMH* normal cubic meter per hour

All of the unit costs, cost rates, and fixed costs of the left-hand column must be the input. The pre-spray setup time is the time needed for tasks which need to be completed before actual spraying, and include items such as powder purchase and robot programming. The setup time per piece is the time it takes to remove a completed piece and install a new base for a subsequent piece. The hourly labor rate includes only salary and fringe benefits directly paid to the operator. The spreadsheet assumes only one worker for all tasks. Yearly administrative cost is the cost for supervision, sales, clerical, etc. devoted to the operation of the CS system considered. The yearly rent is similarly prorated, for example, where the CS system occupies only part of a building and the other production systems occupy the remainder.

The job costs are then calculated, shown in the lower, center column. Powder and gas costs are based upon amounts used and upon unit costs previously determined. The labor cost is simply the labor rate times the sum of all the time needed to complete the job, including setup. Overhead values for a single job are prorated based upon the fraction of time needed to complete the job divided by the available time per year (here 2000 h), divided by the fractional utilization (U). For example,

$$\text{Prorated admin} = (\text{yearly admin})(\text{total job time}) / 2000(\text{U}).$$

11.3 Component Effects

Considering Fig. 11.1 to be a base case, we can estimate the relative importance of each parameter by its variation. There are often trade-offs that can be made between parameter values that can reduce costs. For example, a more expensive powder may allow the use of nitrogen instead of helium. The effects of major cost-affecting parameters are examined below. All of the calculations made are based upon a variation of parameters given by Fig. 11.1.

11.3.1 Gas

By far the largest influence on final cost is the gas used. This can be inferred from the difference in unit price between nitrogen and helium. Adjusting for higher DE and lower feed rate, when the nitrogen used in Fig. 11.1 is switched to helium, the cost increases from US$ 36,562 to US$ 111,329. The cost increase is almost entirely due to the difference in unit costs between nitrogen and helium as can be seen from the comparable cost distributions shown in Fig. 11.2. On first look, it would seem unreasonable to ever use helium; however, there are quality benefits resulting from helium use that are not evident from a manufacturing point of view. Helium can yield improved bond strength and decreased porosity. Nitrogen will sometimes not produce high enough particle velocity needed to allow hard, refractory particles to deposit. System recycle of helium would significantly offset the cost increase described in this example.

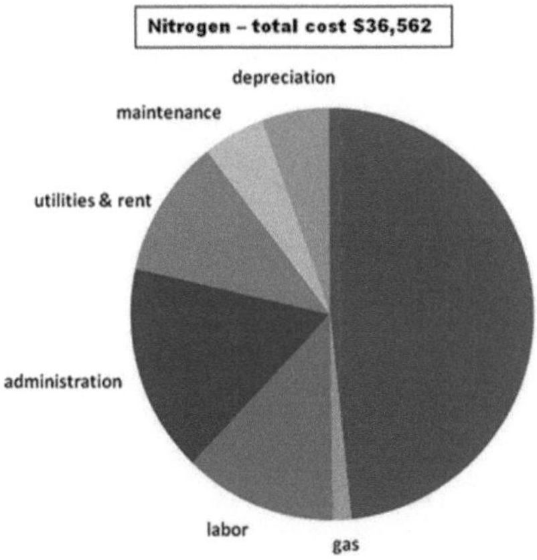

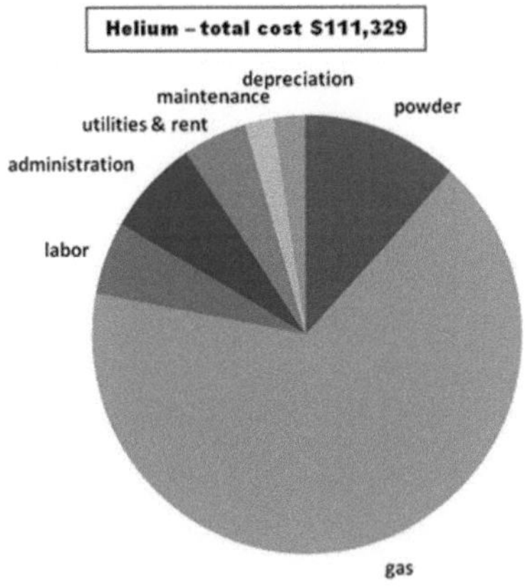

Fig. 11.2 Cost distributions of nitrogen use (*left*) versus helium use (*right*)

11.3.2 Powder Feed

An often overlooked parameter when attempting to minimize cost is the powder feed rate. This is especially true when operating with helium. Clearly, increasing powder feed rate will shorten the time required for completion, which in turn decreases total gas consumption, labor cost, and prorated overhead. For the example above for helium operation, the US$ 111,329 cost was based on a feed rate of 2.5 kg/h, which results in a powder flow equal to 5.3 % of the gas mass flow. Increasing the feed rate to 5 kg/h, without any other changes, would decrease the cost to US$ 66,460. The limit to arbitrary increase of powder feed rate is the carrying capacity of the accelerating gas. Gas flow is relatively unaffected when accelerating powder composing 5 % by mass of the gas mass flow. Gas and hence particle velocities decrease as powder feed rate exceeds 5 %, which in turn adversely affects DE and deposit quality.

11.3.3 Powder Cost

While the unit cost for the purchase of gas does not vary significantly from job to job, the cost of powder can vary between US$ 20/kg and US$ 1000/kg. The powder costs often contain atomization and/or milling costs, so that the forming process "atomize/mill—consolidate by CS" a priori adds costs to the raw material price. For the case described by Fig. 11.1, the effect of only changing the powder unit cost is shown in Fig. 11.3. The total job cost can change tenfold over the possible range

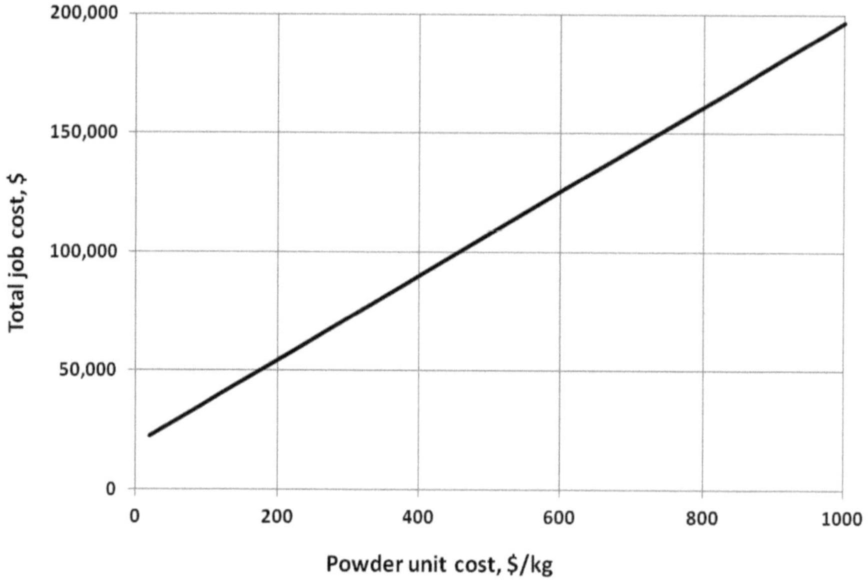

Fig. 11.3 The effect of powder cost on overall cost, based on the example of Fig. 11.1

of powder unit costs. This result weighs in favor of shopping for the lowest powder cost, but the qualities of the powder can directly influence the quality of the deposit and the DE of the specific operation. Determination of the deposit quality and DE among powder candidates must be done by means of test CS runs. Once the results from the test runs are known, the DE's and unit costs can be inserted into the cost spreadsheet and job costs determined. A judgment can then be made with respect to the cost–deposit quality trade–off.

11.3.4 Deposition Efficiency

As described above, powder characteristics directly affect DE. For a given alloy, particle shape and particle size distribution are the principal powder determinants of DE. Particle density is also important when other alloys are included. Besides powder characteristics, operating parameters such as gas pressure and temperature have a large influence on DE. For the case described by Fig. 11.1, the effect of DE on job cost is shown in Fig. 11.4. The figure clearly shows the importance of maximizing DE. Powder characteristics and operating parameters can be adjusted to maximize DE. Computer models are sometimes used to predict DE and to determine an optimum set of parameters, but trial CS runs are generally more accurate and preferred.

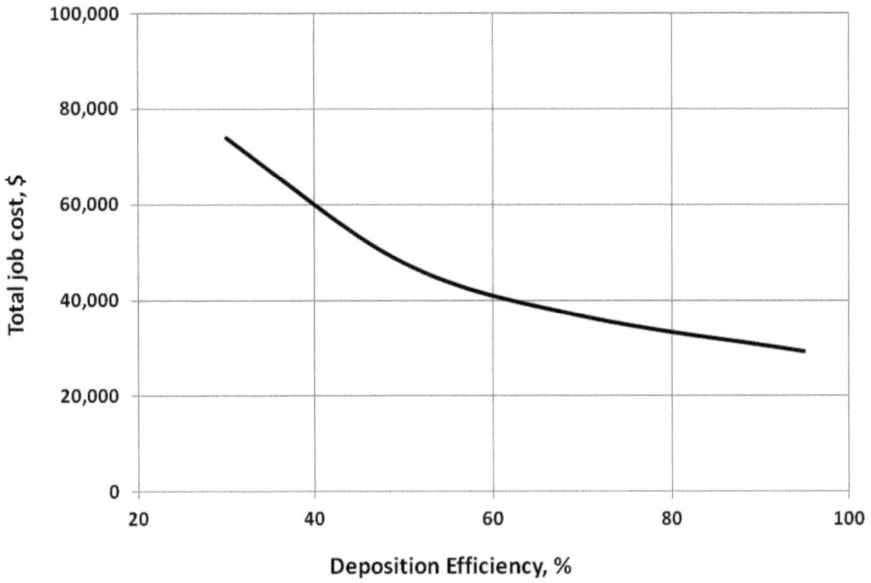

Fig. 11.4 The effect of deposition efficiency

11.3.5 Utilization

Labor and prorated overhead costs are strongly dependent on the time needed to complete the job. Clearly, labor rates, depreciation cost, administration yearly cost all directly affect the bottom line job cost, but the actual time that these services are used for the manufacture of the specific job is what assigns their prorated costs to that job. A second factor, related to the job time, is utilization. Utilization is simply the amount of facility time spent in productive utilization, divided by the total time available. Hundred percent utilization means that the system is in full use throughout the year without any idle time. Figure 11.5 shows how utilization affects costs for the case described by Fig. 11.1. A larger portion of fixed overhead costs must be assumed by each job as utilization decreases, and in this case, job costs can almost double as utilization decreases to below 50 %.

11.3.6 Number of Parts to Be Produced

Given a constant, onetime, setup period (robot programming, purchasing, etc.), the cost per piece obviously decreases as this setup cost is shared with many pieces. Again, considering the case described by Fig. 11.1, the effect of mass production is

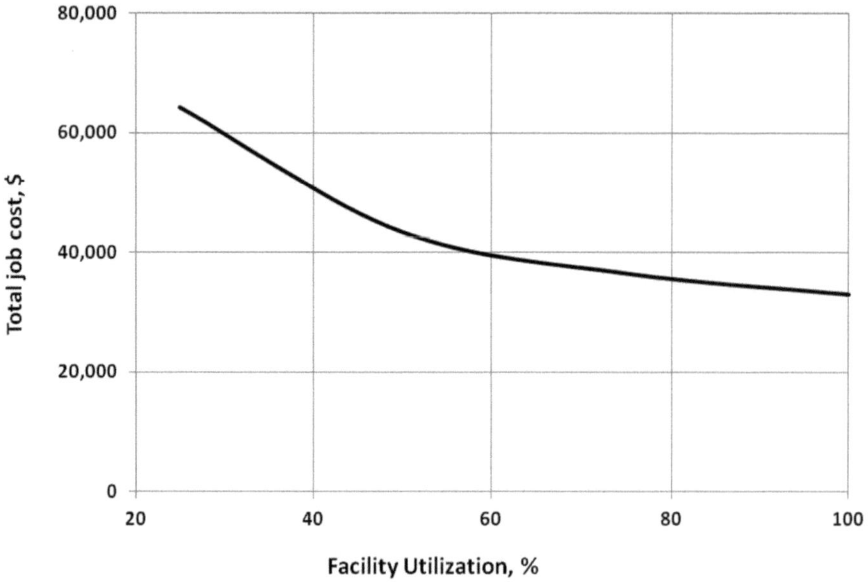

Fig. 11.5. The effect of utilization

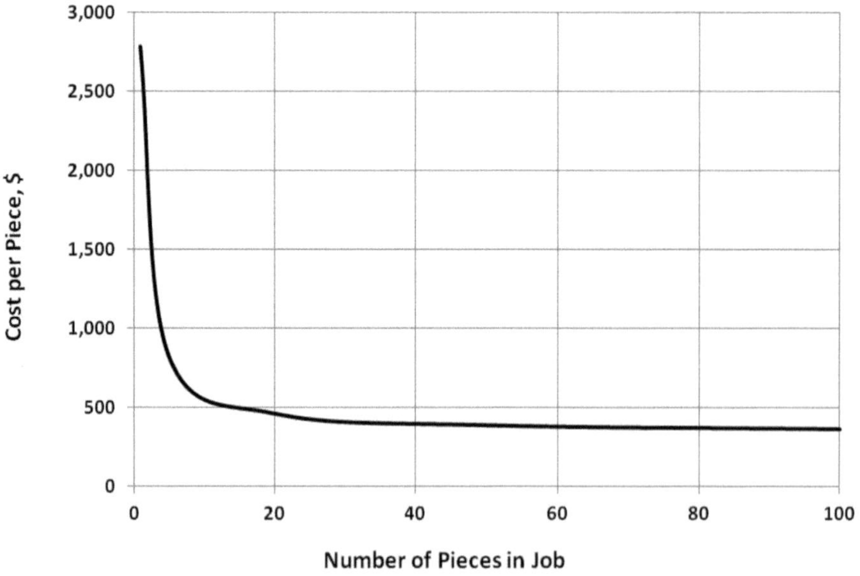

Fig. 11.6. The effect on cost per piece by the number of pieces made

shown in Fig. 11.6. For this example, with 8 h of up-front setup, the cost per piece does not increase significantly until fewer than ten pieces are to be produced. Below ten pieces, the cost per piece increases significantly as up-front costs are shared with fewer pieces.

11.4 Determination of Operating Parameters

CS parameters are often adjusted to provide maximum particle velocity and DE, which generally results in optimum deposition characteristics and maximum cost. A cost spreadsheet allows the cold sprayer to balance deposit quality with cost. Until deposition models are improved, the characterization of DE and deposit quality is best done by means of trial spray runs. A matrix of variables may be gas type, pressure, temperature, and feed rate. The DE would be measured for each run. The measurement of deposition quality could include cross-sectional examination, bond strength, tensile strength, etc., depending on the specific characteristics desired. Costs could be calculated for each run and associated DE. Data generated in this way would then yield how the minimum cost for acceptable quality could be achieved.

For industrial applications, the bonding strength of a CS coating to the substrate, the tensile strength of a CS deposit, its porosity or its ductility, etc. are subject to specification. In the language of mathematics, these properties define *constraints* to the feasible CS processes. The specification has to be met while there is little advantage from exceeding it. Therefore, these properties normally do not serve as a *quality function* for an optimization. The optimization of a CS process should be carried out in order to minimize its total costs (is equal to quality function) while obeying the constraints resulting from coating property specifications.

Decision-Making Example

An example of cost-effectiveness control is as follows. It is desired to coat 20 tubes, 3-cm diameter and 1-m length, with 500 μm of nickel. This yields a coating volume of 24 cm^3/tube. The corresponding cost spreadsheet for helium gas is shown in Fig. 11.7. Table 11.1 can be generated, once the deposition efficiencies and porosities are determined from the test runs and the costs are determined from the spreadsheet. Figure 11.8 shows the porosity and cost values of Table 11.1 for the two gases. A desired porosity of 0.2 % or under would require helium gas and cost US$ 400 per unit. If 0.4 % porosity were acceptable, then nitrogen could be used, and the cost per unit would be at most US$ 300 per unit. The costs are seen to increase as porosity increases for both gases, which seems counterintuitive. Although higher pressures and more gas are used to yield lower porosities, the increasing deposition efficiencies result in decreases in powder usages. The decreased cost of powder more than compensates increased gas costs and results in a net cost savings.

11.5 Cost Model of CS

The calculations discussed above (and performed by the spreadsheet in Fig. 11.1) can be combined to one neat equation. The derivation of the equation also explains the physics of the CS process. This equation contains fewer parameters than the spreadsheet has input cells, for three reasons:

INPUT DATA			CALCULATED VALUES	
nitrogen (0) helium (1)	1		gas flow, NCMH	249.8
throat diameter	2.7	mm	powder to gas, %	4.9
feed rate	2.2	kg/hour	powder per piece, kg	0.25
gas temp after heating	600	degree C	time per piece, hr	0.11
compressed gas pressure	4	MPa	elect per piece, kWhr	6.3
deposit volume	24	cu cm		
number of pieces	20		powder cost	$404
overspray	10	%	gas cost	$3,065
material density	9	gm/cc		
deposition efficiency	94	%	labor cost	$1,072
unit powder cost	80	$/kg		
unit nitrogen cost	0.14	$/kg	administration	$1,430
unit helium cost	30	$/kg	utilities & rent	$953
unit electricity cost	0.15	$/kWhr	maintenance	$429
pre-spray set up time	8	hours	depreciation	$479
set-up time per piece	0.20	hours	production electricity	$19
hourly labor rate	75	$/hr		
equipment capital cost	900,000	$	Cost per Piece	$392
equipment life	15	years		
salvage cost	300,000	$	Total Job Cost	$7,833
interest rate	3	%		
system utilization rate	75	%		
maintenance, % of capital	5	%		
yearly administration	150,000	$		
yearly rent, utilities	100,000	$		

Fig. 11.7 Coating a tube with nickel example. *NCMH* normal cubic meter per hour

Table 11.1 Porosity and cost changes resulting from operational changes

Gas	Pressure (bar)	DE (%)	Porosity (%)	Cost per unit ($)
He	20	80	0.39	447
He	30	90	0.23	407
He	40	94	0.16	392
N2	20	13	1.08	541
N2	30	24	0.58	339
N2	40	34	0.41	280

DE deposition efficiency

The CS deposit mass on one work piece, as well as the number of pieces, is eliminated by referring to a CS deposit unit mass of 1 kg.

Expenses for setup times are neglected because they are independent of the CS process parameters. They can simply be added after using the equation.

The last nine input parameters of the spreadsheet in Fig. 11.1 effectively reduce to an hourly rate which is used in the equation explicitly.

To calculate the hourly rate, the depreciation period is divided into productive and unproductive hours, as visualized in Fig. 11.9. Mobile spray units gain less productive hours than stationary CS systems in a workshop, due to travel times. On top of depreciation, the plant causes running costs for administration, rent, and maintenance. The

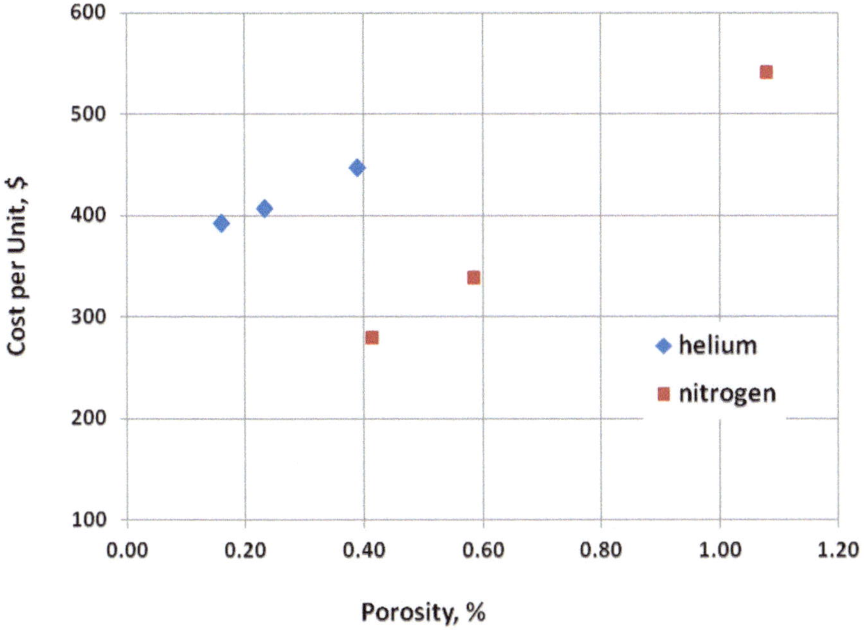

Fig. 11.8 Production cost versus porosity achieved for nitrogen and helium

total of these four costs, divided by the number of productive hours, is an equipment hourly rate adding to the hourly labor rate. The resulting total hourly rate, U_{hr}, covers all expenses for having the facility available. In the cost calculation module of the KSS software, the two hourly rates for equipment and labor can be set directly. In addition, there is a default calculation of the equipment hourly rate from the depreciation period, capacity utilization, and a variety of investment and rent items plus separate maintenance and repair hourly rates covering predefined system components.

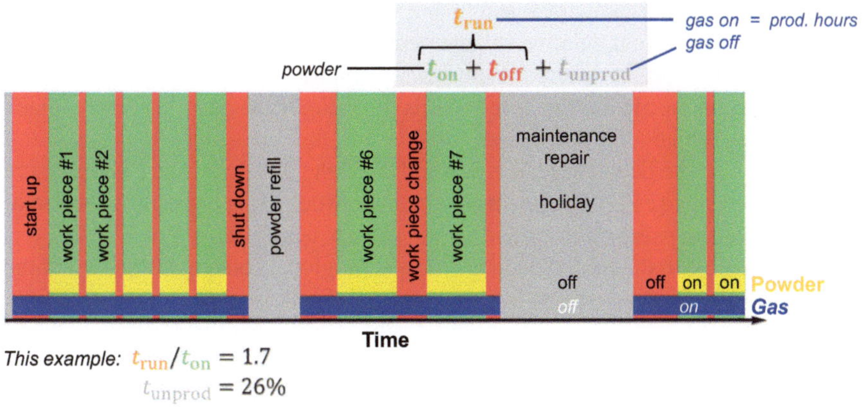

Fig. 11.9 Partitioning of the depreciation period into productive and unproductive hours and visualization of the t_{on}, t_{off}, and t_{run} times

The costs of process consumables (powders, propellant gas, and electric power) all are accounted for per unit amount of CS deposited material (1 kg), rather than per time unit (1 h). The reference unit 1 kg is convenient for estimating future production costs because the mass of CS deposited material per piece of product is known. The costs of equipment and labor are allotted to the reference unit 1 kg by the time t_{run} required for spraying and handling during deposition of that 1 kg material, applying the hourly rate U_{hr}. In case the system components are systematically worn out by certain powders (e.g., irreversible clogging or throat erosion of nozzles), the recurrent replacement costs may be allotted to the respective powder price U_{pwd} since the damage level correlates with the quantity of powder processed.

11.5.1 Generic Cost Function

In this section, a generic expression for the total costs of 1 kg CS deposit, C_{tot}, is presented. All model parameters are summarized in Table 11.2, together with their units. Some of them are explained in the following.

Productive hours are defined as the time when the gas flow is on. Thus, breaks for powder refill, maintenance, etc., do not add to the equipment run time but are covered by the hourly rate U_{hr}.

It may be advantageous not to shut down the gas flow during a change of work pieces, to avoid the delay associated with gas heater shut down and restart. However, the powder feeder would be stopped during work piece change, to save expensive powder. Hence, the equipment run time required for producing 1 kg of CS deposit, t_{run}, is divided into a portion t_{on} where powder is fed, and a powder feeder idle portion t_{off} where only gas is flowing, see Fig. 11.9:

$$t_{run} = t_{on} + t_{off}. \tag{11.2}$$

At turning points of spray tracks, CS deposits tend to pile up to excess thickness. To avoid this, track turning points are often placed outside the work piece edges which results in an extended spray track length or a virtual augmentation of the work piece surface area. The virtual, relative enlargement of the work piece is expressed by an overspray factor, 1+GL, where GL means "geometric loss." For example, in the case of constant nozzle motion speed, GL is the ratio of the cumulated spray track lengths out of, and on, the work piece, see Fig. 11.10.

Let $\dot{m}_{pwd}$ be the powder feeding rate and $\dot{m}_{gas}$ the gas flow rate. Low-pressure gas dynamic spray (LPGDS) systems inject the powder in the expanding section of the Laval nozzle, using atmospheric air at ambient temperature. For LPGDS systems, $\dot{m}_{gas}$ refers to the heated main gas flow through the convergent section of the nozzle. The powder-to-gas mass loading ratio is defined as

$$w = \frac{\dot{m}_{pwd}}{\dot{m}_{gas}}. \tag{11.3}$$

Table 11.2 Nomenclature

a	Speed of sound, m/s
A_{thr}	Nozzle throat area, mm^2
c	Helium mass fraction
c_p	Isobaric specific heat, kJ/(kg·K)
C_{tot}	Total costs of 1 kg deposited material, \$
F_{gas}	Inverse gas flow factor, 3600 m/$\sqrt{\text{K}}$ s
γ	Specific heat ratio
GL	Geometric loss factor
HL	Heat loss factor
M	Mach number
$\dot{m}_{\text{gas}}$	Gas flow rate of heated main gas stream, kg/h
$\dot{m}_{\text{pwd}}$	Powder feeding rate, kg/h
P	Gas stagnation pressure, MPa
R	Specific gas constant, J/(kg·K)
ρ_1	Gas density at nozzle exit, kg/m^3
ρ_{gas}	Gas density, kg/m^3
S_{anc}	Total electric power consumption of anything but gas heating, kW
t_{off}	Total duration of gas flow without powder flow per kilogram of deposited material, h
t_{on}	Total duration of powder flow per kilogram deposited material, h
t_{run}	Total duration of gas flow per kilogram deposited material, h
T	Gas stagnation temperature, K
T_{amb}	Gas inlet temperature, K
U_{elc}	Electrical energy price, \$/kWh
U_{hr}	Total hourly rate, \$/h
U_{gas}	Gas price, \$/kg
U_{pwd}	Powder price, \$/kg
v_1	Gas velocity at nozzle exit, m/s
v_{gas}	Gas velocity, m/s
v_{p}	Particle velocity, m/s
w	Powder-to-gas mass loading ratio
Y_{DE}	Deposition efficiency

The cumulated powder feeder idle time during deposition of 1 kg material, t_{off}, see Fig. 11.9, depends on the application and is an independent model parameter.

At significant temperature differences $T-T_{\text{amb}}$, the gas-heating unit produces thermal losses because of convection of surrounding air and radiation from hot surfaces. This can be taken into account by a temperature-dependent "heat loss" factor HL which may reach values up to 0.4 for nitrogen at $T-T_{\text{amb}} \approx 1000$ K.

Let S_{anc} be the total power consumption of the gas pressurizing and CS control units, helium recovery system, dust collector fan, and ancillary equipment like robots.

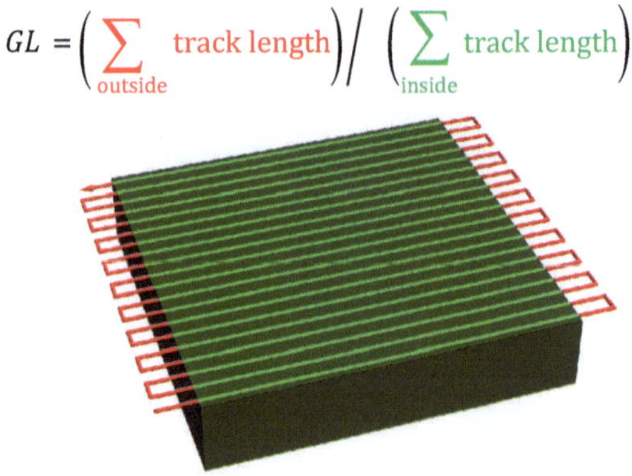

Fig. 11.10 In the case of constant nozzle motion speed, the overspray GL ("geometric loss") is the ratio of the cumulated spray track lengths out of, and on, the work piece

Then the total costs for depositing 1 kg material by CS are given by the generic cost function (Stier 2014):

$$C_{\text{tot}} = \frac{1+\text{GL}}{Y_{\text{DE}}}\left[U_{\text{pwd}} + \frac{1}{w}\frac{t_{\text{run}}}{t_{\text{on}}}\left(U_{\text{gas}} + \frac{U_{\text{hr}}}{\dot{m}_{\text{gas}}} + \frac{1+\text{HL}}{3600}\, c_p\left(T - T_{\text{amb}}\right)U_{\text{elc}}\right)\right] + S_{\text{anc}}t_{\text{run}}U_{\text{elc}}. \quad (11.4)$$

Equation 11.4 applies to all spray powders, propellant gases, any "high pressure CS," "low pressure CS," LPGDS (Maev and Leshchynsky 2008), "vacuum CS" (Fan et al. 2006), aerosol deposition (Akedo 2008), or kinetic metallization system (Gabel 2004), and to all kinds of application, such as coating, restoration, additive manufacturing, near-net forming.

Worked Example 1

The generic cost function Eq. 11.4 is used to calculate the costs per piece for the use case presented in Fig. 11.1. Equation 11.4 returns the costs for depositing 1 kg material. In the use case, the mass of deposited material is 1.125 kg per work piece, as calculated from the deposit volume and material density according to Fig. 11.1. Hence, the costs according to Eq. 11.4 will have to be multiplied by 1.125. For the same reason, the temporal durations occurring in Eq. 11.4 need to be divided by 1.125 when taken from Fig. 11.1. Thus, the time needed to deposit 1 kg is $t_{\text{run}}=0.32$ h/1.125 = 0.286 h. The spreadsheet assumes $t_{\text{on}}=t_{\text{run}}$. Note, the ratio $t_{\text{run}}/t_{\text{on}}$ is independent of the work piece mass.

Further input parameters for Eq. 11.4 are directly read from Fig. 11.1: The gas used is nitrogen, that is, $c_p=1.13$ kJ/kg K (from the literature), $T=773.15$ K, $U_{\text{pwd}}=100$ \$/kg, $U_{\text{gas}}=0.14$ \$/kg, $U_{\text{elc}}=0.15$ \$/kWh, w = 0.0467, $Y_{\text{DE}}=0.7$, and the overspray is GL = 0.1. The gas flow rate is $\dot{m}_{\text{gas}}=94.1$ normal cubic meter per hour (NCMH) × 1.25 kg/m^3 = 117.6 kg/h, using the density of nitrogen at the norm conditions 273.15 K and 101,325 Pa.

Assume a gas inlet temperature $T_{\text{amb}}=293.15$ K, HL = 0.10, and $S_{\text{anc}}=17.5$ kW for ventilation and pressure booster.

The hourly rate U_{hr} is not explicitly given by Fig. 11.1 but can be calculated from the displayed numbers as described above: The CS utility is operated during 2000 h × 75 % = 1500 h per year, taking into account the available time per year (2000 h) and the system utilization rate (0.75). The yearly depreciation cost is calculated by the CRF method, Eq. 11.1, with CRF = 0.0838, and is US$ 50,260. The yearly expenses for administration, rent, and maintenance (= 5 % of capital) are US$ 150,000, US$ 100,000, and US$ 45,000, respectively. The total of the four costs (US$ 345,260), divided by the annual productive hours (1500 h) is the equipment hourly rate, 230 $/h. To this, the hourly labor rate (75 $/h) is added, resulting in U_{hr} = 305 $/h. With these numbers, Eq. 11.4 returns C_{tot} = 251 $/kg. By multiplication with 1.125, the mere CS process costs US$ 282 per work piece.

This price covers all expenses related to the actual spraying, but does not include preparation costs. Obviously, preparation costs cannot be derived from CS process parameters so that they cannot be calculated by a generic cost function. The spread sheet in Fig. 11.1 takes into account such additional costs explicitly: Per work piece, in average 8/100 + 0.2 = 0.28 h are spent for pre-spray setup and setup per piece, respectively. Multiplied by U_{hr}, this time costs another US$ 85, so that the total costs per piece are US$ 367. This total contains US$ 1.79 electricity costs, that is, 0.5 %.

For expensive powders, or expensive gases, electricity costs can be neglected, as suggested by the above example. This allows for a significant simplification of Eq. 11.4:

$$C_{tot} \approx \frac{1+\mathrm{GL}}{Y_{DE}}\left[U_{pwd} + \frac{1}{w}\frac{t_{run}}{t_{on}}\left(U_{gas} + \frac{1}{\dot{m}_{gas}}U_{hr}\right)\right]. \tag{11.5}$$

Equation 11.5 is specific to expensive spray powders (prices $U_{pwd} \geq 100$ $/kg) or expensive gases (with significant helium content) but generally valid otherwise. Against intuition, the gas costs per kilogram deposited material do not depend on the gas flow rate $\dot{m}_{gas}$, but the equipment and labor costs do. This is due to the process duration effect explained above. Using Eq. 11.3, the gas flow rate in Eq. 11.5 can be substituted by the powder feeding rate:

$$C_{tot} \approx \frac{1+\mathrm{GL}}{Y_{DE}}\left[U_{pwd} + \frac{t_{run}}{t_{on}}\left(\frac{U_{gas}}{w} + \frac{U_{hr}}{\dot{m}_{pwd}}\right)\right]. \tag{11.6}$$

Equations 11.5 and 11.6 allow the identification of the main cost factors of CS and to find ways of minimizing those.

CS, kinetic metallization, and aerosol deposition systems are operated at pressures sufficient to produce transonic flow in the nozzle throat, that is, the nozzle throat is choking the gas flow. Hence, the gas flow rate $\dot{m}_{gas}$ is equal to the critical mass flow rate through the nozzle. The critical mass flow rate is known to depend on the nozzle throat orifice cross-sectional area A_{thr} and the gas stagnation properties P and T in the following way:

$$\frac{1}{\dot{m}_{gas}} = F_{gas} \frac{\sqrt{T}}{A_{thr}\, P}. \tag{11.7}$$

Herein,

$$F_{gas} = \frac{\sqrt{R/\gamma}}{3600} \left(\frac{2}{\gamma + 1} \right)^{(\gamma+1)/2(1-\gamma)} \tag{11.8}$$

is the inverse flow factor and depends on the kind of gas. The unit of F_{gas} is $3600\ \text{m}/\sqrt{\text{K}}\,\text{s}$ (the factor 3600 is for unit conversion from s to h). R is the specific gas constant and γ the isentropic exponent of the propellant gas. Equation 11.7 relates the CS costs to the primary process parameters P and T: Substituting Eqs. 11.2 and 11.7 in Eq. 11.5 yields

$$C_{tot} \approx \frac{1+\text{GL}}{Y_{DE}} \left[U_{pwd} + \frac{1}{w} \left(1 + \frac{t_{off}}{t_{on}} \right) \left(U_{gas} + F_{gas} \frac{\sqrt{T}}{A_{thr}\, P} U_{hr} \right) \right]. \tag{11.9}$$

Herein, electricity costs are neglected by assuming expensive powders or gases. For capacity planning, the process duration can be calculated from Eq. 11.2 in conjunction with one of the following three alternative expressions:

$$t_{on} = \frac{1+\text{GL}}{\dot{m}_{pwd} Y_{DE}} = \frac{1+\text{GL}}{\dot{m}_{gas}\, w Y_{DE}} = \frac{1+\text{GL}}{w Y_{DE}} F_{gas} \frac{\sqrt{T}}{A_{thr}\, P}. \tag{11.10}$$

An advantage of the cost function Eq. 11.9 is that it involves variables which are comparably easy to estimate up front for an intended CS application: GL is a ratio of possibly unknown values which is easier to estimate than those values themselves. t_{off} depends more on the article produced than on any final CS process parameters.

Worked Example 2

Equation 11.9 is applied to the use case presented in Fig. 11.1. For nitrogen, F_{gas}=0.0071 in units of $3600\ \text{m}/\sqrt{\text{K}}$ s. P=4 MPa and A_{thr}=5.73 mm^2 (circular cross section). The other input parameters are given in Example 1, t_{off}=0. With these numbers, Eq. 11.9 returns $C_{tot}\approx$ 249 \$/kg, or US\$ 280 per work piece for the mere CS process. Adding the US\$ 85 for setup times gives total costs per piece of US\$ 366. The neglected electricity costs are US\$ 1.79 per work piece, that is, 0.5 %. The process duration for deposition of 1 kg is t_{run}=0.286 h according to Eqs. 11.2 and 11.10, so it will be 0.322 h per piece.

For *P*, *T*, and *w* typical values may be assumed. Y_{DE} may be determined from inexpensive experiments, as described above. The coefficients A_{thr}, F_{gas}, U_{gas}, U_{pwd}, and U_{hr} are known.

11.6 Gases for CS

The requirements to CS propellant gases are that they possess a high speed of sound *a* and that they are neither inflammable, explosive, or toxic nor prohibitively expensive. In addition, nonoxidizing gases are preferred in many applications. Therefore, helium (He), nitrogen (N_2), air, and their mixtures are feasible propellant gases for CS, as well as superheated steam. The technical properties of superheated steam as a CS propellant gas lie between those of N_2 and He while air has thermodynamic properties similar to N_2. A consideration of binary mixtures of N_2 and He will therefore capture the characteristics of all relevant propellant gases for CS.

The relation between the He mass fraction and the He volume (or mole) fraction in a blend of He and N_2 is nonlinear, as shown in Fig. 11.11. Those technical and economic properties of He–N_2 mixtures which are relevant to CS depend more linearly on the He mass fraction *c* than on the He volume fraction: The dependence of the gas flow speed v_{gas} on *c* exhibits less deviation from linearity than the dependence of v_{gas} on the He volume fraction, as shown in Fig. 11.12. The same applies

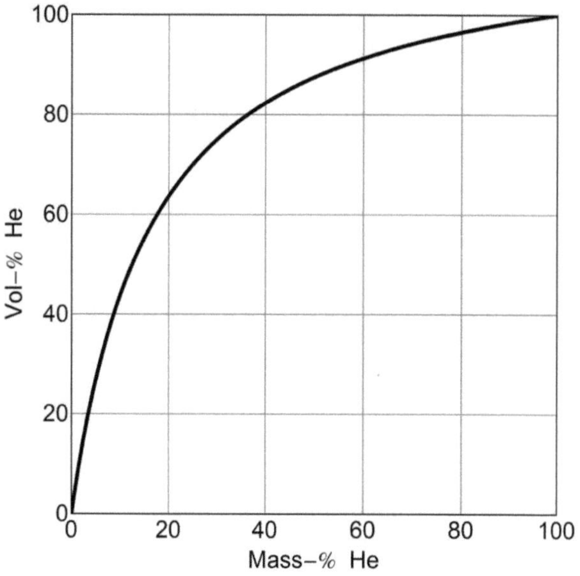

Fig. 11.11 Relation between the He mass fraction *c* and the volume, or mole, fraction of a binary He–N_2 mixture

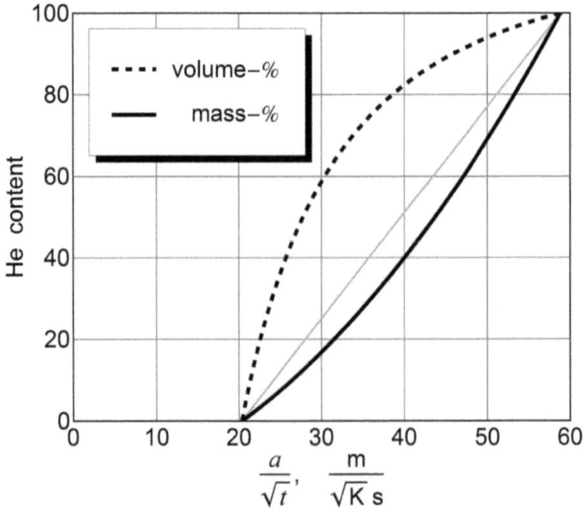

Fig. 11.12 Relation between the sound velocity a at temperature t and the He content for binary He–N_2 mixtures. The gray straight line visualizes a hypothetical linear relation, as a guide to the eye

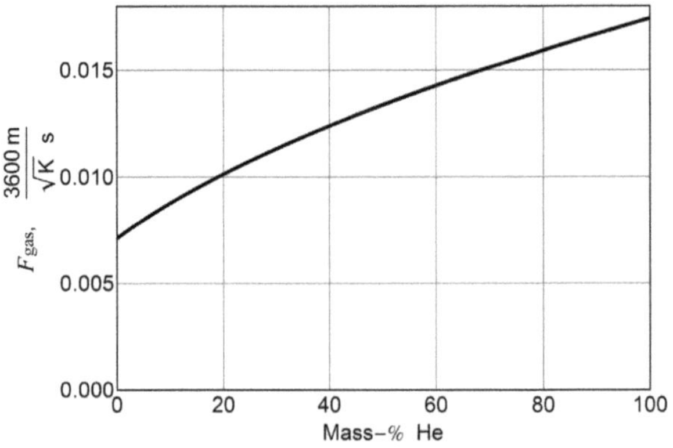

Fig. 11.13 Dependence of the inverse flow factor F_{gas} on the He mass fraction c for binary He–N_2 mixtures. Numeric values of F_{gas} are 0.0071 for N_2 and 0.0174 for He

for the inverse flow factor F_{gas} shown in Fig. 11.13. Furthermore, the highly cost-relevant parameter w refers to mass flow rates (Eq. 11.3). Therefore, the He mass fraction c $(0 \le c \le 1)$ is preferred over the volume (or mole) fraction for the purpose of cost analysis, and the costs of gas per unit mass are preferred over the costs per unit volume. Coarse estimates for the gas price $U_{gas}(c)$ for private industry are shown in Fig. 11.14 for new and recycled He, respectively.

11.7 Cost Factors of CS

Given the powder price and the hourly rate, the costs per kilogram deposited material depend on the application-specific process parameters GL and t_{off}; flow parameters P, T, and w; propellant gas properties F_{gas} and U_{gas}; particle bonding characteristic property Y_{DE}; and equipment (nozzle) parameter A_{thr}.

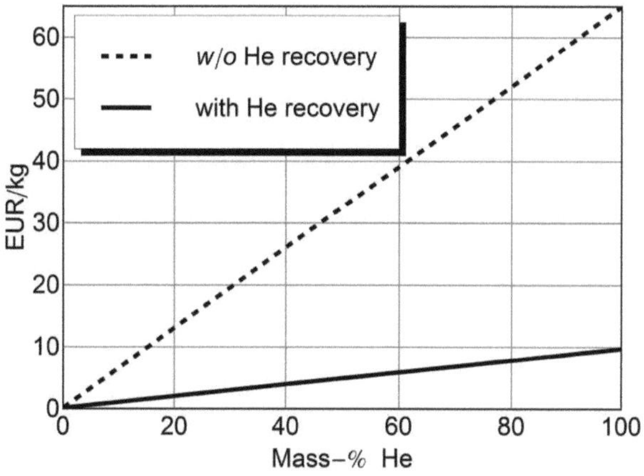

Fig. 11.14 Gas price U_{gas} dependence on the He mass fraction *c*. The solid line refers to a He recovery system with 85 % capture efficiency. The N_2 price refers to a liquid N_2 supply with a pressure boost system (e.g., Linde PRESUS). N_2 from cylinder bulk packs is roughly ten times as expensive which, still, is small compared to the He costs. *w/o* without

Obviously, GL and t_{off} have to be as small as possible, as they represent operation modes producing costs without material deposition. The flow parameters *P*, *T*, and *w* deserve deeper consideration. On first look, Eq. 11.9 suggests that for low process costs both Y_{DE} and *w* ought to be large. Second, the variance of $U_{gas}(c)$ is one order of magnitude larger than that of *w*, so a cost analysis needs to include the effect of U_{gas}, that is, of *c*. Moreover, there is a negative correlation between Y_{DE} and *w*. Therefore, the minimal costs in general will realize a compromise between the powder and gas costs.

Worked Example 3

In the use case of Example 2, nitrogen is replaced by helium, that is, $U_{gas}=30.00$ \$/kg and $F_{gas}=0.0174$ (in units of 3600 m/$\sqrt{K}$ s). Feeding powder at 2.5 kg/h results in a mass loading ratio $w=5.3\,\%$, as calculated from Eqs. 11.3 and 11.7. Assume that the higher gas velocity obtained by helium (see Fig. 11.12) leads to a DE increase from 0.7 to $Y_{DE}=0.95$. Equation 11.9 now returns $C_{tot}\approx 914$ \$/kg, or US\$ 1028 per work piece for the mere CS process. So, the CS process has become a factor 3.7 more expensive than in Example 2, only by using helium ($c=1$) instead of nitrogen ($c=0$). Adding US\$ 85 for setup costs gives a total price of US\$ 1113 per piece.

11.7.1 Gas Stagnation Properties

To analyze the roles of *P* and *T*, consider their effect on the particle velocity and acceleration near the nozzle exit. The accelerating force on particles is proportional to

$$\rho_{gas}\left(v_{gas}-v_{p}\right)\left|v_{gas}-v_{p}\right|, \tag{11.11}$$

where v_p is the particle velocity and ρ_{gas} is the gas density. Assuming 1D, isentropic expansion in a full-flowing nozzle (i.e., at sufficient pressure), the gas velocity at the nozzle exit is

$$v_1 = M\sqrt{\gamma RT / \left(1+\frac{\gamma-1}{2}M^2\right)}, \tag{11.12}$$

where M is the nozzle Mach number. The corresponding gas density is

$$\rho_1 = \frac{P}{RT}\left(1+\frac{\gamma-1}{2}M^2\right)^{1/(1-\gamma)}. \tag{11.13}$$

From Eqs. 11.11 to 11.13, the following conclusions can be drawn:

The velocity v_{gas} originates from T and is independent of P.

The particle acceleration force is proportional to P.

Higher M produce both higher v_{gas} and lower ρ_{gas}. Due to this ambivalence, an optimal nozzle Mach number exists for given gas and powder.

According to Eq 11.9, higher P lead to lower costs. Large stagnation pressures P are, thus, favorable from both the technical and the economic point of view. Therefore, P normally can be set as large as technically possible. In case the powder feeding rate needs to be constrained, it is recommended to reduce the gas flow by shrinking the nozzle throat, rather than by reducing the pressure.

As also can be seen from Eq. 11.9, higher T leads to growing equipment costs. Not even included here is the potential side effect on the hourly rate, that frequent operation at high temperatures may reduce the longevity of the equipment. On the other hand, higher T normally also lead to larger DE. Hence, the overall influence of increasing gas temperature on the process costs is ambivalent. T should, thus, be chosen as large as necessary for sufficient particle deposition and bonding strength.

In most of the presently available CS systems, T is a mixing temperature obtained by injection of cool powder feeding gas (with powder) into the hot main gas flow heated to the displayed temperature. This results in a lower stagnation temperature and a higher gas flow (except, for LPGDS systems where the injection occurs downstream of the nozzle throat). The required flow rate of powder feeding gas is related to the powder feeding rate, so that the mixing temperature T eventually depends on w. If the powder feeding gas has a different He content than the main gas, then even c may depend on w. The limited electrical heating power of an actual CS system may impose a constraint on the possible combinations of T, $\dot{m}_{gas}$, and c. Such effects can be considered when the cost optimization goes into detail.

11.7.2 Mass Loading Ratio

Feeding powder into the gas results in a deceleration of the flow. This particle loading effect on the flow speed imposes an upper bound on w because lower particle impact velocities v_{pi} result in lower DE. At the opposite end, decreasing w will allow it to dominate the costs because C_{tot} is asymptotically proportional to $1/w$ for $w \rightarrow 0$. Mass loading ratios have been reported in the interval from 1 to 30%. Thereby, $w \approx 3 \ldots 5\%$ may be considered typical for CS using N_2, whereas larger ratios may be affordable with He, or gas mixtures containing significant He mass fractions c.

From the cost point of view, U_{gas}/w should be minimal at given powder feeding rate and DE, as can be seen from Eq. 11.6. Depending on the world region, the N_2 supply form, and market conditions, 1 kg He is 75–500 times as expensive as 1 kg N_2, so that the gas price $U_{gas}(c)$ is approximately proportional to the He mass fraction c, see Fig. 11.14. Hence, c/w should be minimal for minimizing C_{tot} at given $\dot{m}_{pwd}$ and Y_{DE}. However, c has influence on both the affordable mass loading ratio and the achievable DE, so that the cost optimal He concentration is not easily determined, in general. Only if pure nitrogen yields high DE and the specified deposit properties, the cost optimal He concentration is certainly $c = 0$.

Worked Example 4

In the use case of Example 3, the powder feeding rate is doubled (5 kg/h) so that $w = 10.6\%$. Everything else remains unaltered, in particular the DE (by assumption). According to Eq 11.9, $C_{tot} \approx 515$ \$/kg which corresponds to US\$ 579 per work piece, for the mere CS process. So, the doubled mass loading ratio results in 44% cost reduction and yields a piece price of US\$ 665 including setup costs.

The CS process is still a factor 2.7 more expensive than in Example 2. For a cost break-even with Example 2, the mass loading ratio would have to be increased to 32% while maintaining a DE of 95%. This is not necessarily realistic. The use of helium does not seem to pay off in this example because a relatively high DE (70%) is already reached using nitrogen. Only if the coating quality obtained with nitrogen was unacceptable, helium would *have to* be used, regardless of the cost increase.

11.8 Cost Optimization of CS

The generic cost function possesses a valley between two regions of high costs: At low c and high w the gas is overloaded with powder which results in flow deceleration and reduced DE. Small values of Y_{DE} result in exceedingly high total costs, see Eq. 11.6. On the contrary, at high c and low w the powder acceleration capacity of the gas is not exhausted which results in exceedingly high gas costs, again see Eq. 11.6. Near a certain optimal ratio of c and w, the total costs become minimal. Then, the acceleration capacity of the gas is fully used while avoiding significant powder loss due to flow deceleration.

In case a He recovery system is used, U_{hr} will increase because of the higher capital costs while U_{gas} will decrease due to the recycling. By recycling of He, the valley of the cost function may become deeper and wider, in spite of increased investment and potentially a reduced number of annual productive hours. This would allow for additional flexibility in choosing the CS process parameters. A He recovery system would pay off in high volume production, even when using gas blends with significant contents of nitrogen. This can be seen from comparing the two summands in the round brackets in Eq. 11.5: For present commercial high flow rate CS systems the term $U_{hr}/\dot{m}_{gas}$ typically assumes values below 10 €/kg. Any reduction of U_{gas} by this order will compensate the increase of the equipment hourly rate, as caused by a He recovery system. In view of He prices around $U_{gas} \approx 65$ €/kg this option appears realistic. Another argument in favor of He recovery is the fact that He is a finite and nonrenewable natural resource.

Worked Example 5

Assume the use case of Example 3 and add US$ 1,100,000 to the equipment capital cost for a He recovery system with 70% capture efficiency. Then, the yearly depreciation cost increases to US$ 142,403 (Eq. 11.1) and maintenance costs US$ 2 M × 5% = US$ 100,000 per year. The total hourly rate thus increases by US$ 98, to U_{hr} = 403 $/h. Due to the gas recycling, $U_{gas} = 30.00 \times (1-0.7) = 9.00$ $/kg. Under these conditions, the total piece price will be US$ 675. Compared with Example 3 where all new He is used, the recycling eventually saves 39% of the costs. In the given example, spraying with recycled He still is more expensive than spraying with nitrogen. However, if the deposit quality obtained with nitrogen was unacceptable, helium recovery would be economically mandatory.

11.9 Concluding Remarks

A framework for CS cost estimation has been presented. The costs have a plain generic structure which applies to all present types of CS systems and kinds of application. Comfortably, this allows an assessment of the economic viability of an intended application before knowing many spray process details.

The principle cost categories of materials, labor, and overhead have been defined and broken down into individual components. The method to determine the cost contribution of each component has been described. These components have been incorporated into a spreadsheet (Fig. 11.1). The spreadsheet utilizes input values, such as DE, powder cost, equipment cost, and coating volume to calculate the cost of the completed products. The calculations of the spreadsheet example are straightforward and can be done by readers either by hand calculation or by computer, for example Windows Excel. Additionally, the authors may be contacted for assistance.

We saw that individual parameters can have significant cost impact and that the judicious combination of these parameters can result in cost savings. The influence of gas on cost is extreme, due to a two order of magnitude difference in gas purchase price between nitrogen and helium. Helium recovery pays off in high volume production. Clearly, the use of nitrogen is favored, but at times, helium must be used for a desired result. Helium–nitrogen blends possess economic potential, and the generic cost function (Eq. 11.9) is useful for determination of the cost optimal helium concentration for a given application. Powder feed rate is often overlooked as a cost driver, but must be carefully understood and controlled to achieve a specified result at a minimum cost. Using high gas stagnation pressures is generally favorable in CS.

There are competing technologies that can be applied to most applications. As CS applications are being developed, it is important to be able to assess the commercial viability of the application. In addition to quality, relative cost is critical for this assessment.

Disclaimer The research reported in this document by Dennis Helfritch was performed in connection with contract/instrument W911QX-14-C-0016 with the U.S. Army Research Laboratory.

The views and conclusions contained in this document are those of TKC Global and the U.S. Army Research Laboratory. Citation of manufacturer's or trade names does not constitute an official endorsement or approval of the use thereof. The U.S. Government is authorized to reproduce and distribute reprints for Government purposes notwithstanding any copyright notation hereon

References

Akedo, J. 2008. Room temperature impact consolidation (RTIC) of fine ceramic powder by aerosol deposition method and applications to microdevices. *Journal of Thermal Spray Technology* 17 (2): 181–198.

Champagne, V. K., ed. 2007. *The cold spray materials deposition process: fundamentals and applications*. Cambridge: Woodhead.

Fan, S.-Q., C.-J. Li, C.-X. Li, G.-J. Liu, L.-Z. Zhang, and G.-J. Yang. 2006. Primary study of performance of dye-sensitized solar cell of nano TiO_2 coating by vacuum cold spraying. *Materials Transactions* 47:1703–1709.

Gabel, H. 2004. Kinetic metallization compared with HVOF. *Advanced Materials & Processes* 162 (5): 47–48.

Helfritch, D., and M. Trexler. 2011. How operating parameters and powder characteristics affect cold spray costs. 1st North American cold spray conference, ed. A. McDonald, Oct 25–27. Windsor, ON, Canada, The Quebec Materials Network.

Karthikeyan, J. 2005. Cold spray technology. *Advanced Materials & Processes*, 163 (3): 33–35.

Maev, R. G., and V. Leshchynsky. 2008. *Introduction to low pressure gas dynamic spray: physics and technology*. Weinheim: Wiley-VCH.

Papyrin, A. N. 2002. Cold spray process for cost-sensitive applications. Proceedings of the TMS 2002 annual meeting, Feb 17–21, 137–149. Seattle: The Minerals, Metals & Materials Society.

Pattison, J., S. Celotto, R. Morgan, M. Bray, and W. O'Neill. 2007. Cold gas dynamic manufacturing: a non-thermal approach to freeform fabrication. *International Journal of Machine Tools & Manufacture* 47:627–634.

Stier, O. 2014. Fundamental cost analysis of cold spray. *Journal of Thermal Spray Technology* 23 (1–2): 131–139.

Chapter 12
Review on Cold Spray Process and Technology US Patents

D. Goldbaum, D. Poirier, E. Irissou, J.-G. Legoux and C. Moreau

12.1 Introduction

In the past decade, the interest in the cold spray (CS) process by the industry has grown significantly with hundreds of new patents released. Over the years, CS process has passed a number of developmental stages, starting from the initial concept to the process optimization in terms of process efficiency, reliability and cost. Better process control and improved efficiency contributed to the integration of CS technology in highly regulated industry for application such as aerospace part repair (Jensen et al. 2013; Calla et al. 2012) or medical implant devices (Kramer 2009). The motivation of the present patent review is to provide an overview of the recent, key developments in the expanding field of the CS technology. A review on CS technology intellectual property covering both the Russian patents and the US patents and patent applications was published in early 2008 (Irissou et al. 2008). Since then, the total number of US patents issued more than doubled. An update on the US patents issued from mid-2007 is provided in this chapter, complementing the previously published review (Irissou et al. 2008).

E. Irissou (✉) · D. Goldbaum · D. Poirier · J.-G. Legoux
Automotive and Surface Transportation, National Research Council of Canada, Boucherville, QC, Canada
e-mail: Eric.Irissou@imi.cnrc-nrc.gc.ca

C. Moreau
Department of Mechanical and Industrial Engineering, Concordia University, Montreal, QC, Canada
e-mail: Christian.moreau@concordia.ca

J. Villafuerte (ed.), *Modern Cold Spray*, DOI 10.1007/978-3-319-16772-5_12

12.1.1 Historical Perspective

The reference to the basic concept of depositing a metal powder on a substrate propelled by a gas flow dates from the early twentieth century, even though the technology that makes CS possible took off only in the early 2000s. Thurston (1902) and Schoop (1915) were first to, separately, develop the idea of solid metal particle deposition on metal using a pressurized gas as the propeller. Some 60 years later, Rocheville (1963) brought the idea of using a de Laval type nozzle to increase the propelling gas velocity with the aim to increase powder particle velocities. The early designs presented important issues in terms of powder feeding control, gas flow saturation and achievable particle velocities, making them unpractical for spraying of viable coatings. It is only in 1980 that scientists in the Institute of Theoretical and Applied Mechanics of Russian Academy of Science in Novosibirsk successfully produced a CS metal deposit by accelerating the metal powder to supersonic velocities in a wind tunnel (Papyrin et al. 2006). The first patents for a workable concept and device were issued in the 1990s. A detailed description on the evolution of the CS system in Russia was provided in a previous intellectual property review on CS process (Irissou et al. 2008).

12.1.2 Search Methodology and General Overview

The current technical review covers exclusively US patents on the CS process. For the sake of this review, CS was defined as an all-solid-state coating process using a high-speed gas jet to propel powder particles towards an article resulting in particle plastic deformation and consolidation upon impact.

Other designations of processes that fit this definition and are found in the literature include: cold gas dynamic spray (CGDS), kinetic spray (KS), supersonic particle deposition (SPD), dynamic metallization (DYMET) or kinetic metallization (KM). For simplicity, the designation CS will be exclusively used for this review and refers to all processes that correspond to the above definitions.

The US patent search was performed with the software ORBIT from Questel (Paris, France) using the above-mentioned keywords and their derivatives. The search returned exactly 300 granted inventions relevant to CS. All these patents were separated into two categories: The first one, called inclusive, refers to patents that include the CS technology as one of the many coating methods that can be employed for the purpose of the invention. The second category, called exclusive, refers to patents that protect an idea that uses exclusively CS as part of the invention.

While the general overview (Sect. 12.2), below, covers the 300 patents of the initial search, the following technical review (Sect. 12.3) includes US patents issued after 2007 and found to fall in the exclusive category only.

12.2 General Overview

The distribution of the US patents with respect to the year of release is presented in Fig. 12.1, starting with the first patent released by Alkhimov et al. (1994). The figure incorporates the patents up to the first 4 months of 2014. The incomplete data for that year explain the apparent decrease in the patents granted for 2014. Figure 12.2 incorporates the number of patents granted each year for the 15 more active assignees/companies. Assignees from different divisions or branches of the same company were assigned to the head company when applicable. It was found that a total of 97 companies or individuals were granted at least one US patent related to the CS technology.

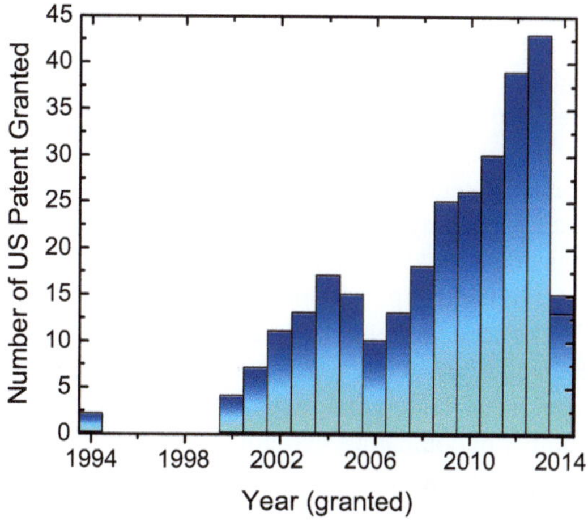

Fig. 12.1 Number of US patents on cold spray technology as a function of year of issuance

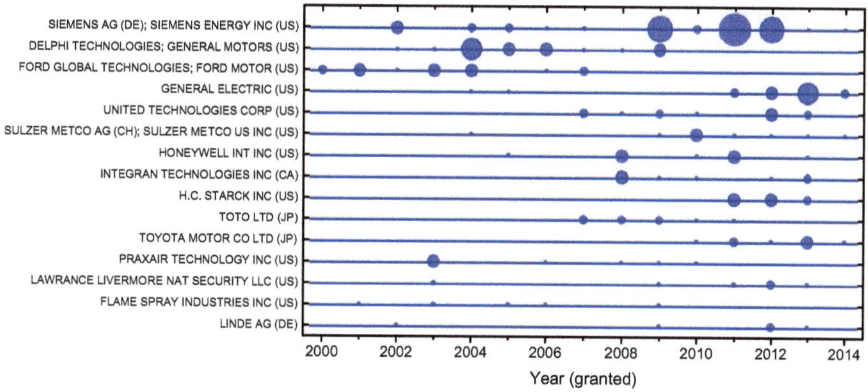

Fig. 12.2 Number of US patents issued each year for the 15 most active assignees for the period ending in April 2014 (*largest bubble* corresponds to eight patents)

From these graphs, we can see two distinct periods.

The first period, defined as being between 2000 and 2007, was marked by an increase in the number of patents that peaked in 2004 following the original Alkhimov et al. US patent (Alkhimov et al. 1994). In this period, the assignees can be divided roughly into three groups of companies. The first group is the automotive sector companies, with Ford and General Motors as the top assignees in this group, that were heavily involved in the development of both equipment and applications with 29 patents combined.

The second group is from the aerospace sector including gas turbine for power generation with Siemens, General Electric, United Technologies and Honeywell as the most active assignees with 14 patents combined. Finally, the third group is composed of thermal spray equipment manufacturer, powder producers and gas suppliers such as Praxair, Sulzer Metco, Flame Spray and Linde as the top assignees with six patents combined. This group represents the largest number of companies; most of them are not shown in Fig. 12.2 because of only few patent granted per assignees. A large majority of the assignees fit into one of these three categories for the period described.

The second period started in 2007 and shows a constantly increasing number of granted patents per year with over 150 patents, half the total patents, released in the last 5 years only. In this period, we notice a phasing down of the more active companies of the first period, General Motors and Ford with only four patents, while other automotive companies such as Toyota (eight patents) are getting involved.

The second group, thermal spray equipment, powders and gas companies, is increasingly involved with new assignees such as H. C. Starck, with a total of 24 patents for the top assignees combined. Again, this group is represented by a large number of companies that hold too few patents to be listed in Fig. 12.2.

The third group, the aerospace sector, became the most active with 53 patents combined. Several other aerospace companies that are not listed in Fig. 12.2 were also granted patents in this period. Contrary to the first period, in the second period, a large number of patents could not be placed into one of these three groups of assignees. Companies from different sectors like energy, chemical, mining, consumer electronics, medical devices, appliances, oil and gas and others were granted, together, a significant portion of the overall patents issued during this period. The broad interest and adoption of the CS technology by a number of industrial sectors, thus, contributes to a substantial increase in the US patents granted since 2007.

12.3 Technical Review

This chapter focuses on US patents exclusive to CS as defined above and released between September 2007 and April 2014. The patents considered were separated into three categories: (1) apparatus and methods, (2) precursor and (3) applications.

12.3.1 *Apparatus and Methods*

The section 'Apparatus and Methods' was divided into four subsections: (1) gun and nozzle design (2) hybrid system, (3) system control and (4) manufacturing design and method.

12.3.1.1 Gun and Nozzle Design

Typical cold spray systems contain a pressurized gun compartment where the gas is preheated, and pressurized. The pressurization chamber requires insulating lining which is reduced in thickness to accommodate for the system weight and size. The conventional design therefore limits the system operating temperature which directly impacts the achievable particle velocity. P. Heinrich *et al.* filed two patents (Heinrich et al. 2009; Heinrich et al. 2012a) with a new pressurized gas heater system. The gas heater consists of an insulated steel container (Fig. 12.3) in which the gas is preheated with heating elements placed near the container walls. In order to optimize heating efficiency a flow distributor element is added in the gas inflow area. The system incorporates a cooling system that maintains the exterior temperature of the container within acceptable limits while minimizing the system size and mass.

The powder is introduced into the preheated and pressured gas within the deLaval type nozzle where it is accelerated to supersonic velocities. Various configurations of the nozzle design and powder inlet locations were explored with the goal of reducing the nozzle wear and improving the control of the particle flow dynamics and deposition. A design patented by Ko et al. (2009) (shown in Fig. 12.4) proposes a movable powder feeder injector (12) which is located in the converging section of the nozzle and can be used to control the particle velocity. The converging section of the nozzle is, also, followed by a buffer chamber (30) which reduces the particle interaction with internal surfaces, preventing clogging.

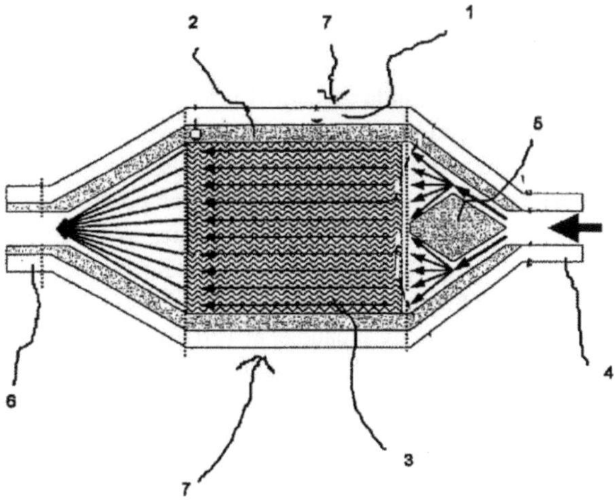

Fig. 12.3 High-pressure gas-heating device. (Heinrich et al. 2012a)

Fig. 12.4 Nozzle with a buffer chamber. (Ko et al. 2009)

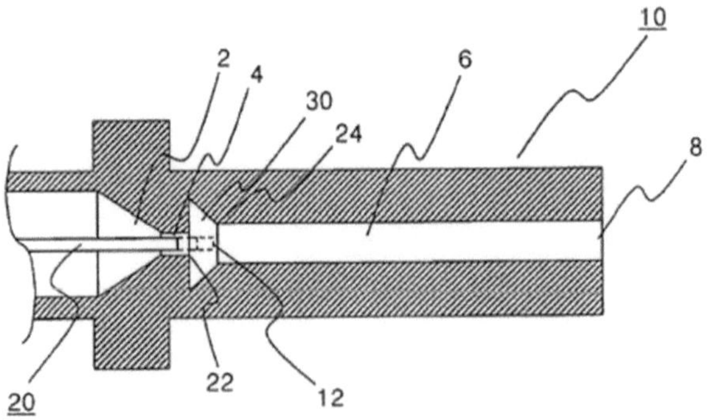

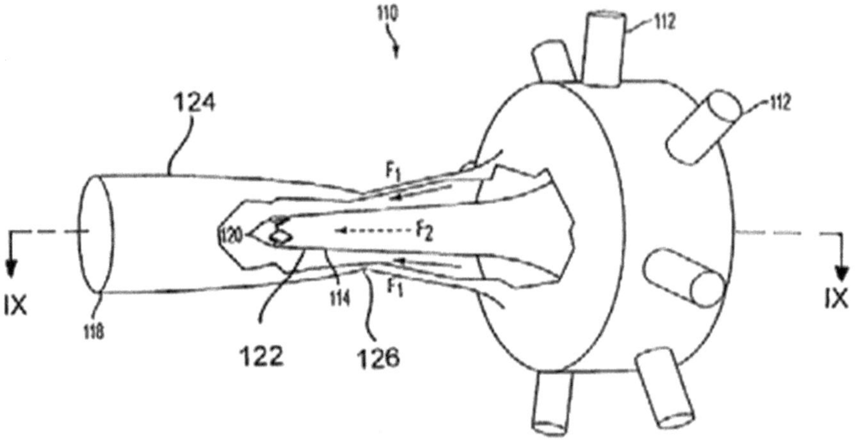

Fig. 12.5 Axial powder injector with chevrons. (Muggli et al. 2013)

In order to ensure homogeneous particle velocity and temperature distribution, Muggli et al. (2013) patented an axial particle injector with a chevron tip located at the end of the convergent section of the nozzle (Fig. 12.5). The chevrons contribute to the particle intermixing within the supersonic effluent stream.

Esfahani and Vanderzwet (2012) also patented a powder injector with a movable piston which can be adjusted with respect to its position at the convergent section of the de Laval nozzle. The configuration is shown in Fig. 12.6. The adjustable member can be used to introduce the feedstock powder in various axial positions, within the convergent section of the nozzle, thus providing multiple throat clearance configurations (Fig. 12.7). Variable powder injection location is also proposed in Maev et al. patent (Maev et al. 2012), further discussed in the *control system* section.

Another design by Jabado et al. (2010) incorporated a dual nozzle structure, with one nozzle arranged within a larger nozzle. The powder is fed through the interior nozzle, while the gas is fed through the exterior nozzle section. The gas envelops the feedstock particle jet resulting in an improved powder flow dynamics with decrease in particle deflection and clogging. The design allows the deposition of very fine particles that are less than 5 μm in size (Jabado et al. 2010).

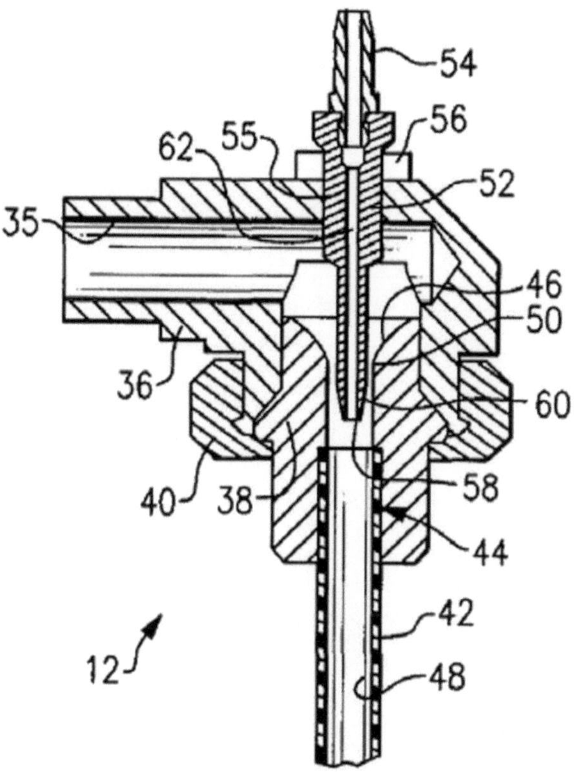

Fig. 12.6 Cold spray nozzle assembly. (Esfahani and Vanderzwet 2012)

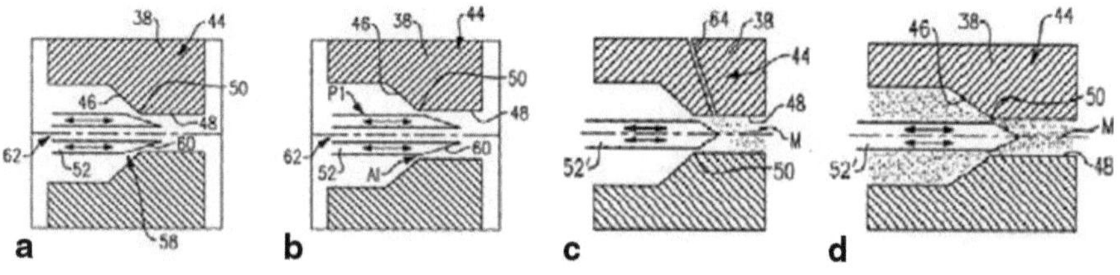

Fig. 12.7 **a** Movable piston with **b** varied axial position configuration with **c** powder feeder located upstream and **d** downstream the adjustment member. (Esfahani and Vanderzwet 2012)

For some materials, powder clogging is a major challenge. The clogging typically occurs in the junction of the convergent/divergent section of the nozzle where the preheated particles impact the walls and block the gas stream. Kay and Karthikeyan (2012) improved the gun design by incorporating a cooling jacket around a polymeric nozzle (6) placed with a housing (Fig. 12.8). The cooling jacket cools the nozzle and prevents the particle adhesion to the nozzle interior walls. The design includes a support to prevent a nozzle deformation upon heating.

In some applications, higher particle temperatures are needed to increase the powder deposition efficiency, the later can be achieved by increasing the time of

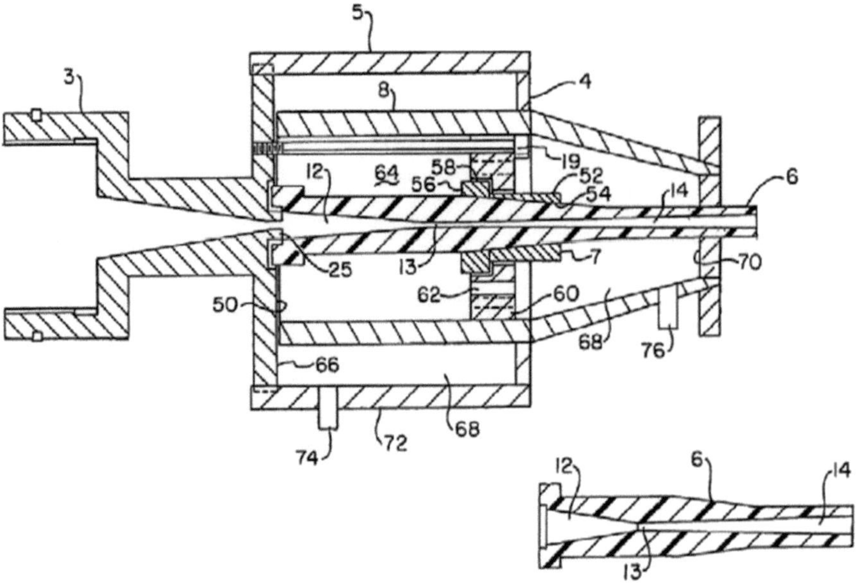

Fig. 12.8 Removable nozzle within a cooling jacket. (Kay and Karthikeyan 2012)

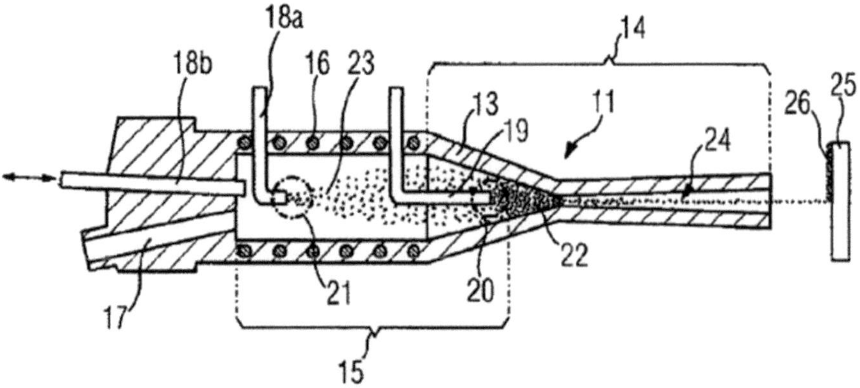

Fig. 12.9 Stagnation chamber with varied powder inlet positions. (Arndt et al. 2012)

the particle interaction with the heated gas jet. For that purpose, Arndt et al. (2012) patented an adjustable powder feeder in the convergent section of the nozzle (shown in Fig. 12.9). The incorporation of the feedstock powder early in the divergent section of the nozzle provides longer course through the stagnation chamber and higher temperature transfer to the feedstock powder material (Arndt et al. 2012).

The system portability in the field can be of importance. A patent from Vanderzwet et al. (2012) presents a portable CS unit with a compact, hand-held CS gun, as shown in Fig. 12.10. The gun is connected, by a cable (22), to a separate unit with gas and powder feeder. The gas, fed through cable (32), is heated by the heating

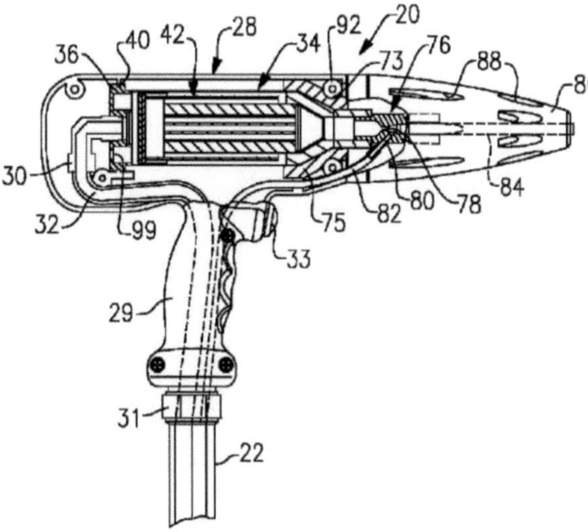

Fig. 12.10 Compact, hand-held CS gun mainly composed of the gun housing *28,* the heater assembly *34,* an insulating cone *75,* a nozzle *76* followed by a tube *(84)* and a shroud *(86)* to prevent the tube *84* from being damaged. (Vanderzwet et al. 2012)

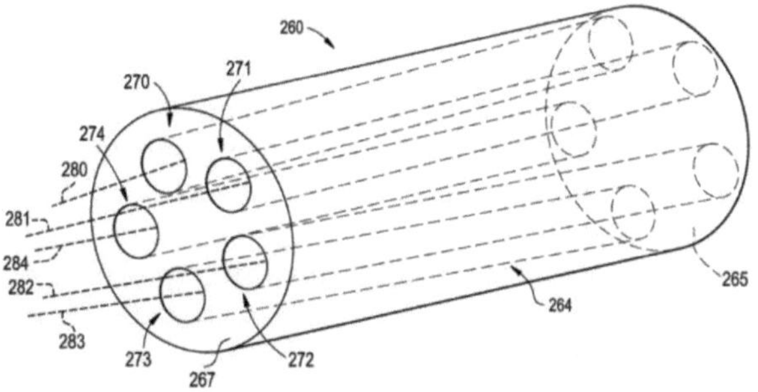

Fig. 12.11 Multi-nozzle configuration. (Calla and Venkatachalapathy 2013)

assembly (34). The powder is introduced by line (80) downstream the divergent section on the nozzle (78).

Finally, a design with multi-nozzle configuration was patented by Calla and Venkatachalapathy (2013) and allows a simultaneous or sequential deposition of different powders. The system, shown in Fig. 12.11, is made of multiple nozzles arranged in parallel. The nozzles can be designed with a convergent/divergent section and can be used to tailor the gas and particle velocities within each individual nozzle for deposition of composite mixture coatings (Calla and Venkatachalapathy 2013).

12.3.1.2 Hybrid Systems

A number of hybrid systems were developed in order to optimize the properties of the CS deposited coatings. Such systems incorporated laser devices which can be used in parallel with the CS system. Calla and Jones (2011) patented a device with auxiliary heating source, shown in Fig. 12.12, with lasers (202) used to preheat the substrate or CS material for enhancing coating adhesion or to anneal the material

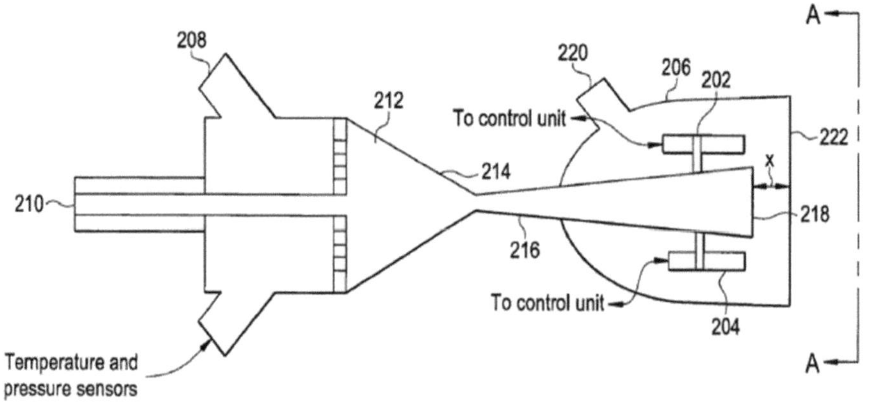

Fig. 12.12 Hybrid cold spray system with laser component. (Calla and Jones 2011)

for better mechanical properties. A laser-guided CS deposition process (Jensen et al. 2011) is also detailed in the *manufacturing design and method section.*

Other modifications to the CS system include a gas envelope housing (206) used to control the oxidation process of the deposited material (Calla and Jones 2011). Hertter et al. (2014) proposed to incorporate the reactive gas jets alongside with the CS jet in order to induce partial oxidation of the particle during CS deposition (Hertter et al. 2014).

Substantial system modifications were proposed by Molz et al. (2009) who re-engineered the CS process to use a plasma spray gun for heating and accelerating the gas instead of the electrical resistors that are used by most of the commercial CS guns. The particles are injected after the converging section of a nozzle (see Fig. 12.13).

Finally, a spray system was developed by Jodoin (2012), which induces the shock waves or compression waves with a shock generator, shown in Fig. 12.14, to project pulses of powder onto the substrate material. The design is claimed to reduce the gas consumption, clogging and to generate less heat at the substrate surface. Additional advantages include superior coating density and uniformity.

12.3.1.3 System Control

In order to optimize the CS process, a number of control systems were introduced into the CS system. Maev et al. (2012) introduced a powder flow rate meter, at the exit of the powder feeder. The powder flow rate is constantly measured and adjusted in order to control the deposition rate and deposition efficiency of the process. For better control, various locations for the powder injector are proposed. Jabado et al. (2009a) proposes a system, shown in Fig. 12.15, with modulated controller for gas stream temperature, pressure and particle density, velocity and temperature. The controller can vary system parameters periodically or aperiodically in order to

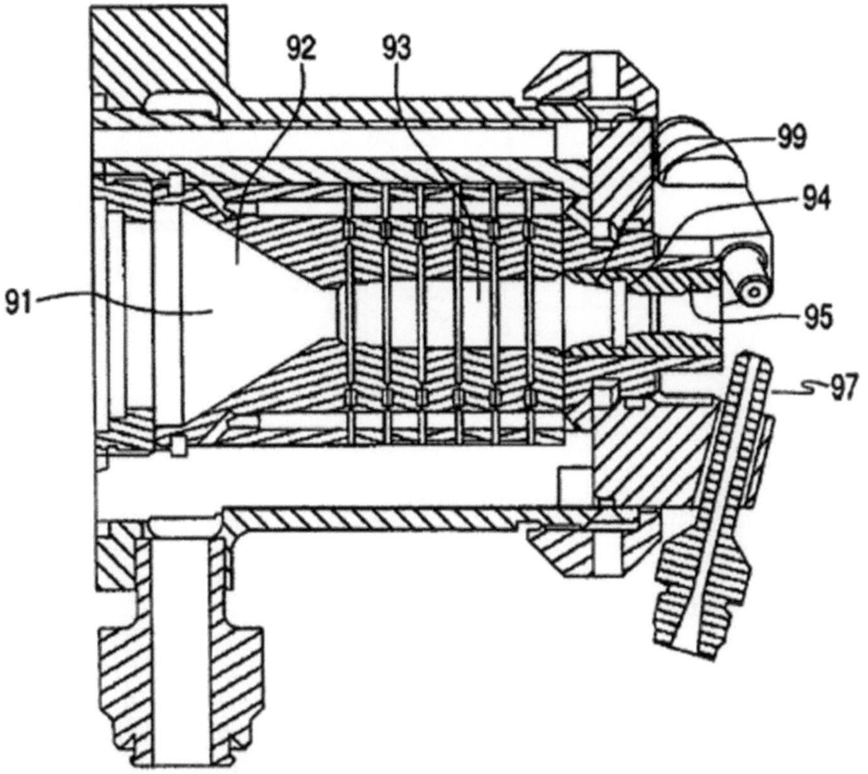

Fig. 12.13 Hybrid plasma/cold spray system. (Molz et al. 2009)

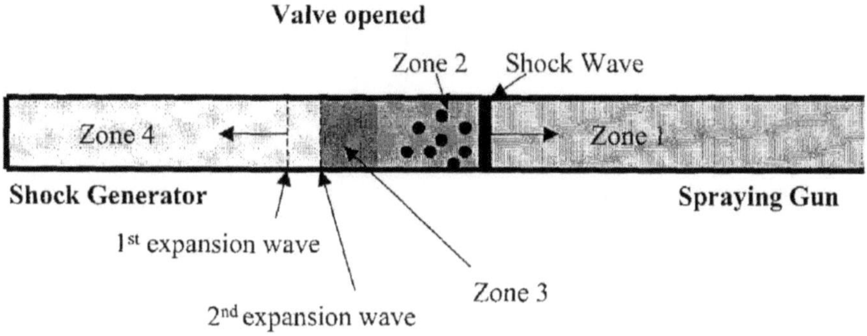

Fig. 12.14 Cold spray system with shock generator. (Jodoin 2012)

produce coating with desired properties. The modulation of process parameters is achieved with pulsed heating; valve-regulated gas flow and powder feeder; piezo-electric pressure generators; wave couple as well as high-pressure valve operated individually or at the same time. The system can be used with submicron particles and is claimed to produce superior coatings (Jabado et al. 2009a).

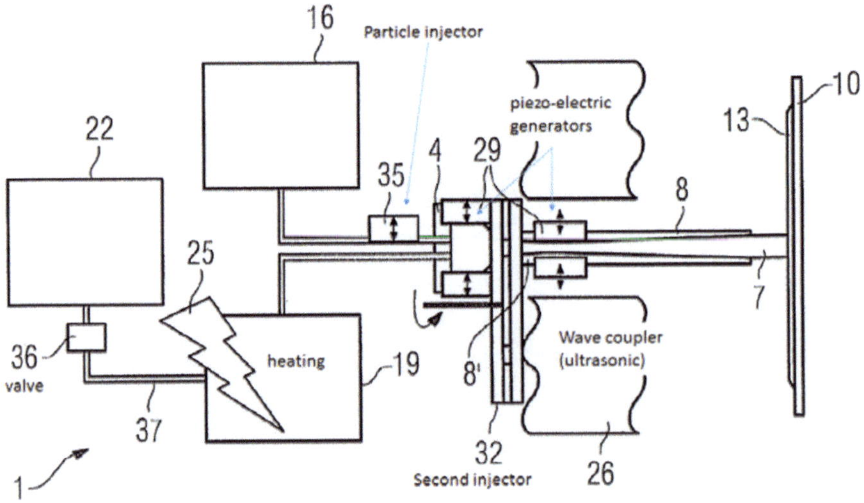

Fig. 12.15 Modulated cold spray process with pulsed system controllers. (Jabado et al. 2009a)

12.3.1.4 Manufacturing Design and Method

System modifications were made to accommodate the CS process in accordance with the manufacturing design and methods. One of the challenges, in any spray process, is the ability to coat inner surfaces. In order to provide solution for this challenge, the nozzle configurations can be modified to meet the application requirements. For the coating applications onto the small diameter bores, Payne (2011) developed a nozzle with a bend, see Fig. 12.16. In order to mitigate the wear and clogging issues, this invention includes a wear and adhesion resistant coating in the inner surface of the bent nozzle in addition to the gas injection inlets used to conform the particle flow to the bent in the nozzle.

Another patent by Venkatachalapathy et al. (2011) optimized the geometry of the gun in order to reduce the length of the convergent/divergent section of the nozzle

Fig. 12.16 Nozzle with a bend. (Payne 2011)

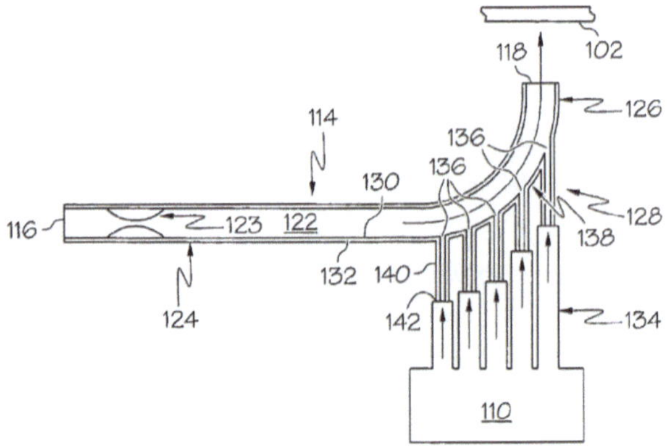

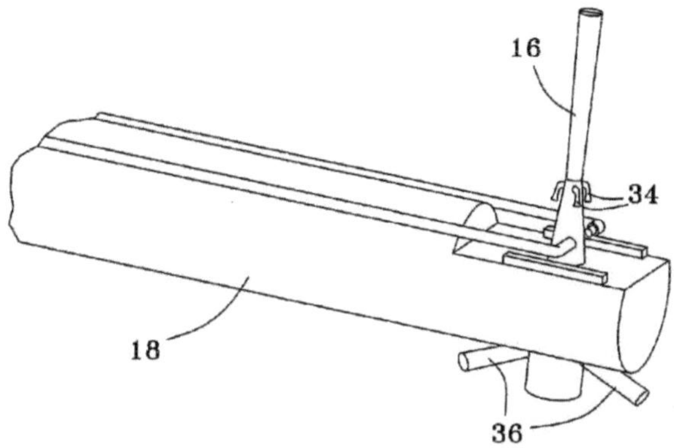

Fig. 12.17 Compact nozzle design. (Venkatachalapathy et al. 2011)

to less than 200 mm (see Fig. 12.17). The design incorporates the powder injector (34) that feeds the powder upstream of the throat in the converging section of the nozzle and can be placed in axial or radial orientation with respect to the body. The compact design increases the nozzle accessibility to the confined locations such as interior wall of the gas turbines and increases overall system manoeuvrability.

A number of modifications were implemented to the CS process to improve deposition resolution. Gambino et al. (2007) patented a method for the powder deposition through a straight bore with a very fine aperture, of less than 25 μm, used for the direct writing of metallic conductor patterns, shown in Fig. 12.18, on the insulating surfaces. A laser-guided CS deposition process was also implemented in precision writing of the metallic structures for applications like strip conductors (Jensen et al. 2011).

Other system adjustments can be made through masking and heating devices used for controlled deposition of CS coatings. Ikejiri (2013) filed a patent on a masking jig, shown in Fig. 12.19 (60). The masking jig contains an opening through which the powder can be deposited. The masking jig is also equipped with a heating element (61) powered by the heating device (50) which is used to heat the vicinity of the opening when in contact with the insulating substrates (10). The insulating

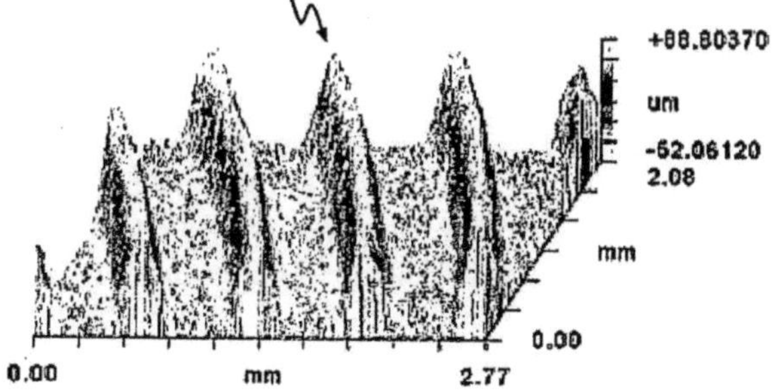

Fig. 12.18 Conductor lines. (Gambino et al. 2007)

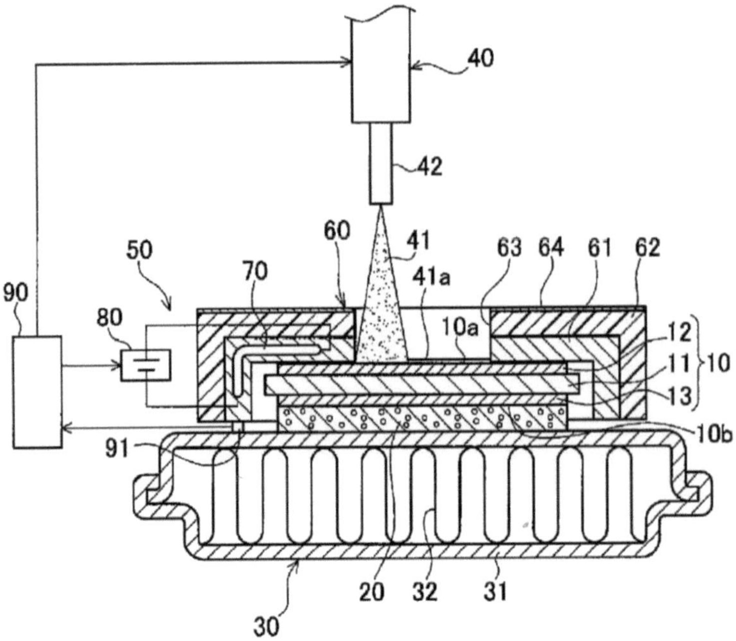

Fig. 12.19 Making jig with heating elements. (Ikejiri 2013)

substrates are made of three layers, one upper (12) and one lower layer (13) made of aluminium and the middle layer made of aluminium nitride (11).

The insulating substrates are placed on the cooling system, and a stress relaxation layer (20) is used to reduce the thermal shock in the insulating substrates. The heating system is reported to be of low cost, requiring only one hearting component with possibility of having multiple openings in the masking jig. The patent claims superior deposition efficiency for copper coating, high coating density and adhesion to the insulating substrates. The patent takes into account the possibility of laser-assisted coating deposition (Ikejiri 2013).

12.3.2 Precursor/Feedstock

12.3.2.1 Feedstock Powder

Conventional CS generally uses powder particles with diameter exceeding 5 μm to ensure sufficient mass inertia to resist the bow shock formed near the substrate surface during spraying. However, for some applications, it is desirable to build a coating presenting nano-features. Two patents filled by Jabado and his team (Jabado et al. 2009b, 2014) propose solutions in this regard. It is proposed to encapsulate nanoparticles in a film that will either disintegrate or be incorporated in the coating upon impact shown in Fig. 12.20a (Jabado et al. 2014) or to bind nanoparticles at

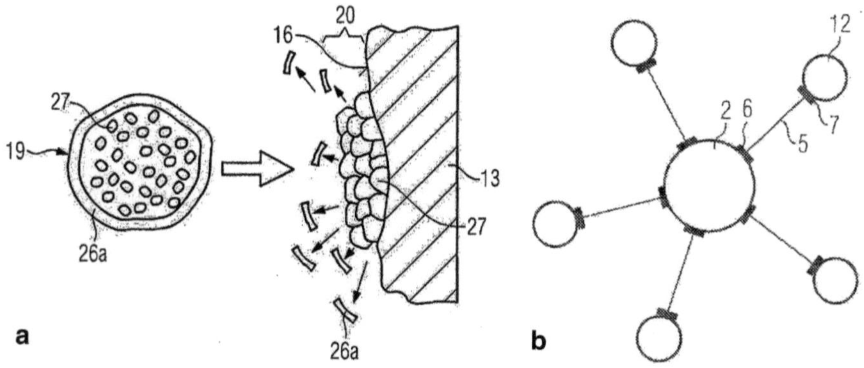

Fig. 12.20 **a** Encapsulated nanoparticles (Jabado et al. 2014). **b** Nanoparticles bound to the microparticles. (Jabado et al. 2009b)

the surface of micronsize particles using coupling molecules such as polymer as can be seen in Fig. 12.20b (Jabado et al. 2009b).

Another patent, by Calla et al., involves the cryomilling of the CS feedstock prior to spraying to produce nano-size grain powders (Anand et al. 2010).

One of the CS challenges is linked to the proper feeding of the feedstock powder, especially for particles tending to agglomerate. Dahl Jensen proposes to feed particles through a supply line in a liquid or solid that will vaporize or sublimate at the mouth of the supply lines, when delivered to the carrier gas stream, thus ensuring better feed rate uniformity. The solid or liquid additive is to be selected so that it takes a gaseous state in the carrier gas stream due to temperature and pressure reduction cause by adiabatic expansion of the carrier gas (Jensen et al. 2010). Finally, the patent (Ajdelsztajn et al. 2013) suggests to improve the sprayability of nickel and other hard and high temperature materials such as stainless steel and titanium alloys by adding an outer particle layer displaying lower melting point and/or softer by techniques such as chemical vapour deposition (CVD) and physical vapour deposition (PVD).

12.3.2.2 New CS Material Systems

In the past years, there was a drive to extend the range of material types and combinations deposited by CS. Polymer ceramics layer was produced by cold spraying of the precursors of a polymer ceramic, also called pre-ceramic polymers, as well as filling materials followed by the appropriate heat treatment, or the conversion reaction which can be performed directly during spraying if the energy input into the cold gas jet is properly adjusted (Krüger and Ullrich 2010). A method was also suggested to form metal matrix composite CS coatings, involving initial wet blending of the metal and ceramic powders (Debiccari et al. 2011). Ko et al. proposed a 1:1 or 3:1 mixture of SiC or alumina into the metallic matrix in order to improve the coatings wear resistance (Ko et al. 2013).

Another patent suggests the use of two different powder sources and sequentially deposits the first and second powders to get a coating graded in composition (Debiccari and Haynes 2013). Methods of interest also include the fabrication of an electrical coil, and especially a superconductor coil, by cold spraying of electrically insulating and electrically conducting powder materials under vacuum (Bohn 2009) as well as coating consolidation through shot peening (Tapphorn and Gabel 2012).

12.3.3 Applications

In recent years, CS process was adapted for the applications in the precision manufacturing of electronic systems to repair, additive manufacturing free forms, welding, brazing and surface protection, medical devices and sputtering targets. The sectors, in which CS process was implemented include: military, automotive, aerospace as well as medical industries.

12.3.3.1 Additive Manufacturing, Free Forms and Repair

High density, superior adhesion, deposition rate and efficiency of the CS process make it attractive for the applications in the near-net-shape production technology and repair. Slattery (2008) patented a method for manufacturing of dense structures such as flanges, ridges and posts on the base material preforms using CS process, shown in Fig. 12.21. In order to improve the mechanical properties of the final components, hot isotactic pressing, heat treatment, aging, quenching, stretching and annealing were incorporated into the manufacturing process.

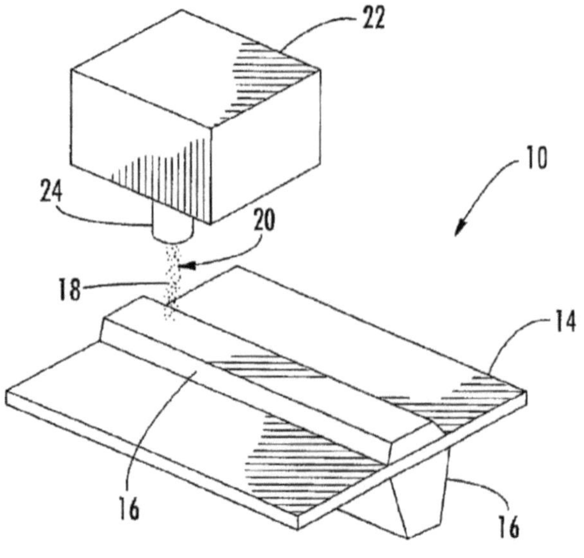

Fig. 12.21 Structural member *16* on a base member *14* to form a preform *10*. (Slattery 2008)

Heinrich et al. (2012b) patented a method by which the CS process can be used to manufacture a hollow pipe. The deposition of the hollow pipe is achieved by spraying the material onto a low adhesion cylindrical substrate. The method relies on the low adhesion of the CS deposited coatings in order to detach the preform from the deposit. Low adhesion was incited by varying the spray angle from 0 to 90° (Heinrich et al. 2012b).

Payne and Garland (2008) filed a patent on the repair of thin-wall housings using the CS process. The patent includes the repair of aluminium castings and other materials with reduced dimensions and worn inner and outer surfaces. The cleaned surface is CS coated with the material of choice and machined to the required dimensions. Additional steps include the heat treatment to restore the mechanical properties followed by the component inspection. Thin-wall components are often found in the aerospace and automotive industries where the weight plays an important role. Thin-wall housings are subject to corrosion, erosion (example fluid pumps) and mechanical wear. Some components, like aluminium castings, can be costly, and high cost savings can be achieved through the CS component repair.

Finally, Ngo et al. (2013) patented a method for the repair of the components having a damaged internally threaded openings. The process consists of machining down the damaged treads with inclusion of internal notches for a better coating adhesion, as can be seen in Fig. 12.22. The machined opening is, then, filled, by CS, with the material of choice and rethreaded to the desired dimensions (43). The patent indicates that a complete CS filling of the notches is not required (Ngo et al. 2013).

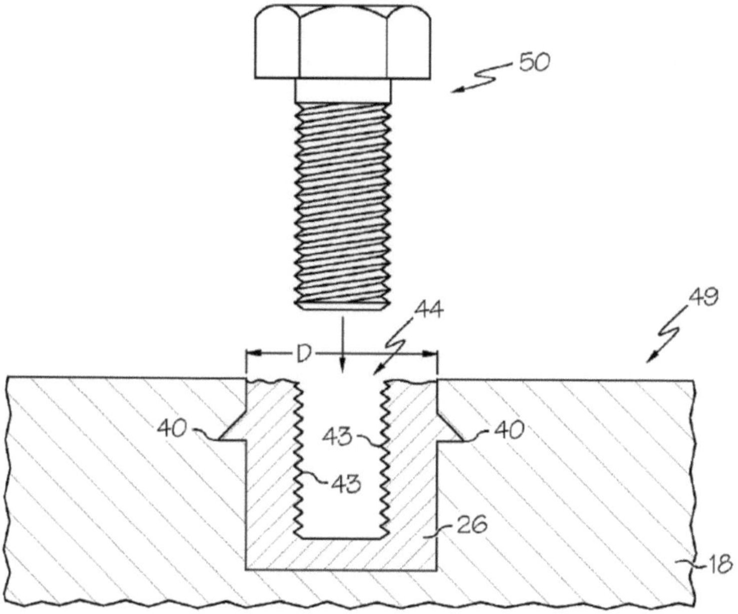

Fig. 12.22 Cold spray repair of the component with damaged internal threaded opening. (Ngo et al. 2013)

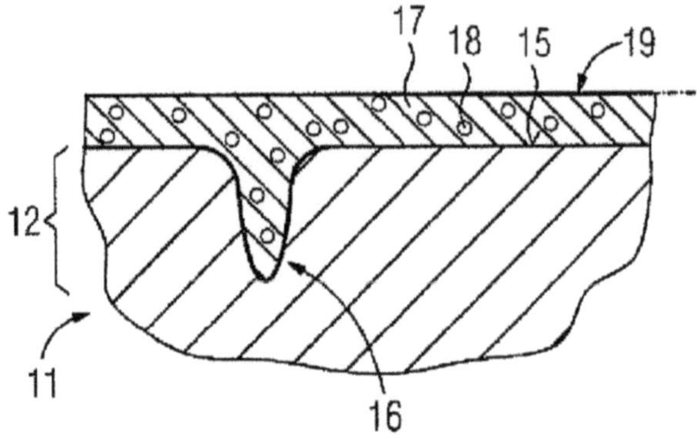

Fig. 12.23 Repair of the turbine blade. (Jensen et al. 2013)

In recent years, CS process was adapted towards the repair and manufacturing of the aerospace and automotive components. A patent on the manufacturing and repair of the turbine rotor was released by Calla et al. (2012). The fabrication process consists of the near-net-shape deposition or repair of the rotor structure on top of the core shaft with CS process. The patent includes many processing steps consisting of heat treatment to relieve the internal stresses and to induce the diffusion bonding between the cold-sprayed splats and the coating–substrate interface. The final step of the process consists of a machining to recover the desired dimension. The materials mentioned in the patent are chromium- and nickel-based alloys deposited with a wide range of powder mixtures and alloy concentrations.

A method for repair of the turbine blades by CS process was recently patented by Jensen et al. (2013). The method consists of repairing the turbine blades that show signs of visible wear, cracking (see Fig. 12.23 (16)) or microstructural degradation with depletion of the key elements. Such elements may be aluminium (18) in MCrAlY alloy used in turbine blades and bond coats. The process consists of tailoring the composition of the repair coating so as to compensate for the depletion of the key element within the metal matrix.

12.3.3.2 Welding and Brazing

The ability of CS to produce metallic coating in intimate contact with the base materials, as well as the low oxygen content of the CS deposited materials, low processing temperatures, high deposition rates, processing speed and excellent deposition control opens the cold process towards the welding and blazing applications. Schmid and Doesburg (2010) patented a method by which the CS process was used to produce solders with mixtures of composite materials with low melting point like Sn-Ag-Cu, Sn-Ni-Cu, Sn-Cu, Sn-Zn and others. CS soldering can be used with powders made to incorporate thermally and electrically conductive fillers (like carbon in graphite, diamond or carbon nanotubes) within a softer metal matrix to impart the thermal and electrical properties of the solder.

Miller et al. released two patents (Miller et al. 2011, 2012b) by which the CS process can be applied to join the clad structures. The process consists of spraying the cladding material between the two joins so as to form a continuous junction. The CS process was patented for the closure of the various joint configurations, shown in Fig. 12.24. The joint can be made of tantalum and can be used to weld steel edges with tantalum welds to produce corrosion-resistant joints (Miller et al. 2011, 2012b).

The application of the CS process extends to the semiconductor seals. Ohno (2013) patented a method by which the CS process is used to deposit a resin-bonding coat made of the metallic powder, like Cu, in between the two semiconductor substrates. Bonding coat provides an anchor for the resin seal which is claimed to contribute to a better seal adhesion at low cost, when compared to conventional techniques.

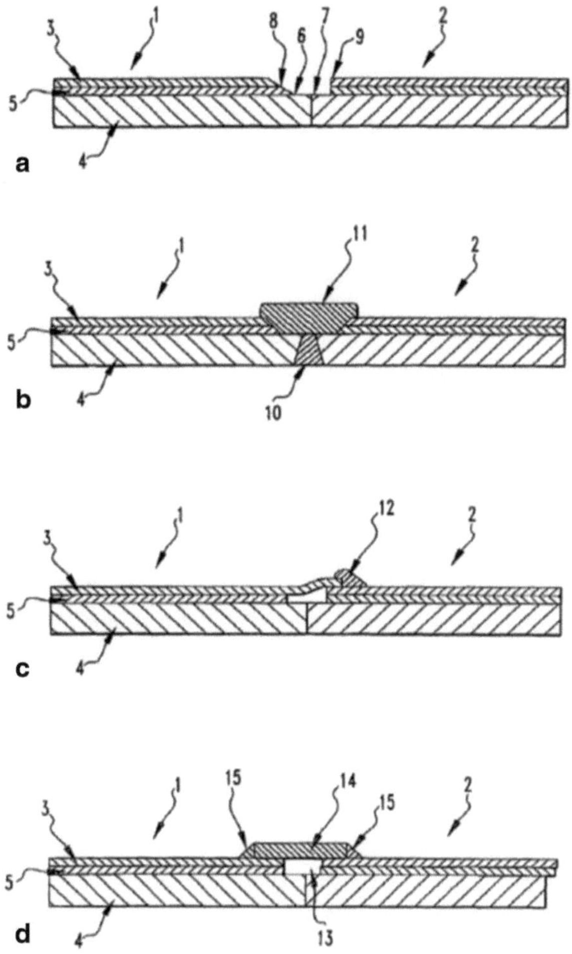

Fig. 12.24 **a** Clad section before joining. **b** Cold spray deposited joint. **c** Cold-sprayed lap joint. **d** Double lap joint. (Miller et al. 2011, 2012b)

Unlike the conventional welding and blazing techniques, CS process is carried under low-temperature processing conditions which limit the microstructural degradation and formation of the detrimental heat-affected zone. Schaeffer et al. (2013) patented a method where welding and brazing of aerospace components, like turbine blades and brackets, are achieved through CS deposition of a titanium aluminide powder with refined microstructure.

12.3.3.3 Protective Coatings

A number of patents were released for the application of the corrosion-resistant CS coatings on the metallic components. Ajdelsztajn et al. (2013) patented a method by which corrosion-resistant coatings made of a metal alloys comprising cobalt and manganese are cold sprayed on components, such as riser tension component, made of iron and its alloys. Similarly, Raybould et al. patented a method for the corrosion/erosion protection and repair of the magnesium components. The corrosion-resistant CS coatings patented for that purpose included aluminium, aluminium alloys, titanium and its alloys as well as hard particle dispersed composites. The protective coatings can be layered, with an aluminium layer for corrosion protection and with titanium or hard particle composite materials in the outer layer for erosion protection (Raybould et al. 2008).

Bunting et al. (2013) developed a method by which a corrosion-damaged gas turbine component can be restored to the original dimensions with added benefit of the decreased oxidation rate of the component. The method consists of CS deposition of the material with superior corrosion resistance to the base material (the materials mentioned include aluminium, magnesium, silicon and their mixtures). The CS process parameters are selected to reduce the material oxidation and therefore consist of low deposition temperatures and non-oxidizing carrier gas like nitrogen or helium.

In a patent filed by Miyamoto and Hirano, an Al-based alloy containing Sn is cold sprayed on a sliding surface to form a bearing with improved performances compared to thermal sprayed coatings, attributed to good distribution of the different phases and low oxide content (Miyamoto and Hirano 2011). Low oxide content and high density of coatings produced by the CS process prove to be also advantageous in conductive liner in the cylindrical blocks in the interior of the combustion engine, an application patented by Miyamoto et al. (2010). Typically such liners are deposited by plasma spray process and result in the molten material oxidation which negatively affects the thermal conductivity of the deposited layer. With CS, the deposition process takes place in solid state which limits the coating oxidation while providing a good coating adhesion to the base material and superior coating thermal conductivity.

In another case, CS is proposed, among other possible methods, to deposit a Sn-containing base layer, such as Babbitt metal, on a ferrous base material to provide a baring surface. A heat input, such as a laser, is used to diffuse Sn and to form a metallurgical bond with the base materials by the formation of a thin (max 10 μm) bonding zone containing $FeSn_2$ (Roeingh and Keller 2009). Alternatively, a tin- and silicon-rich layer can be cold sprayed on an aluminium alloy that does not contain tin (Fujita et al. 2008).

12.3.3.4 Aerospace Applications

New applications of the CS process are being developed in the aerospace industries. Haynes et al. (2012) patented a method for applying abradable coatings on a seal backing material to form an abradable seal between the rotating and stationary components in gas turbines. CS coating made of titanium or nickel-based alloy with varied porosity content was deposited on the backing plate (made of Ti6Al4V) so as to produce a dense seal near the coating base for structural rigidity and with higher porosity near the seal surface for increased coating abradability. The porosity levels were controlled by changing the CS process parameters.

Another application of the CS process includes the deposition of bond coats on turbine engine components. Schlichting and Freling (2012) patented a method for CS deposition of high porosity bond coats shown in Fig. 12.25. The idea is to imbed some of the carrier gas in the coating. In a subsequent heat treatment, the entrapped gas diffuses to form larger pores. The selected materials included MCrAlY, aluminium, titanium, palladium and other transition metals. The process is claimed to produce bond coats with porosity ranging between 25 and 50 %, thus increasing the operating temperatures of the turbine engine.

12.3.3.5 Manufacturing

High density and purity of the cold-sprayed materials make the CS process ideal for the manufacturing of the electrodes and sputtering targets. Three patents, one by Zimmermann et al. (2011) and two by Miller et al. (2012a, 2013), were recently released on the manufacturing and repair of the sputtering targets by the CS process. Patents proposed deposition of high-purity refractory materials, see Fig. 12.26, like

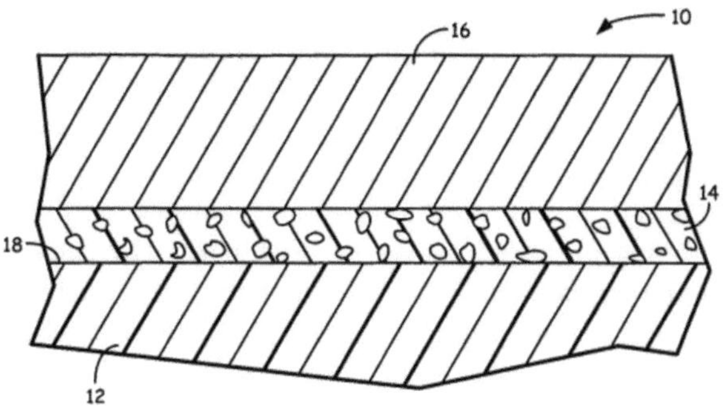

Fig. 12.25 Porous bond coat *14* on a turbine engine component *12* with a thermal barrier top coat *16*. (Schlichting and Freling 2012)

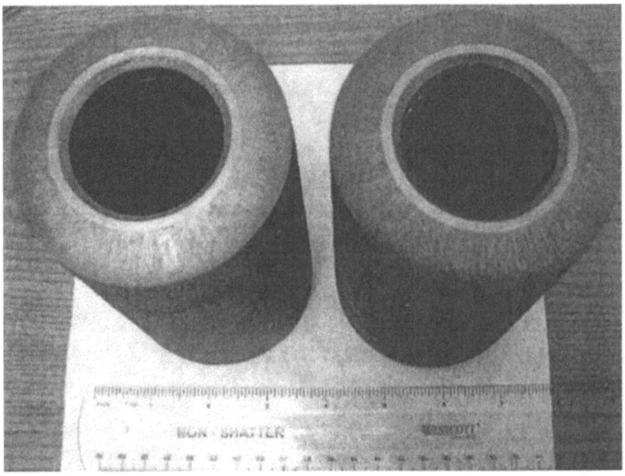

Fig. 12.26 Tantalum tubular preforms. (Miller et al. 2012a, 2013)

niobium, tantalum, tungsten, molybdenum, titanium, zirconium and other metals and their mixtures on the backing plate.

Unlike the conventional thermal spray process techniques, low deposition temperatures of the CS process do not induce the buckling of the backing plate, demonstrate low gas contamination of the refractory materials and do not contribute to the detrimental change in the material microstructure. Barker et al. (2011) patented a method by which high-purity electrocatalytic materials were deposited on an electrode. The CS process was claimed not to alter the characteristics of the feedstock powder material.

CS process can be used for deposition of the materials like carbon, silicon, metal and metal oxides on metallic components. Kalynushkin et al. filed four patents for methods by which such materials were deposited, with sufficient thickness, on the metal collector tapes to produce electrodes (Kalynushkin and Novak 2010, 2011a, b, 2012). A metal collector tape serves as a cathode or anode and, in combination with a separator (see Fig. 12.27), can be used as a fuel cell or as a metal-ceramic membrane in all types of electronic devices (Kalynushkin and Novak 2011a, 2012). Since CS can prevent metallic powder oxidation during deposition, cold-sprayed coatings displaying high thermal conductivity can also be used in heat transfer devices, such as in power modules, process patented by Miyamato and Tsuzuki (2011) and Tsuzuki and Miyamoto (2013).

Kruger and Ullrich (2011) proposed the deposition of the solar cell materials like copper indium selenide (CIS) or YBaCuO in layers on top of textured substrates. The structured texture of substrates was transferred onto the coating layers which was claimed to improve the solar cell efficiency. Another patent by Doye et al. (2012) proposed a method to enhance the photocatalytic properties of materials with the help of the reactive gases. The patent was based on doping of the cold-sprayed titanium dioxide with nitrogen gas activated with radiation source like UV light. The coatings could be produced with a varied porosity contend which increases the surface available for catalysis.

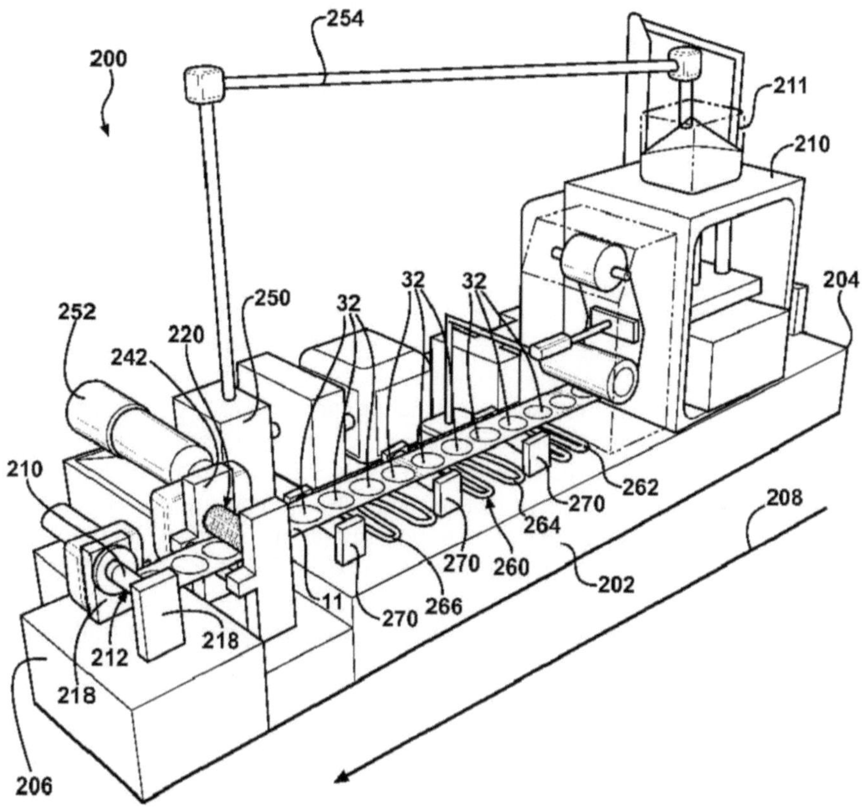

Fig. 12.27 Metal collector for the manufacturing of metal-ceramic membrane. (Kalynushkin and Novak 2011a, 2012)

12.3.3.6 Medical

The control of the coating porosity proves advantageous in medical device applications. Kramer (2009) patented a method for near-net-shape manufacturing of tubes, stocks and substrates sheets for applications in the medical devices such as stents, anastomosis chips, embolic protection filters, graft attachments, ring makes and others. The applications, also, include the CS deposition of high porosity coatings on medical devices with biocompatible materials, polymers and ceramics to increase the device biocompatibility, to serve as a delivery system for the drugs or as therapeutic agents and for the elution of the body. The porous structures can be obtained by spraying porous starter materials or by optimizing the spray parameters. CS deposition of the drugs directly onto the porous medical device was also proposed.

12.4 Concluding Remarks

CS technology, first anticipated by Thurston and Schoop but truly elaborated in Russia following the discovery of Papyrin team in 1984, has gone a long way since its early development. This chapter aimed at reviewing US patents issued between January 2008 and April 2014. Even though great care was taken to present a most extensive review, the authors cannot guarantee that all the existing patents on CS were included.

The judgment of the validity or feasibility of individual patents was considered, by the authors, to be out of the scope of this review. From the patent review presented in this chapter, it can be, however, inferred that CS technology has naturally evolved from an emerging technology with an eclectic growth with fluctuating patent quality (Irissou et al. 2008) to a more mature state with less spectacular breakthroughs but more systematic and controlled development. This technology evolution towards maturity is noticeable through the increasing number of patents on technology applications within an increasing number of industrial sectors, providing detailed apparatus targeting very specific purposes or challenges. It is expected that the technology growth in the niche applications will continue to contribute to the process reliability and consistency in production through the development of designs or sensors for improved process control and robustness as part of the next challenges the technology will face.

Acknowledgments The authors would like to acknowledge the support of Pierre Dion and Dominique Charbonneau (CISTI) for the bibliographic search.

References

Ajdelsztajn, L. 2013. Corrosion resistant riser tensioners, and methods for making. US Patent 8,535,755 B2, 17 Sept.

Ajdelsztajn, L., J. A. Ruud, and T. Hanlon. 2013. Cold spray deposition method. US Patent 8591986 B1.

Alkhimov, A. P., A. N. Papyrin, V. F. Kosarev, N. I. Nesterovich, and M. M. Shushpanov. 1994. Gas-dynamic spraying method for applying a coating. US Patent 5302414, 12 April.

Anand, K., E. Calla, R. Oruganti, S. K. Sondhi, and P. R. Subramanian. 2010. A method of cold spraying with cryo-milled nano-grained particles. European Patent 2 206 568 A2, 17 July.

Arndt, A., U. Pyritz, H. Schiewe, and R. Ullrich. 2012. Method and device for the cold-gas spraying of particles having different solidities and/or ductilities. US Patent 8197895 B2, 12 June.

Barker, M. H., O. Hyvärinen, and K. Osara. 2011. Electrode is sprayed with at least one of the oxides of the transition metals (manganese dioxide) in powder form as a catalytic coating, after which the electrode is ready for use without any separate heat treatments; anode used in the electrolytic recovery of metals. US Patent 7871504 B2, 18 Jan.

Bohn, M. 2009. Method and device for cold gas spraying. US Patent 20090291851, 26 Nov.

Bunting, B. W., A. DeBiccari, C. Vargas, M. D. Kinstler, and D. W. Anderson. 2013. Corrosion protective coating through cold spray. US Patent 8597724 B2, 3 Dec.

Calla, E., and M. G. Jones. 2011. Apparatus, systems, and methods involving cold spray coating. US Patent 8,020, 509, 20 Sept.

Calla, E., and V. Venkatachalapathy. 2013. Multi-nozzle spray gun. US Patent 8544769, 1 Oct.

Calla, E., S. Pabla, and R. Goetze. 2012. Turbine rotor fabrication using cold spraying. US Patent 8261444 B2, 11 Sept.

Debiccari, A., and J. D. Haynes. 2013. Method for creating functionally graded materials using cold spray. European Patent 1712657 B1, 21 Aug.

Debiccari, A., J. D. Haynes, D. A. Hobbs, and J. Karthikeyan. 2011. Cold sprayed metal matrix composites. European Patent 1942209 B1, 3 Aug.

Doye, C., U. Krüger, and U. Pyritz. 2012. Method for producing a coating through cold gas spraying. US Patent 8241702 B2, 14 Aug.

Esfahani, M. K., and D. P. Vanderzwet. 2012. Adjustable cold spray nozzle. US Patent 8282019 B2, 9 Oct.

Fujita, M., E. Inoue, and S. Inami. 2008. Plain bearing and method of manufacturing the same. US Patent 20080206087 A1, 28 Aug.

Gambino, R., R. Greenlaw, S. Kubik, J. Longtin, J. Margolies, and S. Sampath. 2007. Producing a mixture of a metal powder having a flake morphology and gas; accelerating mixture in a subsonic carrier gas jet through a straight-bore tube; directing the subsonic carrier gas jet onto the substrate; manipulating one of the subsonic carrier gas jet and the substrate for forming conductor. US Patent 7208193, 24 April.

Haynes, J. D., A. DeBiccari, and G. Shubert. 2012. Applying abradable material from nozzles onto seal backings, to form an abradable seal between rotating and stationary components of turbines; improving performance. US Patent 8192792 B2, 5 June.

Heinrich, P., H. Kreye, T. Schmidt, and P. Richter. 2009. Cold gas spray gun. US Patent 7637441, 29 Dec.

Heinrich, P., H. Kreye, and T. Schmidt. 2012a High-pressure gas heating device. US Patent 8249439 B2, 21 Aug.

Heinrich, P., P. Richter, H. Höll, and E. Bahr. 2012b. Method for producing a pipe. US Patent 8316916 B2, 27 Nov.

Hertter, M., A. Jakimov, and S. Schneiderbanger. 2014. Gas dynamic cold spraying of oxide-containing protective layers. US Patent 8697184, 15 April.

Ikejiri, T. 2013. Masking jig, substrate heating device, and coating method. US Patent 8414977, 9 April.

Irissou, E., J.-G. Legoux, A. Ryabinin, B. Jodoin, and C. Moreau. 2008. Review on cold spray process and technology: Part I—Intellectual property. *Journal of Thermal Spray Technology* 17 (4): 495–516. doi:10.1007/s11666-008-9203-3.

Jabado, R., J. D. Jensen, U. Krüger, D. Körtvelyessy, V. Lüthen, R. Reiche, M. Rindler, and R. Ullrich. 2009a. Cold spraying installation and cold spraying process with modulated gas stream. US Patent 7631816, 15 Dec.

Jabado, R., J. D. Jensen, U. Krüger, D. Körtvelyessy, V. Lüthen, R. Reiche, M. Rindler, and R. Ullrich. 2009b. Powder for cold spraying processes. US Patent 20090306289, 10 Dec.

Jabado, R., J. D. Jensen, U. Krüger, D. Körtvelyessy, V. Lüthen, R. Reiche, M. Rindler, and R. Ullrich. 2010. Nozzle arrangement and method for cold gas spraying. US Patent 7740905, 22 June.

Jabado, R., J. D. Jensen, U. Krüger, D. Körtvelyessy, V. Lüthen, U. Pyritz, R. Reiche, and R. Ullrich. 2014. Cold gas spraying method. European Patent 1 926 841 B1, 20 Aug.

Jensen, J. D., J. Klingemann, U. Krüger, D. Körtvelyessy, V. Lüthen, R. Reiche, and O. Stier. 2010. Method for feeding particles of a coating material into a thermal spraying process. US Patent 20100098845 A1, 22 April.

Jensen, J. D., U. Krüger, and R. Ullrich. 2011. Cold gas spraying method. US Patent 8021715, 20 Sept.

Jensen, J. D., J. Klingemann, U. Krüger, D. Körtvelyessy, V. Lüthen, R. Reiche, and O. Stier. 2013. Method for repairing a component by coating. US Patent 8343573 B2, 1 Jan.

Jodoin, B. 2012. The use of shock or compression waves to project particles onto surface, for the preparation of coated surfaces that exhibit superior density and uniformity. US Patent 8298612, 30 Oct.

Kalynushkin, Y., and P. Novak. 2010. Electrode for energy storage device and method of forming the same. US Patent 7717968B2, 18 May.
Kalynushkin, Y., and P. Novak. 2011a. Apparatus for forming structured material for energy storage device and method. US Patent 7,951,242 B2, 31 May.
Kalynushkin, Y., and P. Novak. 2011b. Electrode for cell of energy storage device and method of forming the same. US Patent 7972731 B2, 17 Nov.
Kalynushkin, Y., and P. Novak. 2012. Apparatus for forming structured material for energy storage device and method. US Patent 8142569, 27 March.
Kay, A., and J. Karthikeyan. 2012. Spray nozzle assembly for gas dynamic cold spray and method of coating a substrate with a high temperature coating. US Patent 8192799 B2, 5 June.
Ko, K., H. Lee, J. Lee, J. Lee, and Y. Yu. 2009. Nozzle for cold spray and cold spray apparatus using same. US Patent 7621466, 24 Nov.
Ko, K., H. Lee, J. H. Lee, J. Lee, and Y. Yu. 2013. Method of preparing wear-resistant coating layer comprising metal matrix composite and coating layer prepared thereby. US Patent 8486496, 24 Nov.
Kramer, P. A. 2009. Coating a consolidated biocompatible medical device such as a stent by directing a supersonic jet of high pressure gas and particles to form a functionally graded coating to allow direct control of drug elution without an additional polymer topcoat; applying a drug to the coating. US Patent 7514122B2, 7 April.
Krüger, U., and R. Ullrich. 2010. Method for producing ceramic layers. European Patent 1899494 B1, 28 July.
Kruger, U., and R. Ullrich. 2011. Cold gas spraying method. US Patent 8012601 B2, 6 Sept.
Maev, R. G., V. Leshchynsky, and E. E. Strumban. 2012. Gas dynamic spray gun. US Patent 8132740, 13 March.
Miller, S. A., L. N. Shekhter, and S. Zimmerman. 2011. Methods of joining protective metal-clad structures. US Patent 8002169, 23 Aug.
Miller, S. A., O. Schmidt-Park, P. Kumar, R. Wu, S. Sun, and S. Zimmermann. 2012a. Methods of forming sputtering targets. US Patent 8197894, 12 June.
Miller, S. A., L. N Shekhter, and S. Zimmerman. 2012b. Protective metal-clad structures. US Patent 8113413 B2, 14 Feb.
Miller, S. A., O. Schmidt-Park, P. Kumar, R. Wu, S. Sun, and S. Zimmerman. 2013. Methods of rejuvenating sputtering targets. US Patent 8491959 B2, 23 July.
Miyamoto, N., and M. Hirano. 2011. Bearing material coated slide member and method for manufacturing the same. US Patent 7964239 B2, 21 June.
Miyamato, N., and Y. Tsuzuki. 2011. Method of forming a metal powder film a thermal conduction member, power module, vehicle inverter, and vehicle formed thereof. US Patent 8025921, 27 Sept.
Miyamoto, N., M. Hirano, T. Takami, K. Shibata, N. Yamashita, T. Mihara, G. Saito, M. Horigome, and T. Sato. 2010. Component for insert casting, cylinder block, and method for manufacturing cylinder liner. US Patent 7757652 B2, 20 July.
Molz, R. J., D. Hawley, and R. McCullough. 2009. Hybrid plasma-cold spray method and apparatus. US Patent 7582846, 1 Sept.
Muggli, F., M. Heggemann, and R. J. Molz. 2013. Two stage kinetic energy spray device. US Patent 8,590,804 B2, 26 Nov.
Ngo, A. T., S. Yip, C. Macintyre, B. Almond, and C. L. Cahoon. 2013. Methods for structural repair of components having damaged internally threaded openings and components repaired using such methods. US Patent 8601663, 10 Dec.
Ohno, H. 2013. Semiconductor device. US Patent 8,436,461 B2, 7 May.
Papyrin, A., V. Kosarev, S. Klinkov, A. Alkhimov, and V. M. Fomin. 2006. *Cold spray technology*. Oxford: Elsevier Science.
Payne, D. A. 2011. Apparatus for applying cold-spray to small diameter bores. US Patent 7959093, 14 June.
Payne, D. A., and P. E. Garland. 2008. Method of repair of thin wall housings. US Patent 7367488 B2, 6 May.

Raybould, D., M. N. Madhava, V. Chung, T. R. Duffy, and M. Floyd. 2008. Methods for coating a magnesium component. US Patent 7455881 B2, 25 Nov.
Rocheville, C. F. 1963. Device for treating the surface of a workpiece. US Patent 3,100, 724, 13 Aug.
Roeingh, K., and K. Keller. 2009. Plain bearing, method for production and use of a plain bearing of said type. US Patent 0232431 A1, 17 Sept.
Schaeffer, J. C., K. Anand, S. Amancherla, and E. Calla. 2013. Titanium aluminide application process and article with titanium aluminide surface. US Patent 8,475,882 B2, 2 July.
Schlichting, K. W., and M. Freling. 2012. Porous protective coating for turbine engine components. US Patent 8147982 B2, 3 April.
Schmid, R. K., and J. C. Doesburg. 2010. Material and method of manufacture of a solder joint with high thermal conductivity and high electrical conductivity. US Patent 7,758,916 B2, 20 June.
Schoop, M. U. 1915. Method of plating or coating with metallic coatings. US Patent 1128059 A, 9 Feb.
Slattery, K. T. 2008. Mixing hydrogen and particles to form stream with a temperature less than the material's melting temperature; depositing the stream on a base member to form the structural member; and subjecting the member to sub-atmospheric pressure to release the hydrogen; removal from base member; forging; machining. US Patent 7,381,446 B2, 3 June.
Tapphorn, R. M., and H. Gabel. 2012. Technique and process for controlling material properties during impact consolidation of powders. US Patent 8,113,025 B2, 14 Feb.
Thurston, S. H. 1902. Method of impacting one metal upon another. US Patent 706,701 A, 12 Aug.
Tsuzuki, Y., and N. Miyamoto. 2013. Method for manufacturing heat transfer member, power module, vehicle inverter, and vehicle. US 8499825, 6 Aug.
Vanderzwet, D., Z. Baran, and G. Mills. 2012. Gas dynamic cold spray unit. US 8,313,042 B2.
Venkatachalapathy, V., Y. C. Lau, and E. Calla. 2011. Apparatus and process for depositing coatings. US Patent 8,052, 074, 8 Nov.
Zimmermann, S., S. A. Miller, P. Kumar, and M. Gaydos. 2011. Low-energy method for fabrication of large-area sputtering targets. US Patent 7,910,051 B2, 22 March.

MIX
Papier aus verantwortungsvollen Quellen
Paper from responsible sources
FSC® C105338

If you have any concerns about our products,
you can contact us on
ProductSafety@springernature.com

In case Publisher is established outside the EU,
the EU authorized representative is:
Springer Nature Customer Service Center GmbH
Europaplatz 3, 69115 Heidelberg, Germany

Printed by Libri Plureos GmbH
in Hamburg, Germany